Principles and Applications of Climatology

Principles and Applications of Climatology

Editor: Ella Thompson

STATES
ACADEMIC PRESS
www.statesacademicpress.com

States Academic Press,
109 South 5th Street,
Brooklyn, NY 11249, USA

Visit us on the World Wide Web at:
www.statesacademicpress.com

ISBN: 978-1-63989-434-5 (Hardback)

Cataloging-in-Publication Data

Principles and applications of climatology / edited by Ella Thompson.
p. cm.
Includes bibliographical references and index.
ISBN 978-1-63989-434-5
1. Climatology. 2. Atmospheric science. 3. Atmosphere. I. Thompson, Ella.
QC861.3 .P75 2022
551.5--dc23

Table of Contents

Preface

The scientific study of weather patterns averaged over a period of time is called climatology or climate science. It focuses on recording and analyzing weather patterns of the Earth's atmosphere and understanding the conditions that cause them. It is a branch of the atmospheric sciences and a subfield of physical geography. Climatology also comprises aspects of oceanography and biogeochemistry. It is used for the analysis of observations and modeling the physical laws that determine the climate. It encompasses climate variability, mechanisms of climate change, and modern climate change. This book outlines the principles and applications of climatology in detail. It discusses the fundamentals as well as modern approaches of this field. As climatology is emerging at a rapid pace, the contents of this book will help the readers understand the modern concepts and applications of the subject.

All of the data presented henceforth, was collaborated in the wake of recent advancements in the field. The aim of this book is to present the diversified developments from across the globe in a comprehensible manner. The opinions expressed in each chapter belong solely to the contributing authors. Their interpretations of the topics are the integral part of this book, which I have carefully compiled for a better understanding of the readers.

At the end, I would like to thank all those who dedicated their time and efforts for the successful completion of this book. I also wish to convey my gratitude towards my friends and family who supported me at every step.

Editor

1

$\delta^{18}O_{sw}$ estimate for *Globigerinoides ruber* from core-top sediments in the East China Sea

Keiji Horikawa[1*], Tomohiro Kodaira[2], Jing Zhang[1] and Masafumi Murayama[3]

Abstract

The paired analyses of the Mg/Ca ratio and oxygen isotopic composition ($\delta^{18}O_c$) of surface-dwelling planktonic foraminifera have become a widely used method for reconstructing the oxygen isotopic composition of ambient seawater ($\delta^{18}O_{sw}$) as a robust proxy for surface salinity. *Globigerinoides ruber* (*G. ruber*) is a mixed-layer dweller, and its fossil shell is an ideal archive for recording past sea surface water conditions, such as those caused by variability in the East Asian summer monsoon (EASM). Here, we investigate the validity of shell-derived $\delta^{18}O_{sw}$ estimates for *G. ruber* using core-top sediments from the East China Sea (ECS). First, we determined a local $\delta^{18}O_{sw}$–salinity equation for the eastern part of the ECS in July [$\delta^{18}O_{sw} = -7.74 + 0.23 \times salinity$]. Then, we calculated $\delta^{18}O_{sw}$ from core-top $\delta^{18}O_c$ and Mg/Ca values in *G. ruber* using the $\delta^{18}O_c$–temperature equation of Bemis et al. (Paleoceanography 13(2):150–160, 1998) and the Mg/Ca–temperature equation of Hastings et al. (EOS 82:PP12B-10, 2001). The core-top $\delta^{18}O_{sw}$ and salinity were estimated to be in the ranges of −0.2 to +0.39 ‰ and 33.7 to 34.5, respectively, which fall close to the local $\delta^{18}O_{sw}$–salinity regression line. The core-top data showed that the Mg/Ca–temperature calibration by Hastings et al. (EOS 82:PP12B-10, 2001) and the $\delta^{18}O_c$–temperature equation by Bemis et al. (Paleoceanography 13(2):150–160, 1998) are appropriate for calculating $\delta^{18}O_{sw}$ in the ECS. Furthermore, we measured core-top Ba/Ca ratios of *G. ruber* ($Ba/Ca_{G.\ ruber}$), which ranged from 0.66 to 2.82 μmol mol^{-1}. There was not a significant relationship between the salinity and $Ba/Ca_{G.\ ruber}$ ratios due to the highly variable $Ba/Ca_{G.\ ruber}$ data. Given the seawater Ba/Ca data and the published partition coefficient for Ba ($D_{Ba} = 0.15$–0.22), pristine $Ba/Ca_{G.\ ruber}$ ratios at northern Okinawa Trough sites should be less than 0.84 μmol mol^{-1}. Anomalously high $Ba/Ca_{G.\ ruber}$ ratios (>0.84 μmol mol^{-1}) might be attributable to contamination by sedimentary barite adherent on fossil shells. Therefore, further evaluation of the $Ba/Ca_{G.\ ruber}$ ratio as a paleo-salinity proxy requires diethylene triamine pentaacetic acid (DTPA)-cleaned Ba/Ca data that can minimize the influence of barite contamination.

Keywords: Oxygen isotope composition of seawater; Reconstruction of sea surface salinity; Mg/Ca–temperature; *Globigerinoides ruber*; East China Sea

Background

An East Asian summer monsoon (EASM) precipitation zone is formed by the convergence of a westerly flow to the north of the Tibetan Plateau and a southerly monsoon flow over eastern China, which gradually moves northward from South China and reaches northern China in late July (Qian and Lee 2000). This EASM precipitation has caused anomalous climatic events (both wet and dry) on an inter-decadal scale, influencing the agricultural production and economy in Asia (Wang 2006). Paleoclimate records show that anomalous dry periods, when summer monsoons weakened, caused crop failure and led to the collapse of some Chinese dynasties (Yancheva et al. 2007). Therefore, EASM variability, its effect on remote areas, and the mechanisms behind this variability have been investigated intensively as important climate change issues (Wang et al. 2005; Chen et al. 2008; Wang et al. 2008; Chang et al. 2009; Liu et al. 2014).

The EASM brings a large amount of precipitation to South China and the drainage area of the Changjiang River (Yangtze River) and induces a large influx of Changjiang River runoff to the East China Sea (ECS), forming Changjiang diluted water (CDW) in the

* Correspondence: horikawa@sci.u-toyama.ac.jp
[1]Graduate School of Science and Engineering for Research, University of Toyama, Gofuku 3190, Toyama 930-8555, Japan
Full list of author information is available at the end of the article

Changjiang River Estuary (Zhang et al. 1990; Chen et al. 1994) (Fig. 1). The CDW extends farther offshore, driven by the southerly and southwesterly monsoons from June to August, and advection of the CDW causes a significant decrease in the sea surface salinity (SSS) of the northern Okinawa Trough (Chang and Isobe 2003; Lee and Chao 2003). Consequently, the northern Okinawa Trough is considered an oceanic area where EASM-related variability in SSS can be seen (Chang and Isobe 2003).

In such areas, the oxygen isotopic compositions of seawater ($\delta^{18}O_{sw}$) have been estimated from paired analyses

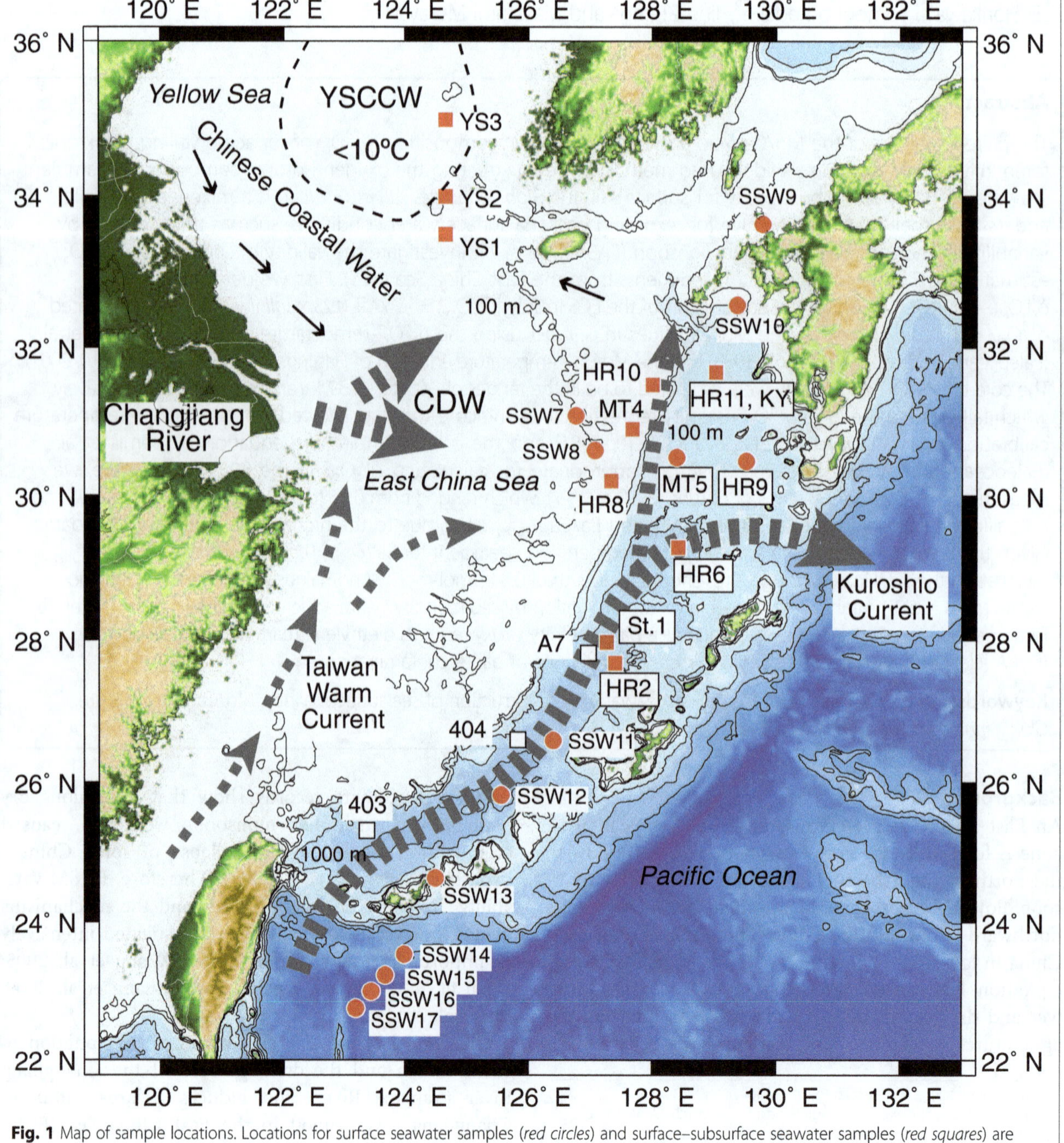

Fig. 1 Map of sample locations. Locations for surface seawater samples (*red circles*) and surface–subsurface seawater samples (*red squares*) are shown together with the present current system (Ichikawa and Beardsley 2002). Surface sediments were taken from HR11, MT5, HR9, HR6, HR2, and St.1. Locations for sediment core at KY, A7, 403, and 404 (Sun et al. 2005; Lin et al. 2006; Chen et al. 2010; Kubota et al. 2010) are also shown. CDW and YSCCW represent the Changjiang diluted water and Yellow Sea Central Cold Water, respectively. The bathymetric contours represent 100, 200, 300, 1000, 2000, and 3000 m

of Mg/Ca temperatures and oxygen isotopes ($\delta^{18}O_c$) in the surface-dwelling foraminifera *Globigerinoides ruber* (*G. ruber*) to reconstruct EASM-related variability in SSS (Sun et al. 2005; Kubota et al. 2010). However, $\delta^{18}O_{sw}$ estimates from paired analyses of Mg/Ca temperatures and $\delta^{18}O_c$ in planktonic foraminifera entail high levels of uncertainty (>0.2 ‰) due to the propagation error of $\delta^{18}O_c$ and Mg/Ca measurements and Mg/Ca–temperature calibration (Schmidt 1999; Rohling 2007). Furthermore, the lack of a local $\delta^{18}O_{sw}$–salinity equation causes additional uncertainties in salinity estimates calculated from $\delta^{18}O_{sw}$ values. Additionally, if the habitat of *G. ruber* shifts seasonally, independent of EASM variability, shell-derived $\delta^{18}O_{sw}$ estimates also change, and interpreting EASM variability will be complicated. Therefore, a reliable reconstruction of SSS with respect to EASM variability requires adequate Mg/Ca–temperature, $\delta^{18}O_c$–temperature, and local $\delta^{18}O_{sw}$–salinity calibration equations and an understanding of the habitat season and depth of *G. ruber* in the ECS.

In this study, we first derive a local $\delta^{18}O_{sw}$–salinity equation for the northern Okinawa Trough to provide a basis for salinity estimates calculated from $\delta^{18}O_{sw}$ for *G. ruber*. Then, using ECS core-top sediments, we calculate the core-top $\delta^{18}O_{sw}$ from Mg/Ca ratios and $\delta^{18}O_c$ for *G. ruber* and compare them with the local $\delta^{18}O_{sw}$–salinity line. We find that the Mg/Ca–temperature calibration by Hastings et al. (2001) and the $\delta^{18}O_c$–temperature equation by Bemis et al. (1998) are appropriate for calculating $\delta^{18}O_{sw}$ in the ECS. Finally, we present the Ba/Ca ratios of *G. ruber* from ECS core-top sediments for the first time and discuss the possibility of using shell Ba/Ca ratios as an independent paleo-salinity proxy in the ECS.

Oceanographic setting

The ECS is a marginal sea in the northwestern Pacific bounded by China, Taiwan, Ryukyu Island, Kyushu, and the Korean Peninsula (Fig. 1). The Yellow Sea is located in the northern part of the ECS, with most (>70 %) of both the Yellow Sea and ECS located above the continental shelf (<200 m water depth). The deep Okinawa Trough (~2000 m water depth) is a back-arc basin that occupies the remaining southeastern part of the ECS (Ichikawa and Beardsley 2002).

The Yellow Sea and the ECS receive the saline and warm Kuroshio water that enters the ECS along the east of Taiwan and flows northeastward along the shelf slope in the ECS (Fig. 1). The annual cycle of Kuroshio volume transport has an estimated maximum of 24 Sv (1 Sv = 1×10^6 m^3 s^{-1}) in the summer and a minimum of 20 Sv in the autumn (Lee et al. 2001). In addition, the Taiwan warm current (TWWC), which originates from the Kuroshio, flows into the ECS from the South China Sea (SCS) through the Taiwan Strait off western Taiwan, with an annual mean northward transport of 0.78 Sv (Jan et al. 2006; Zhang et al. 2014) (Fig. 1). In the autumn and winter, Changjiang River runoff decreases (1.0×10^4 m^3 s^{-1} in January) (Yanagi 1994), and the CDW flows southwestward along the Chinese coast as a narrow band; the saline water (>34) of Kuroshio origin enters the Yellow Sea (Chang and Isobe 2003). However, during the summer, when the Changjiang River supplies a large amount of freshwater to the ECS (4.8×10^4 m^3 s^{-1} in July) (Yanagi 1994), the CDW starts to extend offshore and spread east- and northeastward. Then, the eastward-flowing CDW mixes with the TWWC and the Kuroshio water in the southern and eastern parts of the ECS, respectively (Chang and Isobe 2003) (Fig. 2a).

Since the annually averaged precipitation and evaporation are near equilibrium in the ECS and Yellow Sea and the Changjiang River runoff accounts for 90 % of the total river discharge (with an annual mean of ~3.0×10^4 m^3 s^{-1}), the Changjiang River discharge is considered the dominant freshwater source in the ECS (Beardsley et al. 1985; Chen et al. 1994; Yanagi 1994). This Changjiang River water extends offshore, and therefore, the water mass characteristics of the offshore ECS can be roughly described by the binary mixing of the cooler, less saline CDW and the warmer, more saline Kuroshio water (Zhang et al. 1990; Ichikawa and Beardsley 2002; Zhang et al. 2007). In contrast, there are other local water masses in the coastal ECS and the Yellow Sea, such as the TWWC and the Yellow Sea Central Cold Water (YSCCW) (Fig. 1). Therefore, the water mass characteristics and $\delta^{18}O_{sw}$–salinity relationships of the coastal ECS are more complicated than those of the offshore ECS, as discussed in a later section.

The northern Okinawa Trough, which is the main focus of this study, shows distinct seasonal changes in SSS (e.g., site HR11, SSS = 33.2–34.7) due to the advection of the CDW, in comparison with the central Okinawa Trough (e.g., site HR2, SSS = 34.4–34.8) (Figs. 1 and 2a). Sea surface temperature (SST) distribution also varies seasonally owing to monsoonal winds (Lee and Chao 2003). The climatological mean annual SST is 22.2 °C near site HR11 in the northern Okinawa Trough, with a maximum in August (28.1 °C) and a minimum in February (17.5 °C). In the central Okinawa Trough, the mean annual SST is 25.0 °C near site HR2, with a maximum in August (28.9 °C) and a minimum in February (21.5 °C) (Japan Oceanographic Data Center (JODC), available at http://www.jodc.go.jp/).

Methods

Seawater and sediment samples

Seawater samples were collected using a CTD (conductivity, temperature, and depth) Carousel multisampling

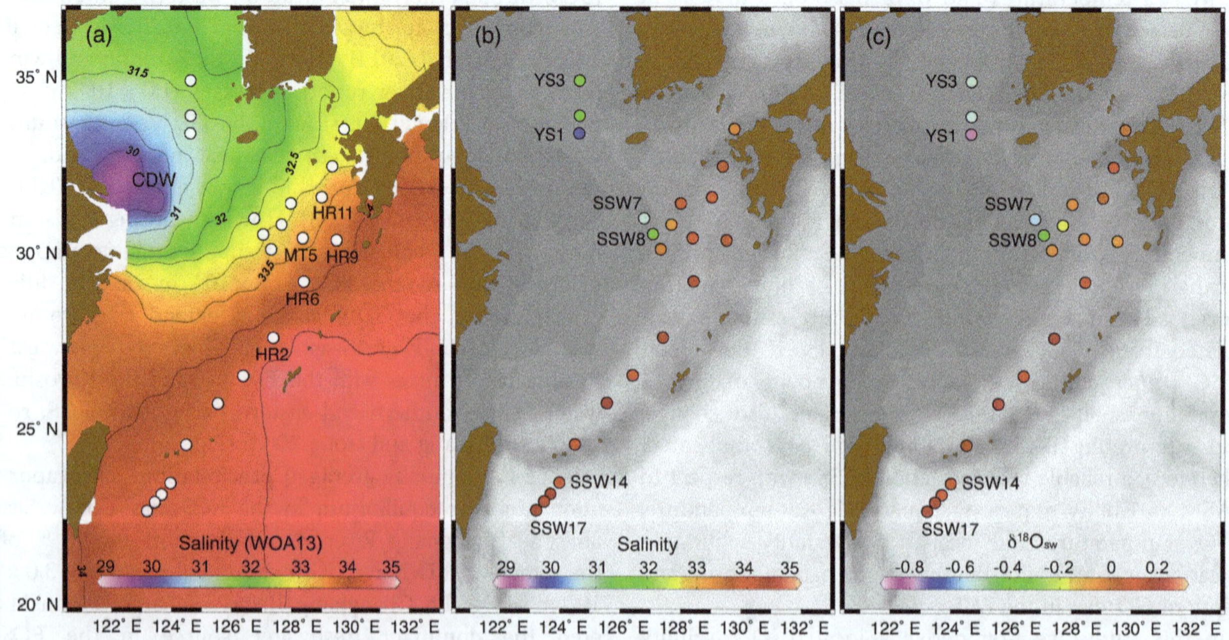

Fig. 2 Surface distribution of salinity and $\delta^{18}O_{sw}$. **a** SSS in July–September averaged statistically for 1955–2012. The data is from the World Ocean Atlas 2013 (Zweng et al. 2013). *Circles* represent locations for surface seawater samples (see Fig. 1 and Table 1). **b** SSS and **c** $\delta^{18}O_{sw}$ of seawater in early July 2013. The lowest SSS and $\delta^{18}O_{sw}$ values were found at YS1 within the CDW

system (Sea-Bird, model SBE 9plus) during the KH-13-4 cruise in early July 2013 (R/V Hakuho Maru) (Table 1). The CTD system has 24 attached Niskin bottles, each with a volume of 12 L. Subsurface seawater samples were taken with the Niskin bottles, and surface water samples were taken using a bucket (~0 m depth) or pumped from below the ship (~5 m depth). All seawater samples were transferred into glass vials for salinity and $\delta^{18}O_{sw}$ measurements, and the samples for the $\delta^{18}O_{sw}$ measurements were stored at 4 °C until laboratory analysis. Salinity was measured with an onboard Autosal laboratory salinometer (Model 8400B, Guildline Instruments Ltd., Canada) and is reported using the practical salinity scale. Sampling vials for salinity were prepared according to Joint Global Ocean Flux Study (JGOFS) protocols. The Autosal was standardized using International Association for the Physical Sciences of the Oceans (IAPSO) standard seawater.

Three sampling stations (YS1–3) are located in the southwest area off the Korean Peninsula, and 19 other stations (MT, HR, and SSW sites) are located in the continental shelf slope area of the ECS and in the Pacific Ocean (Fig. 1). We report salinity and $\delta^{18}O_{sw}$ data from these stations, and the vertical profiles of both datasets were obtained for nine stations (Fig. 1 and Table 1). Surface sediment samples were taken by a multiple corer at six sites along the central to northern parts of the Okinawa Trough from water depths of 709 to 1675 m during the KH-13-4 and KT-12-25 cruises (Fig. 1 and Table 2). Sediment samples were subsampled onboard with a 1 cm resolution and stored in a refrigerator until analysis in the laboratory.

The lithology of the sediment cores (~30 cm core length) was characterized as foraminifer-bearing silty clay or clayey silt with no visible turbidities or erosional surfaces. In general, the central to northern parts of the Okinawa Trough preserve thick late Holocene sediments due to their high sedimentation rates (10–50 cm kyr^{-1}), except in shelf slope regions (Ujiié et al. 2003; Sun et al. 2005; Kubota et al. 2010). All core-top $\delta^{18}O_c$ values of *G. ruber* were isotopically lighter than −2.0 ‰, which is consistent with the late Holocene values in the central to northern Okinawa Trough (Table 2) (Sun et al. 2005; Kubota et al. 2010). Therefore, we conclude that the core-top sediments used in this study represent at least late Holocene sediments. Site HR2 was located at a depth below the present calcite saturation horizon (CSH) in the ECS (~1600 m) (Fig. 6a). The Mg/Ca ratios from this site should be treated with caution regarding the partial dissolution of *G. ruber* shells, which will be discussed in a later section.

Analysis of the oxygen isotopic composition of seawater ($\delta^{18}O_{sw}$)

We measured $\delta^{18}O_{sw}$ with a stable isotope mass spectrometer (PRISM, Micromass UK, Ltd.) at the University of Toyama. Oxygen isotopic analysis was carried out using the automated H_2O–CO_2 equilibrium method

Table 1 Salinity and $\delta^{18}O_{sw}$ data at each station in the ECS

Depth (m)	Temperature (°C)	Salinity	$\delta^{18}O_{sw}$ (‰, VSMOW)	Std.dev (‰)	Site area
YS1 (30 June 2013, 124° 39.66′ E, 33° 29.80′ N)					Yellow Sea–ECS shelf site
0	23.40	29.89	−0.81	0.03	
5	22.95	30.60	−0.67	0.03	
10	22.83	31.51	−0.47	0.02	
20	17.10	32.82	−0.18	0.02	
40	10.17	33.05	−0.12	0.02	
YS2 (30 June 2013, 124° 40.07′ E, 34° 00.40′ N)					Yellow Sea–ECS shelf site
0	22.90	31.37	−0.54	0.03	
5	22.80	31.50	−0.52	0.02	
10	22.76	31.86	−0.43	0.03	
20	17.14	32.35	−0.35	0.04	
30	13.54	32.57	−0.29	0.03	
YS3 (1 July 2013, 124° 40.04′ E, 34° 59.64′ N)					Yellow Sea–ECS shelf site
0	23.90	31.88	−0.53	0.03	
5	23.12	31.86	−0.52	0.03	
10	17.29	31.87	−0.48	0.02	
20	13.73	32.09	−0.44	0.02	
30	7.71	32.21	−0.42	0.02	
HR8 (2 July 2013, 127° 19.84′ E, 30° 12.68′ N)					Yellow Sea–ECS shelf site
0	26.70	33.40	−0.06	0.02	
20	25.50	33.90	0.07	0.02	
40	24.09	34.34	0.15	0.02	
60	21.33	34.58	0.27	0.02	
70	19.44	34.61	0.26	0.02	
MT4 (2 July 2013, 127° 41.25′ E, 30° 54.42′ N)					Yellow Sea–ECS shelf site
0	26.10	33.10	−0.18	0.02	
6	25.81	33.16	−0.19	0.02	
10	25.79	33.37	−0.14	0.04	
30	24.44	33.81	0.00	0.03	
60	21.56	34.19	0.10	0.02	
100	19.08	34.64	0.20	0.02	
HR10 (2 July 2013, 128° 00.21′ E, 31° 30.22′ N)					Yellow Sea–ECS shelf site
0	26.10	33.75	−0.04	0.02	
5	25.87	33.68	−0.08	0.02	
10	25.80	33.82	−0.02	0.02	
20	25.49	34.07	0.05	0.02	
40	24.96	34.20	0.09	0.02	
59	22.91	34.39	0.17	0.04	
80	21.01	34.59	0.22	0.02	
99	19.86	34.65	0.24	0.03	
HR11 (2 July 2013, 129° 01.85′ E, 31° 40.60′ N)					Okinawa Trough site
0	26.90	33.88	0.01	0.02	
49	23.76	34.30	0.15	0.02	

Table 1 Salinity and $\delta^{18}O_{sw}$ data at each station in the ECS *(Continued)*

75	21.96	34.56	0.23	0.02	
99	20.27	34.69	0.24	0.04	
HR6 (8 July 2013, 128° 25.43′ E, 29° 17.42′ N)					Okinawa Trough site
0	29.70	34.22	0.12	0.05	
20	29.57	34.23	0.11	0.02	
50	28.63	34.35	0.23	0.02	
75	27.21	34.52	0.27	0.04	
99	23.69	34.66	0.33	0.04	
HR2 (9 July 2013, 127° 24.03′ E, 27° 40.97′ N)					Okinawa Trough site
0	29.40	34.25	0.18	0.02	
5	29.63	34.24	0.13	0.02	
10	29.36	34.19	0.14	0.03	
20	29.01	34.17	0.14	0.02	
50	26.32	34.40	0.24	0.02	
75	23.86	34.61	0.32	0.02	
100	22.47	34.74	0.35	0.02	
HR9 (7 July 2013, 129° 31.29′ E, 30° 28.26′ N)					Okinawa Trough site
0	28.50	33.73	−0.07	0.03	
MT5 (8 July 2013, 128° 23.41′ E, 30° 31.22′ N)					Okinawa Trough site
0	28.50	33.91	−0.03	0.02	
SSW07 (2 July 2013, 126° 45.99′ E, 31° 04.86′ N)					Yellow Sea–ECS shelf site
5	23.20	30.80	−0.59	0.04	
SSW08 (2 July 2013, 127° 04.33′ E, 30° 37.06′ N)					Yellow Sea–ECS shelf site
5	25.30	31.80	−0.36	0.02	
SSW09 (7 July 2013, 129° 47.01′ E, 33° 38.63′ N)					Off-Kyushu site
5	24.10	33.33	0.00	0.02	
SSW10 (7 July 2013, 129° 22.50′ E, 32° 34.00′ N)					Off-Kyushu site
5	23.60	33.71	0.12	0.05	
SSW11 (9 July 2013, 126° 24.20′ E, 26° 36.00′ N)					Okinawa Trough site
5	28.80	34.00	0.12	0.02	
SSW12 (9 July 2013, 125° 33.84′ E, 25° 49.26′ N)					Okinawa Trough site
5	29.20	34.38	0.21	0.03	
SSW13 (9 July 2013, 124° 30.06′ E, 24° 38.41′ N)					Okinawa Trough site
5	28.80	33.81	0.03	0.02	
SSW14 (10 July 2013, 123° 59.02′ E, 23° 32.27′ N)					Kuroshio site
5	29.30	34.06	0.06	0.02	
SSW15 (10 July 2013, 123° 40.09′ E, 23° 14.07′ N)					Kuroshio site
5	29.30	34.33	0.12	0.09	
SSW16 (10 July 2013, 123° 26.97′ E, 22° 59.89′ N)					Kuroshio site
5	29.40	34.23	0.15	0.02	
SSW17 (10 July 2013, 123° 11.60′ E, 22° 45.00′ N)					Kuroshio site
5	29.40	34.34	0.20	0.02	

Table 2 Details of the core-top sediments and metal/Ca and $\delta^{18}O$ values from *Globigerinoides ruber* (sensu stricto)

Core site	Latitude	Longitude	Water depth (m)	$\delta^{18}O$ (VPDB ‰)		Mg/Ca (mmol mol^{-1})	±SD	Ba/Ca (μmol mol^{-1})	±SD	Mn/Ca (μmol mol^{-1})	±SD
KH13-4-HR2MC	27° 40.79′ N	127° 23.95′ E	1675	−2.93	Ave	4.09	0.18	0.72	0.74	7.88	5.5
					Run #1	4.05		0.66		0.74	
					Run #2	3.94		0.78		6.50	
					Run #3	4.03		1.53		11.58	
					Run #4	4.35		2.25		12.71	
KT12-25-St.1	27° 58′ N	127° 16′ E	1153	−2.52	Ave.	4.20	0.03	0.77	0.01	23.25	21.3
					Run #1	4.22		0.78		8.18	
					Run #2	4.18		0.77		38.31	
KH13-4-HR6MC	29° 17.35′ N	128° 25.18′ E	1065	−2.47	Ave.	3.97	0.25	0.79	0.20	4.37	4.4
					Run #1	3.68		1.04		1.92	
					Run #2	4.13		0.66		9.49	
					Run #3	4.09		0.92		1.72	
KH13-4-HR9MC	30° 28.19′ N	129° 31.06′ E	709	−2.78	Ave.	4.19	0.25	0.81	1.09	53.01	13.7
					Run #1	3.91		0.81		48.40	
					Run #2	4.36		1.11		68.44	
					Run #3	4.30		2.82		42.18	
KH13-4-MT5MC	30° 31.23′ N	128° 23.38′ E	823	−2.27	Ave.	3.61	0.22	0.88	0.11	44.50	7.7
					Run #1	3.41		0.72		51.73	
					Run #2	3.44		0.95		46.37	
					Run #3	3.68		0.97		46.21	
					Run #4	3.89		0.87		33.67	
KH13-4-HR11MC	31° 40.57′ N	129° 01.99′ E	725	−2.42	Ave.	3.30	0.02	0.85	0.06	44.86	37.0
					Run #1	3.31		0.89		18.67	
					Run #2	3.29		0.80		71.05	
A7	27° 49.2′ N	126° 58.7′ E	1264	−2.35		4.06					
MD012403	25° 17′ N	123° 10′ E	1420	−2.51		4.40					
MD012404	26° 38.84′ N	125° 48.75′ E	1397	−2.60		4.24					
KY07-04-01	31° 38.35′ N	128° 56.64′ E	725	−2.50		3.50					

Table 2 Details of the core-top sediments and metal/Ca and $\delta^{18}O$ values from *Globigerinoides ruber* (sensu stricto)

No dissolution-corrected Mg/Ca-derived SST			Dissolution-corrected Mg/Ca-derived SST							
SST[a] (°C)	SST[b] (°C)	SST[c] (°C)	SST[a] (°C)	$\Delta[CO_3^{2-}]$ (μmol kg^{-1})	Corrected Mg/Ca (mmol mol^{-1})	Calcification SST	Shell-derived $\delta^{18}O_{sw}$ (VSMOW ‰)	Calcification salinity	Calcification season	Calcification depth (m)
26.7	24.4	29.4	28.7	−2.3	4.90	28.5	0.22	34.43	July–Sept	0–10
27.0	24.6	29.7	28.0	5.9	4.60	26.6	0.27	34.50	May–Oct	0–30
26.4	24.1	29.0	27.3	7.3	4.30	26.3	0.19	34.41	May–Oct	0–30
27.0	24.6	29.6	27.1	12.9	4.24	27.3	0.01	34.01	June–Sept	0–30
25.3	23.2	27.9	25.7	11.1	3.75	25.1	0.17	34.17	May–Oct	0–30
24.3	22.3	26.9	24.5	12.7	3.37	25.3	−0.20	33.76	June–Sept	0–30
26.6	24.3	29.3	27.9	4.2	4.55	25.9	0.37	34.54	May–Oct	0–30
27.5	25.1	30.2	29.0	1.7	5.01	26.5	0.39	34.43	May and Oct	0–30
27.1	24.7	29.8	28.6	2.1	4.84	26.9	0.21	34.38	May–Oct	0–30
24.9	22.9	27.6	25.2	12.7	3.57	25.6	−0.14	33.68	June–Sept	0–20

Decreases in Mg/Ca ratios due to calcite dissolution were calculated using critical thresholds for the dissolution (14 μmol kg^{-1} $\Delta[CO_3^{2-}]$) and sensitivity of Mg/Ca ratio to $\Delta[CO_3^{2-}]$ (0.05 mmol mol^{-1} per μmol^{-1} kg^{-1}) (Johnstone et al. 2011). Core-top Mg/Ca and $\delta^{18}O$ data from A7, MD012403, MD012404, KY07-04-01 were from Sun et al. (2005), Lin et al. (2006), Chen et al. (2010), and Kubota et al. (2010), respectively

[a]SST = ln(Mg/Ca/0.38)/0.089, Hastings et al. (2001)

[b]SST = ln(Mg/Ca/0.34)/0.102, Anand et al. (2003)

[c]SST = ln(Mg/Ca/0.3)/0.089, Lea et al. (2000)

(Epstein and Mayeda 1953). The stable isotope ratios are given as conventional δ values (‰), and the analytical precision was ±0.02 ‰ (1σ) for $\delta^{18}O$. Oxygen isotopic composition is expressed relative to Vienna Standard Mean Ocean Water (VSMOW).

$\delta^{18}O$ and Mg/Ca analyses of *G. ruber*

Core-top sediments (0–1 cm) were wet washed through a 63 μm sieve and then dried in an oven at 50 °C. To minimize ontogenic and growth rate effects on shell geochemistry, specimens of *G. ruber* (sensu stricto form; 20–30 shells) were handpicked from the 250–355 μm size fraction. Planktonic foraminifera were gently crushed and rinsed three times with ultrapure water and methanol (super special grade, Wako Pure Chemical Industries, Ltd.) to remove adherent clay particles. The shells were then split into two fractions for $\delta^{18}O_c$ and metal/Ca measurements. $\delta^{18}O_c$ values were obtained using a Finnigan MAT 253 mass spectrometer at the Center for Advanced Marine Core Research (CMCR), Kochi University and calibrated in accordance with standard NBS 19. The precision of these measurements was better than ±0.08 ‰ (1σ) for $\delta^{18}O_c$.

The samples for metal/Ca analysis were cleaned according to the procedure developed for trace element analysis (Boyle and Keigwin 1985; Rosenthal et al. 1997). In brief, samples underwent a multistep process consisting of initial rinses in ultrapure water and methanol, followed by treatments with hot reducing and oxidizing solutions, transferred into new acid-leached microcentrifuge tubes (1.5 mL), and finally leached with a dilute ultrapure nitric acid solution (0.001 M HNO_3, TAMAPURE-AA-100 from Tama Chemicals, Ltd.). The sample solution was then dissolved with a Sc-spiked

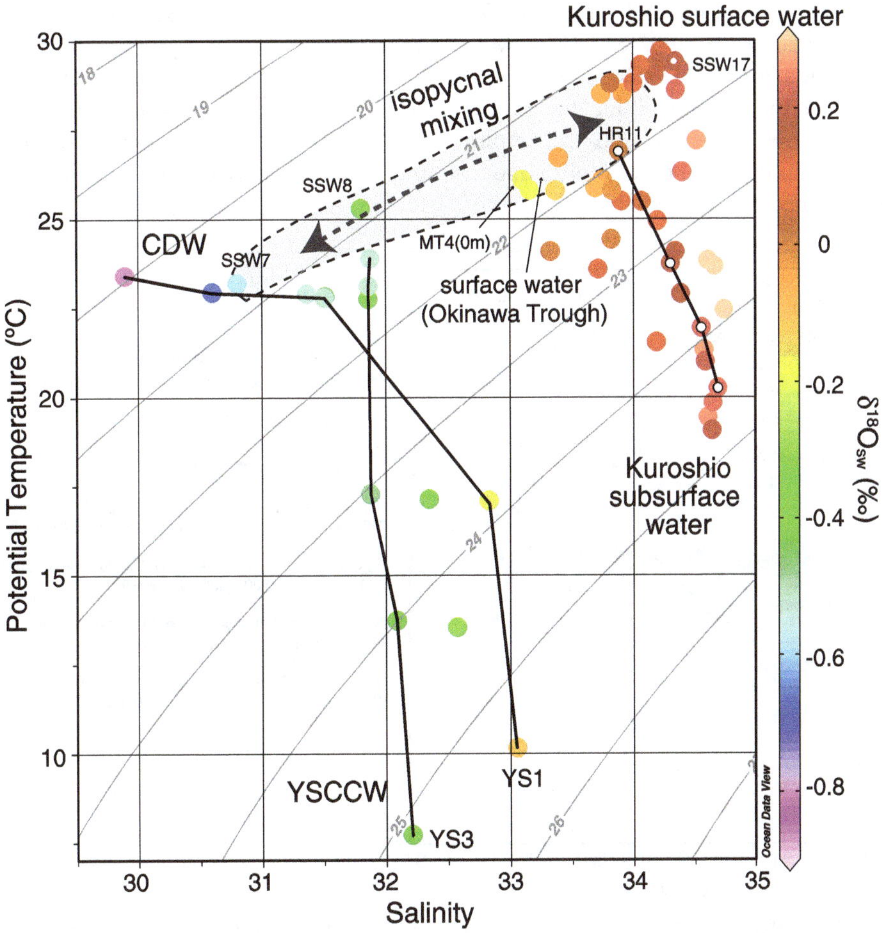

Fig. 3 Potential temperature and salinity (*T–S*) plot in the ECS. The *T–S* diagram shows four distinct water masses in early July: the Kuroshio surface water, Kuroshio subsurface water, YSCCW, and CDW. The *color in circles* indicates the $\delta^{18}O_{sw}$ value of the seawater. The Kuroshio surface and subsurface waters are characterized by higher $\delta^{18}O_{sw}$ values, whereas the CDW is characterized by relatively low $\delta^{18}O_{sw}$ values. Surface water properties in the northern part of the Okinawa Trough can be explained by isopycnal mixing with the CDW and Kuroshio surface water. The potential density curves (σ_θ = 19–27) are also shown. The *T–S* diagram was made using Ocean Data View (Schlitzer R, Ocean Data View, http://odv.awi.de, 2014)

dilute ultrapure nitric acid solution (2 % HNO_3) to obtain a Ca concentration of 10 ± 1 μg g^{-1}. All clean work was conducted in laminar flow benches or a clean room under trace metal clean conditions.

The metal/Ca ratio was determined with a Thermo Scientific ELEMENT 2 sector field inductively coupled plasma mass spectrometer (ICP–MS) at the University of Toyama, operated in a low-resolution mode ($m/\Delta m$ = 300) (Marchitto 2006). We analyzed ^{24}Mg, ^{26}Mg, ^{43}Ca, ^{44}Ca, ^{88}Sr, ^{137}Ba, and ^{138}Ba during the same run to determine Mg/Ca for temperature reconstructions and measured ^{56}Fe (in a mid-resolution mode) and ^{55}Mn to monitor contamination by clay minerals and diagenetic coatings (Table 2). Element counts were converted into molar ratios by the intensity ratio method based on a series of matrix-matched standard solutions. The accuracy and precision of Mg/Ca ratios were confirmed by analyses of the $CaCO_3$ reference materials BAM RS3 and ECRM 752-1. The repeated analyses gave Mg/Ca ratios of 0.786 ± 0.008 (n = 100, 1σ, RSD = 1.0 %) and 3.92 ± 0.06 (n = 24, 1σ, RSD = 1.5 %), respectively, which were within the reported value for BAM RS3 (0.791 ± 0.03) and slightly higher than the reported value for ECRM 752-1 (3.824 ± 0.095) (Greaves et al. 2008). Samples often show high Mn/Ca ratios (>100 μmol mol^{-1}) due to the presence of diagenetic coatings that were not removed during the cleaning process. Such samples were rejected. To further minimize the impact of sample heterogeneity and the analytical error of metal/Ca ratios, we report replicate measurements of metal/Ca ratios on re-picked *G. ruber* shells (Table 2).

Results and discussion

$\delta^{18}O_{sw}$–salinity relationship in the eastern ECS

In early July, surface water samples showed the highest SSS (34.3) at site SSW17 in the Kuroshio water area and the lowest SSS (29.9) at YS1 in the Yellow Sea (Fig. 2b and Table 1). The $\delta^{18}O_{sw}$ ranged from −0.81 to 0.21 ‰, and relatively low $\delta^{18}O_{sw}$ values were observed in the Yellow Sea, the ECS shelf, and the northern part of the Okinawa Trough (Fig. 2c). Spatial distributions of salinity and $\delta^{18}O_{sw}$ values were almost similar to the summer (July–September) salinity distribution averaged statistically over 1955–2012 (Zweng et al. 2013) (Fig. 2a), suggesting that the CDW spread east- and northeastward in the southern Yellow Sea and the northern Okinawa Trough during the sampling period.

A temperature and salinity plot (*T–S* diagram) of the surface and subsurface waters (0–100 m) in the Yellow Sea and the ECS shows four typical water masses: (1) the Kuroshio surface water in the subtropical Pacific Ocean (e.g., SSW14–17), (2) the Kuroshio subsurface

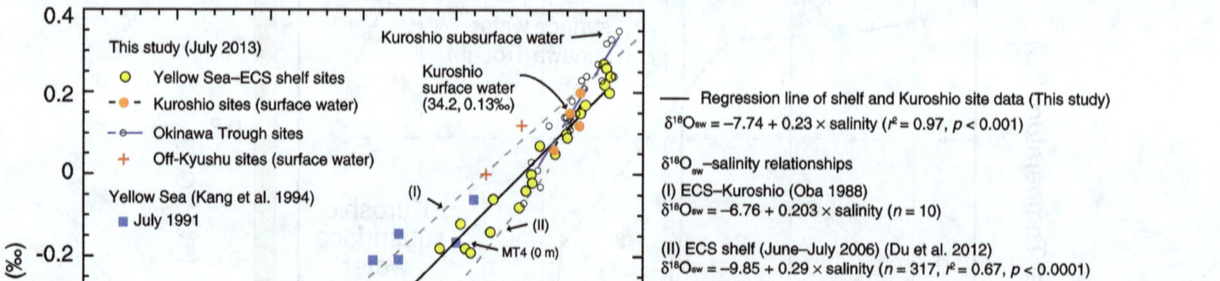

Fig. 4 $\delta^{18}O_{sw}$–salinity plot of shallow seawaters in the ECS. $\delta^{18}O_{sw}$–salinity plot of all shallow seawater (0–100 m) data in this study (early July 2013). The data are classified into four site groups: Yellow Sea–ECS shelf, Kuroshio, Okinawa Trough, and off-Kyushu sites (Table 1). The *black* regression line is derived from the data in the Yellow Sea–ECS shelf and Kuroshio sites. Since the surface waters in the northern Okinawa Trough are a mixture of the CDW and Kuroshio surface water, $\delta^{18}O_{sw}$ and salinity data from northern Okinawa Trough sites fall on this regression line. Regression lines (*dotted lines*) and equations from the literature are also shown. Regression lines III, IV, and V were developed in mainly coastal and shelf waters near the Changjiang Estuary. Although these regression lines represent freshwater end-members that are similar to our regression line, the end-member of the seawater shows lower $\delta^{18}O_{sw}$ values than that of the Kuroshio surface water (34.2 and 0.13 ‰; SSW14–17)

water in the Okinawa Trough (identified by a salinity of ~34.4 and a $\delta^{18}O_{sw}$ of 0.2–0.5 ‰) (Kang et al. 1994; Kim et al. 2005), (3) the YSCCW (<10 °C) in the bottom water (30 m) at YS3, and (4) the less saline CDW in the surface water at YS1 (Fig. 3). The *T–S* diagram also shows that the majority of the shallow water data in the Okinawa Trough fall on the isopycnal mixing line connecting the CDW and Kuroshio surface water (Fig. 3). Indeed, the $\delta^{18}O_{sw}$ values and salinity of the shallow seawater (0–100 m) are strongly correlated, indicating that the data can be explained by a binary mixing of freshwater and saline seawater (Fig. 4).

To derive a local $\delta^{18}O_{sw}$–salinity relationship around the northern Okinawa Trough, sampling sites were classified by geographic and oceanographic characteristics (Table 1). Sites SSW14–17, which are outside the ECS and under the influence of the Kuroshio surface water, were defined as Kuroshio sites. Data from the Yellow Sea and the ECS were divided into three groups: Yellow Sea–ECS shelf sites at depths shallower than 150 m, Okinawa Trough sites deeper than 150 m, and off-Kyushu sites influenced by local freshwater inputs from Kyushu (Fig. 4 and Table 1). Given that the eastward-flowing CDW mixes with the warmer, more saline Kuroshio water near the shelf edge in the summer (Chang and Isobe 2003), salinity and $\delta^{18}O_{sw}$ values in the Kuroshio surface water should reflect the end-member for saline water (Kang et al. 1994). Therefore,

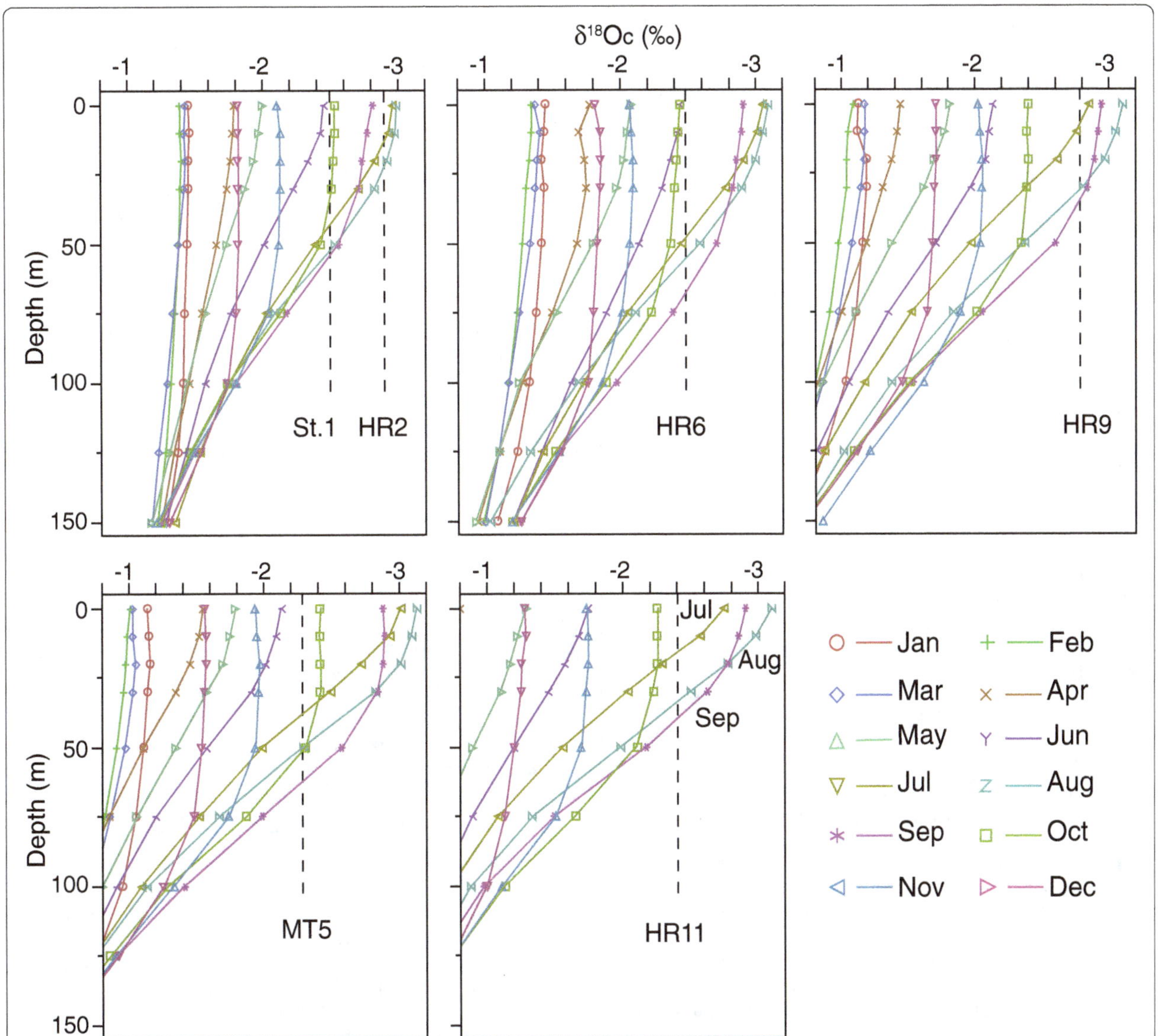

Fig. 5 Comparison of monthly variation in the predicted $\delta^{18}O_c$ values with the core-top $\delta^{18}O_c$ values. The *solid lines* show the monthly variation of the predicted $\delta^{18}O_c$ values within the water column. The *dotted lines* show the core-top $\delta^{18}O_c$ values of *G. ruber*. The predicted $\delta^{18}O_c$ values are calculated from temperature and salinity data (http://www.jodc.go.jp/) using a calcite equilibrium equation (Bemis et al. 1998). See details in the text

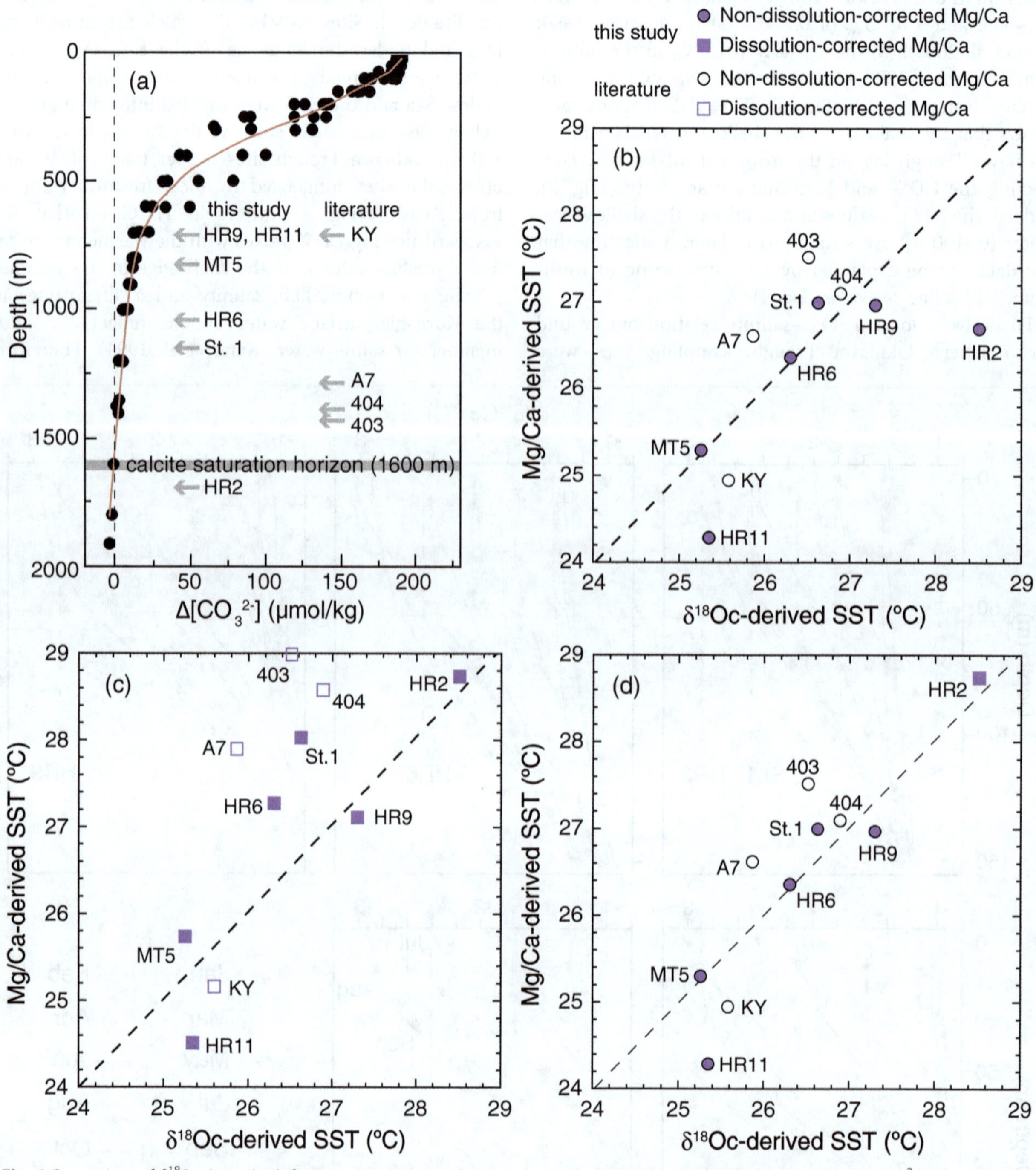

Fig. 6 Comparison of $\delta^{18}O_c$-derived calcification temperature with Mg/Ca-derived temperature. **a** Water column profile of $\Delta[CO_3^{2-}]$ in the ECS. Calculation of the calcite saturation state $\Delta[CO_3^{2-}]$ was accomplished via subtracting the carbonate ion concentration $[CO_3^{2-}]$ at saturation from the in situ $[CO_3^{2-}]$. To compute the in situ $[CO_3^{2-}]$ with the program CO2sys.xls (Lewis and Wallace 1998), we used total alkalinity, TCO_2, pH, and hydrographic water column data in the ECS (NODC Accession 0109919, http://data.nodc.noaa.gov/nodc/archive/metadata/approved/). *Arrows* indicate the water depths of cores analyzed in this study and by Sun et al. (2005), Lin et al. (2006), Chen et al. (2010), and Kubota et al. (2010). **b** $\delta^{18}O_c$-derived calcification temperature and non-dissolution-corrected Mg/Ca-derived temperature plot. **c** $\delta^{18}O_c$-derived calcification temperature and dissolution-corrected Mg/Ca-derived temperature plot. **d** $\delta^{18}O_c$-derived calcification temperature and Mg/Ca-derived temperature plot. The dissolution-corrected Mg/Ca ratio was used for the deepest site, HR2, where $\Delta[CO_3^{2-}] < 0$ μmol kg^{-1}. The other data were non-dissolution-corrected Mg/Ca ratios

taking into consideration this mixing of two water masses in the eastern part of the ECS, we derived the following $\delta^{18}O_{sw}$–salinity equation from the summer Yellow Sea–ECS shelf and Kuroshio site data:

$$\delta^{18}O_{sw} = -7.74\,(\pm 0.4) + 0.23\,(\pm 0.01) \times salinity\,(r^2 = 0.97, p < 0.001, n = 40) \quad (1)$$

The quoted errors of the slope and intercept are 95 % confidence intervals. This equation includes the surface (0–30 m water depth) and subsurface (40–100 m water depth) data. The surface data (0–30 m) alone result in the same equation within the 95 % confidence intervals of Eq. (1).

Equation (1) indicates a projected freshwater end-member of −7.74 ‰ (±0.4 ‰). This value is consistent with the $\delta^{18}O$ values (−8.4 to −7.1 ‰) for Changjiang River water in January and July (Zhang et al. 1990). However, the regression lines of the $\delta^{18}O_{sw}$–salinity relationships developed for the Changjiang River mouth and the western part of the ECS indicate lower $\delta^{18}O_{sw}$ values in the salinity range of 30 to 34 [$\delta^{18}O_{sw} = -8.41 + 0.24 \times$ salinity in January (line IV of Fig. 4) and $\delta^{18}O_{sw} = -7.06 +$ $0.20 \times$ salinity in July (line III of Fig. 4)]. Similar regression lines were developed for the Yellow Sea and the ECS [$\delta^{18}O_{sw} = -8.66 + 0.24 \times$ salinity in winter (line V of Fig. 4) and $\delta^{18}O_{sw} = -10.7 + 0.27 \times$ salinity in summer] (Ye et al. 2014). These regression lines do not pass through the end-member values of the Kuroshio surface water (salinity = 34.2–34.4 and $\delta^{18}O_{sw}$ = 0.1–0.2 ‰), indicating that the Kuroshio surface water was not the source of saline water for these areas (Fig. 4). Since these data were mainly from coastal and shelf waters near the Changjiang Estuary, the less saline, $\delta^{18}O$-depleted TWWC may be the dominant source of saline water rather than the Kuroshio water. Therefore, $\delta^{18}O_{sw}$–salinity relationships in the coastal ECS are highly variable and differ from our equation.

Like our regression line, regression lines that pass through the saline Kuroshio surface water were derived for the ECS shelf covering a broad sampling area for June–July (line II of Fig. 4) (Du et al. 2012), for the southern Yellow Sea in July (Kang et al. 1994), and for the area from the ECS to off the southern Japan Islands (Oba 1990) (line I of Fig. 4). Our regression line occupied an intermediate area between these regression lines. Furthermore, our equation is derived from the data obtained in the eastern part of the ECS where the eastward-flowing CDW mixes with the Kuroshio water in the summer. Therefore, our $\delta^{18}O_{sw}$–salinity equation should be representative of the northern Okinawa Trough during the EASM season.

Even if the freshwater end-member changed from −7.74 to −9.0 ‰ (given the averaged y-intercept of lines II, IV, and V as the potential change in the riverine $\delta^{18}O$), this change in amplitude does not yield a significant difference between the two calibration lines within the salinity range of 33 to 34.2. In addition, since end-member values of the CDW and the Kuroshio surface water may not have changed significantly during the Holocene in comparison with the glacial–interglacial cycles (Wang et al. 2008), our local $\delta^{18}O_{sw}$–salinity equation may provide an approximate relationship between the salinity and $\delta^{18}O_{sw}$ in the eastern part of the ECS, at least during the Holocene.

Core-top $\delta^{18}O_c$ and Mg/Ca ratios for *G. ruber*

The core-top $\delta^{18}O_c$ values for *G. ruber* ranged from −2.9 to −2.3 ‰ in the central to northern parts of the Okinawa Trough (Fig. 1 and Table 2). The lightest value was observed at HR2, the southernmost site, whereas the heaviest value was found at MT5 near the CDW advection area. Core-top Mg/Ca ratios of *G. ruber* ranged from 3.29 to 4.35 mmol mol^{-1} (Table 2). The Mg/Ca value at HR2 (4.09 mmol mol^{-1}, 1675 m water depth), which is the deepest site, was lower than that at the nearby shallower site St.1 (4.20 mmol mol^{-1}, 1153 m), even though HR2 had the lowest $\delta^{18}O_c$ value of *G. ruber* (−2.9 ‰) (Table 2). Since site HR2 is below the CSH (~1600 m) (Fig. 6a), the Mg/Ca ratio at HR2 seems to be lowered due to calcite dissolution (discussed below). Core-top Mg/Ca ratios, including the data from the literature (Sun et al. 2005; Lin et al. 2006; Chen et al. 2010; Kubota et al. 2010), tend to be higher in the southern sites (e.g., HR2 and St.1) and lower in the northern sites (HR11 and MT5) (Table 2), which is consistent with the distribution of the summer SST. In contrast, core-top $\delta^{18}O_c$ values correlate poorly with the Mg/Ca ratios ($r^2 = 0.19$), indicating that $\delta^{18}O_c$ values are strongly influenced by $\delta^{18}O_{sw}$.

Calcification season and calcification temperature of *G. ruber*

The foraminifera shell Mg/Ca response to temperature is biologically modulated and dependent on pH and/or [CO_3^{2-}] (Lea et al. 1999; Russell et al. 2004). Therefore, many species- and basin-specific Mg/Ca–temperature equations have been developed, some of which take into consideration the potential effect of dissolution (Lea et al. 2000; Dekens et al. 2002; Anand et al. 2003; McConnell and Thunell 2005). There is, however, no equation for *G. ruber* in the ECS, and previous studies conducted in the ECS have converted Mg/Ca values of *G. ruber* to SST using a Mg/Ca–temperature calibration equation [T (°C) = ln(Mg/Ca/0.38)/0.089] developed for *G. ruber* based on the SCS core-top sediments by Hastings et al. (2001) (Sun et al. 2005; Lin et al. 2006; Chen et al. 2010; Kubota et al. 2010). Although this calibration equation yields temperatures corresponding to the warm summer months, it has not been fully investigated whether there is consistency between $\delta^{18}O_c$- and Mg/Ca-derived temperatures, even though reconciled estimates of temperatures are crucial for calculating $\delta^{18}O_{sw}$.

First, to calculate calcification temperatures ($\delta^{18}O_c$-derived SSTs), we identified the calcification season and depth based on a comparison of core-top $\delta^{18}O_c$ values with predicted $\delta^{18}O_c$ values. Predicted $\delta^{18}O_c$ values were calculated using the following $\delta^{18}O_c$–temperature equation assuming oxygen isotopic equilibrium in foraminiferal calcite:

$$T(^\circ C) = 14.9 - 4.8\left(\delta^{18}O_c - \left(\delta^{18}O_{sw} - 0.27‰\right)\right) \quad (2)$$

This $\delta^{18}O_c$–temperature equation was determined from culturing experiments of a symbiont-bearing species, *Orbulina universa*, grown under high-light conditions (Bemis et al. 1998). The high-light conditions (>380 μEinstein m^{-2} s^{-1}) in the culturing experiments take into consideration symbiont effects on shell $\delta^{18}O_c$ (i.e., high photosynthetic activity forms ^{18}O-depleted shells). The applicability of this equation to the surface-dwelling, symbiont-bearing species *G. ruber* has been confirmed (Bemis et al. 1998; Benway et al. 2006). Temperature and salinity data averaged statistically for 1930–2003 in a 1° × 1° (latitude × longitude) grid provided by JODC were used for the calculation. Salinity was converted to $\delta^{18}O_{sw}$ values using Eq. (1). The term −0.27 ‰ corrects for the $\delta^{18}O$ difference between VSMOW and Pee Dee Belemnite (PDB).

Figure 5 shows a comparison of the predicted vertical profiles of $\delta^{18}O_c$ values for each month with core-top $\delta^{18}O_c$ values of *G. ruber* at each site. Among the six core-top datasets, HR2 and HR9 showed relatively lighter $\delta^{18}O_c$ values that are approximately equal to the $\delta^{18}O_c$ values in the surface water for July–September. In contrast, the other sites (St.1, HR6, MT5, and HR11) were marked by isotopically heavier values than the peak summer $\delta^{18}O_c$ in the surface water, suggesting a large contribution of *G. ruber* shells formed in deeper waters (>50 m) or in surface water in May, June, and/or October (Fig. 5). Since *G. ruber* (sensu stricto) lives predominantly in the top 20 m of the subtropical North Pacific (Kuroyanagi and Kawahata 2004), calcification of shells in deeper water (>50 m) should be negligible. Furthermore, sediment trap data from the ECS, which unfortunately do not include data for September, show that the fluxes of *G. ruber* from May to October account for 80 % of the total fluxes each year and the fluxes in May alone account for 20 % of the yearly fluxes (Xu et al. 2005). Based on these data, we regard the dominant habitat depth of *G. ruber* in the ECS as the top 0 to 30 m (mixed layer) and contend that the *G. ruber* shell geochemistry in the ECS records the weighted mean surface ocean conditions from May to October.

To calculate the calcification temperatures and salinity of the seawater in which *G. ruber* shells were formed, we computed seasonal average (May–October or July–September) $\delta^{18}O_c$ values and compared these predicted values to core-top $\delta^{18}O_c$ values (Table 2). The calcification seasons and depths at HR2 and HR11 were identified as July–September and 0–10 m and June–September and 0–30 m, respectively. The $\delta^{18}O_c$-derived SSTs at HR2 and HR11 were calculated as 28.5 and 25.3 °C, respectively (Fig. 6 and Table 2). The calculated calcification SSTs at all sites ranged from 25.1 to 28.5 °C in the central to northern Okinawa Trough (Table 2). It is noteworthy that the calcification season of *G. ruber* was not necessarily uniform across the ECS; data from sites HR2 and HR9 clearly show a warmer seasonal bias (close to the peak summer SST), in contrast to other site data (Figs. 1 and 5). The core-top $\delta^{18}O_c$ data shows that Eq. (2) provides better estimates for calcification depth, season, and temperature that agree with sediment trap and plankton tow data (Kuroyanagi and Kawahata 2004; Xu et al. 2005).

Although we applied the $\delta^{18}O_c$–temperature equation by Bemis et al. (1998) to the core-top data, there are other $\delta^{18}O_c$–temperature relationships that yield various estimates for calcification temperature. For instance, $\delta^{18}O_c$–temperature equations by Kim and O'Neil (1997) and Mulitza et al. (2003) yield an offset of approximately 1 °C to the calcification temperatures calculated by Eq. (2). If these equations are applied to our core-top data with the assumption of $\delta^{18}O$ equilibrium, calcification temperatures at HR2 and HR9 will exceed the peak summer SST. Therefore, we should consider $\delta^{18}O$ disequilibrium in foraminiferal calcite (i.e., vital effect) for at least these sites. However, since the vital effect for *G. ruber* has not been determined precisely (0 to −1.0 ‰) (Niebler et al. 1999), applying these equations will introduce significant uncertainties in calcification temperature estimates.

Assessment of *G. ruber* Mg/Ca-derived temperatures

G. ruber is one of the species of planktonic foraminifera most susceptible to dissolution (Thunell and Honjo 1981). Partial dissolution of *G. ruber* shells can decrease Mg/Ca ratios owing to the higher solubility of Mg-rich calcite (Dekens et al. 2002). The ECS is the most undersaturated marginal sea with respect to calcite, and the CSH is shallow (~1600 m) (Fig. 6a). In our dataset, site HR2 is below the CSH, and the published core-top data from site A7, MD012404 (hereafter 404), and MD012403 (hereafter 403) were also derived from bottom water conditions with low calcite saturation states of <5 μmol kg^{-1} ($\Delta[CO_3^{2-}] = [CO_3^{2-}]_{in\ situ} - [CO_3^{2-}]_{saturation}$) (Fig. 6a; see caption of Fig. 6a for more details). According to data from core-top *G. ruber* tests of a depth transect in the Ontong Java Plateau, Mg/Ca ratios of *G. ruber* started to decrease with decreasing $\Delta[CO_3^{2-}]$ (−0.05 mmol mol^{-1} per μmol^{-1} kg^{-1}) from an initial $\Delta[CO_3^{2-}]$ of 14 μmol kg^{-1} (Johnstone et al. 2011). This finding is a cause for concern that ECS Mg/Ca data obtained under bottom water conditions with low $\Delta[CO_3^{2-}]$ may be influenced by dissolution (Fig. 6a).

Therefore, we assessed the impact of calcite dissolution on Mg/Ca ratios in the ECS by applying critical thresholds for the dissolution (14 μmol kg^{-1} $\Delta[CO_3^{2-}]$) and the sensitivity of the Mg/Ca ratio to $\Delta[CO_3^{2-}]$ (0.05 mmol mol^{-1} per μmol^{-1} kg^{-1}) (Johnstone et al. 2011). We calculated $\Delta[CO_3^{2-}]$ for the bottom waters at each core site and estimated the dissolution-corrected Mg/Ca ratios for all core sites (Table 2). Figure 6 shows a comparison of the $\delta^{18}O_c$-derived SSTs with non-dissolution-corrected and dissolution-corrected Mg/Ca-derived SSTs (Fig. 6b, c). Both Mg/Ca ratios were converted to SSTs using the SCS calibration equation of Hastings et al. (2001).

The shallowest site, HR9, where the $\Delta[CO_3^{2-}]$ value is 12.9 μmol kg^{-1}, showed good agreement with the non-dissolution-corrected Mg/Ca- and $\delta^{18}O_c$-derived SSTs (Fig. 6b). It is noteworthy that core-top data at depths ranging from 1000 to ~1400 m (HR6, St.1, A7, 404, and 403) also showed agreement with the non-dissolution-corrected Mg/Ca- and $\delta^{18}O_c$-derived SSTs, even though the $\Delta[CO_3^{2-}]$ values were lower than the critical threshold for dissolution (10–20 μmol kg^{-1}) (Regenberg et al. 2009; Johnstone et al. 2011) (Fig. 6b). The deepest site, HR2 (1675 m water depth), showed that the non-dissolution-corrected Mg/Ca-derived SST underestimates the $\delta^{18}O_c$-derived SST by 1.8 °C (Fig. 6b), and the dissolution-corrected Mg/Ca-derived SST agrees well with the $\delta^{18}O_c$-derived SST (Fig. 6c). However, this dissolution correction induced a larger offset to the $\delta^{18}O_c$-derived SSTs for other sites (1000 to ~1400 m water depths) (Fig. 6c).

The Mg/Ca–temperature calibration equation of Hastings et al. (2001) was developed for *G. ruber* from SCS core-top samples recovered from depths above 2000 m. As in the ECS, the deep waters in the SCS are characterized by low $\Delta[CO_3^{2-}]$ with the CSH at a water depth of 2500 m (Regenberg et al. 2009). Presumably, samples recovered from depths of 1500 to 2000 m (<15 μmol kg^{-1} $\Delta[CO_3^{2-}]$) for the calibration equation might have been influenced by partial dissolution under low $\Delta[CO_3^{2-}]$. Therefore, we consider that the SCS calibration equation potentially involves a correction for calcite dissolution under low $\Delta[CO_3^{2-}]$. Consequently, the correction for calcite dissolution is required only for samples obtained below the CSH in the ECS.

Thus, in this study, we integrated data of dissolution-corrected (for HR2 from below the CSH) and non-dissolution-corrected (for other cores above the CSH) Mg/Ca values to obtain Mg/Ca-derived SSTs (Fig. 6d). Although HR11 and 403 are still offset from the $\delta^{18}O_c$-derived SST by 1 °C, the other core-top data showed good agreement with both SST estimates ($r^2 = 0.83$, $p < 0.001$, except HR11 and 403) (Fig. 6d). This SCS calibration equation reconciles Mg/Ca-derived and $\delta^{18}O_c$-derived SSTs in the ECS, both of which reflect SSTs in the warmer season months. However, other calibration equations (Lea et al. 2000; Anand et al. 2003; McConnell and Thunell 2005) do not yield temperatures corresponding to the warmer season months and do not agree well with $\delta^{18}O_c$-derived SSTs (Table 2).

$\delta^{18}O_{sw}$ estimate for *G. ruber* in core-top sediments

Core-top $\delta^{18}O_{sw}$ values were calculated from Mg/Ca-derived SSTs and core-top $\delta^{18}O_c$ values using Eq. (2). The core-top data of the shell-derived $\delta^{18}O_{sw}$ and the salinity, including the published data from KY, A7, 404, and 403 (Sun et al. 2005; Lin et al. 2006; Chen et al. 2010; Kubota et al. 2010), are plotted in Fig. 7a. Since the estimation of the shell-derived $\delta^{18}O_{sw}$ is affected by the uncertainty in the $\delta^{18}O_c$ (±0.11 ‰) and the Mg/Ca-derived temperature (±1 °C, corresponding to approximately ±0.26 ‰) (Schmidt 1999; Rohling 2007), the reconstructed $\delta^{18}O_{sw}$ introduces a large propagation error of ±0.28 ‰ (1σ). Although the regression line of the core-top $\delta^{18}O_{sw}$ data has a steeper gradient and a lower *y*-intercept, given 1σ error bars of $\delta^{18}O_{sw}$ estimates the data points fall primarily on the local $\delta^{18}O_{sw}$–salinity regression line of the eastern part of the ECS (Fig. 7a). Indeed, sites KY and HR11 in the northern Okinawa Trough were marked by the lowest shell-derived $\delta^{18}O_{sw}$ values, whereas St.1 and A7 in the central Okinawa Trough showed the highest shell-derived $\delta^{18}O_{sw}$ values. Sites HR9 and MT5 had intermediate $\delta^{18}O_{sw}$ values between the above two datasets (Fig. 7a). The core-top $\delta^{18}O_{sw}$ values were roughly distributed as expected from the peak summer SSS at each site.

Although site MT5 is in an area with a lower summer SSS than that of site HR9, the shell-derived $\delta^{18}O_{sw}$ value at MT5 (0.17 ‰) was higher than that at HR9 (0.01 ‰) (Fig. 7a). One potential explanation is that *G. ruber* shells at HR9 were formed in the surface waters during the peak summer season when SSS decreases distinctly, whereas *G. ruber* at MT5 was not abundant in the peak summer season, as inferred from the $\delta^{18}O_c$ data (Fig. 5). These data tell us that the difference of 0.16 ‰ in the shell-derived $\delta^{18}O_{sw}$ may be caused by changes in the seasonal habitat of *G. ruber*.

This core-top investigation shows that realistic $\delta^{18}O_{sw}$ values can be reconstructed by the Mg/Ca–temperature calibration of Hastings et al. (2001) and the $\delta^{18}O_c$–

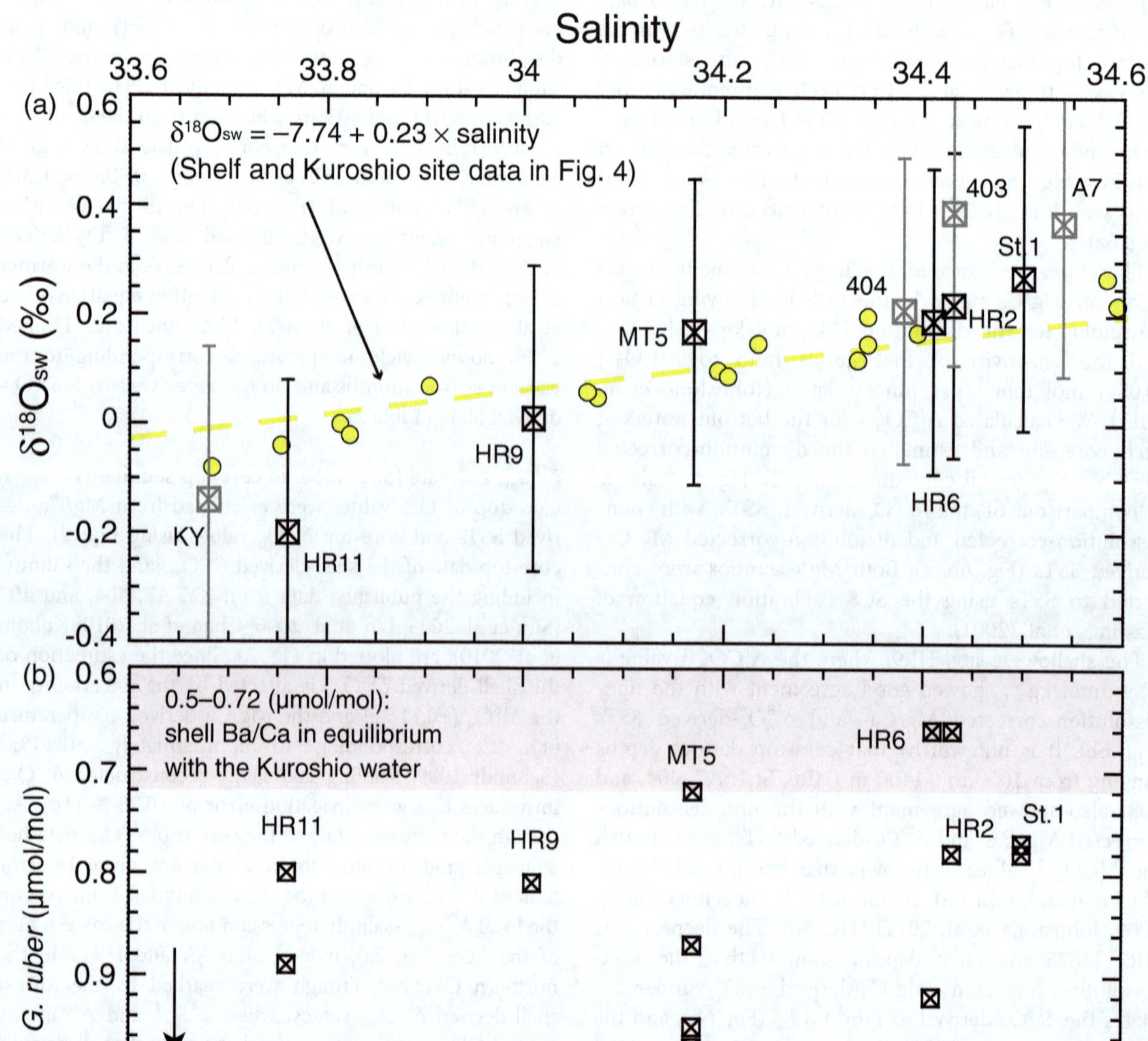

Fig. 7 Salinity, shell-derived $\delta^{18}O_{sw}$, and Ba/Ca ratios of core-top *G. ruber* tests. **a** Shell-derived $\delta^{18}O_{sw}$ and calcification salinity plot of core-top samples in this study and the literature. Shell-derived $\delta^{18}O_{sw}$ values are calculated using the Mg/Ca-derived temperature and core-top $\delta^{18}O_c$ of *G. ruber*. The data from KY, A7, 403, and 404 (*gray squares*) are from Kubota et al. (2010), Sun et al. (2005), Lin et al. (2006), and Chen et al. (2010), respectively. *Error bars* show the propagation error of ±0.28 ‰ (1σ). Seawater data (*green circles*) and their regression line (green broken line, Fig. 4) are also shown with the core-top data. Core-top data fall close to the local $\delta^{18}O_{sw}$–salinity regression line. **b** Ba/Ca$_{G.\ ruber}$ ratios and calcification salinity plot of all core-top samples in this study. The replicated analyses of Ba/Ca$_{G.\ ruber}$ ratios at each core site show high variability, resulting in the range from 0.66 to 2.82 μmol mol^{-1}. The calculated shell Ba/Ca ratio in equilibrium with the Kuroshio water is 0.5–0.72 μmol mol^{-1}. Lower values at HR2 and HR6 are within this range. Higher shell Ba/Ca ratios than this Kuroshio water range might be attributable to barite contamination or calcification in low saline, Ba-enriched CDW

temperature equation of Bemis et al. (1998), although $\delta^{18}O_{sw}$ estimates introduce a large error. Furthermore, this finding indicates that robust reconstruction of EASM-related variability in SSS might be possible by conducting multiple replicates of paired $\delta^{18}O_c$ and Mg/Ca analyses to decrease the large uncertainty. In addition, the utility of potential paleo-salinity proxies, such as the foraminiferal Ba/Ca ratio and the organic compound-specific δD, should be investigated further (Rohling 2007).

Possibility of using Ba/Ca ratios of *G. ruber* as a proxy for surface salinity

In general, the desorption of Ba^{2+} from suspended sediments in rivers results in a very high riverine [Ba^{2+}] relative to seawater. If there is a single river source, conservative mixing of river water and ambient seawater produces a linear inverse correlation between salinity and [Ba^{2+}] (Bahr et al. 2013). Ba^{2+} incorporation into living planktonic foraminifera shells is linearly dependent on the [Ba^{2+}] of the water, with a constant partition coefficient for Ba ($D_{Ba} = 0.15$) (Lea and Spero 1994; Hönisch et al. 2011). *G. ruber* populations can thrive in low-salinity surface water influenced by river discharge (Schmuker and Schiebel 2002; Arnold and Parker 2003). This ecological preference of *G. ruber* and the incorporation of Ba into foraminifera shells makes Ba/Ca ratios in fossil *G. ruber* a proxy for the influence of river discharge on salinity (Weldeab et al. 2007; Schmidt and Lynch-Stieglitz 2011).

In the case of the ECS, the Changjiang River accounts for approximately 90 % of the total river discharge into the ECS, with large quantities of Ba^{2+} (~360 nmol L^{-1}) (Beardsley et al. 1985; Qu et al. 1993; Yanagi 1994). Based on our preliminary [Ba^{2+}] data in the ECS measured by an isotope dilution method (Klinkhammer and Chan 1990), the surface water at SSW8 (30.80 salinity) showed a high [Ba^{2+}] of 57.0 nmol L^{-1} (Ba/Ca ratio of 6.31 μmol mol^{-1}), and the Kuroshio surface water (SSW14–17, average 34.2) showed a low [Ba^{2+}] of 32.9 nmol L^{-1} (Ba/Ca of 3.27 μmol mol^{-1}). The [Ba^{2+}] data for the Kuroshio water were consistent with previously reported data from Kuroshio waters (29.9–33.5 nmol L^{-1}) (Sugiyama et al. 1984). Other surface samples from the northern Okinawa Trough fall within these [Ba^{2+}] and Ba/Ca ranges, showing a linear relationship between $Ba/Ca_{seawater}$ and salinity (Horikawa et al., personal communication).

Our Ba/Ca ratios in *G. ruber* (hereafter $Ba/Ca_{G.\ ruber}$) from core-top sediments ranged from 0.66 to 2.82 μmol mol^{-1} (Fig. 7b). The lowest Ba/Ca value of 0.66 μmol mol^{-1} was observed at sites HR2 and HR6 in the central Okinawa Trough. Given the $Ba/Ca_{seawater}$ ratio of 3.27 μmol mol^{-1} in the Kuroshio surface water and the likely D_{Ba} of 0.15–0.22 (Hall and Chan 2004; Hönisch et al. 2011), the calculated shell Ba/Ca ratio in equilibrium with the Kuroshio water should be 0.5 to 0.72 μmol mol^{-1}. Indeed, HR2 and HR6 values were within the expected range for Kuroshio surface water. In contrast, HR11, where the lowest SSS among the studied sites was observed, had relatively higher $Ba/Ca_{G.\ ruber}$ ratios (0.80–0.89 μmol mol^{-1}) (Fig. 7b). Given the $Ba/Ca_{seawater}$ ratio of 3.8 μmol mol^{-1} (at HR10, 0 m) and the D_{Ba} of 0.15 to 0.22, the shell Ba/Ca ratio in equilibrium should be 0.57 to 0.84 μmol mol^{-1}. The core-top $Ba/Ca_{G.\ ruber}$ ratios observed at HR11 were consistent with the upper range of expected values, probably reflecting the influence of the Ba-enriched CDW (i.e., low-salinity waters).

However, all core-top $Ba/Ca_{G.\ ruber}$ data at six sites do not show a significant relationship with the salinity, but replicate data at each site showed highly variable $Ba/Ca_{G.\ ruber}$ ratios, some of which exceeded the values at site HR11 (Fig. 7b). Although we are unable to adequately explain such highly variable $Ba/Ca_{G.\ ruber}$ ratios at this time, one likely reason is sample heterogeneity. In this study, to obtain averaged $Ba/Ca_{G.\ ruber}$ ratios for each site, we re-picked 20–30 shells of *G. ruber* more than twice after picking the samples labeled run #1 (Table 2). Although the samples from run #1 gave reasonable $Ba/Ca_{G.\ ruber}$ ratios (0.66–0.89 μmol mol^{-1}) at each site except HR6, the samples from runs #2 to #4 yielded anomalously high $Ba/Ca_{G.\ ruber}$ ratios compared with the samples from run #1. Given that there were few well-preserved, clean *G. ruber* tests for these sites and that replicate measurements of Mn/Ca and Mg/Ca ratios were within an acceptable range, we argue that samples from runs #2 to #4 may have involved impure *G. ruber* tests in which barite crystallized on fossil shell surfaces.

Finally, based on the available dataset of $Ba/Ca_{seawater}$ in the ECS, we propose that pristine $Ba/Ca_{G.\ ruber}$ ratios at northern Okinawa Trough sites would be less than 0.84 μmol mol^{-1}. A possible reason for higher core-top $Ba/Ca_{G.\ ruber}$ ratios (>0.84 μmol mol^{-1}) is contamination by sedimentary barite adherent on fossil shells. Our preliminary data suggest that $Ba/Ca_{G.\ ruber}$ data as a paleo-salinity proxy should be derived from well-preserved, clean *G. ruber* tests or diethylene triamine pentaacetic acid (DTPA)-cleaned *G. ruber* tests that can minimize the influence of barite contamination (Lea and Boyle 1991; Hall and Chan 2004). Since the $Ba/Ca_{G.\ ruber}$ records from the northern Okinawa Trough sediments may potentially identify high river discharge events related to stronger EASMs, assessment of foraminiferal Ba/Ca should be continued in the ECS.

Conclusions

The main findings of this study are as follows:

(1) Shallow seawater samples (0–100 m) taken from the ECS in early July showed a strong correlation between $\delta^{18}O_{sw}$ and salinity. The *T–S* diagram in the surface waters in the ECS shows that the eastward-flowing CDW mixes with the saline Kuroshio waters. Based on this finding, we derived the following $\delta^{18}O_{sw}$–salinity equation from the Yellow Sea–ECS shelf and Kuroshio site data:

$$\delta^{18}O_{sw} = -7.74(\pm 0.4) + 0.23(\pm 0.01) \times salinity(r^2 = 0.97, p < 0.001, n = 40).$$

This local $\delta^{18}O_{sw}$–salinity equation might be representative of the northern Okinawa Trough during the EASM season.

(2) We found that the dominant habitat depth of *G. ruber* is within the top 0 to 30 m in the ECS. The calcification season of *G. ruber* is mainly during the warm summer months (May–October) but may not necessarily be uniform across the ECS. We confirmed that the Mg/Ca–temperature calibration by Hastings et al. (2001) yields temperatures corresponding to the warmer season months, as expected from sediment trap data. The Mg/Ca-derived SSTs agreed with calcification temperatures calculated by Bemis et al. (1998). Site HR2, where the bottom water is undersaturated with calcite, required a correction for calcite dissolution.

(3) We found that core-top data for shell-derived $\delta^{18}O_{sw}$ and salinity fall primarily on our local $\delta^{18}O_{sw}$–salinity regression line giving 1σ error of $\delta^{18}O_{sw}$ estimates. The Mg/Ca–temperature calibration by Hastings et al. (2001) and the $\delta^{18}O_c$–temperature equation by Bemis et al. (1998) should be appropriate for calculating $\delta^{18}O_{sw}$ in the ECS.

(4) $Ba/Ca_{G.\ ruber}$ from core-top sediments ranged from 0.66 to 2.82 μmol mol^{-1}. There was not a significant relationship between salinity and $Ba/Ca_{G.\ ruber}$ due to the highly variable $Ba/Ca_{G.\ ruber}$ data. Given the seawater Ba/Ca data and the published partition coefficient for Ba (D_{Ba} = 0.15–0.22), pristine $Ba/Ca_{G.\ ruber}$ ratios at northern Okinawa Trough sites should be less than 0.84 μmol mol^{-1}. One possible reason for higher $Ba/Ca_{G.\ ruber}$ ratios (>0.84 μmol mol^{-1}) is contamination by sedimentary barite adherent on fossil shells. Further evaluation of the $Ba/Ca_{G.\ ruber}$ ratio as a paleo-salinity proxy requires DTPA-cleaned Ba/Ca data that can minimize the influence of barite contamination.

Abbreviations

EASM: East Asian summer monsoon; ECS: East China Sea; CDW: Changjiang diluted water; SSS: sea surface salinity; TWWC: Taiwan warm current; SCS: South China Sea; YSCCW: Yellow Sea Central Cold Water; SST: sea surface temperature; JODC: Japan Oceanographic Data Center; CTD: conductivity, temperature, and depth; JGOFS: Joint Global Ocean Flux Study; IAPSO: International Association for the Physical Sciences of the Oceans; CSH: calcite saturation horizon; VSMOW: Vienna Standard Mean Ocean Water; CMCR: Center for Advanced Marine Core Research; PDB: Pee Dee Belemnite; DTPA: diethylene triamine pentaacetic acid.

Competing interests

The authors declare that they have no competing interests.

Authors' contributions

KH conceived and designed the study. TK carried out the measurements of the seawater $\delta^{18}O$, foraminiferal $\delta^{18}O_c$, and metal/Ca ratios under the supervision of Professors MM and JZ. All authors analyzed and discussed the data. All authors contributed toward writing the manuscript and read and approved the final manuscript.

Authors' information

KH is an associate professor at the University of Toyama. He is a paleoceanographer and a geochemist who has worked on SST reconstruction around Japan using the Mg/Ca–temperature proxy and the analysis of deep-sea water mass circulation using neodymium isotopes. JZ is also a geochemist and a professor at the University of Toyama with considerable experience in geochemical analysis. TK is a Ph.D. student supervised by KH and JZ. MM is a professor at the Center for Advanced Marine Core Research (CMCR), Kochi University.

Acknowledgements

We acknowledge the tremendous support of the KH-13-4 and KT-12-25 shipboard scientists and staff. This study was performed under the cooperative research program of the Center for Advanced Marine Core Research (CMCR), Kochi University (#13B050). We thank the chief editors Ryuji Tada, Hodaka Kawahata, Yasufumi Iryu, and two anonymous reviewers for their significant help in improving this manuscript.

Author details

[1]Graduate School of Science and Engineering for Research, University of Toyama, Gofuku 3190, Toyama 930-8555, Japan. [2]Graduate School of Science and Engineering for Education, University of Toyama, Gofuku 3190, Toyama 930-8555, Japan. [3]Center for Advanced Marine Core Research, Kochi University, B200 Monobe, Nankoku 783-8502, Japan.

References

Anand P, Elderfield H, Conte MH (2003) Calibration of Mg/Ca thermometry in planktonic foraminifera from a sediment trap time series. Paleoceanography 18(2):1050. doi:10.1029/2002PA000846

Arnold A, Parker W (2003) Biogeography of planktonic Foraminifera. In: Gupta BKS (ed) Modern Foraminifera. Springer, Netherlands, pp 103–122. doi:10.1007/0-306-48104-9_7

Bahr A, Schönfeld J, Hoffmann J, Voigt S, Aurahs R, Kucera M, Flögel S, Jentzen A, Gerdes A (2013) Comparison of Ba/Ca and $\delta^{18}O_{WATER}$ as freshwater proxies: A multi-species core-top study on planktonic foraminifera from the vicinity of the Orinoco River mouth. Earth Planet Sci Lett 383:45–57. doi:10.1016/j.epsl.2013.09.036

Beardsley RC, Limeburner R, Yu H, Cannon GA (1985) Discharge of the Changjiang (Yangtze River) into the East China Sea. Cont Shelf Res 4(1–2):57–76. doi:10.1016/0278-4343(85)90022-6

Bemis BE, Spero HJ, Bijima J, Lea DW (1998) Reevaluation of the oxygen isotopic composition of planktonic foraminifera: Experimental results and revised paleotemperature equations. Paleoceanography 13(2):150–160

Benway HM, Mix AC, Haley BA, Klinkhammer GP (2006) Eastern Pacific Warm Pool paleosalinity and climate variability: 0–30 kyr. Paleoceanography 21(3):PA3008. doi:10.1029/2005PA001208

Boyle EA, Keigwin LD (1985) Comparison of Atlantic and Pacific paleochemical records for the last 215,000 years: changes in deep ocean circulation and chemical inventories. Earth Planet Sci Lett 76(1–2):135–150

Chang P-H, Isobe A (2003) A numerical study on the Changjiang diluted water in the Yellow and East China Seas. J Geophys Res 108(C9):3299. doi:10.1029/2002JC001749

Chang Y-P, Chen M-T, Yokoyama Y, Matsuzaki H, Thompson WG, Kao S-J, Kawahata H (2009) Monsoon hydrography and productivity changes in the East China Sea during the past 100,000 years: Okinawa Trough evidence (MD012404). Paleoceanography 24(3):PA3208. doi:10.1029/2007PA001577

Chen C, Beardsley RC, Limeburner R, Kim K (1994) Comparison of winter and summer hydrographic observations in the Yellow and East China Seas and adjacent Kuroshio during 1986. Cont Shelf Res 14(7–8):909–929. doi:10.1016/0278-4343(94)90079-5

Chen F, Yu Z, Yang M, Ito E, Wang S, Madsen DB, Huang X, Zhao Y, Sato T, Birks HJB, Boomer I, Chen J, An C, Wünnemann B (2008) Holocene moisture evolution in arid central Asia and its out-of-phase relationship with Asian monsoon history. Quat Sci Rev 27(3–4):351–364. doi:10.1016/j.quascirev.2007.10.017

Chen MT, Lin XP, Chang YP, Chen YC, Lo L, Shen CC, Yokoyama Y, Oppo DW, Thompson WG, Zhang R (2010) Dynamic millennial-scale climate changes in the northwestern Pacific over the past 40,000 years. Geophys Res Lett 37(23):L23603. doi:10.1029/2010GL045202

Dekens PS, Lea DW, Pak DK, Spero HJ (2002) Core top calibration of Mg/Ca in tropical foraminifera: Refining paleotemperature estimation. Geochem Geophys Geosyst 3(4):1–29. doi:10.1029/2001GC000200

Du J, Chen M, Cao J, Qiu Y, Tong J, Ma Q, Yang J (2012) Oxygen isotope in seawater and its hydrological implication in the southern Yellow Sea and the East China Sea. Oceanol Limnol Sin 43(6):1057–1066 (in Chinese with English abstract)

Epstein S, Mayeda T (1953) Variation of O18 content of waters from natural sources. Geochim Cosmochim Acta 4(5):213–224. doi:10.1016/0016-7037(53)90051-9

Greaves M, Caillon N, Rebaubier H, Bartoli G, Bohaty S, Cacho I, Clarke L, Cooper M, Daunt C, Delaney M, deMenocal P, Dutton A, Eggins S, Elderfield H, Garbe-Schoenberg D, Goddard E, Green D, Groeneveld J, Hastings D, Hathorne E, Kimoto K, Klinkhammer G, Labeyrie L, Lea DW, Marchitto T, Martinez-Boti MA, Mortyn PG, Ni Y, Nuernberg D, Paradis G et al (2008) Interlaboratory comparison study of calibration standards for foraminiferal Mg/Ca thermometry. Geochem Geophys Geosyst 9(8):Q08010. doi:10.1029/2008gc001974

Hall JM, Chan LH (2004) Ba/Ca in Neogloboquadrina pachyderma as an indicator of deglacial meltwater discharge into the western Arctic Ocean. Paleoceanography 19(1):PA1017. doi:10.1029/2003PA000910

Hastings DW, Whitko A, Kienast M, Steinke S (2001) A comparison of three independent paleotemperature estimates from a high resolution record of deglacial SST records in the tropical South China Sea. Eos 82:PP12B–10

Hönisch B, Allen KA, Russell AD, Eggins SM, Bijma J, Spero HJ, Lea DW, Yu J (2011) Planktic foraminifers as recorders of seawater Ba/Ca. Mar Micropaleontol 79(1–2):52–57. doi:10.1016/j.marmicro.2011.01.003

Ichikawa H, Beardsley R (2002) The Current System in the Yellow and East China Seas. J Oceanogr 58(1):77–92. doi:10.1023/A:1015876701363

Jan S, Sheu DD, Kuo H-M (2006) Water mass and throughflow transport variability in the Taiwan Strait. J Geophys Res 111(C12):C12012. doi:10.1029/2006JC003656

Johnstone HJH, Yu J, Elderfield H, Schulz M (2011) Improving temperature estimates derived from Mg/Ca of planktonic foraminifera using X-ray computed tomography–based dissolution index, XDX. Paleoceanography 26(1):PA1215. doi:10.1029/2009PA001902

Kang D-J, Chung CS, Kim SH, Hong GH, Kim K-R (1994) Oxygen isotope characteristics of seawaters in the Yellow Sea. La mer 32:279–284

Kim S-T, O'Neil JR (1997) Equilibrium and nonequilibrium oxygen isotope effects in synthetic carbonates. Geochim Cosmochim Acta 61(16):3461–3475. doi:10.1016/S0016-7037(97)00169-5

Kim K-R, Cho Y-K, Kang D-J, Ki J-H (2005) The origin of the Tsushima Current based on oxygen isotope measurement. Geophys Res Lett 32(3):L03602. doi:10.1029/2004GL021211

Klinkhammer GP, Chan LH (1990) Determination of barium in marine waters by isotope dilution inductively coupled plasma mass spectrometry. Anal Chim Acta 232(0):323–329. doi:10.1016/S0003-2670(00)81249-0

Kubota Y, Kimoto K, Tada R, Oda H, Yokoyama Y, Matsuzaki H (2010) Variations of East Asian summer monsoon since the last deglaciation based on Mg/Ca and oxygen isotope of planktic foraminifera in the northern East China Sea. Paleoceanography 25(4):PA4205. doi:10.1029/2009pa001891

Kuroyanagi A, Kawahata H (2004) Vertical distribution of living planktonic foraminifera in the seas around Japan. Mar Micropaleontol 53(1–2):173–196. doi:10.1016/j.marmicro.2004.06.001

Lea DW, Boyle EA (1991) Barium in planktonic foraminifera. Geochim Cosmochim Acta 55(11):3321–3331

Lea DW, Spero HJ (1994) Assessing the reliability of paleochemical tracers: Barium uptake in the shells of planktonic foraminifera. Paleoceanography 9(3):445–452. doi:10.1029/94PA00151

Lea DW, Mashiotta TA, Spero HJ (1999) Controls on magnesium and strontium uptake in planktonic foraminifera determined by live culturing. Geochim Cosmochim Acta 63(16):2369–2379

Lea DW, Pak DK, Spero HJ (2000) Climate impact of Late Quaternary equatorial Pacific sea surface temperature variations. Science 289(5485):1719–1724. doi:10.1126/science.289.5485.1719

Lee H-J, Chao S-Y (2003) A climatological description of circulation in and around the East China Sea. Deep-Sea Res PT II 50(6–7):1065–1084. doi:10.1016/S0967-0645(03)00010-9

Lee TN, Johns WE, Liu C-T, Zhang D, Zantopp R, Yang Y (2001) Mean transport and seasonal cycle of the Kuroshio east of Taiwan with comparison to the Florida Current. J Geophys Res 106(C10):22143–22158. doi:10.1029/2000JC000535

Lewis E, Wallace DWR (1998) CO2SYS–Program Developed for CO_2 System Calculations ORNL/CDIAC-105 (Carbon Dioxide Information Analysis Center, Oak Ridge National Laboratory, U.S. Department of Energy, Oak Ridge, TN)

Lin Y-S, Wei K-Y, Lin I-T, Yu P-S, Chiang H-W, Chen C-Y, Shen C-C, Mii H-S, Chen Y-G (2006) The Holocene Pulleniatina Minimum Event revisited: Geochemical and faunal evidence from the Okinawa Trough and upper reaches of the Kuroshio current. Mar Micropaleontol 59(3–4):153–170. doi:10.1016/j.marmicro.2006.02.003

Liu Z, Wen X, Brady EC, Otto-Bliesner B, Yu G, Lu H, Cheng H, Wang Y, Zheng W, Ding Y, Edwards RL, Cheng J, Liu W, Yang H (2014) Chinese cave records and the East Asia Summer Monsoon. Quat Sci Rev 83:115–128. doi:10.1016/j.quascirev.2013.10.021

Marchitto TM (2006) Precise multielemental ratios in small foraminiferal samples determined by sector field ICP-MS. Geochem Geophys Geosyst 7(5):Q05P13. doi:10.1029/2005GC001018

McConnell MC, Thunell RC (2005) Calibration of the planktonic foraminiferal Mg/Ca paleothermometer: Sediment trap results from the Guaymas Basin, Gulf of California. Paleoceanography 20(2):PA2016. doi:10.1029/2004PA001077

Mulitza S, Boltovskoy D, Donner B, Meggers H, Paul A, Wefer G (2003) Temperature:δ^{18}O relationships of planktonic foraminifera collected from surface waters. Palaeogeogr Palaeocl 202(1–2):143–152. doi:10.1016/S0031-0182(03)00633-3

Niebler HS, Hubberten HW, Gersonde R (1999) Oxygen isotope values of planktic foraminifera: A tool for the reconstruction of surface water stratification. In: Wefer G (ed) Fischer G. Use of Proxies in Paleoceanography, Springer Berlin Heidelberg, pp 165–189. doi:10.1007/978-3-642-58646-0_6

Oba T (1990) Paleoceanographic information obtained by isotopic measurement of individual foraminiferal specimens. In: Wang P, Lao Q, He Q (eds) Proceedings of the First International Conference on Asian Marine Geology. China Ocean Press, Beijing, pp 169–180

Qian W, Lee D-K (2000) Seasonal march of Asian summer monsoon. Int J Climatol 20:1371–1386

Qu CH, Chen CZ, Yang JR, Wang LZ, Lu YL (1993) Geochemistry of dissolved and particulate elements in the major rivers of China (The Huanghe, Changjiang, and Zhunjiang rivers). Estuaries 16(3):475–487. doi:10.2307/1352595

Regenberg M, Steph S, Nürnberg D, Tiedemann R, Garbe-Schönberg D (2009) Calibrating Mg/Ca ratios of multiple planktonic foraminiferal species with δ^{18}O-calcification temperatures: Paleothermometry for the upper water column. Earth Planet Sci Lett 278(3–4):324–336. doi:10.1016/j.epsl.2008.12.019

Rohling EJ (2007) Progress in paleosalinity: Overview and presentation of a new approach. Paleoceanography 22(3):PA3215. doi:10.1029/2007PA001437

Rosenthal Y, Boyle EA, Labeyrie L (1997) Last glacial maximum paleochemistry and deepwater circulation in the Southern Ocean: Evidence from foraminiferal cadmium. Paleoceanography 12(6):787–796

Russell AD, Hönisch B, Spero HJ, Lea DW (2004) Effects of seawater carbonate ion concentration and temperature on shell U, Mg, and Sr in cultured planktonic foraminifera. Geochim Cosmochim Acta 68(21):4347–4361. doi:10.1016/j.gca.2004.03.013

Schmidt GA (1999) Error analysis of paleosalinity calculations. Paleoceanography 14(3):422–429. doi:10.1029/1999PA900008

Schmidt MW, Lynch-Stieglitz J (2011) Florida Straits deglacial temperature and salinity change: Implications for tropical hydrologic cycle variability during the Younger Dryas. Paleoceanography 26(4):PA4205. doi:10.1029/2011PA002157

Schmuker B, Schiebel R (2002) Planktic foraminifers and hydrography of the eastern and northern Caribbean Sea. Mar Micropaleontol 46(3–4):387–403. doi:10.1016/S0377-8398(02)00082-8

Sugiyama M, Matsui M, Nakayama E (1984) Direct determination of barium in sea water by inductively coupled plasma emission spectrometry. J Oceanogr 40:295–302

Sun Y, Oppo DW, Xiang R, Liu W, Gao S (2005) Last deglaciation in the Okinawa Trough: Subtropical northwest Pacific link to Northern Hemisphere and tropical climate. Paleoceanography 20(4):PA4005. doi:10.1029/2004PA001061

Thunell RC, Honjo S (1981) Calcite dissolution and the modification of planktonic foraminiferal assemblages. Mar Micropaleontol 6(2):169–182. doi:10.1016/0377-8398(81)90004-9

Ujiié Y, Ujiié H, Taira A, Nakamura T, Oguri K (2003) Spatial and temporal variability of surface water in the Kuroshio source region, Pacific Ocean, over the past 21,000 years: evidence from planktonic foraminifera. Mar Micropaleontol 49(4):335–364

Wang B (2006) The Asian Monsoon. Springer Praxis, Chichester

Wang Y, Cheng H, Edwards RL, He Y, Kong X, An Z, Wu J, Kelly MJ, Dykoski CA, Li X (2005) The Holocene Asian Monsoon: Links to solar changes and North Atlantic climate. Science 308(5723):854–857. doi:10.1126/science.1106296

Wang Y, Cheng H, Edwards RL, Kong X, Shao X, Chen S, Wu J, Jiang X, Wang X, An Z (2008) Millennial- and orbital-scale changes in the East Asian monsoon over the past 224,000 years. Nature 451(7182):1090–1093

Weldeab S, Lea DW, Schneider RR, Andersen N (2007) 155,000 years of west African monsoon and ocean thermal evolution. Science 316(5829):1303–1307. doi:10.1126/science.1140461

Xu X, Yamasaki M, Oda M, Honda MC (2005) Comparison of seasonal flux variations of planktonic foraminifera in sediment traps on both sides of the Ryukyu Islands, Japan. Mar Micropaleontol 58(1):45–55. doi:10.1016/j.marmicro.2005.09.002

Yanagi T (1994) Material transport in the Yellow/East China Seas. Bull Coast Oceanogr 31:239–256

Yancheva G, Nowaczyk NR, Mingram J, Dulski P, Schettler G, Negendank JFW, Liu J, Sigman DM, Peterson LC, Haug GH (2007) Influence of the intertropical convergence zone on the East Asian monsoon. Nature 445(7123):74–77

Ye F, Deng W, Xie L, Wei G, Jia G (2014) Surface water $\delta^{18}O$ in the marginal China seas and its hydrological implications. Est Coast Shelf Sci 147:25–31. doi:10.1016/j.ecss.2014.05.033

Zhang J, Letolle R, Martin JM, Jusserand C, Mouchel JM (1990) Stable oxygen isotope distribution in the Huanghe (Yellow River) and the Changjiang (Yangtze River) estuarine systems. Cont Shelf Res 10(4):369–384. doi:10.1016/0278-4343(90)90057-S

Zhang L, Liu Z, Zhang J, Hong GH, Park Y, Zhang HF (2007) Reevaluation of mixing among multiple water masses in the shelf: An example from the East China Sea. Cont Shelf Res 27(15):1969–1979. doi:10.1016/j.csr.2007.04.002

Zhang W-Z, Chai F, Hong H-S, Xue H (2014) Volume transport through the Taiwan Strait and the effect of synoptic events. Cont Shelf Res. doi:10.1016/j.csr.2014.07.010

Zweng MM, Reagan JR, Antonov JI, Locarnini RA, Mishonov AV, Boyer TP, Garcia HE, Baranova OK, Johnson DR, Seidov D, Biddle MM (2013) World Ocean Atlas 2013, Volume 2: Salinity. NOAA Atlas NESDIS 74. Silver Spring, MD

2

Diurnal pattern of rainfall in Cambodia: its regional characteristics and local circulation

Kumiko Tsujimoto[1*], Tetsu Ohta[2], Kentaro Aida[3], Katsunori Tamakawa[4] and Monichoth So Im[5]

Abstract

We analyze the hourly rainfall data of 30 rain gauges in Cambodia from 2010 to 2015 to focus on the diurnal pattern of rainfall and its regional characteristics, with the underlying mechanisms inferred from the observed data. The observed annual rainfall in inland Cambodia ranges from 1087 to 1528 mm on station-average. Approximately 5–20% of the annual rainfall occurs during the pre-monsoon season, 50–78% during the summer monsoon season, and 12–36% during the post-monsoon season. During the pre-monsoon season, rainfall is dominant on the coast and over the Cardamom Mountains, with a maximum in the afternoon. The rainfall amount is smaller around the Tonle Sap Lake. During the summer monsoon season, rainfall is larger in the northern region and smaller in the western region in inland Cambodia, in both amount and proportion to annual rainfall. The rainfall amount on the coast is distinctively large. The diurnal rainfall maximum occurs in the early afternoon in the Cardamom Mountains, in the afternoon on the plain at the southwestern side of the Tonle Sap Lake, in the evening on the wide area of the northeastern side of the lake, and in the early morning on the coast. The clear regional characteristics in the diurnal rainfall pattern suggest significant effects of local features, even during the Asian summer monsoon season. During the post-monsoon season, rainfall is larger on the southwestern side of the Tonle Sap Lake with dominant nocturnal rainfall. These diurnal patterns are, however, not clear on some days, and analysis of the synoptic-scale atmospheric condition suggests the effect of the large-scale low-pressure system and disturbances on the appearance of the clear diurnal rainfall pattern. The effect of land–lake and mountain–valley circulations on forming the diurnal rainfall pattern is also implied from ground-observed meteorological data, although further numerical studies are required to examine the detailed mechanisms. The study of local effects on rainfall with consideration of the land-surface dynamics may aid flood and drought management in Cambodia by facilitating a greater understanding of its rainfall pattern.

Keywords: Diurnal pattern of rainfall, Local circulation, Land breeze, Sea/lake breeze, Katabatic wind, Anabatic wind, Asian summer monsoon, Cambodia, Indochina Peninsula, Tonle Sap Lake

Introduction

Study area

Cambodia is located in the southern portion of the Indochina Peninsula, in the lower Mekong River basin (Fig. 1). In the center of the country is a huge lake, the Tonle Sap (Fig. 2), which has an area of about 2500 km^2 at the end of the dry season (around April–June) and expands to about five times this size at the end of the rainy season (around October–November) in connection with the seasonal rise of the water level of the Mekong River (Masumoto et al. 2007). The seasonally expanding area of the lake is indicated by the white hatched area in Fig. 2. Responding to the change in lake area are the surrounding land-cover characteristics and soil moisture, which also change drastically within a year. Topographically, most of the land to the west of the Mekong River in Cambodia is basin-shaped flat terrain except for the Cardamom Mountains to the southwest facing the Gulf of Thailand, with the highest elevation at 1813 m.

Since it is located on the Indochina Peninsula, Cambodia is influenced by both the Asian summer monsoon (ASM), or Indian monsoon, and the Asian winter monsoon (AWM) (Chang et al. 2005). The ASM is established in early to middle May, associated with strong convection and the change of direction of the

* Correspondence: tsujimoto@okayama-u.ac.jp
[1]Graduate School of Environmental and Life Science, Okayama University, 3-1-1 Tsushima-naka, Kita-ku, Okayama 700-8530, Japan
Full list of author information is available at the end of the article

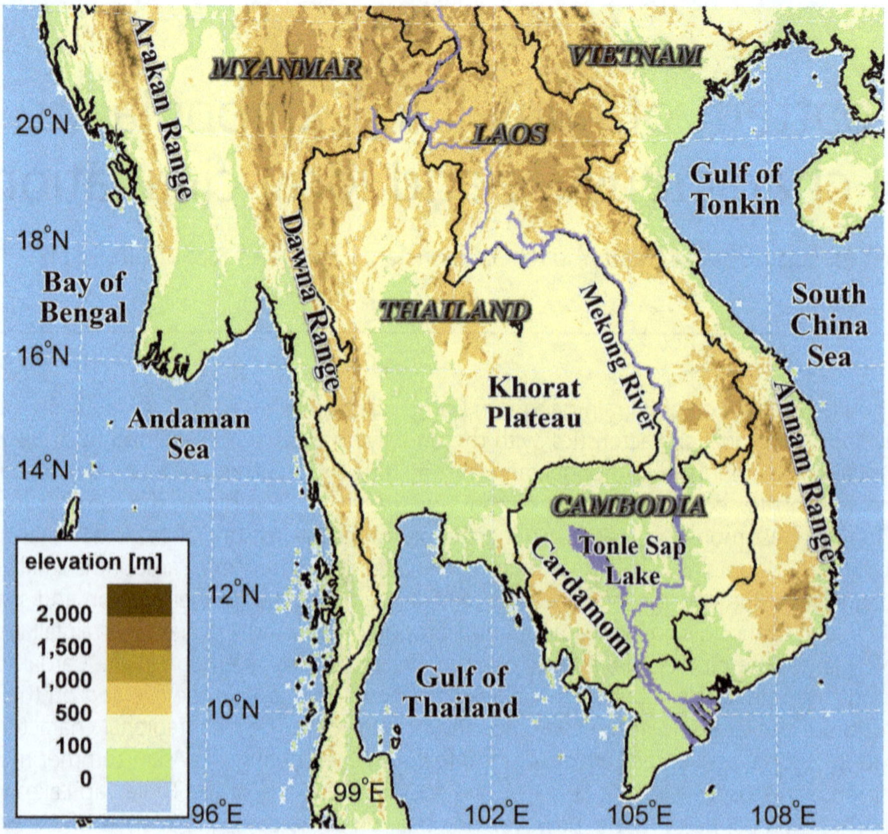

Fig. 1 Topography of the Indochina Peninsula and the location of Cambodia. The topographical map is drawn based on the Global 30 Arc-Second Elevation (GTOPO30) data provided by the US Geological Survey

prevailing wind over the Bay of Bengal, the Indochina Peninsula, and the South China Sea (Zhang et al. 2002). In the boreal winter season, the entire Indochina is subject to the dry season from mid-December until mid-April (Yen et al. 2011).

Rainfall characteristics in the Southeast Asian monsoon region

The onset and withdrawal dates of the rainy season and their spatial distribution over and around the Indochina Peninsula were analyzed by Matsumoto (1997) using rain-gauge data in 1975–1987. He found that the rainy season starts earlier in the inland region of Thailand from late April to early May and withdraws later in the central-eastern coast of Vietnam in late November. Based on his results, the onset date over the southern Indochina Peninsula (around Cambodia) is around 28 April to 8 May, and the withdrawal date is around 31 October to 10 November.

Takahashi et al. (2010a) analyzed summer (May–September) rainfall over the Indochina Peninsula using both satellite and ground observations and found the lowest rainfall over ridges and the leeward side of the high mountains of the Arakan, Dawna, and Annam Ranges and the highest rainfall offshore in the eastern Bay of Bengal, in the eastern Gulf of Thailand, and at the windward side of the Annam Range of the eastern Khorat Plateau.

The spatial variability of rainfall has been especially observed in the diurnal precipitation cycles of different regions. Hirose and Nakamura (2005) examined the spatial and diurnal variations of rainfall over Asia using satellite data, considering precipitation at multiple scales, to find small precipitation systems occurring most frequently around the early afternoon over most land areas, and large systems associated with the terrain occurring mostly in the evening.

The diurnal cycle of convection tends to be stronger over land during summer, with maxima in the late afternoon or early evening under a dominant influence of daytime boundary-layer heating (Chaboureau et al. 2004). Tsujimoto et al. (2008) analyzed the diurnal cycle of rainfall in Cambodia and showed clear afternoon rainfall peaks at two stations in Cambodia, in the city of Phnom Penh and over a paddy field near Phnom Penh, and a weaker afternoon peak over the Tonle Sap Lake near the city of Siem Reap (see Fig. 2). As Wallace (1975) has summarized, daytime rainfall peaks are

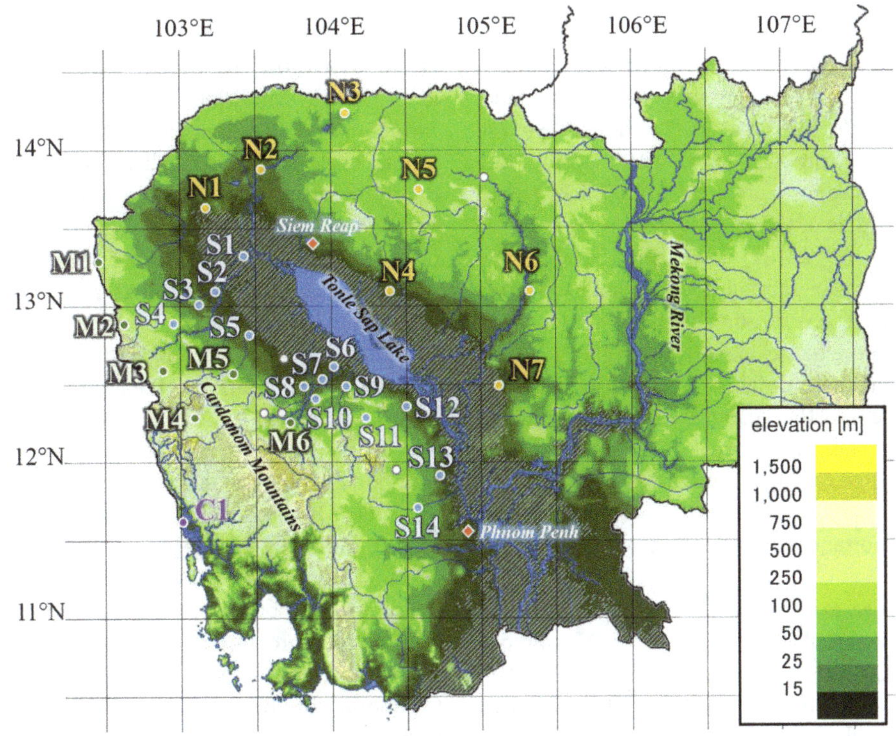

Fig. 2 Map of Cambodia and location of rain gauges with their symbols (N1–N7 in yellow for northern stations, S1–S14 in blue for southern stations, M1–M6 in green for mountainous stations, and C1 in purple for the coastal station). Four white circles show the excluded stations for the analysis due to frequent missing data. Station S7 is equipped with an AWS. The color shades indicate the elevation, and the white hatched area around the Tonle Sap Lake is its seasonal expansion area where it is under water from the middle of the rainy season (around August) to the beginning of the dry season (around December). The base map is adapted from The Atlas of Cambodia (Save Cambodia's Wildlife 2006)

mainly controlled by thermodynamic processes affecting the atmospheric stability. Sensible heat released from the ground surface under daytime solar radiation warms the lower boundary layer, which decreases the low-level atmospheric stability, leaving favorable conditions for moist convection during the afternoon and thus contributing to the afternoon maxima of rainfall observed at many places over land.

However, the rainfall peak tends to occur at night or in the early morning over some regions. For example, over the southern and central Tibetan Plateau, afternoon rainfall is pronounced over mountains, while late evening to early morning rainfall is dominant over valleys and lakes (Fujinami et al. 2005; Singh and Nakamura 2009). Over the southward slope of the Himalayas, late evening to early morning rainfall is more pronounced (Bhatt and Nakamura 2005). Over tropical Asia in Bangladesh, Thailand, Vietnam, and Malaysia, Ohsawa et al. (2001) analyzed the spatial distribution of summer rainfall (June–August) from rain gauges as well as the convective activity observed by satellites and found that late night–early morning rainfall was stronger in intensity.

Based on the summary by Wallace (1975), apart from the abovementioned thermodynamic processes, dynamical processes may also explain the shift of rainfall maxima to other hours of the day, such as the convergence of flow within the planetary boundary layer. In addition, radiative cooling from cloud tops and low-level warm advection are thermodynamic processes that enhance nighttime rainfall maxima (Wallace 1975). Moreover, the downward propagation of atmospheric convective systems associated with rainfall peaks along mountain slopes has been reported at various places. Using rain gauges in China, Yu et al. (2007) showed the eastward propagation of the diurnal phase of rainfall systems along the Yangtze River Valley in summer, with midnight maxima in the upper valley, early morning maxima in the middle valley, and late afternoon maxima in the lower valley. Satomura (2000) numerically simulated the eastward propagation of rainfall initiated at the leeward side of the Dawna Range in western Indochina in the late afternoon with water vapor supplied from the Andaman Sea. As a result, there was a nighttime maximum of precipitation in the inland areas of northeast Thailand. Okumura et al. (2003) analyzed the diurnal cycle of precipitation in rainy seasons using radar echo data observed in northern Thailand to show moving mesoscale systems with a phase delay corresponding to the distance from mountains. Satomura et al. (2011) used the radar echo area observed in Lao PDR and analyzed the propagation speed of its diurnal variation in relation to the large-scale wind in the lower troposphere. Takahashi et al. (2010a) examined the 3 h mean diurnal cycle of 10-year climatological rainfall occurring from May to September to show that, around the Tonle Sap Lake in Cambodia, evening rainfall begins at approximately 16:00–19:00 Local Time (LT) and becomes strongest at 19:00–22:00 LT, before weakening in the middle of the night.

As Takahashi et al. (2010a) have pointed out, the observed diurnal cycle of rainfall is a "visualized" result of the interaction of several mechanisms under both large- and local-scale effects. Thus, the observation and interpretation of regional rainfall characteristics in terms of the diurnal cycle and the spatial variability are useful for understanding the local circulation, especially in regions where the observational network is sparse.

Limitations and importance of studying rainfall in Cambodia

Although many studies exist on the climatology and rainfall in the Indochina Peninsula (Nieuwolt 1981; Matsumoto 1997; Satomura 2000; Ohsawa et al. 2001; Okumura et al. 2003; Xie et al. 2006; Tsujimoto et al. 2008; Takahashi et al. 2010a; Satomura et al. 2011; Takahashi 2013; Takahashi 2016), limitations of the automatic rain-gauge network in Cambodia have led preceding studies of rainfall in Indochina to use rain-gauge data from mostly only neighboring countries and/or satellite data. As mentioned by Takahashi et al. (2010a), rain-gauge observations were inadequate for analyzing the diurnal cycle of rainfall over the Indochina Peninsula because of their limited availability, whereby gauges have only been located in populated and low-altitude areas until very recently. Even regarding floods and droughts, which frequently affect Cambodia and result in enormous damage (Torti 2012; National Committee for Disaster Management and United Nations Development Programme 2014; Chhinh and Millington 2015), such events are usually classified in terms of the inundated area, the water level of rivers and lakes, and damage to households, people, and agricultural production. Hence, the spatio-temporal distribution of rainfall itself over Cambodia has yet to be clarified in detail. Even though the inundation area of the Tonle Sap Lake is highly affected by the water level of the Mekong River due to its hydrological system described above, and thus, the rainfall amount in the upper Mekong River basin and the water level of the Mekong can explain the floods and droughts in Cambodia to a large extent, deepening the understanding of rainfall that directly falls onto Cambodia, including both large-scale and locally induced rainfall, will help the development of strategies for mitigating water-related disasters such as floods and droughts.

In addition, while surface properties such as vegetation and soil moisture are considered important for the formation and distribution of rainfall (Xie et al. 2006), these effects are not yet well understood. Takahashi et al. (2010b) examined the potential impact of land-surface conditions on regional climate over the Indochina Peninsula associated with the diurnal precipitation cycle using a high-resolution regional climate model. Such a numerical study was also conducted for Bangladesh, a wet tropical region, to examine the precipitation sensitivity to soil moisture in different seasons (Sugimoto and Takahashi 2017). The relationship between soil moisture and precipitation in Bangladesh was recognized by satellite observation as well, with higher afternoon precipitation intensity over drier land surfaces over a flat terrain (Ono and Takahashi 2016).

With the unique hydrological characteristics of land in Cambodia, the rainfall pattern would be highly affected by surface heterogeneity, in both time and space, resulting in rainfall characteristics within Cambodia having regional characteristics. Clear land–lake circulation, with a daytime lake breeze (onshore flow) and nighttime land breeze (offshore flow) in the presence of a linear convective system at its front, has been recognized in Cambodia by both the analysis of observations and numerical simulations (Tsujimoto and Koike 2008 Tsujimoto and Koike 2013). Although the effect of mountain–valley circulation interacting with land–lake circulation is often discussed in studies on the diurnal cycle of rainfall around the lake (e.g., Verburg and Antenucci 2010; McGloin et al. 2014), the surrounding area of the Tonle Sap Lake is almost flat (Fig. 2) compared with other lakes in the world; therefore, the effect of land–lake circulation would clearly appear, even though the effects of mountains also exist. Tsujimoto and Koike (2012) analyzed rain-gauge data in Cambodia together with global reanalysis data and found that about 30–40% of post-monsoon rainfall was induced by the local circulation within Cambodia.

However, because of the lack of multi-year hourly rainfall data at various places in Cambodia, rainfall distribution associated with local-scale heterogeneity of the land, as well as the diurnal, seasonal, and intra-seasonal variations, has not been explored. To rectify this situation, we installed 16 automatic rain gauges in September 2009 and rehabilitated a further 19 abandoned rain gauges for establishment of the ground truth of rainfall distribution in Cambodia. We concentrate on the surrounding area of the Tonle Sap Lake west of the Mekong River, with the aim of detecting the influence of the Tonle Sap Lake.

Scope of this study

In this study, we aim to quantify the rainfall amount in Cambodia, i.e., the annual and seasonal amounts at various places from 2010 to 2015. We also aim to clarify the diurnal rainfall pattern and its regional characteristics for the pre-monsoon, summer monsoon, and post-monsoon seasons to highlight locally induced rainfall. The intra-seasonal variability of the clarity of the diurnal rainfall cycle is discussed as well to examine the influence of synoptic-scale phenomena on locally induced rainfall. Quality-controlled hourly rainfall data from 30 rain gauges are used to quantify the diurnal patterns of rainfall over Cambodia. We also refer to data from a ground meteorological observation station to examine the possible mechanisms underlying the observed diurnal patterns of rainfall in each region in each season.

Methods/Experimental

Rainfall data

The locations of the automatic rain gauges from which we gather rainfall data in Cambodia are shown in Table 1 and Fig. 2, along with each station's symbol. At station S7 in the city of Pursat, surface meteorological variables are observed in addition to rainfall by an Automatic Weather Station (AWS), which is explained in the next subsection.

Rain-gauge locations were carefully selected to be away from tall trees and buildings to carry out observation at open and wide spaces and to be safe against theft and interference by humans, animals, vehicles, and so on. At locations where shrubs were growing, we asked local residents and/or landowners to keep them cut during the observation period. The setting environment is different among stations, as summarized in Table 1. Some stations are inside the observation sites of the local meteorological office or grounds of the local district/commune/police office, and others are inside gardens of private houses or paddy fields. The gauge at station S1 was installed on a roof over a private house, since the ground there is completely inundated as a part of the lake during the rainy season. The height of the rain-gauge orifice to the land or water surface there is around 4–8 m and changes within a year, following the seasonal change of the water level of the lake.

Each rain gauge was equipped with a tipping bucket and a data logger. There are three types of gauges (A, B, and C as indicated in Table 1) with different specifications. The type-A rain gauge is a No. 34-T (RA-1) gauge manufactured by Ota Keiki Seisakusho Co., Ltd., and data are recorded to the Onset Computer Corporation's HOBO event data logger (H07-002-04). It measures at a 0.5 mm resolution with a receiving orifice diameter of 200 mm. All the type-A rain gauges had been abandoned for observation as of June 2009, and therefore, we rehabilitated these rain gauges to restart observation where possible. On restarting the observation, we checked the performance of tipping buckets and data loggers by pouring known amounts of water; we also carefully cleaned the buckets, nets, and data loggers and changed the desiccants and batteries to new ones. The type-B rain gauge consists of an RG3-M gauge and a HOBO pendant event data logger (UA-003-64) produced by the Onset Computer Corporation. Its measuring resolution is 0.2 mm, and the receiving orifice diameter is 154 mm. All the type-B rain gauges were installed by the authors in September 2009, especially concentrating on the poorly gauged area in the northeastern side of the lake. At two stations (S4 and M3), the type-B rain gauges were installed next to the type-A gauges to check the instrumental difference on the observed rainfall amount. The type-C gauge is an A-TK-2 rain gauge manufactured by Takeda Keiki Kogyo Co., Ltd., constituting one component of the AWS that was also installed by the authors. It has a gauge resolution of 1.0 mm with a receiving orifice diameter of 200 mm. Rainfall data were recorded, along with other meteorological variables, to a Campbell Scientific, Inc. data logger (CR1000). After the rehabilitation and installation of rain gauges in September 2009, we serviced stations two to three times per year to obtain continuous, highly reliable, and quality-controlled data for our detailed investigation of the rainfall in Cambodia.

The quality control of data was applied in two stages. First, we inspected the physical status of each station at every visit and carefully cleaned the gauges and removed any tall grass surrounding the station. Any rainfall counts recorded by mistake during maintenance work were removed. For data loggers found to function abnormally, or where there existed some artificial disturbance to the observation (i.e., construction work, interference by humans or animals, or excess dust, insects, or leaves affecting the gauge or tipping bucket), we assumed that observations from that station from the previous to the current data collection dates are missing. Second, in graphs of the rainfall distribution in which exceptionally high or low data occurred, we assumed that observations around that period are also missing. While this standard is not quantitative, the reference to local information allows us to treat doubtful data as missing.

Following the removal of questionable data, we converted the recorded data to hourly and daily datasets. For definitions of date and hour, we used Cambodian LT (UTC + 7 h). We defined hourly rainfall at hour h as accumulated rainfall amount observed from h:00:00 to h:59:59. Note that this definition is different from that for the normal gauge observation (h – 1:00:01 to h:00:00), and here we set the definition to match that of the satellite precipitation product, such as Global Satellite Mapping of Precipitation (GSMaP) (Earth Observation Research Center of Japan Aerospace Exploration Agency 2017).

Table 1 Specification of each station and its gauge used in the analysis

Symbol	Location name	Province	Latitude [°N]	Longitude [°E]	Altitude [m]	Setting environment	Gauge type	Gauge height [m]
N1	Preah Net Preah	Banteay Meancheay	13.63	103.19	18	2	B	0.50
N2	Chrouy Neang Nuong	Siem Reap	13.86	103.52	26	2	B	0.50
N3	Anlung Veng	Oddar Meancheay	14.23	104.08	58	2	B	0.5
N4	Kampong Kdei	Siem Reap	13.12	104.34	21	2	B	0.50
N5	Srayang	Preah Vihear	13.72	104.55	138	2	B	0.50
N6	Mean Rith	Kampong Thom	13.10	105.28	41	3	B	0.50
N7	Kampong Thmar	Kampong Thom	12.50	105.15	15	1	B	0.50
S1	Bac Prea	Battambang	13.31	103.40	8	4	B	-
S2	Battambang	Battambang	13.09	103.21	15	1	B	0.50
S3	Phnom Sam Pov	Battambang	13.02	103.11	15	5	B	1.38
S4	Ratanak Mondol	Battambang	12.89	102.97	58	2	A B	1.5 0.5
S5	Muang Russey	Battambang	12.78	103.45	18	2	A	1.49
S6	Kandieng	Pursat	12.61	103.99	21	3	A	1.67
S7	Pursat	Pursat	12.54	103.91	20	1	C	0.60
S8	Dambat	Pursat	12.49	103.81	24	3	A	1.58
S9	Koh Chum	Pursat	12.50	104.05	28	5	A	1.48
S10	Bactra	Pursat	12.43	103.86	57	3	A	1.46
S11	Bamnak	Pursat	12.30	104.18	69	3	A	1.50
S12	Boribo	Kampong Chhnang	12.39	104.48	17	2	A	1.50
S13	Kampong Tralach	Kampong Chhnang	11.94	104.71	21	3	A	1.50
S14	Peam	Kampong Chhnang	11.74	104.54	48	3	A	1.50
M1	Phnom Proek	Battambang	13.22	102.40	125	2	A	1.50
M2	Pailin	Pailin	12.87	102.56	219	1	B	0.26
M3	Samlot	Battambang	12.61	102.86	116	2	A B	1.50 0.50
M4	Veal Veng	Pursat	12.31	103.11	237	2	A	1.49
M5	Bassac Reservoir	Battambang	12.57	103.34	49	3	A	1.46
M6	Peam	Pursat	12.28	103.72	42	3	A	1.46
C1	Koh Kong	Koh Kong	11.61	102.99	5	1	B	0.50
–	Preah Vihear	Preah Vihear	13.80	104.98	54	1	B	0.50
–	Rovieng	Pursat	12.32	103.55	63	3	B	1.53
–	Svay Donkeo	Battambang	12.67	103.64	15	3	A	1.50
–	Taing Louch	Pursat	12.32	103.64	42	3	A	1.55
–	Takab	Kampong Chhnang	12.02	104.45	64	3	A	1.50

The setting environment is indicated by a classification number: 1 for inside the observation site of the local meteorological office; 2 for inside the grounds of a local district/commune/police office; 3 for inside the garden of a private house; 4 for on a roof over a private house; and 5 for inside a paddy field. The measuring resolution is 0.5, 0.2, and 1.0 mm for gauge types A, B, and C, respectively. The gauge height is from the ground surface to the top of the receiving orifice. Five stations with the symbol "–" are excluded from the analysis due to frequent missing data

More than one missing point per hour resulted in that hour being regarded as a "data-missing hour," and more than one missing hour per day resulted in that day being regarded as a "data-missing day." At each station, a year with more than 10 missing days was regarded as a "data-missing year." Five stations were found to have more than four data-missing years and were thus excluded from our analysis. These stations are indicated by white circles in Fig. 2 and by the symbol "–" in Table 1. With this procedure, we have set severe criteria for the purpose of analyzing rainfall characteristics with only high-quality data.

At stations where both type-A and type-B rain gauges are available, we compared the observed rainfall amounts from both gauges. The comparison periods are from 1 January 2010 to 31 December 2015 (2191 days) for S4 and from 1 January 2010 to 29 July 2012 (921 days) for M3. The bias (type A minus type B) in hourly rainfall amount averaged throughout the comparison periods is 0.0346 mm for S4 and 0.0344 mm for M3. The mean rainfall amounts in hours when both types of gauges record rainfall equal to or more than 1.0 mm (the threshold to discuss rainfall frequency in the following sections) are 3.8 and 3.3 mm in S4 for type A and type B, respectively, and 3.5 and 2.9 mm in M3 for type A and type B, respectively. The frequency of rainfall equal to or more than 1.0 mm h^{-1} recorded by both rain gauges is 1.7%, whereas that recorded only by type A is 0.77% and only by type B is 0.14% in S4. In M3, the value is 2.3, 0.89, and 0.27%, respectively. Although there exist some differences and type-A rain gauges tend to record larger and frequent rainfall than type-B gauges, both types of gauges are used in the following analysis without any bias correction applied. At S4 and M3, we used type-A gauges.

Ground weather data

Meteorological data observed at the AWS at S7 (see Table 1 and Fig. 2) were used as a reference for local circulation. Among the observed elements, here, we used data on wind direction and wind speed at a 10 m height, air temperature and relative humidity at a 7 m height, and air pressure at a 1.5 m height, in addition to the rainfall amount. The meteorological elements except for rainfall were observed with a 1 s sampling interval, and we used the preceding 10 min means as hourly data. In this paper, air humidity is discussed in terms of specific humidity, which is calculated from relative humidity, air temperature, and air pressure.

Global reanalysis data

The Japanese 55-year Reanalysis data (JRA-55) with a horizontal resolution of 1.25° and 37 pressure levels were used to check the synoptic-scale atmospheric fields during the target period. Further details of the JRA-55 data are provided by Kobayashi et al. (2015) and Harada et al. (2016).

Analysis

To aid the description of the results, we divided the surrounding area of the Tonle Sap Lake through the long axis of the lake, which runs from the northwest to the southeast. The rain gauges on the northeastern side of the lake are designated beginning with the letter N. For the southwestern side of the lake, we divided the rain gauges into three further groups: the lakeshore plain and the lower hillslope (stations beginning with the letter S), in the Cardamom Mountains (M), and on the coast facing the Gulf of Thailand (C). The number of rain-gauge stations for the region N, S, M, and C is 7, 14, 6, and 1, respectively.

The target period for data analysis is from 1 January 2010 to 31 December 2015.

Results

Synoptic-scale atmospheric condition

Figure 3 shows the 6-year-averaged monthly mean synoptic-scale atmospheric fields at 1000 hPa according to JRA-55 data. The 6-hourly data in each month are used to calculate the averages. From March to April (Fig. 3a, b), the monthly mean wind is easterly over the South China Sea and southeasterly over the Gulf of Thailand. From May to September (Fig. 3c, d, e), the entire Cambodia is subject to a large-scale southwesterly wind. From October to November (Fig. 3f, g), the northeasterly wind prevails over the Gulf of Tonkin and the northern South China Sea. From December to February (Fig. 3h, i), the large-scale northeasterly wind becomes stronger with the intrusion of drier air. Thus, in a simple definition on a monthly basis, the summer monsoon season is from May to September, with the pre-monsoon season in March and April and the post-monsoon season in October and November. The dry season is December, January, and February.

Annual and seasonal rainfall

Figure 4 shows the annual rainfall observed at each rain gauge in inland Cambodia from 2010 to 2015, where "data-missing years" have been excluded. The number of available stations, N, and the station-averaged annual rainfall amount are shown for each year. In each bar chart, the rainfall amount from May to September (summer monsoon season) is shown in blue, that from March to April (pre-monsoon season) is in orange, that from October to November (post-monsoon season) is in green, and that in January, February, and December (dry season) is in purple. At the one coastal station in C1, the annual rainfall amount is obtained only in 2011 and it is 3923 mm, with 2825 mm in May–September and 551 mm in October–November.

The rainfall amount is distinctively large on the southwestern coast facing the Gulf of Thailand (C1) with values of nearly 4000 mm. Except for this station, i.e., over inland Cambodia, the station-averaged annual rainfall ranges from 1087 in 2015 to 1528 mm in 2011. The difference is 441 mm, corresponding to 33% of the 6-year average.

The 6-year-averaged rainfall at stations in inland Cambodia shows that the rainfall amount from May to September (summer monsoon season) is 877 mm on average and varies from 519 to 1350 mm. It is larger in the northern region (N5, N6; around Preah Vihear and Kampong Thom provinces or the Stung Sen River basin)

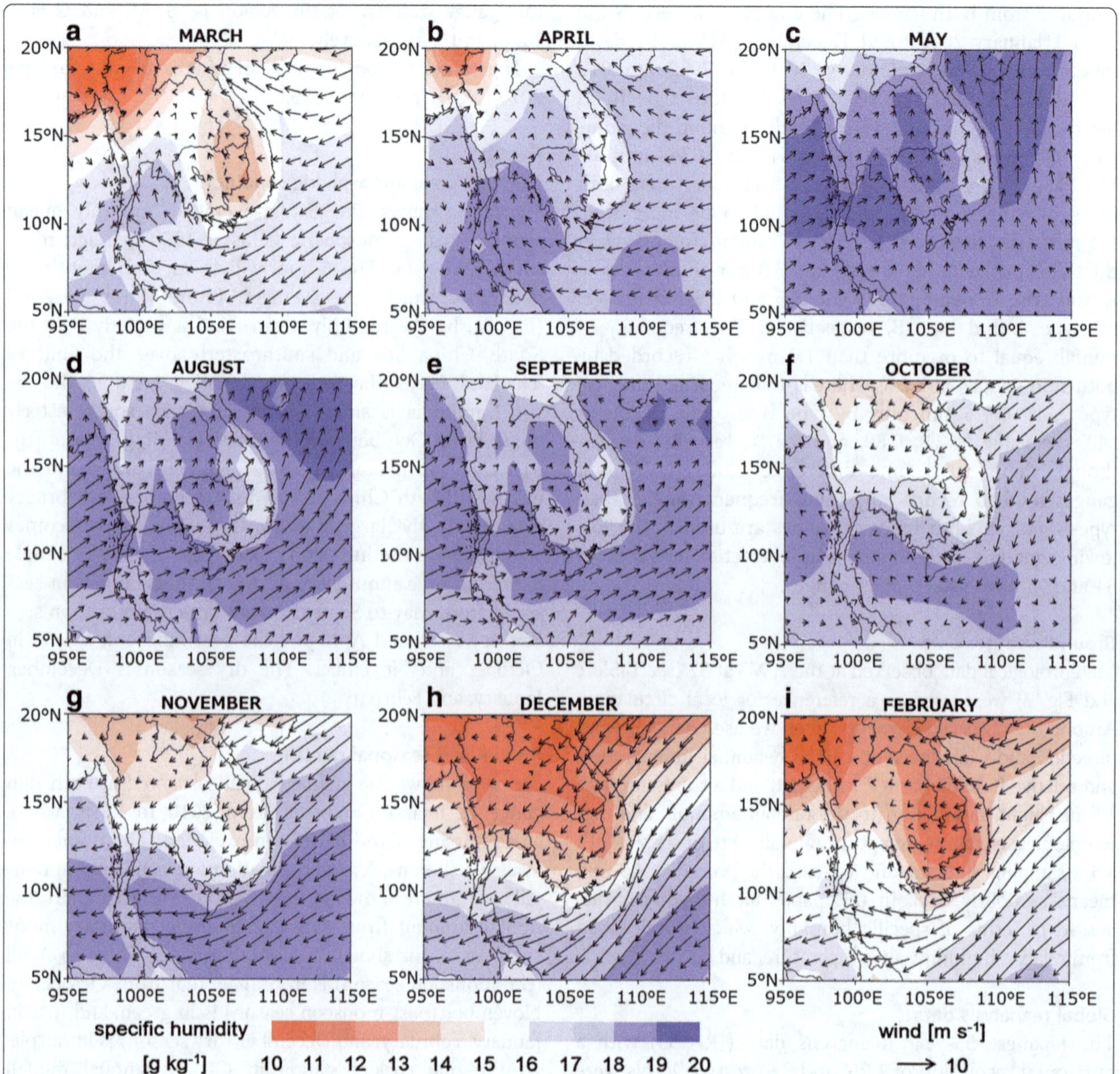

Fig. 3 Wind and specific humidity at 1000 hPa over the Indochina Peninsula. Monthly-mean data for March (**a**), April (**b**), May (**c**), August (**d**), September (**e**), October (**f**), November (**g**), December (**h**), and February (**i**) as 6-year (2010–2015) averages are shown. Data are from JRA-55 historical reanalysis

at around 1230–1350 mm and smaller in the western region (S3, M2, S5, N1, S1; around Battambang province or the Stung Sangkae River basin) at around 520–700 mm. Its proportion to the annual rainfall is 50–78%, with a high proportion mostly in the northern region (N6, N2, N5, N3, S12, N7, and N4 in descending order) and a low proportion in the western region (S5, M2, M1, S3, S4, M4, and M5 in ascending order).

The rainfall amount from March to April (pre-monsoon season) is 137 mm on average and varies from 63 to 278 mm. It contributes about 5–20% of the annual rainfall, with a high proportion in the Cardamom Mountains (M4, S3, M3, and M2 in descending order; near the Thai border) and a low proportion mostly in the southern area (N2, S6, S12, and S14 in ascending order; around Pursat and Kampong Chhnang provinces).

The rainfall amount from October to November (post-monsoon season) is 299 mm on average and varies from 181 to 457 mm. It is larger around the northern ridges of the Cardamom Mountains and the western lakeshore plain (S5, S6, M6, S7, M5, S2, S14, S11, S9, M2, and M3 in descending order; around Pursat province) and smaller at the northern side of the lake (N3, N6, and N2 in ascending order; around Kampong Thom, Siem Reap, and Oddar Meanchey provinces). It consists

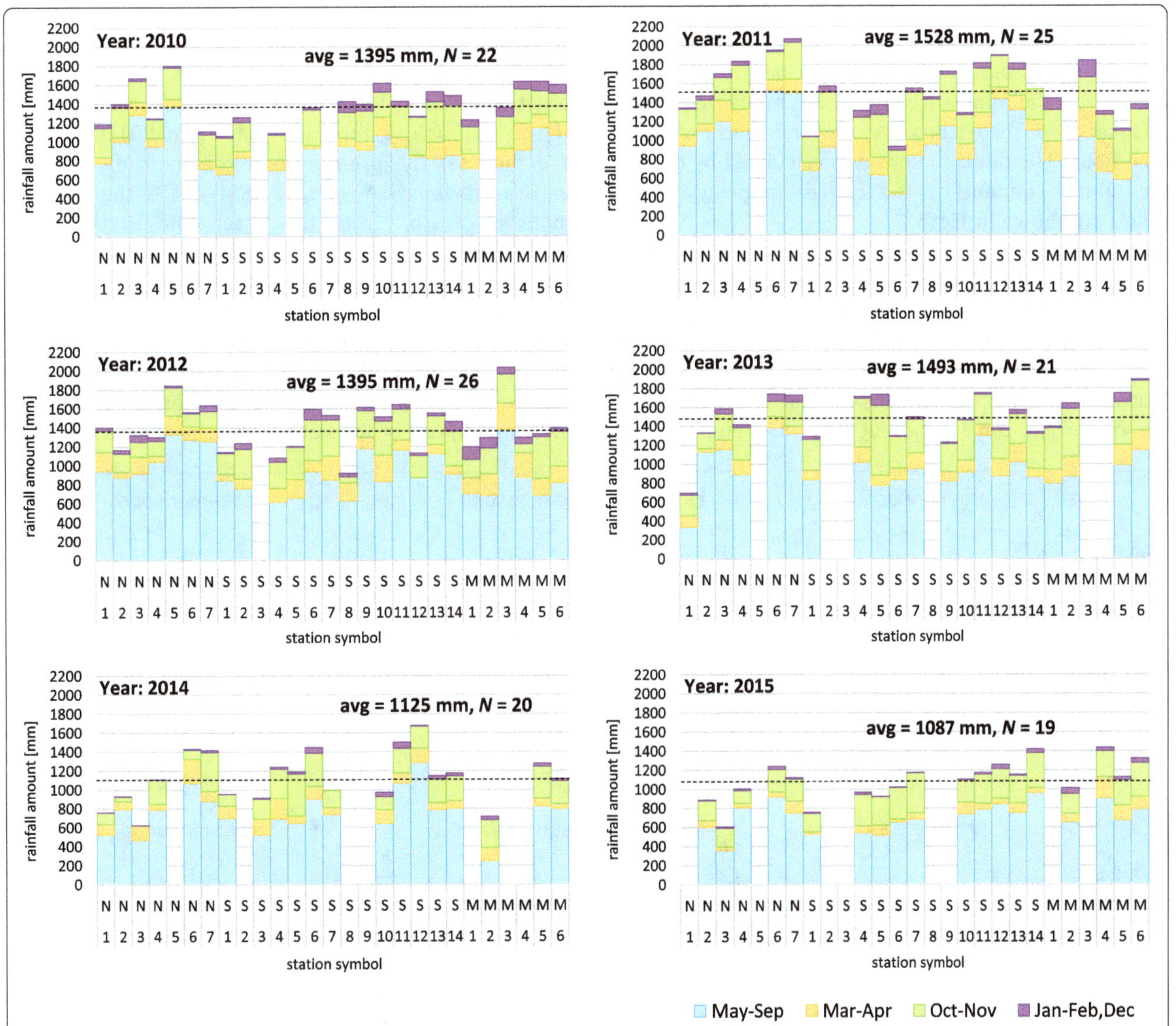

Fig. 4 Annual rainfall observed at each rain gauge in inland Cambodia from 2010 to 2015. Rainfall amounts from May to September (summer monsoon season, in blue), in March and April (pre-monsoon season, in orange), in October and November (post-monsoon season, in green), and in January, February, and December (dry season, in purple) are shown. The station symbol of each station is shown on the horizontal axis. Data in "data-missing years" have been excluded, and the number of available rain gauges, *N*, and annual rainfall amount averaged over available inland rain gauges (except for station C1) are shown in each panel with a broken black line

of 12–36% of the annual rainfall, with a high proportion at the wide area of the western lakeshore plain in the lower hillslopes of the Cardamom Mountains (S5, S6, M2, S7, M6, S4, S2, and M5 in descending order; around Battambang and Pursat provinces) and a low proportion at the northern side of the lake (N6, N3, N2, M4, and N5 in ascending order; around Kampong Thom, Preah Vihear, Siem Reap, and Oddar Meanchey provinces).

Although an altitudinal increase in the rainy season rainfall has been reported in northern Thailand (Kuraji et al. 2001), the terrain around the Tonle Sap Lake is flat with a mild mountain slope (Fig. 2), and 23 out of 28 gauges used in this analysis are located at the area with an altitude lower than 100 m (Table 1). The highest station, M4, is located at a 237 m altitude. The altitudinal difference of gauges is much smaller than in the study by Kuraji et al. (2001), in which the rain gauges were distributed at various altitudes at 282–2535 m. Thus, we consider that the effect of the altitudinal increase in rainfall is not large, if any, in the rainfall observed in this study.

Diurnal cycle of rainfall

The diurnal cycle of rainfall is discussed mainly in terms of the rainfall frequency by defining any hour with rainfall amount equal to or more than 1.0 mm as "rainy." At each hour of the day, we calculated the percentage of the number of "rainy" stations against the total number of stations for each region, N, S, M, and C, and calculated its season-averaged value for the pre-monsoon, summer monsoon, and post-monsoon seasons. In addition to the rainfall frequency, the rainfall amount is calculated as the station-averaged rainfall amount at each hour. The rainfall intensity is calculated by dividing the rainfall amount by the rainfall frequency. These values are shown in Fig. 5.

For these calculations, we defined these three seasons not on a monthly basis but more exactly by defining the onset and withdrawal dates of the ASM for each year. Among various studies on the definition of the ASM onset, we followed the definition used by Kiguchi and Matsumoto (2005), which considers the onset date to be the day when zonal winds at both 850 and 700 hPa become positive and last for at least 20 days. Similarly, the withdrawal date was defined in this study as the day when zonal winds at both levels become negative and last for at least 20 days. These dates are shown in Table 2 for each year from 2011 to 2015. We then set the pre-monsoon season from 1 March to the previous day of the ASM onset, the summer monsoon season from the onset date to the withdrawal date, and the post-monsoon season from the next day of the withdrawal to 30 November.

We further calculated their 6-year averages for each season for each region to show the averaged diurnal cycle with its seasonal and regional characteristics (Fig. 5). To discuss the spatial variability within region S, Fig. 6 shows the diurnal cycle of the rainfall frequency at each station,

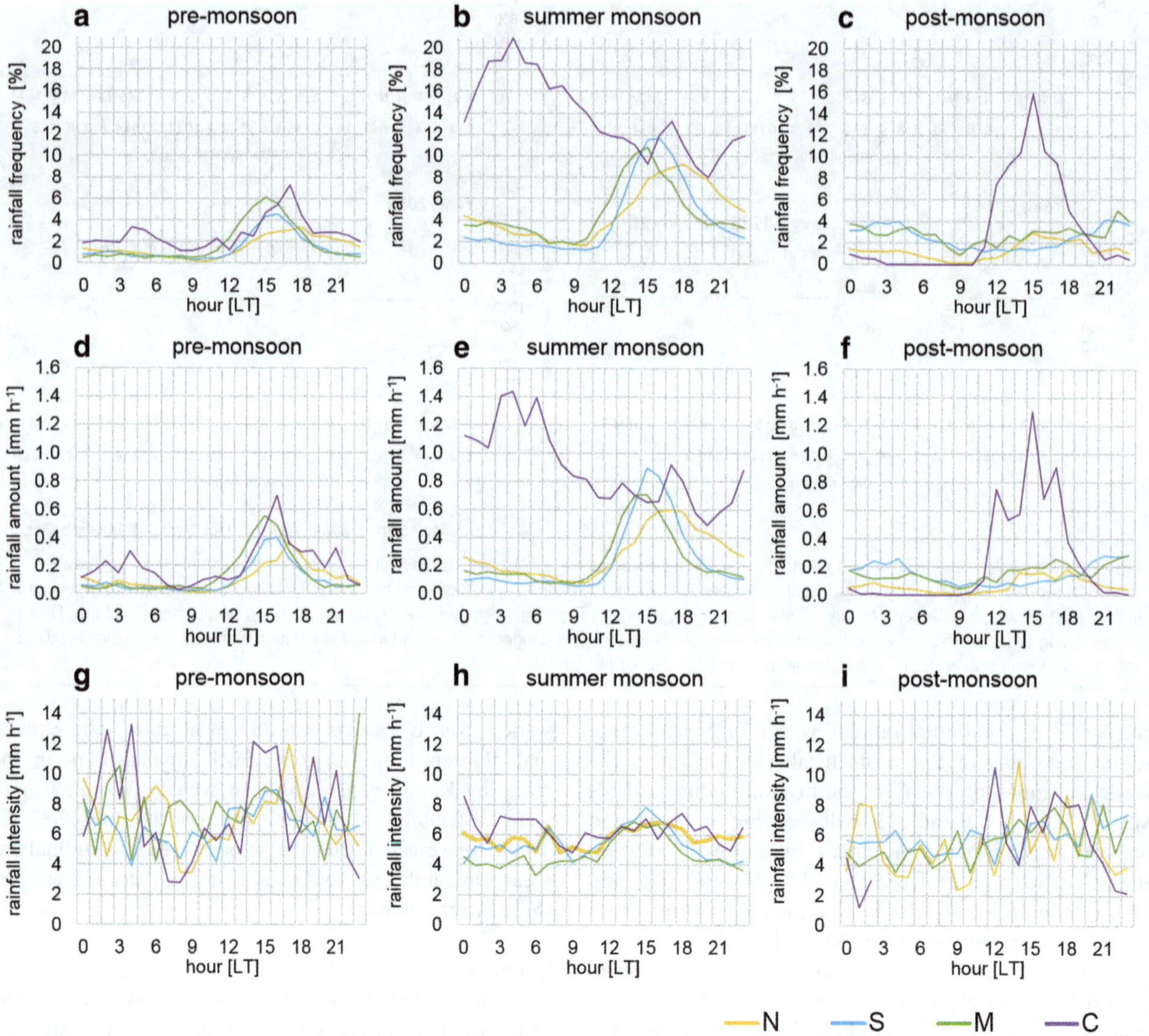

Fig. 5 Diurnal pattern of rainfall on rainfall frequency (**a**–**c**), rainfall amount (**d**–**f**), and rainfall intensity (**g**–**i**). The left panels (**a**, **d**, **g**) are for the pre-monsoon season, the middle panels (**b**, **e**, **h**) are for the summer monsoon season, and the right panels (**c**, **f**, **i**) are for the post-monsoon season. The values are the 6-year-averaged stats for regions N (yellow), S (blue), M (green), and C (purple)

Table 2 Onset and withdrawal dates of the ASM defined for each year in 2010–2015

Year	2010	2011	2012	2013	2014	2015
Onset	22 May	3 June	3 June	7 June	4 June	17 June
Withdrawal	13 November	17 October	8 October	29 October	12 October	26 October

S1 (over the lake), S9 (lower floodplain), S10 (upper floodplain), and S14 (away from the mountain hillslopes).

During the pre-monsoon season, rainfall is frequent in the afternoon over all the four regions (Fig. 5a). The peak frequency is clear at around 15:00–16:00 LT in region S and 15:00 LT in region M. Region C has the first peak at 17:00 LT and the second weaker peak at 04:00 LT. The rainfall frequency in region N is high in the late afternoon with a peak at 18:00 LT. The rainfall intensity is around 4–12 mm h^{-1} (Fig. 5g), showing higher values than other seasons (Figs. 5h, i). The rainfall intensity is especially high in the early morning in region C and in the afternoon in regions C and N.

In the summer monsoon season, more than half of the rainfall over the coastal region C occurs in the early morning, whereas most of the rainfall over inland Cambodia (regions N, S, M) occurs in the afternoon (Fig. 5e). In terms of rainfall frequency, a clear afternoon peak at around 15:00–16:00 LT in region S and a slightly earlier peak at around 14:00–15:00 LT in region M are observed (Fig. 5b). Among the stations in region S, the peak frequency is lower and less clear over the lake (S1) compared with the inland stations (S9, S10, S14) (Fig. 6). The diurnal patterns over the mountain hillslopes (S9, S10) and at the station away from the mountain hillslopes (S14) are similar in Fig. 6, suggesting that the rainfall in region S are mostly independent systems from those in mountains in region M. In region N, the rainfall frequency has a peak during evening around 18:00 LT. In region C, rainfall is frequent at night with a maximum frequency in the early morning at 04:00 LT. The clear regional characteristics of the rainfall pattern within Cambodia imply that, even during the summer monsoon season, most of the rainfall in Cambodia is not simply determined by the strength of the ASM but is highly controlled by local features.

During the post-monsoon season, the peak of the rainfall frequency is in the afternoon at 15:00 LT in regions N and C while it is from night to early morning (around 21:00–4:00 LT) in regions M and S (Fig. 5c). More than half of the post-monsoon rainfall occurs at night in region S, and there exist two diurnal peaks of the rainfall amount in the afternoon and at midnight in region M (Fig. 5f). Since the post-monsoon rainfall in regions M and S is larger than that in region N as described in the previous subsection, this suggests that the regions which receive relatively large post-monsoon rainfall have the favorable condition for nocturnal rainfall.

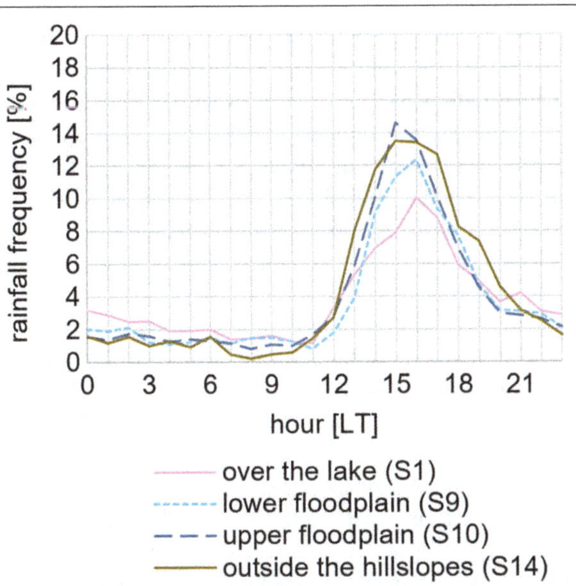

Fig. 6 Same as Fig. 5b but for each station at S1 (over the lake), S9 (lower floodplain), S10 (upper floodplain), and S14 (away from the mountain hillslopes)

Intra-seasonal variability of the diurnal rainfall cycle

In addition to the season-averaged diurnal rainfall cycle (Fig. 5), the intra-seasonal or day-to-day variability of the clarity of the diurnal rainfall cycle is examined associated with the synoptic-scale atmospheric condition. Since the intra-seasonal variability becomes unclear in multi-year averages, two example years, 2011 and 2015, are selected from the results presented in Fig. 5 to draw the date–hour cross-section for the rainfall frequency, shown in Figs. 7 and 8, respectively. In these figures, the areal-averaged geopotential height at 850 hPa (Φ) and zonal winds (u) at 850 and 700 hPa from JRA-55 over Cambodia (102–108°E and 10–15°N) at 7:00 LT (0:00 UTC) are shown in each day as indicators of the synoptic-scale atmospheric condition. The peak-frequency hours recognized in Fig. 5 in each region and season are shown with vertical yellow shades. In contrast, the days with prominent rainfall with an ambiguous diurnal cycle or in different hours are shown with horizontal orange broken lines. The horizontal red lines are drawn on 1 March, the onset and withdrawal dates of the ASM shown in Table 2, and 30 November.

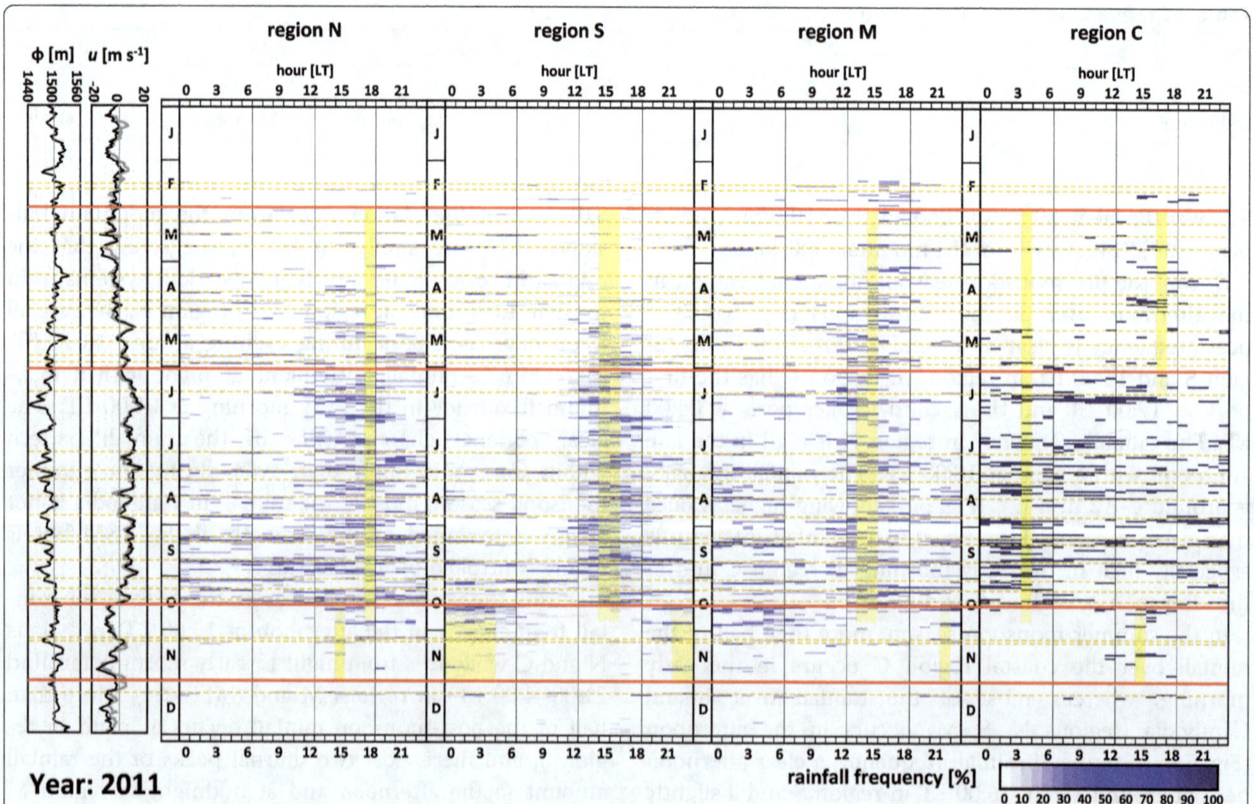

Fig. 7 Diurnal pattern of rainfall frequency for each region N, S, M, and C as the date–hour cross-section in 2011. The percentage of the number of "rainy" stations against the total number of stations in each region is shown with a blue shade. At region C, there is only one gauge, and thus the percentage is either 0 or 100. The horizontal red lines indicate 1 March, the onset and withdrawal dates of the summer monsoon season, and 30 November. The vertical yellow shades indicate hours with peak rainfall frequency in each season in each region, whereas the horizontal orange broken lines indicate days with prominent rainfall with no diurnal cycle or in different hours. The geopotential height (Φ) at 850 hPa and zonal winds (*u*) at 700 hPa (gray line) and 850 hPa (black line) as the areal averages over Cambodia (102–108°E and 10–15°N) at 7:00 LT (0:00 UTC) from JRA-55 are also shown for each day in the left panels

The basic diurnal patterns described in the previous subsection are recognized in both years. However, in 2011, the summer monsoon rainfall in regions N and M is frequent in the morning especially during the latter period (Fig. 7), in addition to the characteristic afternoon rainfall recognized in the season-averaged values (Fig. 5b, e). The geopotential height at 850 hPa is low throughout the summer monsoon season in this year, and its depressions often correspond to the broken orange lines in Fig. 7, which show the days without or with different diurnal patterns of rainfall. On the contrary, summer monsoon rainfall is dominant only in the afternoon in most of the days in 2015 (Fig. 8) when the entire Cambodia is mostly covered by high pressure, with a few days having morning rainfall in regions N and M corresponding to low pressure.

In the early post-monsoon season, the alteration of rainy hours in region S from afternoon to midnight is suggested in the season-averaged discussion in the previous subsection, and it is clear in 2011 along with the withdrawal of the ASM. However, the characteristic midnight to early morning rainfall there is not obvious in 2015 when Cambodia is under high pressure.

In summary, the diurnal pattern of rainfall shown in the previous subsection does not always appear, both with days having rainfall in other hours of a day and with days having no rainfall at the expected hours, and it likely occurs under the influence of the synoptic atmospheric condition.

Diurnal cycle of wind direction by ground observation

The diurnal patterns of wind direction and specific humidity observed at the AWS are shown in Fig. 9 as 6-year-averaged values for April (pre-monsoon season), August (summer monsoon season), and October (post--monsoon season), to examine possible mechanisms underlying the observed diurnal patterns of rainfall as seasonal averages from the viewpoint of local circulation.

In April (Fig. 9a), the wind direction shows a clear diurnal cycle, with northeasterly to easterly wind during the daytime and southwesterly wind during the

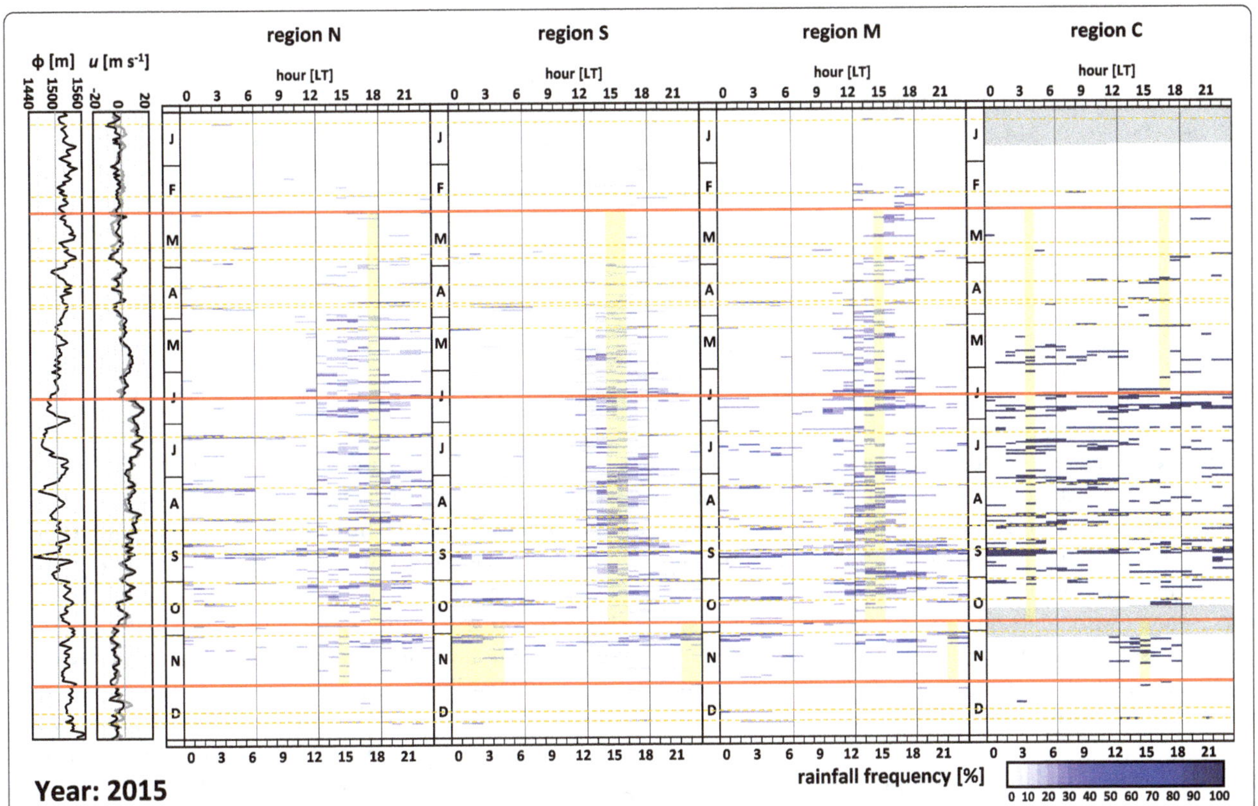

Fig. 8 Same as Fig. 7 but for 2015. Days in the gray-hatched area in region C are "data-missing days"

nighttime. Since the lake is located northeast to the AWS, the observed winds at the AWS would be the land–lake breezes resulting from the presence of the Tonle Sap Lake. The alteration from land breeze to lake breeze occurs at about 7:00 LT and from lake breeze to land breeze at around 17:00–19:00 LT. During the daytime lake breeze hours, the specific humidity decreases.

In August (Fig. 9b), the observed wind is northwesterly with 1.4–1.6 m s^{-1} during the daytime and southerly to southwesterly during the nighttime. By considering the daytime vector balance with the large-scale wind in JRA-55 data (2.3 m s^{-1} from the southwest at 1000 hPa at the nearest point), the existence of the local atmospheric driving force at about 2.1 m s^{-1} blowing from the direction at 12° (northerly with a slight easterly component) is suggested to change the wind direction at the local scale. Thus, we can reasonably infer the existence and the effect of the lake breeze originating from the Tonle Sap Lake during the daytime at this location.

In October (Fig. 9c), a clear diurnal cycle of the wind direction is detected at the AWS, with a southwesterly land breeze during the nighttime and a northeasterly lake breeze during the daytime. The alteration between land breeze and lake breeze in wind direction is clearer than in April and occurs at 09:00 and 18:00 LT.

Discussion

Rainfall amount and its regional difference

The gauge-observed rainfall amount during the summer monsoon season (877 mm on average in inland Cambodia) and its spatial distribution coincide with the results by Takahashi et al. (2010a), which showed that summer monsoon rainfall was smaller around the Tonle Sap Lake (less than 760 mm in 5 months) and slightly larger at the northeastern side of the lake. Under the prevailing ASM, the western region is on the leeward side of the Cardamom Mountains, and the northeastern region is on the windward side of the Annam Range. The gauge-observed spatial distribution of the rainfall in inland Cambodia thus matches the satellite-observed results by Takahashi et al. (2010a), which showed larger rainfall on the leeward side and smaller rainfall on the windward side of mountain ranges.

Further study would be required to analyze the effect of the lake/land breeze circulation in addition to the abovementioned topographical effect for making an observed difference of rainfall amount between regions N and S.

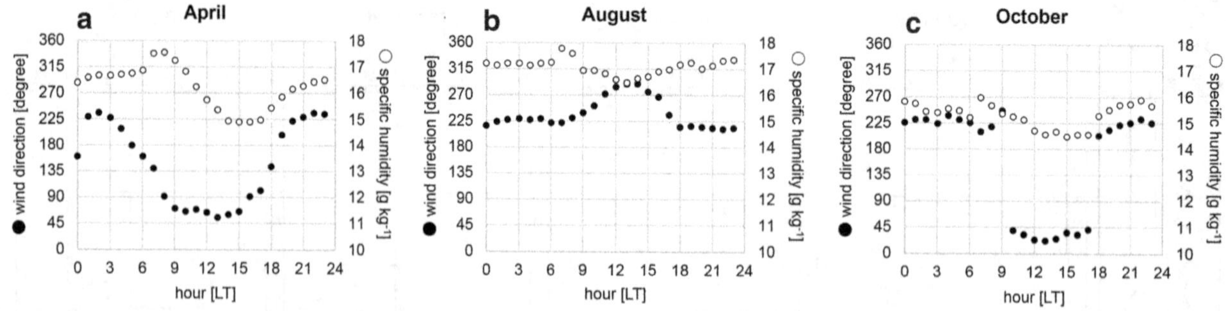

Fig. 9 Diurnal patterns of wind direction (filled circles) and specific humidity (open circles) observed at the AWS at S7 (Fig. 2) in April (**a**), August (**b**), and October (**c**) as 6-year monthly averages

Contribution of pre- and post-monsoon rainfall

Our ground observation shows that around 22–50% of annual rainfall in inland Cambodia is brought from January to April and from October to December, i.e., not in the months under the prevailing ASM. This result corresponds with Matsumoto's (1997) study, which suggests the substantial rainfall in April and October. The result also suggests the existence of factors other than the ASM to influence the rainfall in Cambodia.

Intra-seasonal variability and synoptic-scale influences

Kiguchi and Matsumoto (2005) and Kiguchi et al. (2016) showed that the passage of the synoptic-scale eastward-moving trough produced the intermittent rainfall in Thailand during the pre-monsoon season. Yokoi et al. (2007) showed the predominance of intra-seasonal variations in rainfall over the Indochina Peninsula during the summer monsoon season. Takahashi and Yasunari (2006) and Takahashi et al. (2015) showed that the active and break phases of rainfall in Thailand during the summer monsoon season were brought by the westward propagation of precipitation systems from the western North Pacific. For the post-monsoon season, Nguyen-Thi et al. (2012) showed that approximately 20% of the rainfall in the central Vietnam coastal area was brought by tropical cyclones, and Chen et al. (2012) showed that the heavy rainfall in Vietnam was basically by the westward-moving disturbances and not due to the northeasterly monsoon itself. Takahashi (2013) investigated clouds and rainfall over the Indochina Peninsula in the post-monsoon season and showed that their characteristics changed within this season, with well-developed cloud-precipitation systems and large rainfall during October and November and low clouds with little rainfall in December, January, and February. He discussed that the first half of the AWM period was probably associated with tropical cyclone activity, although diurnal variation in cloud activity was still recognized, and the low clouds in the latter half would be associated with the orographically forced upward motions.

These previous studies showed that there exist synoptic-scale to intra-seasonal variations for rainfall phenomena over the Indochina Peninsula. Our results in Figs. 7 and 8 show the characteristic diurnal patterns in each region and season as well as the intra-seasonal variations, including days with no clear or different diurnal patterns. Since days with prominent rainfall in different hours correspond to days with lower geopotential height and/or stronger zonal wind, the large-scale low-pressure system and/or disturbances are considered to affect rainfall in Cambodia, as suggested by previous studies.

In this study, the frequently observed typical diurnal patterns of rainfall for each season and region are highlighted for further discussion.

Season-averaged diurnal pattern of rainfall in this region

Ohsawa et al. (2001) showed that the late night to early morning maximum over tropical Asia from June to August was found at stations with large rainfall amounts. Takahashi et al. (2010b) showed that heavy rainfall in the early morning was found around the coasts over the eastern Gulf of Thailand and the Bay of Bengal, comprising nearly half of the total rainfall there. In our observation, station C1 located onshore to the western Gulf of Thailand recorded distinctively larger summer monsoon rainfall than other stations in inland Cambodia (Fig. 4), and more than half of the rainfall occurred in the early morning (Fig. 5e). There is no significant difference in the rainfall intensity over regions (Fig. 5h), and large summer monsoon rainfall in the coastal region C is mostly by the higher rainfall frequency there. The late night to early morning rainfall is also dominant in regions S and M during the post-monsoon season, where the total rainfall amount is larger. The rainfall intensity is not significantly larger in these regions than in other regions.

In this observation, the diurnal rainfall peak in the summer monsoon season shifts from region C (at 04:00 LT) to regions M (14:00–15:00 LT), S (15:00–16:00 LT), and N (18:00 LT), whereas that in the post-monsoon season shifts from regions N and C (15:00 LT) to regions M and S (21:00–04:00 LT). For northern Thailand, Okumura et al. (2003) showed that the phase shift of the diurnal rainfall cycle moved leeward from mountains corresponding to the wind direction of the prevailing monsoon. The abovementioned observed phase shift in this study agrees with their study considering the southwesterly flow in the summer monsoon season and the northeasterly flow in the post-monsoon season. However, the observed diurnal rainfall peak in the pre-monsoon season moves from regions M (15:00 LT) and S (15:00–16:00 LT) through region C (17:00 and 04:00 LT) to region N (18:00 LT), and this phase shift cannot be explained in relation to the distance to mountains and the prevailing large-scale southeasterly flow.

Although Takahashi (2016) showed by satellite data that precipitation ended in the evening in April and May and lasted until early morning from July to September over the Indochina Peninsula, such characteristics have been recognized only at coastal station C1. Most of the rainfall in inland Cambodia is observed to occur in the afternoon in both the pre-monsoon and the summer monsoon seasons, except for the dominant evening rainfall in the summer monsoon season in region N.

Local circulation

The observed diurnal cycle of rainfall can be partly explained by the convergence of the large-scale and local-scale flows, as discussed by Houze et al. (1981). Although further study with numerical simulations is needed to examine the mechanisms of the diurnal rainfall cycle in this region, here we summarize a working hypothesis inferred from data from our dense gauge observations, focusing on the viewpoint from the effect of the land–lake circulation and the low-level wind convergence.

During the pre-monsoon season, the low-level large-scale wind over Cambodia is southerly with a slight easterly component (Fig. 3b). During the daytime at station C1, both the sea breeze from the Gulf of Thailand and the anabatic wind to the Cardamom Mountains are southwesterly, so that no clear low-level winds converging to a large-scale southerly are expected. In this situation, the orographic lifting of moist air from the sea caused by the Cardamom Mountains may determine the observed afternoon rainfall. As pointed out by Xie et al. (2006) regarding the abundant rainfall under the ASM over the Indochina Peninsula, moist air is forced to rise over the mountains and thereby initiate intense convection on the windward side of the orographic lifting. The increased atmospheric instability by sensible heat from the drier ground in this season and also by the low-level intrusion of moist air from the sea contributes to the afternoon rainfall maxima and the abundant rainfall on the coast and over the Cardamom Mountains.

During the summer monsoon season, dominant evening rainfall in region N characterizes the difference from afternoon rainfall in regions M and S. Contrary to the southwestern side of the lake, at the northeastern side of the lake (in region N), no clear low-level convergence is expected between the prevailing southwesterly and the daytime lake breeze. Instead, nighttime land breeze converges with the ASM at the northeastern side. Although detection of the land breeze from the AWS data is difficult in August when the land breeze there is expected to flow from almost the same direction with the prevailing ASM, the alteration to land breeze in April is around 17:00–19:00 LT in Fig. 9a and matches the timing of the rainfall peak observed in region N. Therefore, we speculate that the evening rainfall in region N would be driven by the low-level wind convergence with the ASM at the onset of the land breeze.

The possible mechanism for the afternoon rainfall in regions M and S would be the daytime convergence of the southwesterly ASM and the northeasterly lake breeze at the southwestern side of the lake as well as the increased atmospheric instability by sensible heat from the ground. The convergence with the anabatic wind as well as the orographic lifting of the moist lake air over the northeastern slopes of the Cardamom Mountains would also contribute to the afternoon rainfall in region M. On the coast in region C, the ASM converges with the nocturnal northeasterly land breeze and the katabatic wind from the southwestern slope of the Cardamom Mountains, and this would contribute to the late night to early morning rainfall there.

Xie et al. (2006) suggested separate convection lines for dominant rainfall at the southwestern coast in Cambodia, on the coast at the foothills of the Cardamom Mountains, and on the western slope of the Annam Range. However, for our target area west of the Mekong River, the Annam Range is about 400 km away, and thus, the nighttime katabatic wind is unable to reach this area by evening when the rainfall maximum is observed. Therefore, the observed peak in the evening may result from an independent rainfall system from that initiated by the Annam Range.

During the post-monsoon season, the low-level large-scale southwesterly wind withdraws almost entirely, and dry air intrudes to the Indochina Peninsula with a northeasterly AWM flow (Fig. 3h). The situation in region N during the post-monsoon season resembles that in region S during the summer monsoon season, as the daytime lake breeze (southwesterly) converges with the large-scale northeasterly AWM. However, the post-monsoon rainfall amount in region N is even smaller than that in region S, suggesting the importance of the role of the ASM in transporting abundant moisture from the sea as well as the role of the Cardamom Mountains in initiating rainfall. In fact, even during the post-monsoon season, at the coastal station (C1) where the AWM converges with a daytime sea breeze and the anabatic wind to the southwestward slope of the Cardamom Mountains, the rainfall amount is large, with the peak occurring in the afternoon. In addition, over region S, the rainfall amount is larger than in other regions in inland Cambodia and most frequent at night when the land breeze and the katabatic wind from the northeastward slope of the Cardamom Mountains converge with the AWM. As Tsujimoto and Koike (2013) demonstrated, a clear nighttime thermal contrast between the lake and the surrounding lands enhanced by the high lake surface temperature is also important for the strong nighttime land breeze circulation associated with the nocturnal rainfall over the southwestern plain.

Conclusions

Automatic rain-gauge data in Cambodia from 35 stations, including one AWS, have been collected since September 2009, from which the quality-controlled data of 30 of these stations are used to analyze the rainfall characteristics in Cambodia for the period 2010–2015. The analysis focuses on the regional characteristics of the diurnal patterns of rainfall and the underlying mechanisms with reference to the observed surface meteorological data. Although further examination with atmospheric models is required, our discussion of the observed rainfall characteristics and examination of the potential underlying mechanisms can provide a reference for future studies.

The annual rainfall amount in inland Cambodia is 1087–1528 mm on station-average and is much larger on the coast at nearly 4000 mm. Over inland Cambodia, approximately 5–20% of the annual rainfall (63–278 mm) occurs in March and April (pre-monsoon season), 50–78% (519–1350 mm) occurs from May to September (summer monsoon season), and 12–36% (181–457 mm) occurs in October and November (post-monsoon season).

During the pre-monsoon season, rainfall is dominant on the coast and over the Cardamom Mountains, with a maximum in the afternoon. Orographic lifting as well as atmospheric instability by larger sensible heat from the drier ground is likely the driving mechanism.

During the summer monsoon season, the rainfall amount and its proportion to the annual rainfall are larger in the northern region (around Kampong Thom, Preah Vihear, and Siem Reap provinces) and smaller in the western region (around Battambang province) in inland Cambodia. The rainfall amount on the coast is distinctively large. Clear regional characteristics in the diurnal precipitation patterns are recognized, and they suggest the significant effects of local features even during the ASM season. The diurnal rainfall maximum occurs (i) in the early afternoon in the Cardamom Mountains, (ii) in the afternoon on the plain at the southwestern side of the Tonle Sap Lake, (iii) in the evening on the wide area at the northeastern side of the lake, and (iv) and in the early morning on the coast. The possible mechanism for each rainfall is (a) daytime low-level convergence of the southwesterly ASM and the northeasterly lake breeze at the southwestern side of the lake as well as increased atmospheric instability by sensible heat from land (for (i) and (ii)), (b) daytime low-level convergence of the ASM with the anabatic wind, as well as orographic lifting of the moist lake air, over the northeastern slopes of the Cardamom Mountains (for (i)), (c) low-level convergence of the ASM with the evening northeasterly land breeze at the northeastern side of the lake (for (iii)), and (d) low-level convergence of the ASM with the nocturnal northeasterly land breeze enhanced by the katabatic wind from the southwestern slope of the Cardamom Mountains at the coastal area (for (iv)).

The rainfall amount during the post-monsoon season is larger on the southwestern side of the lake, i.e., around the northern ridges of the Cardamom Mountains and the western lakeshore plain, compared with the other regions in inland Cambodia, with a diurnal maximum at night and in the early morning. It probably results from the low-level convergence of the northeastern AWM with the southwestern land breeze enhanced by the katabatic wind from the northeastern slope of the Cardamom Mountains. Since the rainfall amount is smaller at the northeastern side of the lake where the AWM converges with the daytime lake breeze, the importance of the role of mountains on rainfall is proposed.

The diurnal pattern of rainfall in each season and in each region is, however, not clear on some days, and analysis of the synoptic-scale atmospheric condition suggests the effect of the large-scale low-pressure system and disturbances on the appearance of the clear diurnal rainfall pattern. Further numerical studies are required to examine the effect of the local circulation and the interaction with the synoptic-scale atmospheric condition.

We have demonstrated the diurnal patterns of rainfall in Cambodia with their distinct regional and seasonal characteristics. Clear regional differences suggest the importance of the effect of local features, especially topography, in the formation and development of rainfall in the pre-monsoon, summer monsoon, and post-monsoon seasons, in addition to the synoptic-scale conditions suggested by many previous studies. The study of local effects on rainfall with consideration of the land-surface dynamics may aid flood and drought management in Cambodia.

Abbreviations

ASM: Asian summer monsoon; AWM: Asian winter monsoon; AWS: Automatic Weather Station; JRA-55: The Japanese 55-year Reanalysis data; LT: Local Time in Cambodia (UTC + 7 h)

Acknowledgements

We greatly appreciate the kind support from H.E. Mr. Pich Veasna, H.E. Mr. Long Saravuth, Mr. Mao Hak, Mr. Oum Ryna, Mr. Sreng Sotha, Mr. Khum Vany, Mr. Preap Sameng, Mr. Sok Sokhon, and Mr. Mao Soksan of the Ministry of Water Resources and Meteorology, Cambodia, for facilitation of study in Cambodia. We also thank Prof. Toshio Koike for valuable comments during the 7-year observation. We used digital elevation data from the U.S. Geological Survey and global reanalysis data by the Japan Meteorological Agency. The JRA-55 data are used through the Data Integration and Analysis System (DIAS). We are grateful to the editor Prof. Jun Matsumoto and the anonymous reviewers for their helpful and constructive comments that greatly contributed to improving the manuscript from its earlier version.

Funding

Parts of this research were supported by the Japan Society for the Promotion of Science (JSPS) KAKENHI Grant Numbers JP25820222 and JP16K06503, and also by the Space Applications for Environment (SAFE) and the Precipitation Measuring Mission (PMM) projects by Japan Aerospace Exploration Agency (JAXA), and by the Data Integration and Analysis System (DIAS) by Ministry of Education, Culture, Sports, Science and Technology (MEXT).

Authors' contributions

KuT was the principal investigator leading the installation of the observation instrument, designing data analysis, and discussing on the interaction with local circulation. TO conducted data analysis with KuT. KA and KaT conducted fieldwork in Cambodia together with KuT. MSI led the fieldwork with his Cambodian colleagues and provided local information on both observation status and locally known rainfall characteristics. All authors read and approved the final manuscript.

Competing interests

The authors declare that they have no competing interests.

Author details

[1]Graduate School of Environmental and Life Science, Okayama University, 3-1-1 Tsushima-naka, Kita-ku, Okayama 700-8530, Japan. [2]Independent Consultant, Okayama, Japan. [3]Center for Research in Isotopes and Environmental Dynamics (CRiED), University of Tsukuba, 1-1-1 Tennodai, Tsukuba, Ibaraki 305-8577, Japan. [4]International Centre for Water Hazard and Risk Management (ICHARM) of the Public Works Research Institute, 1-6 Minamihara, Tsukuba, Ibaraki 305-8516, Japan. [5]Department of Meteorology, Ministry of Water Resources and Meteorology, 364 Monivong Boulevard, Chamkarmorn, Phnom Penh, Cambodia.

References

Bhatt BC, Nakamura K (2005) Characteristics of monsoon rainfall around the Himalayas revealed by TRMM precipitation radar. Mon Weather Rev 133:149–165. https://doi.org/10.1175/mwr-2846.1

Chaboureau JP, Guichard F, Redelsperger JL, Lafore JP (2004) The role of stability and moisture in the diurnal cycle of convection over land. Q J R Meteorol Soc 130:3105–3117. https://doi.org/10.1256/qj.03.132

Chang C-P, Wang Z, McBride J, Liu C-H (2005) Annual cycle of Southeast Asia—maritime continent rainfall and the asymmetric monsoon transition. J Clim 18:287–301. https://doi.org/10.1175/jcli-3257.1

Chen T-C, Tsay J-D, Yen M-C, Matsumoto J (2012) Interannual variation of the late fall rainfall in central Vietnam. J Clim 25:392–413. https://doi.org/10.1175/jcli-d-11-00068.1

Chhinh N, Millington A (2015) Drought monitoring for rice production in Cambodia. Climate 3:792–811. https://doi.org/10.3390/cli3040792

Earth Observation Research Center (EORC), Japan Aerospace Exploration Agency (JAXA) (2017) Global Rainfall Map in Near Real Time (GSMaP_NRT) and Gauge-calibrated Rainfall Product (GSMaP_Gauge) data format description. http://sharaku.eorc.jaxa.jp/GSMaP/document/DataFormatDescription.pdf. Accessed 5 Feb 2018

Fujinami H, Nomura S, Yasunari T (2005) Characteristics of diurnal variations in convection and precipitation over the southern Tibetan Plateau during summer. SOLA 1:49–52. https://doi.org/10.2151/sola.2005-014

Harada Y, Kamahori H, Kobayashi C, Endo H, Kobayashi S, Ota Y, Onoda H, Onogi K, Miyaoka K, Takahashi K (2016) The JRA-55 reanalysis: representation of atmospheric circulation and climate variability. J Meteor Soc Japan Ser II 94: 269–302. https://doi.org/10.2151/jmsj.2016-015

Hirose M, Nakamura K (2005) Spatial and diurnal variation of precipitation systems over Asia observed by the TRMM Precipitation Radar. J Geophys Res Atmos 110:D05106. https://doi.org/10.1029/2004JD004815

Houze RA Jr, Geotis SG, Marks FD Jr, West AK (1981) Winter monsoon convection in the vicinity of north Borneo. Part I: structure and time variation of the clouds and precipitation. Mon Weather Rev 109:1595–1614. https://doi.org/10.1175/1520-0493(1981)109<1595:WMCITV>2.0.CO;2

Kiguchi M, Matsumoto J (2005) The rainfall phenomena during the pre-monsoon period over the Indochina Peninsula in the GAME-IOP Year, 1998. J Meteor Soc Japan. Ser. II 83:89–106. https://doi.org/10.2151/jmsj.83.89

Kiguchi M, Matsumoto J, Kanae S, Oki T (2016) Pre-monsoon rain and its relationship with monsoon onset over the Indochina Peninsula. Front Earth Sci 4:42. https://doi.org/10.3389/feart.2016.00042

Kobayashi S, Ota Y, Harada Y, Ebita A, Moriya M, Onoda H, Onogi K, Kamahori H, Kobayashi C, Endo H, Miyaoka K, Takahashi K (2015) The JRA-55 reanalysis: general specifications and basic characteristics. J Meteor Soc Japan Ser. II 93: 5–48. https://doi.org/10.2151/jmsj.2015-001

Kuraji K, Punyatrong K, Suzuki M (2001) Altitudinal increase in rainfall in the Mae Chaem watershed. Thailand J Meteor Soc Japan Ser II 79:353–363. https://doi.org/10.2151/jmsj.79.353

Masumoto T, Tsujimoto K, Somura H (2007) Hydro-meteorological observation and analysis of observed data at Tonle Sap Lake and its environs, urban and paddy areas. Technical Report of the National Institute for Rural Engineering 206:219–236 In Japanese with English abstract

Matsumoto J (1997) Seasonal transition of summer rainy season over Indochina and adjacent monsoon region. Adv Atmos Sci 14:231–245. https://doi.org/10.1007/s00376-997-0022-0

McGloin R, McGowan H, McJannet D (2014) Effects of diurnal, intra-seasonal and seasonal climate variability on the energy balance of a small subtropical reservoir. Int J Climatol 35:2308–2325. https://doi.org/10.1002/joc.4147

National Committee for Disaster Management, United Nations Development Programme (2014) Cambodia disaster loss and damage analysis report 1996–2013. http://www.kh.undp.org/content/cambodia/en/home/library/environment_energy/cambodia-disaster-loss-and-damage-analysis-report-1996 2013.html of subordinate document. Accessed 29 Jun 2017

Nguyen-Thi HA, Matsumoto J, Ngo-Duc T, Endo N (2012) A climatological study of tropical cyclone rainfall in Vietnam. SOLA 8:41–44. https://doi.org/10.2151/sola.2012-011

Nieuwolt S (1981) The climates of continental Southeast Asia. In: Takahashi K, Arakawa H (eds) Climates of southern and western Asia. World survey of climatology 9. Elsevier, Amsterdam

Ohsawa T, Ueda H, Hayashi T, Watanabe A, Matsumoto J (2001) Diurnal variations of convective activity and rainfall in tropical Asia. J Meteor Soc Japan Ser. II 79:333–352. https://doi.org/10.2151/jmsj.79.333

Okumura K, Satomura T, Oki T, Khantiyanan W (2003) Diurnal variation of precipitation by moving mesoscale systems: radar observations in northern Thailand. Geophys Res Lett 30:2073. https://doi.org/10.1029/2003GL018302

Ono M, Takahashi HG (2016) Seasonal transition of precipitation characteristics associated with land surface conditions in and around Bangladesh. J Geophys Res Atmos 121:11,190–111,200. https://doi.org/10.1002/2016JD025218

Satomura T (2000) Diurnal variation of precipitation over the Indo-China PPeninsula: two-dimensional numerical simulation. J Meteor So Japan Ser II 78:461–475. https://doi.org/10.2151/jmsj1965.78.4_461

Satomura T, Yamamoto K, Sysouphanthavong B, Phonevilay S (2011) Diurnal variation of radar echo area in the middle of Indochina. J Meteor Soc Japan Ser. II 89A:299–305. https://doi.org/10.2151/jmsj.2011-A20

Save Cambodia's Wildlife (2006) The atlas of Cambodia: national poverty and environment maps. Phnom Penh

Singh P, Nakamura K (2009) Diurnal variation in summer precipitation over the central Tibetan Plateau. J Geophys Res Atmos 114:D20107. https://doi.org/10.1029/2009JD011788

Sugimoto S, Takahashi HG (2017) Seasonal differences in precipitation sensitivity to soil moisture in Bangladesh and surrounding regions. J Clim 30:921–938. https://doi.org/10.1175/jcli-d-15-0800.1

Takahashi HG (2013) Orographic low-level clouds of Southeast Asia during the cold surges of the winter monsoon. Atmos Res 131:22–33. https://doi.org/10.1016/j.atmosres.2012.07.005

Takahashi HG (2016) Seasonal and diurnal variations in rainfall characteristics over the tropical Asian monsoon region using TRMM-PR data. SOLA 12A:22–27. https://doi.org/10.2151/sola.12A-005

Takahashi HG, Fujinami H, Yasunari T, Matsumoto J (2010a) Diurnal rainfall pattern observed by Tropical Rainfall Measuring Mission Precipitation Radar (TRMM-PR) around the Indochina Peninsula. J Geophys Res Atmos 115: D07109. https://doi.org/10.1029/2009jd012155

Takahashi HG, Fujinami H, Yasunari T, Matsumoto J, Baimoung S (2015) Role of tropical cyclones along the monsoon trough in the 2011 Thai flood and interannual variability. J Clim 28:1465–1476. https://doi.org/10.1175/jcli-d-14-00147.1

Takahashi HG, Yasunari T (2006) A climatological monsoon break in rainfall over Indochina—a singularity in the seasonal march of the Asian summer monsoon. J Clim 19:1545–1556. https://doi.org/10.1175/jcli3724.1

Takahashi HG, Yoshikane T, Hara M, Takata K, Yasunari T (2010b) High-resolution modelling of the potential impact of land surface conditions on regional climate over Indochina associated with the diurnal precipitation cycle. Int J Climatol 30:2004–2020. https://doi.org/10.1002/joc.2119

Torti J (2012) Floods in Southeast Asia: a health priority. J Global Health 2:020304. https://doi.org/10.7189/jogh.02.020304

Tsujimoto K, Koike T (2008) Mechanism of locally-induced convection development and its effects on vapor transportation over Tonle Sap Lake area. Ann J Hydraulic Eng Japan Soc Civil Eng 52:247–252. https://doi.org/10.2208/prohe.52.247 In Japanese with English abstract

Tsujimoto K, Koike T (2012) Requisite conditions for post-monsoon rainfall in Cambodia by looking through 2009 rainfall data. J Hydrosci Hydraul Eng 30:1–14

Tsujimoto K, Koike T (2013) Land-lake breezes at low latitudes: the case of Tonle Sap Lake in Cambodia. J Geophys Res Atmos 118:6970–6980. https://doi.org/10.1002/jgrd.50547

Tsujimoto K, Masumoto T, Mitsuno T (2008) Seasonal changes in radiation and evaporation implied from the diurnal distribution of rainfall in the Lower Mekong. Hydrol Process 22:1257–1266. https://doi.org/10.1002/hyp.6935

Verburg P, Antenucci JP (2010) Persistent unstable atmospheric boundary layer enhances sensible and latent heat loss in a tropical great lake: Lake Tanganyika. J Geophys Res Atmos 115:D11109. https://doi.org/10.1029/2009JD012839

Wallace JM (1975) Diurnal variations in precipitation and thunderstorm frequency over the conterminous United States. Mon Weather Rev 103:406–419. https://doi.org/10.1175/1520-0493(1975)103<0406:dvipat>2.0.co;2

Xie S-P, Xu H, Saji NH, Wang Y, Liu WT (2006) Role of narrow mountains in large-scale organization of Asian Monsoon convection. J Clim 19:3420–3429. https://doi.org/10.1175/jcli3777.1

Yen M-C, Chen T-C, Hu H-L, Tzeng R-Y, Dinh DT, Nguyen TTT, Wong CJ (2011) Interannual variation of the fall rainfall in Central Vietnam. J Meteor Soc Japan Ser II 89A:259–270. https://doi.org/10.2151/jmsj.2011-A16

Yokoi S, Satomura T, Matsumoto J (2007) Climatological characteristics of the intraseasonal variation of precipitation over the Indochina Peninsula. J Clim 20:5301–5315. https://doi.org/10.1175/2007jcli1357.1

Yu R, Zhou T, Xiong A, Zhu Y, Li J (2007) Diurnal variations of summer precipitation over contiguous China. Geophys Res Lett 34:L01704. https://doi.org/10.1029/2006GL028129

Zhang Y, Li T, Wang B, Wu G (2002) Onset of the summer monsoon over the Indochina Peninsula: climatology and interannual variations. J Clim 15:3206–3221. https://doi.org/10.1175/1520-0442(2002)015<3206:ootsmo>2.0.co;2

3

Seasonal modulation of the Asian summer monsoon between the Medieval Warm Period and Little Ice Age

Youichi Kamae[1,2*], Toshi Kawana[3], Megumi Oshiro[4] and Hiroaki Ueda[1]

Abstract

Instrumental and proxy records indicate remarkable global climate variability over the last millennium, influenced by solar irradiance, Earth's orbital parameters, volcanic eruptions and human activities. Numerical model simulations and proxy data suggest an enhanced Asian summer monsoon during the Medieval Warm Period (MWP) compared to the Little Ice Age (LIA). Using multiple climate model simulations, we show that anomalous seasonal insolation over the Northern Hemisphere due to a long cycle of orbital parameters results in a modulation of the Asian summer monsoon transition between the MWP and LIA. Ten climate model simulations prescribing historical radiative forcing that includes orbital parameters consistently reproduce an enhanced MWP Asian monsoon in late summer and a weakened monsoon in early summer. Weakened, then enhanced Northern Hemisphere insolation before and after June leads to a seasonally asymmetric temperature response over the Eurasian continent, resulting in a seasonal reversal of the signs of MWP–LIA anomalies in land–sea thermal contrast, atmospheric circulation, and rainfall from early to late summer. This seasonal asymmetry in monsoon response is consistently found among the different climate models and is reproduced by an idealized model simulation forced solely by orbital parameters. The results of this study indicate that slow variation in the Earth's orbital parameters contributes to centennial variability in the Asian monsoon transition.

Keywords: Asian monsoon, Medieval Warm Period, Little Ice Age, Meridional thermal gradient, Orbital forcing

Introduction

The last millennium, from 850 A.D. to the present, is a key period in understanding Earth's climate response to external forcing and its internal variability on decadal-to-centennial timescales (e.g., Crowley 2000; Jones et al. 2009; PAGES 2k Consortium 2013; Smerdon and Pollack 2016). A variety of instrumental and proxy records indicate that the global climate has experienced remarkable fluctuations during this period due to large volcanic eruptions (e.g., Miller et al. 2012; Schurer et al. 2014; Yang and Jiang 2015; Liu et al. 2016), variations in total solar irradiance (TSI) (Gray et al. 2010; Lean 2010), Earth's orbital parameters (Berger and Loutre 1991; Kaufman et al. 2009), greenhouse gases (GHGs) (e.g., Schmidt et al. 2011), and internal climate variability. On a centennial timescale, anthropogenic global warming accompanying rapid GHG increases since the 19th century is the most remarkable climate fluctuation (e.g., Bradley et al. 2003; Jones and Mann 2004; Ammann et al. 2007; Phipps et al. 2013; Masson-Delmotte et al. 2014). In addition, predominant global climate perturbation from the Medieval Warm Period (MWP) of 950–1250 A.D. to the Little Ice Age (LIA) of 1400–1700 A.D. (e.g., Mann et al. 2009) has also attracted much attention because it is the most recent example of long-term, naturally-generated global climate variability (e.g., Liu et al. 2011; Man et al. 2012; Otto-Bliesner et al. 2016; Shi et al., 2016a, b).

* Correspondence: kamae.yoichi.fw@u.tsukuba.ac.jp
[1]Faculty of Life and Environmental Sciences, University of Tsukuba, 1-1-1 Tennoudai, Tsukuba, Ibaraki 305-8506, Japan
[2]Scripps Institution of Oceanography, University of California San Diego, 9500 Gilman Drive, MC 206, La Jolla, CA 92093-0206, USA
Full list of author information is available at the end of the article

Historical variations in the Asian monsoon have important links to environmental changes and the cultural development of human societies over South and East Asia (Clift and Plumb 2008; Fu et al. 2008). Past monsoon variability has been addressed both from paleoenvironmental reconstructions based on proxy records and from numerical climate simulations (e.g., Kutzbach et al. 2008; Zhang et al. 2008; Battisti et al. 2014; Masson-Delmotte et al. 2014; Fallah and Cubasch 2015; Yan et al. 2015). Variations in summertime temperature over the continents in the Northern Hemisphere (NH) in response to seasonal insolation anomalies brought on by changes in orbital parameters (including the longitude of perihelion precession) are an essential factor in past monsoon variability (e.g., COHMAP members 1988; Mohtadi et al. 2016). The land–sea contrast in tropospheric temperature favored by summertime insolation is further amplified by latent heating of the monsoon rainfall that establishes deep-tropospheric monsoon circulation (e.g., Li and Yanai 1996; Webster et al. 1998; Dai et al. 2013; Wu et al. 2015; Zhang et al. 2016). On a millennial timescale, past monsoon variability has largely been controlled by precession variability (e.g., Kutzbach et al. 2008; Merlis et al. 2013; Mohtadi et al. 2016) and amplified by obliquity variability (Bosmans et al. 2015; Wu et al. 2016) as it changed the large-scale thermal contrast between land and sea. Oxygen isotope-based reconstructions of Indian and East Asian monsoon intensity show coherent variability on a millennial timescale (e.g., Battisti et al. 2014; Kathayat et al. 2016; Mohtadi et al. 2016), suggesting that orbital forcing has a strong effect on continental-scale monsoon variability.

On a centennial timescale, orbital forcing also played an important role in long-term differences of the intertropical convergence zone and associated monsoon activity between the MWP and LIA (e.g., Sachs et al. 2009; Oppo et al. 2009; Schneider et al. 2014; Yan et al. 2015). Major characteristics of the monsoon during these two periods have been reconstructed by proxy records that are broadly available across Asia, especially in China (e.g., Chen et al. 2015). Historical documents and speleothem records suggest that precipitation over northern China associated with the East Asian summer monsoon increased significantly during the MWP and decreased during the LIA (Ren and Zhang 1996; Wang et al. 2001; Qian et al. 2003; Tan 2007). Oxygen isotope data retrieved from Indian stalagmite records also suggest an enhanced Indian summer monsoon during the MWP (Sinha et al. 2011, 2015; Berkelhammer et al. 2014). The coherent enhancement of the East Asian and Indian summer monsoon during the MWP, as compared to the LIA suggests that orbitally forced summertime insolation variability is also a key factor for centennial-scale monsoon variability during the last millennium.

Monsoon variability over the last millennium has also been addressed by numerical simulations based on single (Fan et al., 2009b; Liu et al. 2009, 2011; Man et al. 2012; Polanski et al. 2014) or multiple climate models (Fallah and Cubasch 2015; Rojas et al. 2016; Shi et al., 2016a, b; Yang and Jiang 2017) estimated as part of the Paleoclimate Modeling Intercomparison Project Phase 3 (PMIP3) (Otto-Bliesner et al. 2009). The climate models reproduced, in large part, the enhanced Asian summer monsoon and increased precipitation over northern China (35°N–42°N, 100°E–120°E) during the MWP (e.g., Man et al. 2012; Shi et al., 2016b), consistent with the proxy evidence. The enhanced Asian summer monsoon during the MWP, as compared to the LIA, is associated with larger land–sea temperature contrasts (e.g., Man et al. 2012; Shi et al., 2016a, b) induced by strong summertime insolation (Fig. 1; detailed in the Methods section). These large-scale features of the Asian summer monsoon during the MWP are consistent among multiple climate model simulations (Shi et al., 2016a) with nonnegligible inter-model spread in regional rainfall patterns (Shi et al., 2016b).

In addition to seasonal-mean features, seasonal transitions (changes in onset and withdrawal timing) are also important aspects of paleo-monsoon variability (e.g.,

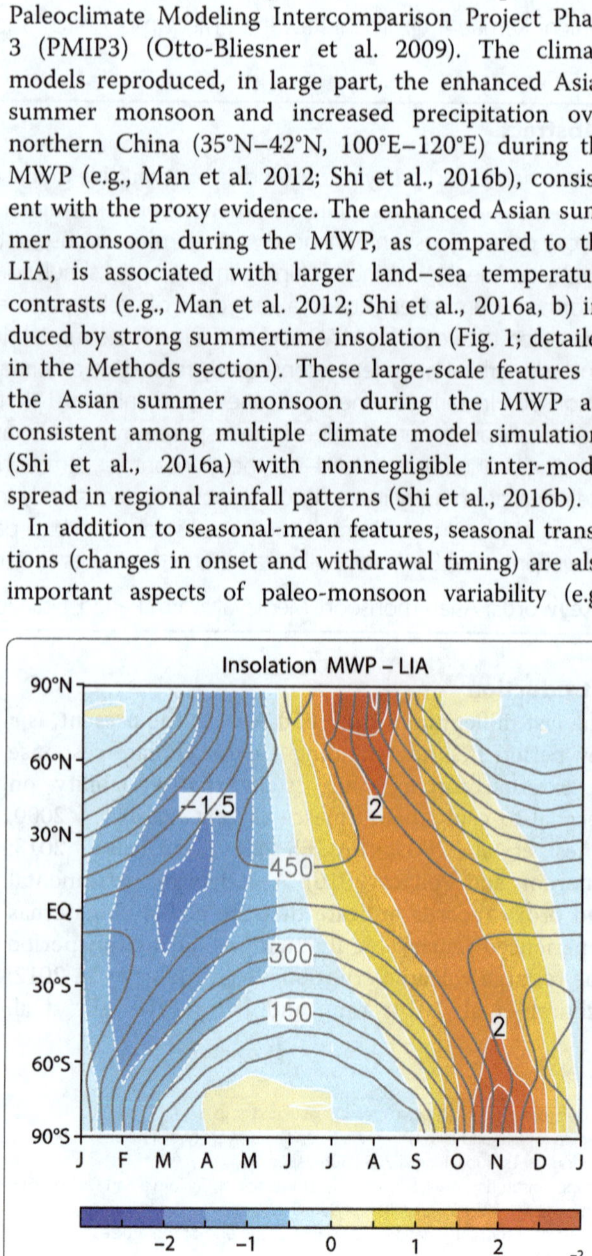

Fig. 1 Seasonal variation of insolation (W m^{-2}) at the top of the atmosphere averaged over 10 PMIP3 models. Shading represents the anomaly between the MWP (950–1250 A.D.) and LIA (1400–1800 A.D.; hereafter ML anomaly). *Contours* represent climatology in the LIA

Ueda et al. 2011; Shi 2016; Kong et al. 2017). The peak NH insolation anomaly due to the Earth's orbit variability is not necessarily concurrent with the summer solstice. As for the difference between the MWP and LIA, time-varying precession (detailed in the Methods section) leads to a positive NH insolation anomaly with its peak in August to September as shown in Fig. 1 (Kaufman et al. 2009). In contrast, a negative insolation anomaly is found in May to June, around the timing of the monsoon onset, implying a seasonally asymmetric response of the Asian summer monsoon to orbital forcing (e.g., Shi 2016; Kong et al. 2017) during the last millennium.

This study reveals that the slow variation in orbital parameters from the MWP to the LIA results in an inverse response of the Asian monsoon between early and late summer, as shown by simulations of the last millennium using multiple climate models and sensitivity experiments by an atmosphere–ocean coupled general circulation model (AOGCM). The results indicate that centennial Asian monsoon variability shown in previous works is not seasonally uniform, but rather characterized by changes in the seasonal transition of the monsoon between the MWP and LIA, which is suggestive for past, present and future climate variability.

Methods/Experimental

PMIP3 multiple model ensemble

We analyzed climate simulations of the last millennium (850–1850 A.D.) of 10 AOGCMs (hereafter denoted as past1000) from PMIP3 (Otto-Bliesner et al. 2009) and the Coupled Model Intercomparison Project phase 5 (CMIP5) (Taylor et al. 2012). Table 1 summarizes the models used in this study. Although simulation outputs from 11 models are available from the CMIP5 Earth System Grid Federation data server as of 01 February 2017, we excluded output from the MIROC-ESM due to climate drift in its simulation (Sueyoshi et al. 2013).

Table 1 List of PMIP3 models used in this study. The TSI and volcanic columns indicate forcing data sets prescribed in the models. V11: Vieira et al. (2011), S09: Steinhilber et al. (2009), C00: Crowley (2000), JM04: Jones and Mann (2004), DB09: Delaygue and Bard (2009), G08: Gao et al. (2008), C08: Crowley et al. (2008)

Model	Atmospheric resolution	TSI	Volcanic
BCC-CSM1-1	2.813° × 2.791°, L26	V11	G08
CCSM4	1.25° × 0.942°, L26	V11	G08
CSIRO-Mk3L-1-2	5.625° × 3.186°, L18	S09	C08
FGOALS-gl	5.0° × 4.103°	C00, JM04	C00, JM04
FGOALS-s2	2.813° × 2.791°, L26	V11	G08
GISS-E2-R	2.5° × 2.0°, L40	V11	G08
HadCM3	3.75° × 2.5°, L19	S09	C08
IPSL-CM5A-LR	3.75° × 1.875°, L39	V11	G08
MPI-ESM-P	1.875° × 1.865°, L47	V11	C08
MRI-CGCM3	1.125° × 1.125°, L48	DB09	G08

External forcing factors for the past1000 runs include time-varying TSI, volcanic eruptions, GHGs, land use, and orbital parameters. Details of the experimental setup are found in Schmidt et al. (2011, 2012). Note that the models employed different TSI and volcanic forcing data as listed in Table 1 (see Schmidt et al. 2011, 2012 for details). This difference contributes partly to inter-model spread among simulation outputs. Varying land uses over time such as gradually increasing cropland and pasture area (Schmidt et al. 2012) prescribed in these simulations can affect regional climate variability, including surface temperature and monsoon systems (Takata et al. 2009; He et al. 2014). Historical GHG concentrations (CO_2, CH_4 and N_2O) are derived from ice cores in Antarctica (Schmidt et al. 2011). Orbital parameters that vary over time, including eccentricity, obliquity and precession result in low-frequency variability in the seasonal-latitudinal distribution of insolation as shown in Fig. 1 (Kaufman et al. 2009; Schmidt et al. 2011). Table 2 summarizes the orbital parameters averaged during the MWP and LIA. The negative early summer insolation anomaly in the NH during the MWP relative to the LIA is largely due to a shift of the perihelion (Schmidt et al. 2011). In this study, we compare long-term mean climates of the MWP and LIA. Note that we determined the MWP climatology in FGOALS-gl by averaging the period from 1000 to 1250 A.D. because only outputs after 1000 A.D. are provided by this model.

Figure 1 shows the 10-model mean insolation anomaly (MWP minus LIA; hereafter ML) at the top of the atmosphere. The seasonal NH insolation anomaly due to orbital forcing, especially precession, is remarkable (−1.5–3 W m^{-2}) despite its slow variation on a centennial timescale (e.g., the difference in perihelion is 7.28°; Table 2). A large positive anomaly in August and September (1.5–2.5 W m^{-2}), in contrast to a small anomaly in June (Fig. 1), enhances the land–sea thermal gradient and the Asian summer monsoon (Man et al. 2012). In contrast, a broad negative insolation anomaly is found in the NH in the winter-to-spring period (peaking at −1.5–−0.5 W m^{-2} in March to April), resulting in the strong seasonal reversal of the NH insolation anomaly. This seasonally asymmetric forcing potentially

Table 2 Orbital parameters for MRI-CGCM2.3 orbital forcing runs

	Eccentricity	Obliquity	Precession
Medieval Warm Period	0.017059	23.556°	87.52°
Little Ice Age	0.016884	23.498°	95.20°

affects the seasonal transition of the Asian summer monsoon (e.g., Kong et al. 2017).

Idealized orbital forcing experiments in an AOGCM

The Asian monsoon simulated in the past 1000 runs is influenced by the time-varying external forcing listed above. The seasonally asymmetric orbital forcing (Fig. 1) may have an especially strong effect on the Asian monsoon (Ueda et al. 2011; Man et al. 2012; Kong et al. 2017). To evaluate the contribution of the orbital forcing to the ML anomaly in the Asian monsoon, we conducted idealized simulations using an AOGCM, the Meteorological Research Institute (MRI) Coupled General Circulation Model version 2.3 (MRI-CGCM2.3) (Yukimoto et al. 2006a).

MRI-CGCM2.3 has widely been used for future climate projections (Yukimoto et al. 2006b) and for paleoclimate simulations including the mid-Holocene, 6 ka (Kitoh et al. 2007; Braconnot et al. 2007), Last Glacial Maximum (LGM), 21 ka (Ueda et al. 2011), and Pliocene, 3.3–3.0 Ma (Kamae and Ueda 2012; Kamae et al. 2016). Ueda et al. (2011) used this model to examine the seasonal transition of the Asian monsoon during the LGM. This model is capable of reproducing general features of the Asian summer monsoon including its seasonal transition (Inoue and Ueda 2009). Its atmospheric model is a spectral model with T42 spatial resolution (approximately 2.8°) that uses 30 vertical layers. The ocean component is a Bryan-Cox-type global ocean model with a horizontal resolution of 2.5° longitude and 2.0°–0.5° latitude and 23 layers. We integrated the model without any flux adjustments. The CO_2 concentration and solar constant were set to 280 ppmv and 1365 W m^{-2}, respectively. We set the other boundary conditions as presented in Table 1 of Kamae et al. (2016) for their pre-industrial control simulations, and integrated the model for 500 years to spin up the model in the same way as Kamae et al. (2016) had done for their pre-industrial control run. Next, we changed the orbital parameters as shown in Table 2 and integrated the model for another 55 years for the MWP and LIA cases. In this paper, we call these two simulations "orbital forcing runs." We performed an ensemble of five member simulations for each with slightly different initial conditions derived from the pre-industrial run with 5-year intervals. We determined the climatology as an average over the last 50 years for the individual runs (250 years in total). Note that this study only addresses the transient response of the Asian monsoon to radiative forcing on a multi-decade timescale. Effects of long-term deep-ocean temperature variations and overturning circulations on the Asian monsoon during the MWP and LIA are not examined in this study.

Results and Discussion

Seasonally asymmetric response in PMIP3 model simulations

In this section, we compare the seasonal evolutions of the Asian monsoons during the MWP and LIA simulated in multiple PMIP3 models. Figures 2 and 3 show the ML rainfall and atmospheric circulation anomalies over the Asian monsoon region averaged for April–May–June (AMJ) and August–September–October (ASO). For spatial patterns, we chose these months to compare general features of late spring to early summer (around the monsoon onset) period, with its negative insolation anomaly (Fig. 1) and late summer to early fall (around monsoon withdrawal), with its positive insolation anomaly, and to emphasize the seasonal asymmetry in the atmospheric circulation and rainfall responses. Note that the timing of seasonal monsoon transitions varies by region (e.g., onset

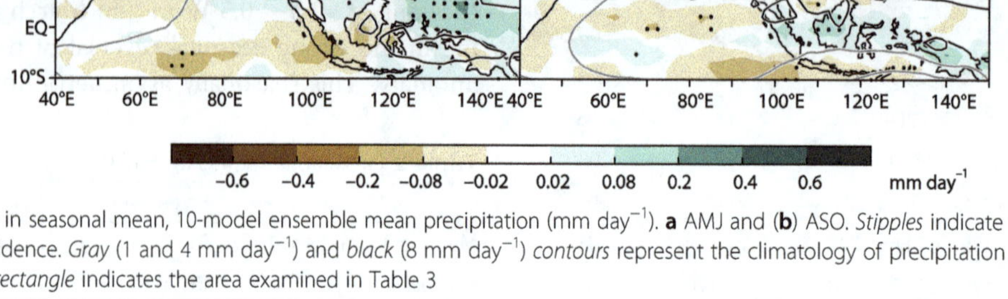

Fig. 2 ML anomaly in seasonal mean, 10-model ensemble mean precipitation (mm day^{-1}). **a** AMJ and (**b**) ASO. *Stipples* indicate the areas with 95% statistical confidence. *Gray* (1 and 4 mm day^{-1}) and *black* (8 mm day^{-1}) *contours* represent the climatology of precipitation during the LIA. The *dashed purple rectangle* indicates the area examined in Table 3

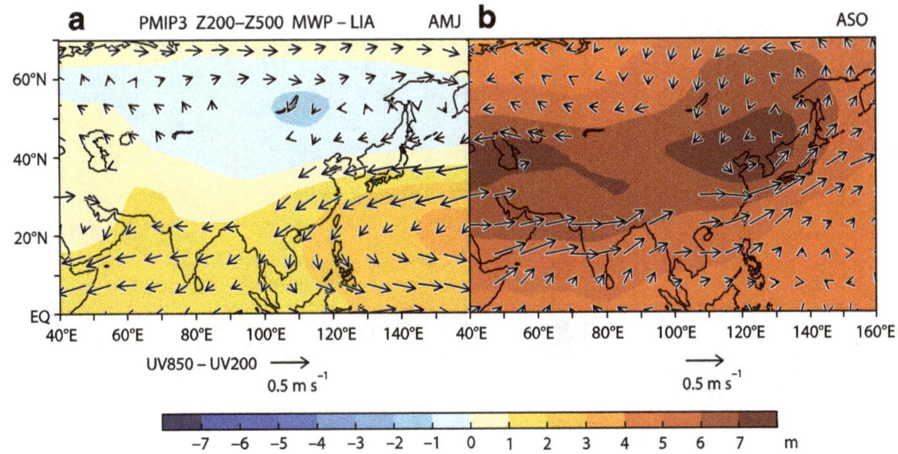

Fig. 3 Similar to **a** Fig. 2a and **b** Fig. 2b, but for tropospheric thickness (m) between the 200 and 500 hPa levels. *Vectors* indicate vertical wind shear (m s^{-1}) between 850 and 200 hPa levels

over the South China Sea in mid-May is 1 month earlier than that over Korea and Japan in early to mid-June; e.g., Tao and Chen 1987; Wang and Lin, 2002; Ueda et al. 2009). In this study, our main focus is continental-scale Asian monsoon variability including the South and East Asian monsoons. We also show seasonal evolution of the ML anomalies in individual months (detailed below).

In the 10-model ensemble mean, ASO rainfall is more active over Asia during the MWP than during the LIA (Fig. 2b), similar to June–July–August (JJA) mean anomalies shown in previous studies (Man et al. 2012; Shi et al., 2016a). The positive ML anomaly, which has high inter-model correspondence, can be found broadly over the Indian subcontinent, northeastern India, Southwest China and Eastern China. The mid-to-upper tropospheric thickness (200–500 hPa; Fig. 3b), an index for mid-to-upper tropospheric temperature (e.g., Dai et al. 2013; Kamae et al. 2017; see the Introduction section), shows general warming over the Eurasian continent, especially over the middle latitudes. In the PMIP3 multi-model mean, the summertime climatological position of the upper-tropospheric South Asian High (Shi et al., 2016a) and its seasonal evolution (intensification and northward shift from spring to summer, followed by zonal expansion in summer; e.g., Yanai and Wu 2006) are well reproduced (figure not shown). The spatial pattern of vertical wind shear (850–200 hPa) indicates enhanced low-level (upper-level) monsoon westerly (easterly) over the Arabian Sea, Indian subcontinent, Bay of Bengal, Southeast Asia, South China Sea, and Eastern China (Fig. 3b). Tropospheric temperature and associated vertical wind shear are widely used as indices for Asian monsoon intensity (Webster and Yang 1992; Li and Yanai 1996; Kawamura 1998; Kamae et al. 2014). These anomalies correspond well with the climatological structure of the Asian monsoon (Fig. 6a in Kamae et al. 2017), indicating a general enhancement of the Asian summer monsoon. The spatial patterns of tropospheric warming and associated anomalous atmospheric circulation and rainfall are partly dependent on which model is used (Shi et al., 2016a, b) especially over the western North Pacific (detailed below), but mid-latitude Eurasian continental warming and the enhanced monsoon are consistent among the different climate model simulations as shown in Figs. 2b and 3b (Man et al. 2012; Shi et al., 2016a, b).

In contrast to the enhanced late summer monsoon, AMJ-mean precipitation (Fig. 2a) shows a negative anomaly over Southeast Asia and Southern China and a positive anomaly over the western tropical Pacific. Figure 4a shows the climatological seasonal variation of rainfall averaged over 15°N–35°N and its ML anomaly. The model captures an earlier climatological monsoon onset over Southeast Asia (90°E–120°E in May; Tao and Chen 1987; Wang and Lin, 2002) than over the Arabian Sea (60°E–80°E in June to July). The negative AMJ rainfall anomaly over Southeast Asia is distinct from the positive ASO anomaly (70°E–120°E; Figs. 2b, 4a). Table 3 summarizes ML anomalies in seasonal rainfall increment (ASO minus AMJ) averaged over the Asian region shown by the rectangles in Figs. 2 and 4a (15°N–35°N, 70°E–120°E). Although inter-model spread of the JJA-mean ML anomaly in the Asian monsoon rainfall is remarkable (Shi et al., 2016b), the seasonal asymmetry in the ML rainfall anomaly is robust among the PMIP3 models (0.09 ± 0.03 mm day^{-1}). Note that an East Asian-Western North Pacific monsoon index, determined by a meridional gradient of low-level zonal wind (20°N–30°N, 110°E–140°E minus 5°N–15°N, 90°E–130°E)

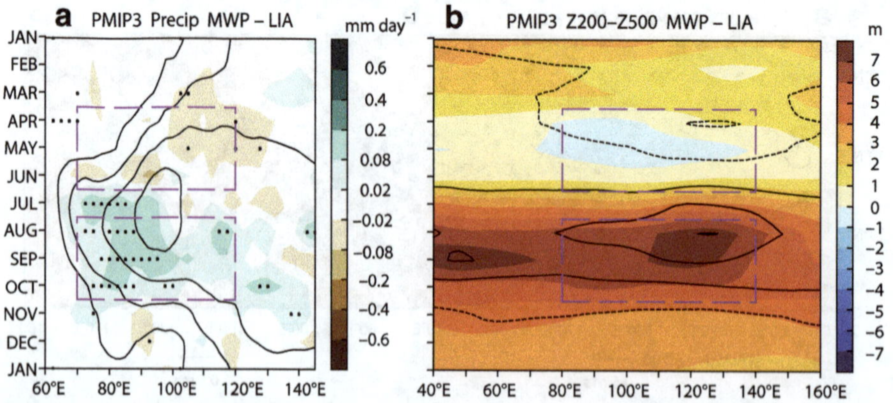

Fig. 4 Seasonal evolutions of ML anomalies in precipitation and tropospheric thickness. **a** Precipitation (mm day^{-1}) averaged over 15°N–35°N. *Contours* represent climatology (1, 2, 4, 8 mm day^{-1}) during the LIA. *Stipples* indicate the areas with 95% statistical confidence. *Dashed purple rectangles* indicate the months and area examined in Table 3. **b** Thickness (m) between 200 and 500 hPa levels averaged over 25°N–60°N. *Contours* represent the difference between 25°N–60°N and EQ–25°N (−6, −4, −2, 0, 2, 4, 6 m)

(Wang et al. 2008), shows a large intermodel spread (0.05 ± 0.28 m s^{-1}). The rainfall anomaly over East Asia is also highly dependent on which model is assumed (Shi et al., 2016b). The model diversity may be related to a large intermodel spread in the ML anomaly in the Pacific Ocean (Shi et al., 2016b). The possible importance of the Pacific Ocean to the East Asian monsoon anomaly is addressed in the Discussion section.

Seasonally asymmetric responses can also be found in monsoon circulation and continental-scale temperature patterns. Figure 3a shows AMJ-mean tropospheric thickness and vertical wind shear. In contrast to ASO (Fig. 3b), a negative anomaly with its peak at 50°N is broadly found over the Eurasian continent. An anomalous meridional temperature gradient associated with this cooling results in a low-level easterly wind anomaly over East and South Asia (5°N–45°N). This easterly wind anomaly is indicative of a delay in the seasonal migration of the Asian summer monsoon (reversal of wind direction). Figure 4b shows the meridional tropospheric temperature gradient (25°N–60°N minus EQ–15°N) over Asia. Except for the AMJ period, the NH mid-latitude tropospheric temperature is higher during the MWP than during the LIA. Figure 5a shows the simulated annual-mean ML anomaly in sea surface temperature (SST). In the PMIP3 ensemble mean, the surface and troposphere over the NH middle latitudes are generally warmer during the MWP than in the LIA in response to external forcing including GHGs and orbital forcing (Schmidt et al. 2011, 2012). Despite the generally warmer conditions, a widespread cool anomaly is found in the AMJ period (Figs. 3a, 4b), consistent with the anomalies in circulation and precipitation over Asia having opposite signs to those during ASO (Figs. 2, 3, 4). This continental-scale structure of the tropospheric temperature anomaly is a robust feature among the PMIP3 models (Table 3). Note that the East Asian-Western North Pacific monsoon index (Wang et al. 2008) is largely consistent with the South Asian monsoon (stronger monsoons in ASO than in AMJ) but shows a large inter-model spread (Table 3). More details on the inter-model divergence are presented in the Discussion section.

Effects of orbital forcing

In the previous section, we showed that there is consistent seasonal reversal of sign in ML anomalies in precipitation, atmospheric circulation, and continental-scale tropospheric temperature among the PMIP3 models.

Table 3 Seasonal differences (ASO minus AMJ) and 95% confidence intervals for precipitation (mm day^{-1}), tropospheric thickness (between 200 and 500 hPa levels; m) over Asia, and an East Asian-Western North Pacific monsoon index, the meridional gradient of zonal wind at the 850 hPa level (Wang et al. 2008) during the MWP relative to the LIA. PMIP3 (orbital forcing run) row represents a 10-model ensemble mean (5-member ensemble mean for the MRI-CGCM2.3 orbital forcing run)

	PMIP3	Orbital forcing run
Precip (mm day^{-1})15°N–35°N, 70°E–120°E	0.09 ± 0.03	0.06 ± 0.04
Z200–Z500 (m) 25°N–60°N, 80°E–140°E	6.39 ± 1.82	6.52 ± 1.60
Z200–Z500 (m) EQ–25°N, 80°E–140°E	3.18 ± 1.02	3.81 ± 1.66
Z200–Z500 (m) 25°N–60°N minus EQ–25°N, 80°E–140°E	3.22 ± 1.14	2.70 ± 2.16
U850 in 20°N–30°N, 110°E–140°E minus U850 in 5°N–15°N, 90°E–130°E (m s^{-1})	0.05 ± 0.28	0.24 ± 0.69

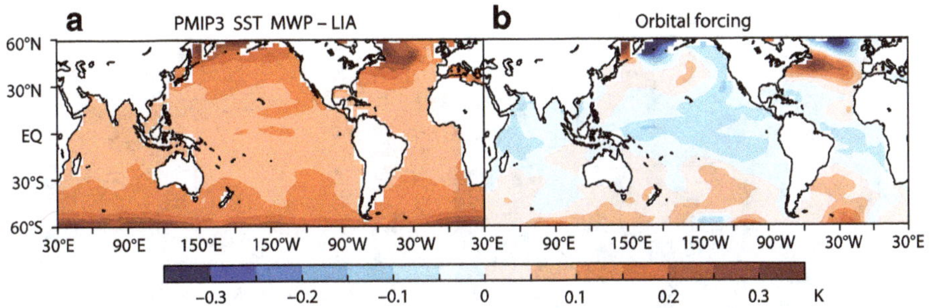

Fig. 5 ML anomaly in annual-mean SST (K). **a** PMIP3 10-model ensemble mean. **b** MRI-CGCM2.3 orbital forcing runs

These seasonally asymmetric responses are difficult to understand as a forced response to seasonally uniform forcing that includes TSI and GHGs (e.g., Ueda et al. 2006; Endo and Kitoh 2014). Here, the seasonal NH insolation anomaly due to changes in orbital parameters (Table 2) has opposite signs in AMJ and ASO (Fig. 1), consistent with the Asian monsoon response. In this section, results of the orbital forcing runs conducted in MRI-CGCM2.3 are compared with the PMIP3 ensemble to examine the physical relationship between orbital parameters and the Asian monsoon.

Figure 6 shows the results of the five-member ensemble simulations with prescribed MWP and LIA orbital parameters. The ML rainfall anomaly is largely negative over Southeast Asia and Southern China (15°N–35°N, 90°E–120°E) in AMJ (Fig. 6a). The AMJ negative anomaly is indicative of a delay in the seasonal migration of the Asian summer monsoon, similar to the PMIP3 results (Fig. 2a). In ASO, positive anomalies are found over the Indian subcontinent, Southeast Asia, and southern China (Fig. 6b). However, negative anomalies are also found over the Bay of Bengal and the Yellow River Basin, distinct from the PMIP3 ensemble mean (detailed below). The seasonal sign reversal (from negative to positive) of the rainfall anomaly averaged over the Asian region (Fig. 6) is qualitatively consistent with the PMIP3 model ensemble (Table 3). Figures 7 and 8 show the seasonal responses of tropospheric thickness and rainfall. An negative tropospheric temperature anomaly in AMJ and a positive one in ASO are broadly found over the mid-latitude Eurasian continent, similar to the PMIP3 outcomes. Associated easterly anomalies over northeastern, eastern and southern China and southern India are found in AMJ, in contrast to an ASO westerly anomaly over Japan, northeastern and eastern China, and western India. The signs of the rainfall and tropospheric temperature anomalies over Asia reverse by season from negative to positive (Fig. 8a, b), largely similar to the PMIP3 results (Figs. 3, 4). This suggests that orbital forcing is an important factor in the differing monsoon responses of the MWP and LIA.

The simulated responses to orbital forcing are somewhat different from the PMIP3 multimodel mean. The Asian rainfall anomaly (negative during AMJ and positive during ASO; Figs. 6, 8a) is generally weaker and less spatially uniform than that estimated by the PMIP3 (Figs. 2, 4a), contributing to less statistical significance (Table 3). The simulated ML anomaly in the tropical SST is generally negative, while the PMIP3 multimodel mean shows a positive anomaly (Fig. 5). Tropospheric

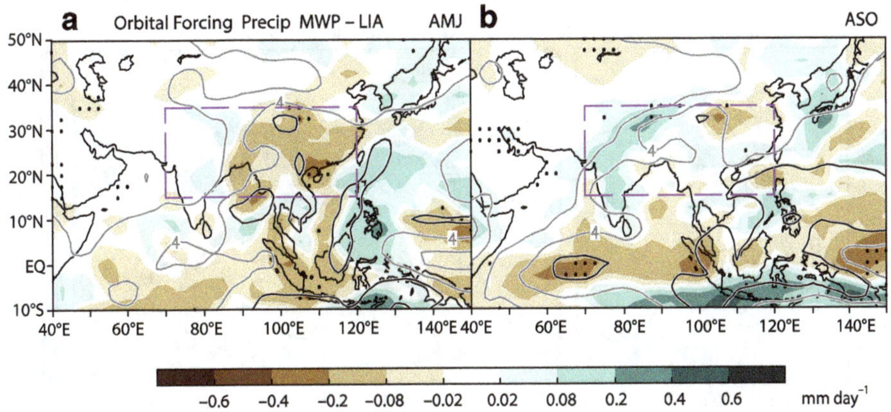

Fig. 6 Similar to **a** Fig. 2a and **b** Fig. 2b, but for MRI-CGCM2.3 orbital forcing runs

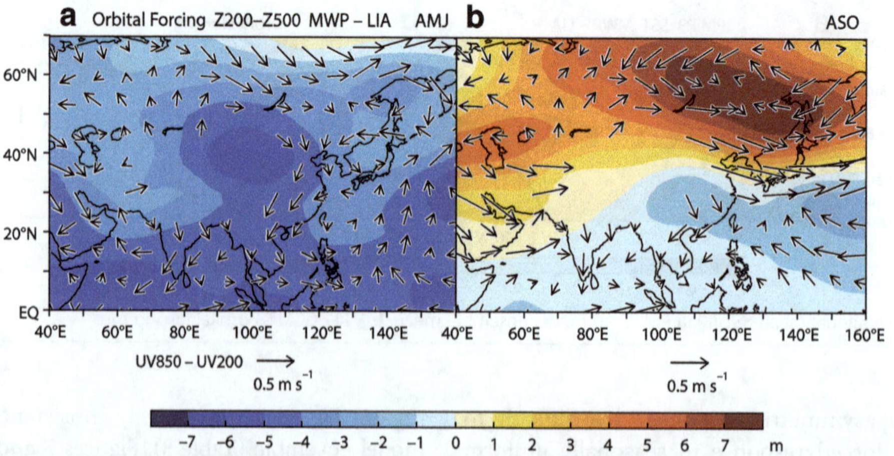

Fig. 7 Similar to **a** Fig. 6a and **b** Fig. 6b, but for tropospheric thickness (shading; m) between 200 and 500 hPa levels and vertical wind shear (vector; ms^{-1}) between 850 and 200 hPa levels

temperature has a larger negative anomaly in AMJ and a smaller positive ASO anomaly, and is less zonally uniform (Figs. 6, 7, and 8b) than in the PMIP3 (Figs. 3, 4b). In the next section, we discuss the limitations of the sensitivity experiments we conducted to quantify effects of orbital forcing on Asian monsoon variability over the last millennium.

Possible importance of other external forcing factors

The idealized orbital forcing run examined in the previous section reproduces seasonal asymmetry in the Asian monsoon rainfall and circulation anomalies similar to the PMIP3 model ensemble mean. However, these two (the PMIP3 and orbital forcing run) also show distinct regional responses. The PMIP3 model ensemble mean produces larger ML temperature and rainfall anomalies (warmer and wetter during the MWP than the LIA in both AMJ and ASO; Figs. 2, 3, 4, 5a) than the orbital forcing run (Figs. 5b, 6, 7, 8). Especially, the ML tropical tropospheric temperature anomaly is broadly negative (EQ–20°N) in the orbital forcing run (Fig. 7), resulting in a low-latitude circulation anomaly (e.g. westerly over northern India in AMJ) that is inconsistent with the PMIP3 (Fig. 3). The Asian rainfall anomaly is less zonally uniform than the PMIP3 estimates (Figs. 2, 6). These differences may be ascribed to less ML difference in the global-mean temperature between the MWP and LIA in the orbital forcing run than in the PMIP3 case. Phipps et al. (2013) also showed that the ML warming anomaly in the last-millennium orbital forcing run is smaller than in all forcing runs. The PMIP3 multimodel mean produces a larger global SST warming anomaly during the MWP than in the LIA, particularly over the middle and high latitudes (Fig. 5a), in contrast to the tropical-wide negative SST anomaly that the orbital forcing run estimates (Fig. 5b). This SST difference may make a substantial contribution to the difference in the mid-to-upper tropospheric temperatures, where we see a cooler troposphere in the orbital forcing run than the PMIP3 as in Figs. 3 and 7 (Ueda et al. 2006; Kamae et al. 2015). In

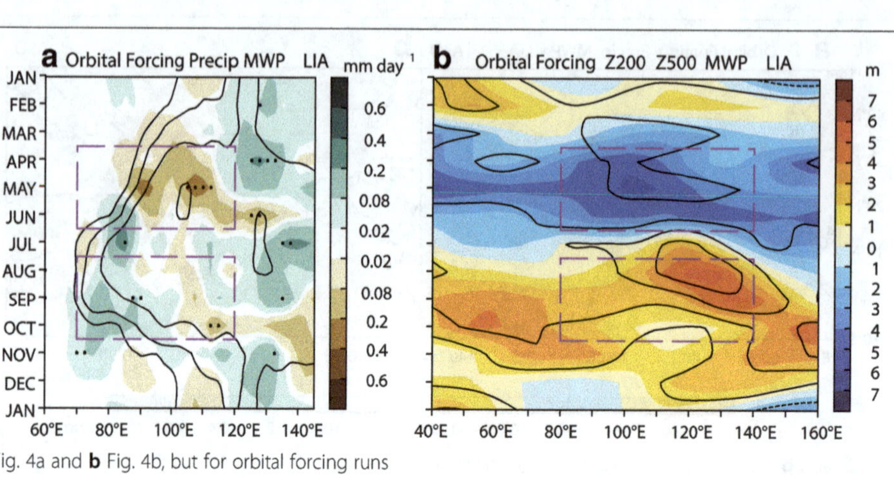

Fig. 8 Similar to **a** Fig. 4a and **b** Fig. 4b, but for orbital forcing runs

contrast to increased summer insolation (Fig. 1), winter insolation inversely decreases over the NH and SH, resulting in limited global-mean annual-mean insolation anomalies. This limited global-mean forcing cannot simply explain the globally-warmer climate during the MWP than in the LIA (Fig. 5a). Under the PMIP3 framework, GHGs, volcanic aerosols, TSI, and land cover all varied over time in the climate model simulations in addition to the orbital forcing. For example, the CO_2 concentration during the MWP is higher than in the LIA, especially after 1600 A.D. (Schmidt et al. 2011), contributing partly to global warming and the larger summertime tropospheric warming over the Eurasian continent than over the surrounding oceans as shown in Fig. 3b (Kamae et al. 2014). In addition, TSI and large volcanic eruptions are also important factors in the ML climate anomaly (e.g., Schurer et al. 2014; Otto-Bliesner et al. 2016). Phipps et al. (2013) showed that the last-millennium temperature variation, in response to orbital forcing, is smaller than that in response to all forcing factors. Orbital forcing is substantially important for the seasonally asymmetric Asian monsoon response (Fig. 8) although it is a minor factor in the global ML climate anomaly (Fig. 5). The effect of volcanic eruptions on the SST anomaly is also discussed in the next subsection.

Role of sea surface temperature pattern

The PMIP3 multi-model mean reproduces general sea surface warming during the MWP than during the LIA (Fig. 5a). We should note that the spatial pattern of the ML SST anomaly and associated regional anomalies in the Asian and Western North Pacific monsoon (rainfall and atmospheric circulation) are highly dependent on the models used, as can be seen in Table 3 as per Shi et al. (2016b), who compared ML SST anomalies simulated in the individual PMIP3 models and found that (1) generally, mid-to-high latitude warming is larger than in the tropics; and (2) inter-model spread in the regional SST anomaly is substantially large over the equatorial Pacific and the mid-latitude North Pacific. A majority of the PMIP3 models simulate global ocean warming with limited east-west gradients in the equatorial Pacific (Mann et al. 2009; Shi et al., 2016b). In contrast, one PMIP3 model, MRI-CGCM3, reproduces a La Niña-like SST anomaly, largely consistent with proxy evidence (Mann et al. 2009; Shi et al., 2016b). This equatorial Pacific SST anomaly contributes to the simulated East Asian rainfall anomaly (Shi et al., 2016b). These results indicate that the reproducibility of the SST pattern over the tropical Pacific and the western North Pacific is essential in simulations of the Asian monsoon for the last millennium.

It is also noteworthy that large uncertainties still exist in the centennial variability of the equatorial Pacific SST associated with ENSO activity and tropical volcanic eruptions during the last millennium (e.g., Adams et al. 2003). Mann et al. (2009) revealed that positive tropical radiative forcing due to high solar irradiance and inactive volcanism are responsible for the more La Niña-like mean state during the MWP as compared to the LIA, inconsistent with a more El Niño-like condition suggested in Yan et al. (2011). In the near future, we plan to conduct a set of atmospheric simulations prescribing tropical SST patterns (i.e., La Niña-like and El Niño-like patterns) to evaluate the relative importance of the SST patterns to regional and seasonal Asian rainfall and the effect of orbital forcing-induced land–sea warming contrast.

Data–model and model–model comparisons

A variety of proxy records for the last millennium Asian monsoon could improve our understanding of the long-term variation in the Asian monsoon's seasonal transition. However, it is not easy to procure such records because temporally high-resolution (i.e., sub-seasonal) information is needed to evaluate data–model consistency in the seasonal monsoon transition. One of promising data source for this purpose is meteorological information in historical documents including diaries. Adamson and Nash (2014) reconstructed regional monsoon behavior over western India for the period 1781–1860 from diaries and other historical documents. They converted qualitative information in the historical documents into a quantitative monthly rainfall index. Wang and Zhang (1992) reviewed several studies trying to reconstruct the historical variation of monthly rainfall and onset/withdrawal dates of the Meiyu rainy season in the Lower Yangtze region in the 18th century. Daily meteorological information recorded in historical documents is beneficial for reconstructing the variability of the Asian summer monsoon. However, such meteorological records were rarely continued for even several decades, let alone centuries, so they are not broadly effective in evaluating long-term monsoon variability. Further efforts to reconstruct past climate variability over Asia by combining a variety of proxy data including tree-ring records (e.g., Cook et al. 2013; Fan et al., 2009a; Fritts 2001; Takahashi et al. 2011), lake-freezing records (e.g., Arakawa 1954, Sharma et al. 2016), and phenological records (e.g. Aono and Omoto 1994; Mikami 2008) may improve our understanding of the long-term variability of the Asian monsoon.

In this study, the effect of orbital forcing is evaluated using only a single climate model. The derived results suffer from model-dependency problems including the effects of model biases. More systematic efforts are needed to better evaluate each particular forcing effect and the results' dependency on models. Previous studies have only evaluated the effects of respective forcing on continental-scale, NH-mean, or global-mean climate features (e.g., Phipps et al. 2013; Schurer et al. 2013, 2014; Otto-Bliesner et al. 2016). Monsoon responses to historical land use changes (e.g., Takata et al. 2009), aerosol forcing (e.g., Bollasina et al. 2011; Polson et al. 2014), and TSI (e.g., Shi et al. 2014) should be further evaluated in a multi-model framework. Further modeling work may contribute to better understanding of the forced monsoon response and its internal variability on interannual, decadal, centennial, and millennial timescales.

Conclusions

We examined long-term variation in the seasonal transition of the Asian monsoon during the last millennium using ten climate model simulations. The models reproduce an enhanced late summer MWP Asian monsoon but weakened early summer Asian rainfall compared to the LIA. The cooling-to-warming transition over the Eurasian continent before and after June with limited variation over the oceans results in the seasonally asymmetric deep-tropospheric circulation and rainfall responses over Asia. The robust seasonal asymmetry in the Asian monsoon response among the climate models is qualitatively reproduced by an AOGCM simulation forced by orbital parameters, suggesting that the seasonal-latitudinal distribution of the insolation anomaly is important to the centennial variability of the seasonal transition of the Asian monsoon.

Orbital forcing cannot explain the annual-mean climate anomaly between the MWP and LIA, implying that other forcing factors, including GHGs, have large contributions. Tropical SST response to orbital forcing in particular differs from the multiple model simulations. This inconsistency contributes to the tropospheric temperature and circulation anomaly over Asia. The east–west SST pattern in the Pacific shows remarkable inter-model variance, contributing substantially to the mid-latitude terrestrial climate anomaly between the MWP and LIA (Shi et al., 2016b). The La Niña-like ML SST anomaly in the Pacific suggested by proxy records was not successfully reproduced in the majority of climate model simulations. Uncertainty in centennial tropical SST variability and its contribution to the monsoon during the last millennium still remains a challenge. Further efforts in data–model synthesis including comparisons of simulation outputs with paleoenvironment reconstructions and paleoclimate data assimilation (e.g., von Storch et al. 2000; Fang and Li 2016) are needed to advance our understanding of the last millennium's monsoon variability.

Abbreviations

AMJ: April–May–June; AOGCM: Atmosphere–ocean coupled general circulation model; ASO: August–September–October; CMIP5: Coupled Model Intercomparison Project phase 5; GHGs: Greenhouse gases; JJA: June–July–August; LGM: Last Glacial Maximum; LIA: Little Ice Age; ML: MWP minus LIA; MRI: Meteorological Research Institute; MRI-CGCM2.3: MRI Coupled General Circulation Model version 2.3; MWP: Medieval Warm Period; NH: Northern Hemisphere; PMIP3: Paleoclimate Modelling Intercomparison Project Phase 3; SH: Southern Hemisphere; SST: Sea surface temperature; TSI: Total solar irradiance

Acknowledgements

We thank two anonymous reviewers whose suggestions helped to greatly improve the manuscript. The MRI made the MRI-CGCM2.3 available under a cooperative agreement. We acknowledge the World Climate Research Programme's Working Group on Coupled Modeling, which is responsible for the CMIP, and thank the climate modeling groups listed in Table 1 for producing and making their model output available. For CMIP5/PMIP3, the US Department of Energy's Program for Climate Model Diagnosis and Intercomparison provided coordinating support, and led the development of the software infrastructure in partnership with the Global Organization for Earth System Science Portals.

Funding

This work was supported by JSPS KAKENHI Grant Numbers 17 K14388 and 17 K01223.

Authors' contributions

YK drafted the manuscript. YK, MO, and HU proposed the topic and conceived and designed the study. TK and MO analyzed the multi-model data and helped in interpretation. YK conducted the sensitivity experiments and analyzed the simulated results. All authors read and approved the final manuscript.

Authors' information

MO's current affiliation is Weathernews Inc.

Competing interests

The authors declare that they have no competing interest.

Author details

[1]Faculty of Life and Environmental Sciences, University of Tsukuba, 1-1-1 Tennoudai, Tsukuba, Ibaraki 305-8506, Japan. [2]Scripps Institution of Oceanography, University of California San Diego, 9500 Gilman Drive, MC 206, La Jolla, CA 92093-0206, USA. [3]College of Geoscience, School of Life and Environmental Sciences, University of Tsukuba, 1-1-1 Tennoudai, Tsukuba, Ibaraki 305-8506, Japan. [4]Graduate School of Life and Environmental Sciences, University of Tsukuba, 1-1-1 Tennoudai, Tsukuba, Ibaraki 305-8506, Japan.

References

Adams JB, Mann ME, Ammann CM (2003) Proxy evidence for an el Niño-like response to volcanic forcing. Nature 426:274–278

Adamson GCD, Nash DJ (2014) Documentary reconstruction of monsoon rainfall variability over western India, 1781–1860. Clim Dyn 42:749–769

Ammann CM, Joos F, Schimel DS, Otto-Bliesner BL, Tomas RA (2007) Solar influence on climate during the past millennium: results from transient simulations with the NCAR climate system model. Proc Natl Acad Sci U S A 104:3713–3718

Aono Y, Omoto Y (1994) Estimation of temperature at Kyoto since the 11th century using flowering data of cherry trees in old documents. J Agr Meteorol Jpn 49:263–272

Arakawa H (1954) Fujiwhara on five centuries of freezing dates of Lake Suwa in the Central Japan. Arch Meteorol Geophys Bioklimatol Ser B 6:152–166

Battisti DS, Ding Q, Roe GH (2014) Coherent pan-Asian climatic and isotopic response to orbital forcing of tropical insolation. J Geophys Res Atmos 119: 11997–12020

Berger A, Loutre MF (1991) Insolation values for the climate of the last 10 million of years. Quat Sci Rev 10:297–317

Berkelhammer M, Sinha A, Mudelsee M, Cheng H, Yoshimura K, Biswas J (2014) On the low-frequency component of the ENSO–Indian monsoon relationship: a paired proxy perspective. Clim Past 10:733–744

Bollasina MA, Ming Y, Ramaswamy V (2011) Anthropogenic aerosols and the weakening of the south Asian summer monsoon. Science 334:502–505

Bosmans JHC, Hilgen FJ, Tuenter E, Lourens LJ (2015) Obliquity forcing of low-latitude climate. Clim Past 11:1335–1346

Braconnot P, Otto-Bleisner B, Harrison S, Joussaume S, Peterschmitt J-Y, Abe-Ouchi A, Crucifix M, Driesschaert E, Fichefet T, Hewitt CD, Kagayama M, Kitoh A, La né A, Loutre M-F, Marti O, Merkel U, Ramstein G, Valdes P, Weber SL, Yu Y, Zhao Y (2007) Results of PMIP2 coupled simulations of the mid-Holocene and last glacial maximum, part 1: experiments and large-scale features. Clim Past 3:261–277

Bradley RS, Briffa KR, Cole J, Osborn TJ (2003) The climate of the last millennium. In: Alverson KD, Bradley RS, Pedersen TS (eds) Paleoclimate, global change and the future. Springer, Heidelberg, pp 105–141

Chen J, Chen F, Feng S, Huang W, Liu J, Zhou A (2015) Hydroclimatic changes in China and surroundings during the medieval climate anomaly and little ice age: spatial patterns and possible mechanisms. Quat Sci Rev 107:98–111

Clift PD, Plumb RA (2008) The Asian monsoon: causes, history & effects. Cambridge University Press, Cambridge

COHMAP members (1988) Climatic changes of the last 18,000 years: observations and model simulations. Science 241:1043–1052

Cook ER, Krusic PJ, Anchukaitis KJ, Buckley BM, Nakatsuka T, Sano M (2013) Tree-ring reconstructed summer temperature anomalies for temperate East Asia since 800 CE. Clim Dyn 41:2957–2972

Crowley TJ (2000) Causes of climate change over the past 1000 years. Science 289:270–277

Crowley TJ, Zielinski G, Vinther B, Udisti R, Kreutz K, Cole-Dai J, Castellano E (2008) Volcanism and the little ice age. PAGES News 16:22–23

Dai A, Li H, Sun Y, Hong L-C, Lin H, Chou C, Zhou T (2013) The relative roles of upper and lower tropospheric thermal contrasts and tropical influences in driving Asian summer monsoons. J Geophys Res Atmos 118:7024–7045

Delaygue G, Bard E (2009) Solar forcing based on be-10 in Antarctica ice over the past millennium and beyond. EGU Gen Assembly Conf Abstr 11:6943

Endo H, Kitoh A (2014) Thermodynamic and dynamic effects on regional monsoon rainfall changes in a warmer climate. Geophys Res Lett 41:1704–1710

Fallah B, Cubasch U (2015) A comparison of model simulations of Asian mega-droughts during the past millennium with proxy reconstructions. Clim Past 11:253–263

Fan Z-X, Bräuning A, Yang B, Cao K-F (2009a) Tree ring density-based summer temperature reconstruction for the central Hengduan Mountains in southern China. Glob Planet Change 65:1–11

Fan F, Mann ME, Amman CM (2009b) Understanding changes in the Asian summer monsoon over the past millennium: insights from a long-term coupled model simulation. J Clim 22:1736–1748

Fang M, Li X (2016) Paleoclimate data assimilation: its motivation, progress and prospects. Sci Chi Earth Sci 59:1817–1826

Fritts HC (2001) Tree rings and climate, 2nd edn. Blackburn Press, Caldwell

Fu C, Freney JR, Stewart JWB (eds) (2008) Changes in the human-monsoon system of East Asia in the context of global change. World Scientific Publishing, Singapore

Gao C, Robock A, Ammann C (2008) Volcanic forcing of climate over the past 1500 years: an improved ice core-based index for climate models. J Geophys Res Atmos 113:D23111. doi:10.1029/2008JD010239

Gray LJ, Beer J, Geller M, Haigh JD, Lockwood M, Matthes K, Cubasch U, Fleitmann D, Harrison G, Hood L, Luterbacher J, Meehl GA, Shindell D, van Geel B, White W (2010) Solar influences on climate. Rev Geophys 48:RG4001

He F, Vavrus SJ, Kutzbach JE, Ruddiman WF, Kaplan JO, Krumhardt KM (2014) Simulating global and local surface temperature changes due to Holocene anthropogenic land cover change. Geophys Res Lett 41:623–631

Inoue T, Ueda H (2009) Evaluation for the seasonal evolution of the summer monsoon over the Asian and western North Pacific sector in the WCRP CMIP3 multi-model experiments. J Meteorol Soc Jpn 87:539–560

Jones PD, Mann ME (2004) Climate over past millennia. Rev Geophys 42:RG2002. doi:10.1029/2003RG000143

Jones PD, Briffa KR, Osborn TJ, Lough JM, van Ommen TD, Vinther BM, Luterbacher J, Wahl ER, Zwiers FW, Mann ME, Schmidt GA, Ammann CM, Buckley BM, Cobb KM, Esper J, Goosse H, Graham N, Jansen E, Kiefer T, Kull C, Küttel M, Mosley-Thompson E, Overpeck JT, Riedwyl N, Schulz M, Tudhope AW, Villalba R, Wanner H, Wolff E, Xoplaki E (2009) High-resolution palaeoclimatology of the last millennium: a review of current status and future prospects. The Holocene 19:3–49

Kamae Y, Ueda H (2012) Mid-Pliocene global climate simulation with MRI-CGCM2.3: set-up and initial results of PlioMIP experiments 1 and 2. Geosci Model Dev 5:793–808

Kamae Y, Watanabe M, Kimoto M, Shiogama H (2014) Summertime land–sea thermal contrast and atmospheric circulation over East Asia in a warming climate–part I: past changes and future projections. Clim Dyn 43:2553–2568

Kamae Y, Shiogama H, Watanabe M, Ishii M, Ueda H, Kimoto M (2015) Recent slowdown of tropical upper tropospheric warming associated with Pacific climate variability. Geophys Res Lett 42:2995–3003

Kamae Y, Yoshida K, Ueda H (2016) Sensitivity of Pliocene climate simulations in MRI-CGCM2.3 to respective boundary conditions. Clim Past 12:1619–1634

Kamae Y, Li X, Xie S-P, Ueda H (2017) Atlantic effects on recent decadal trends in global monsoon. Clim Dyn. doi:10.1007/s00382-017-3522-3

Kathayat G, Cheng H, Sinha A, Spötl C, Edwards RL, Zhang H, Li X, Yi L, Ning Y, Cai Y, Lui WL, Breitenbach SFM (2016) Indian monsoon variability on millennial-orbital timescales. Sci Rep 6:24374

Kaufman DS, Schneider DP, McKay NP, Ammann CM, Bradley RS, Briffa KR, Miller GH, Otto-Bliesner BL, Overpeck JT, Vinther BM, Abbott M, Axford Y, Bird B, Birks HJB, Bjune AE, Briner J, Cook T, Chipman M, Francus P, Gajewski K, Geirsdóttir Á, Hu FS, Kutchko B, Lamoureux S, Loso M, Mac Donald G, Peros M, Porinchu D, Schiff C, Seppä H, Thomas E (2009) Recent warming reverses long-term Arctic cooling. Science 325:1236–1239

Kawamura R (1998) A possible mechanism of the Asian summer monsoon-ENSO coupling. J Meteorol Soc Jpn 76:1009–1027

Kitoh A, Motoi T, Murakami S (2007) El Niño-southern oscillation simulation at 6000 years before present with the MRI-CGCM2.3: effect of flux adjustment. J Clim 20:2484–2499

Kong W, Swenson L, Chiang J (2017) Seasonal transitions and the westerly jet in the Holocene east Asian summer monsoon. J Clim. doi:10.1175/JCLI-D-16-0087.1

Kutzbach JE, Liu XD, Liu ZY, Chen GS (2008) Simulating the evolution response of global monsoons to orbital forcing over the past 280,000 years. Clim Dyn 30: 567–579

Lean JL (2010) Cycles and trends in solar irradiance and climate. WIREs Clim Change 1:111–122

Li CF, Yanai M (1996) The onset and interannual variability of the Asian summer monsoon in relation to land-sea thermal contrast. J Clim 9:358–375

Liu J, Wang B, Ding Q, Kuang X, Soon W, Zorita E (2009) Centennial variations of the global monsoon precipitation in the last millennium: results from ECHO-G model. J Clim 22:2356–2371

Liu J, Wang B, Wang H, Kuang Z, Ti R (2011) Forced response of the east Asian summer rainfall over the past millennium: results from a coupled model simulation. Clim Dyn 36:323–336

Liu F, Chai J, Wang B, Liu J, Zhang X, Wang Z (2016) Global monsoon precipitation responses to large volcanic eruptions. Sci Rep 6:24331

Man W, Zhou T, Jungclaus JH (2012) Simulation of the east Asian summer monsoon during the last millennium with the MPI earth system model. J Clim 25:7852–7866

Mann ME, Zhang Z, Rutherford S, Bradley RS, Hughes MK, Shindell D, Ammann C, Faluvegi G, Ni F (2009) Global signatures and dynamical origins of the little ice age and medieval climate anomaly. Science 326:1256–1260

Masson-Delmotte V, Schulz M, Abe-Ouchi A, Beer J, Ganopolski A, González Rouco JF, Jansen E, Lambeck K, Luterbacher J, Naish T, Osborn T, Otto-Bliesner B, Quinn T, Ramesh R, Rojas M, Shao X, Timmermann A (2014) Information from Paleoclimate archives. In: Stocker TF et al (eds) Climate change 2013: the physical science basis. Contribution of working group I to the fifth assessment report of the intergovernmental panel on climate change. Cambridge University Press, Cambridge, pp 383–464

Merlis TM, Schneider T, Bordoni S, Eisenman I (2013) The tropical precipitation response to orbital precession. J Clim 26:2010–2021

Mikami T (2008) Climatic variation in Japan reconstructed from historical documents. Weather 63:190–193

Miller GH, Geirsdóttir Á, Zhong Y, Larsen DJ, Otto-Bliesner BL, Holland MM, Bailey DA, Refsnider KA, Lehman SJ, Southon JR, Anderson C, Björnsson H, Thordarson T (2012) Abrupt onset of the little ice age triggered by volcanism and sustained by sea-ice/ocean feedbacks. Geophys Res Lett 39:L02708

Mohtadi M, Prange M, Steinke S (2016) Palaeoclimatic insights into forcing and response of monsoon rainfall. Nature 533:191–199

Oppo DW, Rosenthal Y, Linsley BK (2009) 2,000-year-long temperature and hydrology reconstructions from the indo-Pacific warm pool. Nature 460: 1113 1116

Otto-Bliesner BL, Joussaume S, Braconnot P, Harrison SP, Abe-Ouchi A (2009) Modeling and data syntheses of past climates. EOS Trans AGU. doi:10.1029/2009EO110013

Otto-Bliesner B, Brady E, Fasullo J, Jahn A, Landrum L, Stevenson S, Rosenbloom N, Mai A, Strand G (2016) Climate variability and change since 850 C.E.: an ensemble approach with the community earth system model (CESM). Bull Am Meteorol Soc 97:735–754

PAGES 2k Consortium (2013) Continental-scale temperature variability during the last two millennia. Nat Geosci 6:339–346

Phipps SJ, McGregor HV, Gergis J, Gallant AJE, Neukom R, Stevenson S, Ackerley D, Brown JR, Fischer MJ, van Ommen TD (2013) Paleoclimate data–model comparison and the role of climate forcings over the past 1500 years. J Clim 26:6915–6936

Polanski S, Fallah B, Befort DJ, Prasad S, Cubasch U (2014) Regional moisture change over India during the past millennium: a comparison of multi-proxy reconstructions and climate model simulations. Glob Planet Change 122:176–185

Polson D, Bollasina M, Hegerl GC, Wilcox LJ (2014) Decreased monsoon precipitation in the northern hemisphere due to anthropogenic aerosols. Geophys Res Lett 41:6023–6029

Qian W, Hu Q, Zhu Y, Lee D-K (2003) Centennial-scale dry–wet variations in East Asia. Clim Dyn 21:77–89

Ren G, Zhang L (1996) Climate of the medieval warm period in Maili region as inferred from fossil pollen record (in Chinese). Clim Env Res 1:81–86

Rojas M, Arias PA, Flores-Aqueveque V, Seth A, Vuille M (2016) The south American monsoon variability over the last millennium in climate models. Clim Past 12:1681–1691

Sachs JP, Sachse D, Smittenberg RH, Zhang Z, Battisti DS, Golubic (2009) Southward movement of the Pacific intertropical convergence zone AD 1400-1850. Nat Geosci 2:519–525

Schmidt GA, Jungclaus JH, Ammann CM, Bard E, Braconnot P, Crowley TJ, Delaygue G, Joos F, Krivova NA, Muscheler R, Otto-Bliesner BL, Pongratz J, Shindell DT, Solanki SK, Steinhilber F, Vieira LEA (2011) Climate forcing reconstructions for use in PMIP simulations of the last millennium (v1.0). Geosci Model Dev 4:33–45

Schmidt GA, Jungclaus JH, Ammann CM, Bard E, Braconnot P, Crowley TJ, Delaygue G, Joos F, Krivova NA, Muscheler R, Otto-Bliesner BL, Pongratz J, Shindell DT, Solanki SK, Steinhilber F, Vieira LEA (2012) Climate forcing reconstructions for use in PMIP simulations of the last millennium (v1.1). Geosci Model Dev 5:185–191

Schneider T, Bischo T, Haug GH (2014) Migrations and dynamics of the intertropical convergence zone. Nature 513:45–53

Schurer AP, Hegerl GC, Mann ME, Tett SFB, Phipps SJ (2013) Separating forced from chaotic climate variability over the past millennium. J Clim 26:6954–6973

Schurer AP, Tett SFB, Hegerl GC (2014) Small influences of solar variability on climate over the past millennium. Nat Geosci 7:104–108

Sharma S, Magnuson JJ, Batt RD, Winslow JK, Aono Y (2016) Direct observations of ice seasonality reveal changes in climate over the past 320–570 years. Sci Rep 6:25061

Shi Z (2016) Response of Asian summer monsoon duration to orbital forcing under glacial and interglacial conditions: implication for precipitation variability in geological records. Quat Sci Rev 139:30–42

Shi F, Li JP, Wilson RJS (2014) A tree-ring reconstruction of the south Asian summer monsoon index over the past millennium. Sci Rep 4:6739

Shi Z, Xu T, Wang H (2016a) Sensitivity of Asian climate change to radiative forcing during the last millennium in a multi-model analysis. Glob Planet Change 139:195–210

Shi J, Yan Q, Jiang D, Min J, Jiang Y (2016b) Precipitation variation over eastern China and arid central Asia during the past millennium and its possible mechanism: perspectives from PMIP3 experiments. J Geophys Res Atmos 121:11989–12004

Sinha A, Berkelhammer M, Stott L, Mudelsee M, Cheng H, Biswas J (2011) The leading mode of Indian summer monsoon precipitation variability during the last millennium. Geophys Res Lett 38:L15703. doi:10.1029/2011gl047713

Sinha A, Kathayat G, Cheng H, Breitenbach SFM, Berkelhammer M, Mudelsee M, Biswas J, Edwards RL (2015) Trends and oscillations in the Indian summer monsoon rainfall over the last two millennia. Nat Commun 6:6309

Smerdon JE, Pollack HN (2016) Reconstructing Earth's surface temperature over the past 2000 years: the science behind the headlines. WIREs Clim Change 7: 746–771

Steinhilber F, Beer J, Fröhlich C (2009) Total solar irradiance during the Holocene. Geophys Res Lett 36:L19704. doi:10.1029/2009GL040142

von Storch H, Cubasch U, Gonzalez-Rouco JF, Jones JM, Voss R, Widmann M, Zorita E (2000) Combining paleoclimatic eviedence and GCMs by means of data assimilation though upscaling and nudging (DATUN). In: Proc 11th symposium on global climate change studies. AMS Long Beach, CA

Sueyoshi T, Ohgaito R, Yamamoto A, Chikamoto MO, Hajima T, Okajima H, Yoshimori M, Abe M, O'ishi R, Saito F, Watanabe S, Kawamiya M, Abe-Ouchi A (2013) Set-up of the PMIP3 paleoclimate experiments conducted using an earth system model, MIROC-ESM. Geosci Model Dev 6:819–836

Takahashi K, Okuhara I, Tokumitsu Y, Yasue K (2011) Responses to climate by tree-ring widths and maximum latewood densities of two *Abies* species at upper and lower altitudinal distribution limits in central Japan. Trees 25:745–753

Takata K, Saito K, Yasunari T (2009) Changes in the Asian monsoon climate during 1700–1850 induced by preindustrial cultivation. Proc Natl Acad Sci U S A 106:9586–9589

Tan M (2007) Climatic differences and similarities between Indian and east Asian monsoon regions of China over the last millennium: a perspective based mainly on stalagmite records. Int J Speleol 36:75–81

Tao S, Chen L (1987) A review of recent research on the east Asian summer monsoon in China. In: Chang C-P, Krishnamurti TN (eds) Monsoon meteorology. Oxford University Press, Oxford, pp 60–92

Taylor KE, Stouffer RJ, Meehl GA (2012) An overview of CMIP5 and the experiment design. Bull Am Meteorol Soc 90:485–498

Ueda H, Iwai A, Kuwako K, Hori ME (2006) Impact of anthropogenic forcing on the Asian summer monsoon as simulated by 8 GCMs. Geophys Res Lett 33: L06703. doi:10.1029/2005GL025336

Ueda H, Ohba M, Xie S-P (2009) Important factors for the development of the Asian-Northwest Pacific summer monsoon. J Clim 22:649–669

Ueda H, Kuroki H, Ohba M, Kamae Y (2011) Seasonally asymmetric transition of the Asian monsoon in response to ice age boundary conditions. Clim Dyn 37:2167–2179

Vieira LEA, Solanki SK, Krivova NA, Usoskin I (2011) Evolution of the solar irradiance during the Holocene. Astr Astrophys 531:A6. doi:10.1051/0004-6361/201015843

Wang B, Lin H (2002) Rainy season of the Asian-Pacific summer monsoon. J Clim 15:386–398

Wang PK, Zhang D (1992) Recent studies of the reconstruction of east Asian monsoon climate in the past using historical literature of China. J Meteorol Soc Jpn 70:423–446

Wang S, Gong D, Zhu J (2001) Twentieth-century climatic warming in China in the context of the Holocene. The Holocene 11:313–321

Wang B, Wu Z, Li J, Liu J, Chang C-P, Ding Y, Wu G (2008) How to measure the strength of the east Asian summer monsoon. J Clim 21:4449–4463

Webster PJ, Yang S (1992) Monsoon and ENSO: selectively interactive systems. Quart J Roy Meteorol Soc 118:877–926

Webster PJ, Magaña VO, Palmer TN, Shukla J, Tomas RA, Yanai M, Yasunari T (1998) Monsoons: processes, predictability, and the prospects for prediction. J Geophys Res 103(C7):14451–14510

Wu G, He B, Liu Y, Bao Q, Ren R (2015) Location and variation of the summertime upper-troposphere temperature maximum over South Asia. Clim Dyn 45:2757–2774

Wu C-H, Lee S-Y, Chiang JCH, Hsu H-H (2016) The influence of obliquity in the early Holocene Asian summer monsoon. Geophys Res Lett 43:4524–4530

Yan H, Sun L, Wang Y, Huang W, Qiu S, Yang C (2011) A record of the southern oscillation index for the past 2,000 years from precipitation proxies. Nat Geosci 4:611–614

Yan H, Wei W, Soon W, An Z, Zhou W, Liu Z, Wang Y, Carter RM (2015) Dynamics of the intertropical convergence zone over the western Pacific during the little ice age. Nat Geosci 8:315–320

Yanai M, Wu G-X (2006) Effects of the Tibetan plateau. In: Wang B (ed) The Asian monsoon. Springer, Heidelberg, pp 513–549

Yang K, Jiang D (2015) Interannual climate variability of the past millennium from simulations. Atmos Ocean Sci Lett 8:160–165

Yang K, Jiang D (2017) Interannual climate variability change during the medieval climate anomaly and little ice age in PMIP3 last millennium simulations. Adv Atmos Sci 34:497–508

Yukimoto S, Noda A, Kitoh A, Hosaka M, Yoshimura H, Uchiyama T, Shibata K, Arakawa O, Kusunoki S (2006a) Present-day and climate sensitivity in the meteorological research institute coupled GCM version 2.3 (MRI-CGCM2.3). J Meteorol Soc Jpn 84:333–363

Yukimoto S, Noda A, Uchiyama T, Kusunoki S, Kitoh A (2006b) Climate changes of the twentieth through twenty-first centuries simulated by the MRI-CGCM2.3. Pap Meteorol Geophys 56:9–24

4

Tropical cyclone influence on the long-term variability of Philippine summer monsoon onset

Hisayuki Kubota[1*], Ryuichi Shirooka[2], Jun Matsumoto[2,3], Esperanza O. Cayanan[4] and Flaviana D. Hilario[4]

Abstract

The long-term variability of Philippine summer monsoon onset from 1903 to 2013 was investigated. The onset date is defined by daily rainfall data at eight stations in the northwestern Philippines. Summer monsoons tended to start earlier in May after the mid-1990s. Other early onset periods were found during the 1900s, 1920s, and 1930s, and an interdecadal variability of summer monsoon onset was identified. Independent surface wind data observed by ships in the South China Sea (SCS) revealed prevailing westerly wind in May during the early monsoon onset period. To identify atmospheric structures that trigger Philippine summer monsoon onset, we focused on the year 2013, conducting intensive upper-air observations. Tropical cyclone (TC) Yagi traveled northward in the Philippine Sea (PS) in 2013 and triggered the Philippine monsoon onset by intensifying moist low-level southwesterly wind in the southwestern Philippines and intensifying low-level southerly wind after the monsoon onset in the northwestern Philippines. The influence of TC was analyzed by the probability of the existence of TC in the PS and the SCS since 1951, which was found to be significantly correlated with the Philippine summer monsoon onset date. After the mid-1990s, early monsoon onset was influenced by active TC formation in the PS and the SCS. However, the role of TC activity decreased during the late summer monsoon periods. In general, it was found that TC activity in the PS and the SCS plays a key role in initiating Philippine summer monsoon onset.

Keywords: Summer monsoon onset, Tropical cyclone, Long-term variability, Interdecadal variability, PALAU, Philippines

Introduction

The Philippines is an archipelago located in the western rim of the western tropical Pacific. There are distinct summer monsoons (local name *Habagat*) on the western side and winter monsoons (local name *Amihan*) on the eastern side of the country associated with the seasonal shift of major wind direction (Flores and Balagot 1969). The seasonal change of wind direction characterizes the climate phenomenon referred to as a "monsoon" (Ramage 1971). The thermal contrast between land and ocean contributes to this seasonal variation of winds in Asia (Murakami and Matsumoto 1994). During summer, moist southwesterly to westerly wind blows from the Arabian Sea through the Bay of Bengal, the South China Sea (SCS), and the Philippine Sea (PS), bringing seasonal rainfall to India, the Indochina Peninsula, and the Philippines (Wang 2006). Therefore, seasonal wind (monsoon) is closely linked to the seasonal variation of rainfall; namely, the Asian summer monsoon system is characterized by seasonal wind and rainfall. The Asian summer monsoon has several subsystems. Over the oceanic region of the PS including the Philippine region, the western North Pacific (WNP) summer monsoon dominates (Murakami and Matsumoto 1994).

Agriculture and other water resources on land rely strongly on when the seasonal rainfall starts and the amount of seasonal rainfall. Therefore, the starting date of the summer monsoon is a critical issue. The onset of the WNP summer monsoon, including that in the Philippine region, has been the focus of several studies. Rainfall data were used to define the Asian monsoon onset (Matsumoto 1997; Akasaka et al. 2007). Some studies

* Correspondence: hkubota@ep.sci.hokudai.ac.jp
[1]Faculty of Science, Hokkaido University, Kita 10 Nishi 8, Kita-ku, Sapporo 060-0810, Japan
Full list of author information is available at the end of the article

(Tanaka 1992; Lau and Yang 1997; Wang and LinHo 2002) used satellite data of outgoing longwave radiation (OLR) to define an index of convective activity. Kubota et al. (2005) used satellite-based wind data to identify the change of wind direction to define the monsoon onset. These studies determined that the Philippine summer monsoon starts around May to early June with the arrival of moist low-level westerly wind and the start of active convection in climatology. However, both monsoon activity and onset date have interannual and interdecadal variability.

The summer monsoon activity over the WNP has interannual variability associated with the El Niño-Southern Oscillation (ENSO; Ropelewski and Halpert 1987). Rainfall tends to increase during the summer (June–August) of the El Niño developing year (Lyon et al. 2006; Kubota and Wang 2009). However, rainfall decreases because of the eastward shift of the tropical cyclone (TC) track in autumn (September–November; Saunders et al. 2000; Lyon et al. 2006; Lyon and Camargo 2008; Kubota and Wang 2009). The summer monsoon activity of the subsequent El Niño decaying year becomes weak, a phenomenon which is associated with the ENSO influence on sea surface temperature (SST) anomalies in the Indian Ocean (IO; Xie et al. 2009).

The summer monsoon onset date over the WNP tends to be delayed during the El Niño decaying year, as revealed by the interannual variability of summer monsoon activity (Tanaka 1997). The decadal variability of summer monsoon onset in the Philippines and SCS includes early onset during the late twentieth century, as reported by Akasaka (2010) and Kajikawa and Wang (2012). Akasaka (2010) suggested that the early approach of an easterly wave might trigger an early monsoon onset in the Philippines. Kajikawa and Wang (2012) noted that the enhancement of intraseasonal variability of convection and TC activity is important for early monsoon onset in the SCS after the mid-1990s. Both studies highlighted the potential impact of recent sea surface temperature (SST) warming in the SCS. However, both studies focused on recent 30- to 50-year variability, but SST in the SCS has undergone multi-time-scale variability over the past 138 years (Yang et al. 2015). The data used in these monsoon onset studies are not extensive enough to determine how recently observed early monsoon onsets are connected to other longer time-scale variability.

Recently, we collected Philippine rainfall data from the late nineteenth century (Chowdary et al. 2012; Villafuerte et al. 2014) and found that the northwestern Philippine station rainfall has been strongly correlated with the Pacific-Japan (PJ) pattern for more than 100 years (Kubota et al. 2016). The PJ pattern is a dominant teleconnection of interannual variability in the WNP monsoon and East Asian summer monsoon (Nitta 1987) and is correlated to the preceding boreal winter ENSO (Chang et al. 2000a, b; Wang et al. 2000). The occurrence of TC over the WNP during summer is also correlated with the PJ pattern. The relationship between the PJ pattern and ENSO and TC activity was modulated on the interdecadal time scale. It has not been discussed how the interdecadal modulation of the WNP summer monsoon and its related phenomena, such as TC, affect the variability of the Philippine monsoon onset date. In this study, we investigate the long-term variability of the Philippine summer monsoon onset date using rainfall data covering more than 100 years.

The atmospheric structures that cause Asian summer monsoon onset have been investigated using a reanalysis dataset that includes the Philippine region (Wang 2006). The quality of the three-dimensional reanalysis dataset strongly depends on the availability of upper-air observations (Hattori et al. 2016). However, in the Philippines, upper-air observations were no longer carried out throughout the year after the US Clark Air Base was closed because of the eruption of Mt. Pinatubo in 1991. Here, we report the results of intensive upper-air observations in Laoag, Cebu Mactan, and Puerto Princesa in the Philippines from May to August 2013 (Fig. 1). The year 2013 was the first summer that comprehensive upper-air observations were conducted in the Philippines since 1991 (Fig. 2). Upper-air observations that include intensive observations provide essential data for compiling a vertical profile of the atmosphere and are assimilated to create the reanalysis dataset. The inclusion of intensive observations in the reanalysis dataset allows us to capture qualitatively which atmospheric structures trigger Philippine summer monsoon onset.

Methods

Continuous rainfall observations in the Philippines were started in Manila in 1865 by Jesuits in the Manila Observatory (Deppermann 1939; Udias 1996). The Manila Observatory expanded the number of weather stations in the Philippines in the 1880s. However, many of these observational data are now missing. The observations of the Observatorio Meteorológico de Manila made by Spanish meteorologists were collected from 1891 to 1900; unfortunately, the quality of the rainfall data is questionable, except for the data in Manila. The Monthly Bulletin of the Philippine Weather Bureau was published from January 1901 to August 1940. In 1903, the Philippine Weather Bureau expanded the number of weather stations in the Philippines, again under the direction of the Jesuits of the Manila Observatory, along with American meteorologists (Sola 1903). From September 1940, weather observations were obtained by Japanese meteorologists. Daily rainfall data exist for September and October 1942; September, October, and

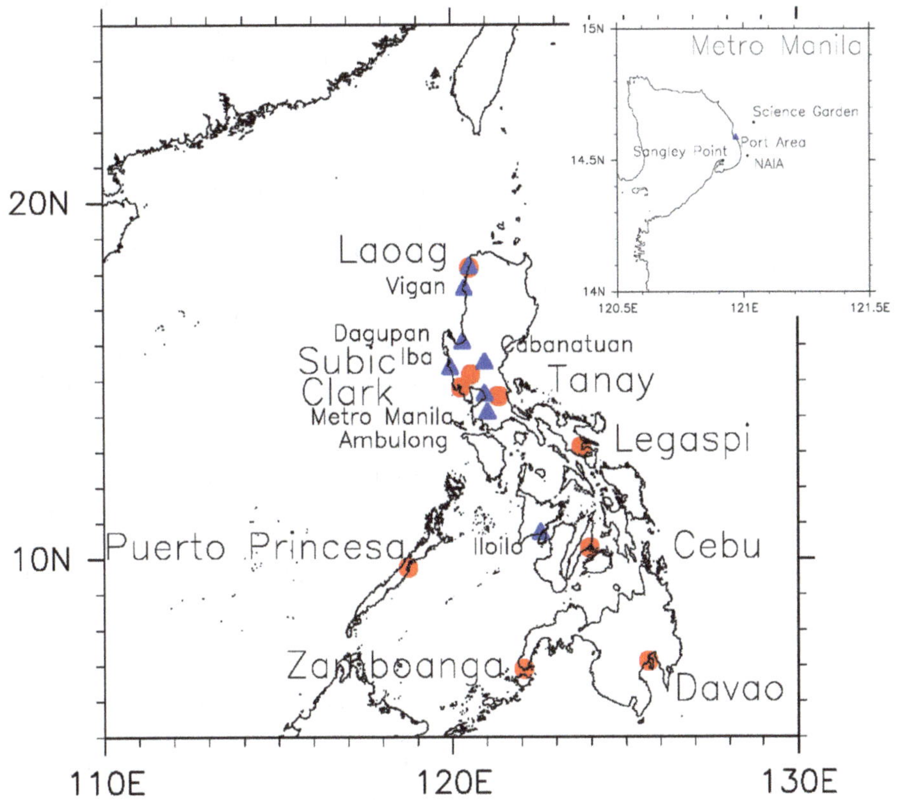

Fig. 1 Locations of observation stations. Upper-air observation stations (red dots with large station name) and surface rainfall observation stations (blue triangles with small station name) used to define Philippine monsoon onset

December 1943; and January 1944 (Kobayashi and Yamamoto 2013). After 1949, the Philippine Atmospheric Geophysical and Astronomical Services Administration (PAGASA) operated the observations in the Philippines. The homogeneity of the station rainfall data was tested by the R-based RHtests_dlyPrcp software (Wang and Feng 2013). If any discontinuity was detected, we adjusted the data based on the quantile-matching algorithm of the RHtests_dlyPrcp software (Wang et al. 2010).

PAGASA defined Philippine summer monsoon onset using eight stations in the northwestern Philippines (Fig. 1; Cayanan 2010). When more than half of the stations satisfied the following criteria during May–July, PAGASA defined it as a summer monsoon onset:

1. Five-day total rain being 25 mm or more;
2. At least three rainy days, each having 5 mm or more in a 5-day period.
3. The first day of the 5-day period satisfying both 1 and 2 is taken as the beginning of the onset.

They defined Metro Manila as a combination of four stations: Science Garden, Port Area, NAIA (Ninoy Aquino International Airport), and Sangley Point. When three of the four stations satisfied the monsoon onset criteria, Metro Manila was considered to have satisfied the criteria. In this study, we follow these Philippine summer monsoon onset criteria using the rainfall data at eight stations. Port Area is the oldest station in Manila. The other three stations of NAIA, Science Garden, and Sangley Point started observations in 1951, 1961, and 1974, respectively. Therefore, we used Port Area as representative of Metro Manila. The official PAGASA summer monsoon onset date was determined by considering not only the rainfall criteria but also other factors like wind data after 1979. Since the other factors may change over time because of the development of observation technology, it is difficult to reproduce these factors for the past. In this study, we used the same criteria as PAGASA with rainfall data from eight stations to identify the monsoon onset. As a result, the summer monsoon onset dates may differ between our study and the official PAGASA announcements.

The International Comprehensive Ocean-Atmosphere Dataset (ICOADS; Woodruff et al. 2011) based on monthly ship observations dataset over the SCS is used for the period from 1901 to 2007 (Yang et al. 2015). These data are available at 2° resolution along the SCS in the meridional direction, where the monthly SST data cover more than 80% of the total months during the

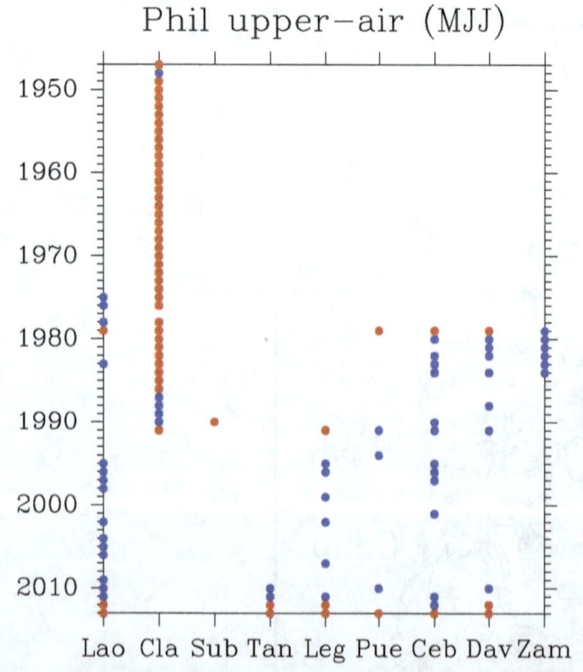

Fig. 2 Availability of upper-air observations. More than half of the period of May to July, upper-air observation is conducted at 00Z or at both 00Z and 12Z—i.e., once or twice a day, are colored in blue and red respectively. The measurement stations Laoag (Lao), Clark air base (Cla), Subic (Sub), Tanay (Tan), Legaspi (Leg), Puerto Princesa (Pue), Cebu (Ceb), Davao (Dav), and Zamboanga (Zam) are shown along the x-axis

whole period (Fig. 1 of Yang et al. 2015). The ICOADS surface wind speed has a noticeable upward trend due to an increase in anemometer height during the twentieth century (Tokinaga and Xie 2011). We use zonal wind data and also consider the wind direction along the SCS in this study, which is not affected by the artificial change in the observation data. The TC best track data from the Joint Typhoon Warning Center (JTWC) and the Japan Meteorological Agency (JMA) are used. The ENSO index is defined using Niño 3.4 and the Hadley Center Global Sea Ice Sea Surface Temperature (HadISST) data (Rayner et al. 2003).

PAGASA carried out upper-air observations at eight stations (from the north: Laoag, Subic, Tanay, Legaspi, Puerto Princesa, Cebu Mactan, Davao, and Zamboanga) in the Philippines (Fig. 1). Data are obtained from the Integrated Global Radiosonde Archive (Durre et al. 2006). Upper-air observations in the Philippines were started in 1947 by the US Clark Air Base staff (Fig. 2). Unfortunately, continuous observations stopped in 1991 because of the eruption of Mt. Pinatubo. Weather balloons were launched once a day at other upper-air observation stations. However, they were interrupted frequently because of an insufficient supply of sensors until 2012. Upper-air observations in the Subic and Zamboanga stations were performed only for a short period. We conducted intensive upper-air observations in Laoag, Puerto Princesa, and Cebu Mactan in collaboration with PAGASA during the Pacific Area Long-term Atmospheric observation for the Understanding of climate change (PALAU) 2013 observation project. At the station of Cebu Mactan, upper-air observations were conducted twice a day from May 1 to August 29, 2013, except for during June 1–30, 2013, when they were conducted four times a day. At the Laoag and Puerto Princesa stations, observations were conducted twice a day from June 1 to July 31, 2013. The Japanese 55-year reanalysis dataset (JRA55; Ebita et al. 2011) on a 1.25° grid was used to obtain three-dimensional structures of wind and moisture. Upper-air observation provides the most useful data for developing a vertical profile of the atmosphere, and these upper-air observations, including our intensive observations, are assimilated into the reanalysis dataset using the Global Telecommunications System.

Results

Climatology of the Philippine monsoon

Figure 3a, b shows the climatology of monthly rainfall distributions along the western stations of the Philippines during 1977–2009. Rainfall increased from June to October north of 10° N, associated with the summer monsoon. Rainfall at all western stations increased or decreased at the same time. However, rainfall at the eastern Philippine stations increased in October from the northern station of Aparri, propagating southward in November and reaching a maximum in January at the southern station of Hinatuan, associated with the winter monsoon (Fig. 3c, d) of the winter monsoon occurred in December in northern Luzon Island and tended to be delayed moving southward. At the southern station in Hinatuan, withdrawal of the winter monsoon occurred in March. However, a classic Philippine monsoon study showed the winter monsoon continued until March at all stations on the eastern side of the Philippines by considering the northeasterly wind direction (Flores and Balogot 1969). Northeasterly wind prevailed throughout winter from north to south at the eastern stations (e.g., Aparri, Legaspi, and Surigao). However, rainfall amounts reached a minimum in February and March in Aparri in northern Luzon. This indicates that the winter monsoon changed its structure in the second half of the winter monsoon season in Luzon Island in the northern Philippines, with less rainfall under northeasterly monsoon wind conditions. Flores and Balogot (1969) defined a unified climatology of the Philippine winter monsoon structure but did not mention the difference in meridional rainfall structures. The southern-most station, Davao, did not show a clear

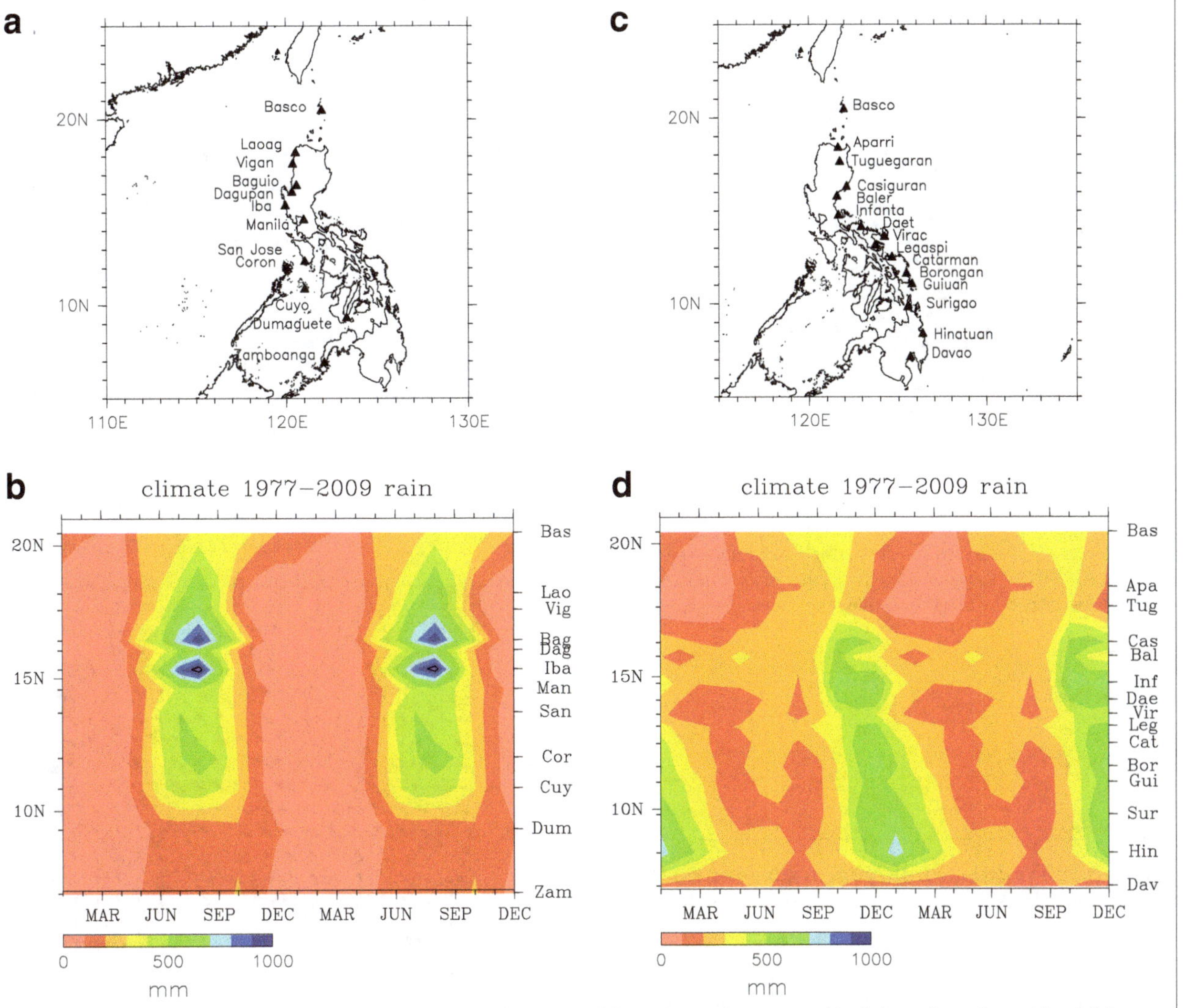

Fig. 3 Climatology of the western and eastern sides of the Philippines rainfall. **a**, **c** Station locations and **b**, **d** climatology of monthly rainfall distribution along the western and eastern sides of the Philippines during the period 1977–2009

winter rainfall maximum. The structure of summer monsoon onset is the focus of the following section.

Long-term variability of Philippine summer monsoon onset

An increase in rainfall in the northwestern Philippines associated with moist southwesterly wind indicates the onset of a Philippine summer monsoon. Here, we use rainfall data from eight stations in the northwestern Philippines to define the Philippine summer monsoon onset, according to the PAGASA rainfall criteria, and expand the study period to the past 100 years. Figure 4 shows daily rainfall during 1901–2013 averaged over the eight northwest Philippine stations used to define the summer monsoon onset from 1903 onward. Rainfall increased around May to June and decreased around October to November. The summer monsoon onset dates are indicated by triangles. During 1941–1950, rainfall data were missing from most of the stations, primarily due to the occurrence of the Second World War.

Figure 5a shows a time series of the summer monsoon onset dates from 1903 to 2013. After the mid-1990s, the onset date tended to be earlier, and the monsoons started in May. Such early onset dates after the mid-1990s were similarly reported in Akasaka (2010) and Kajikawa and Wang (2012). The long-term dataset that we used in this study further reveals other early onsets that occurred in the 1900s and from the 1920s to the 1930s, suggesting an interdecadal variability of the Philippine summer monsoon onset. After the start of the Philippine summer monsoon, the low-level wind direction shifts to westerly over the SCS (e.g., Puerto Princesa in the SCS, Fig. 8a). This suggests that when the summer monsoon starts in May, a monthly averaged westerly

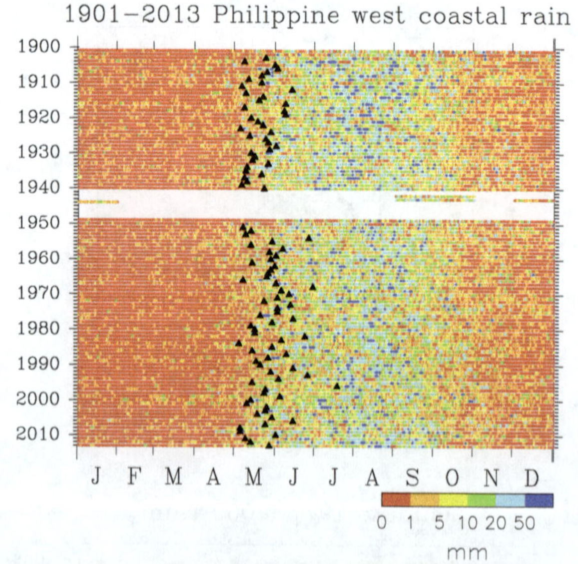

Fig. 4 Daily rainfall averaged at eight northwest Philippine stations during 1901–2013. Triangles indicate Philippine summer monsoon onset dates defined by rainfall data using the criteria (1–3) listed in the text

wind prevails near the surface (Fig. 6a). However, when the summer monsoon starts in June, the surface zonal wind component remains easterly or very weak in May (Fig. 6b). Figure 5b shows the monthly surface zonal wind along the SCS in May from 1901 to 2007. After the 1990s, strong westerly wind appears around 5° N–14° N, associated with early summer monsoon onset. However, during the 1950s to the early 1990s, the surface zonal wind was weakly westerly or easterly. This is consistent with the results of late summer monsoon onset in June defined by the rainfall data. It is worth noting that the westerly wind was also strong around 10° N around 1910 and during the 1920s and 1930s in May, which is consistent with the early summer monsoon onset seen in Fig. 5a.

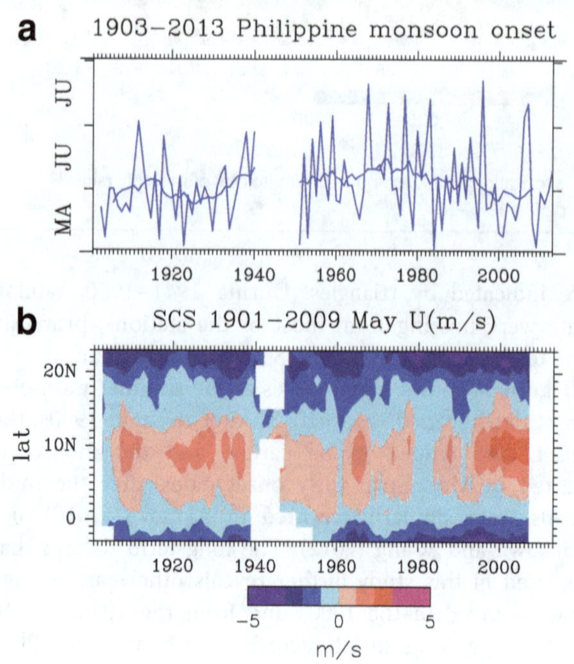

Fig. 5 Philippine summer monsoon onset date and surface zonal wind in May along the South China Sea. Time series of **a** Philippine summer monsoon onset date from 1903 to 2013 and **b** monthly surface zonal wind in May along the South China Sea from 1901 to 2007. Thick lines indicate an **a** 11-year or **b** 5-year running mean

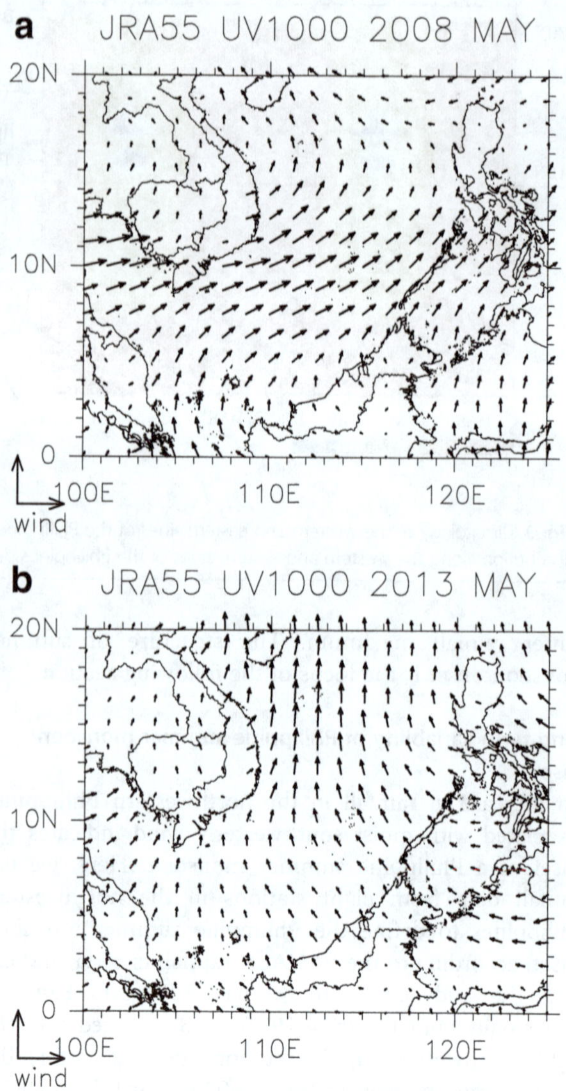

Fig. 6 Monthly averaged horizontal wind at 1000 hPa in May in **a** 2008 and **b** 2013. Unit vectors represent 12.5 m/s

1Earlier studies have shown the strong influence of the ENSO on Philippine summer monsoon rainfall, which increases during the summer of El Niño developing years (Lyon et al. 2006; Kubota and Wang 2009) and decreases during the summer of the following El Niño decaying year (Xie et al. 2009). The correlation between Philippine summer monsoon rainfall and the ENSO also shows interdecadal variability (Chowdary et al. 2012). However, the Philippine summer monsoon onset date was delayed during El Niño decaying years only from around the 1970s to the 1980s (Fig. 7). Tanaka (1997) mentioned the delay of WNP summer monsoon onset during this high-correlation period around the 1970s to the 1980s. The correlations in other periods were insignificant. In summary, the summer monsoon onset date has been weakly correlated to the ENSO during the past 100 years. The PAGASA official summer monsoon onset date has also a weak correlation with the ENSO since the 1990s (not shown).

Case study of 2013 Philippine summer monsoon onset

What caused the interdecadal variability of Philippine summer monsoon onset? Unfortunately, the accuracy of atmospheric structures around the Philippines in the reanalysis data has decreased since 1991 due to a lack of upper-air observations. Here, we use intensive upper-air observations in 2013 in the Philippines to discuss the atmospheric factors responsible for triggering summer monsoon onset and then compare the case of 2013 to the past in the following subsections.

We focus on the change in atmospheric structure during Philippine summer monsoon onset in 2013 using intensive upper-air observations, which are included in the reanalysis data. Using the locally defined rainfall-based criteria for declaring the summer monsoon onset (see the "Methods" section), the Philippine summer monsoon started on June 9, 2013. Figure 8a shows the time series of upper-air observations and rainfall at Puerto Princesa in the southwestern Philippines before and after summer monsoon onset. The layers of relative humidity greater than 70% in the troposphere gradually expand from the lower troposphere over 7 days before the monsoon onset and are associated with rainfall events. Wind direction clearly shifts from northeasterly to southwesterly below 8 km after the monsoon onset on June 9. The arrival of moist southwesterly wind in the lower troposphere is associated with the large-scale seasonal migration of low-level convergence and activated the Asian summer monsoon in the Philippines (Wang 2006). However, different atmospheric structures were present before and after the monsoon onset in the northern Philippines. Figure 8b shows the time series of upper-air observations and rainfall at Laoag in the northwestern Philippines. It was very dry in the mid and upper troposphere before the monsoon onset. The layers of relative humidity greater than 70% suddenly increased up to 11 km and became wet and unstable from June 11. The shifts in wind direction before and after the monsoon onset differed from

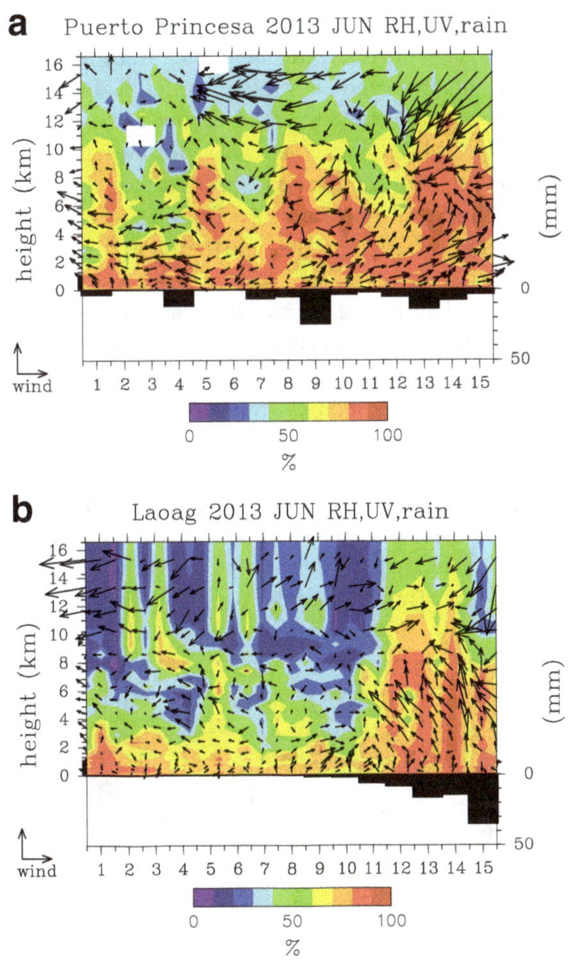

Fig. 8 Atmospheric profile during June 1–15, 2013. Vertical profiles of relative humidity (shading), horizontal wind (vector), and daily rainfall (bars) at **a** Puerto Princesa and **b** Laoag during June 1–15, 2013. Unit vectors represent 12.5 m/s

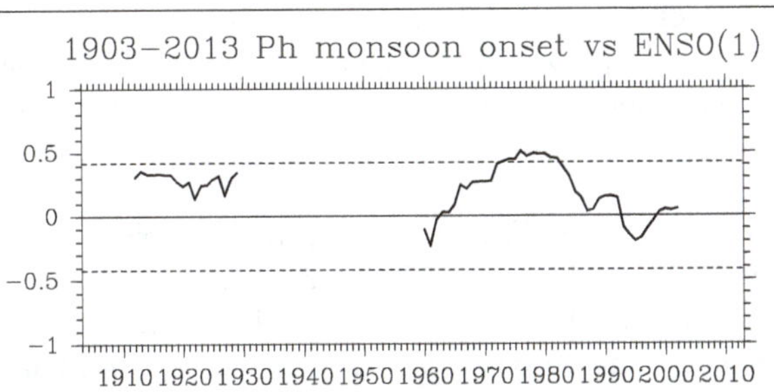

Fig. 7 Correlation between ENSO and Philippine monsoon onset date. Twenty-one-year running correlation between Niño 3.4 SST averaged during the preceding December to February and Philippine monsoon onset date. Dashed lines indicate the 95% confidence level

those in the southwestern Philippines. Low-level southerly wind prevailed after the monsoon onset. In this study, we defined the summer monsoon onset using the northwestern Philippines rainfall data. Summer monsoon rainfall increased at the same times in the western Philippines (Fig. 3b). However, the time required for the moistening process to cause the lower to middle troposphere to become wet and unstable differs between the northern and southern Philippines.

The vertical profile of relative humidity and horizontal wind from the JRA55 reanalysis data, which were assimilated by upper-air observation data, are plotted in Fig. 9 including the same location of Puerto Princesa and Laoag during the same period as in Fig. 8. Wind direction changes associated with monsoon onset are consistent with upper-air observations. However, the JRA55 data showed unrealistically high relative humidity in the wet layers, especially above 12 km and below 1 km, during the analysis period in both regions. This represents wet bias in the upper troposphere and near the surface in the JRA55 dataset. However, lower troposphere JRA55 data well capture the increase of relative humidity and are consistent with in situ observational data. The horizontal distribution of the wind and water vapor at 850 hPa (about 1.5 km) from the JRA55 data are discussed below.

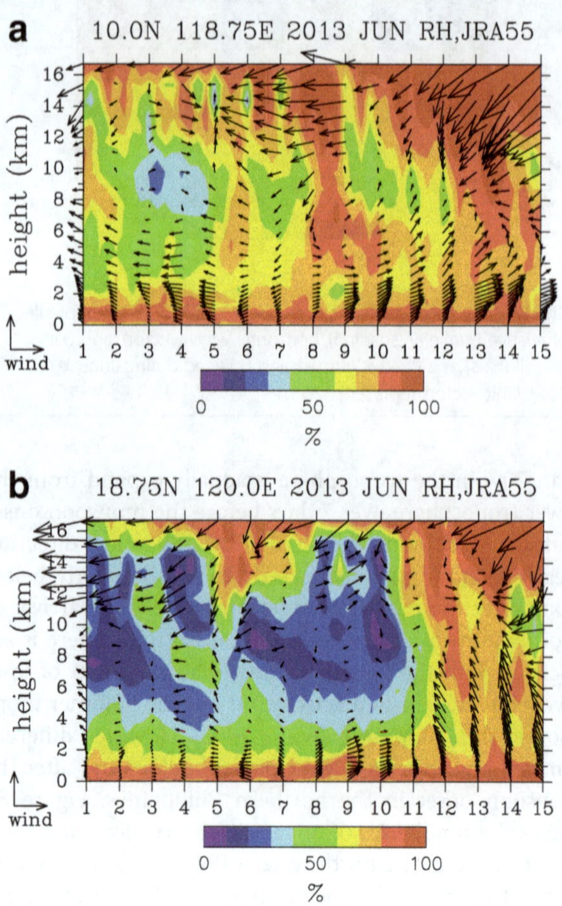

Fig. 9 Atmospheric profile by JRA55 during June 1–15, 2013. Same as Fig. 8 but for JRA55 relative humidity and wind at **a** 10.0° N, 118.75° E (including Puerto Princesa) and **b** 18.75° N, 120.0° E (including Laoag)

The horizontal structures of water vapor mixing ratio and wind at 850 hPa before and after the monsoon onset are shown in Fig. 10. Tropical cyclone (defined as an intensity of tropical storm exceeding 34 kt maximum wind speed and hereafter referred to as TC) Yagi (1303) was generated at 12 UTC (coordinated universal time) on June 8, 2013, and traveled northward in the PS (Fig. 10a–e). As TC Yagi developed during June 8–9, moist southwesterly wind from the SCS passed over the Philippines and fed into TC Yagi (Fig. 10b, c). While TC Yagi traveled northward, the wind direction shifted from westerly to southerly in the northern Philippines. Figure 11 presents a cross section along the west Philippines (117.5° E–120.0° E, 5° N–20° N,) for the period June 5–14. Water vapor mixing ratio increased in the southern part of the cross section as the westerly wind intensified and moisture was transported northward, concurrently with the wind direction shift to southerly in the northern part of the cross section, after the monsoon onset. This suggests that in situ observations in Puerto Princesa at the southern Philippines station and in Laoag at the northern station, where we carried out intensive upper-air observations, can capture representative atmospheric structure changes in the south and north Philippines before and after the summer monsoon onset. The TC induced the moist southwesterly wind and activated the Philippine summer monsoon. The arrival of a TC is one of the major contributors to Philippine summer monsoon rainfall (Kubota and Wang 2009; Chen et al. 2010). Even if a TC does not make landfall in the Philippines, it can still enhance southwesterly flow and activate Philippine summer monsoon rainfall (Cayanan et al. 2011).

Existence of tropical cyclones during the Philippine summer monsoon onset

As demonstrated in the previous section, the passage of a TC over the PS played an important role in the Philippine summer monsoon onset in 2013. We then investigated whether a TC existed within the area comprising the PS and the SCS (100° E–140° E, 0° N–20° N; the area bounded by dashed lines in Fig. 10c) during the Philippine summer monsoon from 1951 to 2013. We also included tropical depressions that developed into TC within 2 days after the summer monsoon onset date. The existence of TC during the summer monsoon onset from 1951 to 2013 is shown in Fig. 12 (labeled "T" in Fig. 12). To explore the influence of TC on Philippine summer monsoon onset, we obtain the probability of TC existence (PTCe) during summer monsoon onset, expressed as

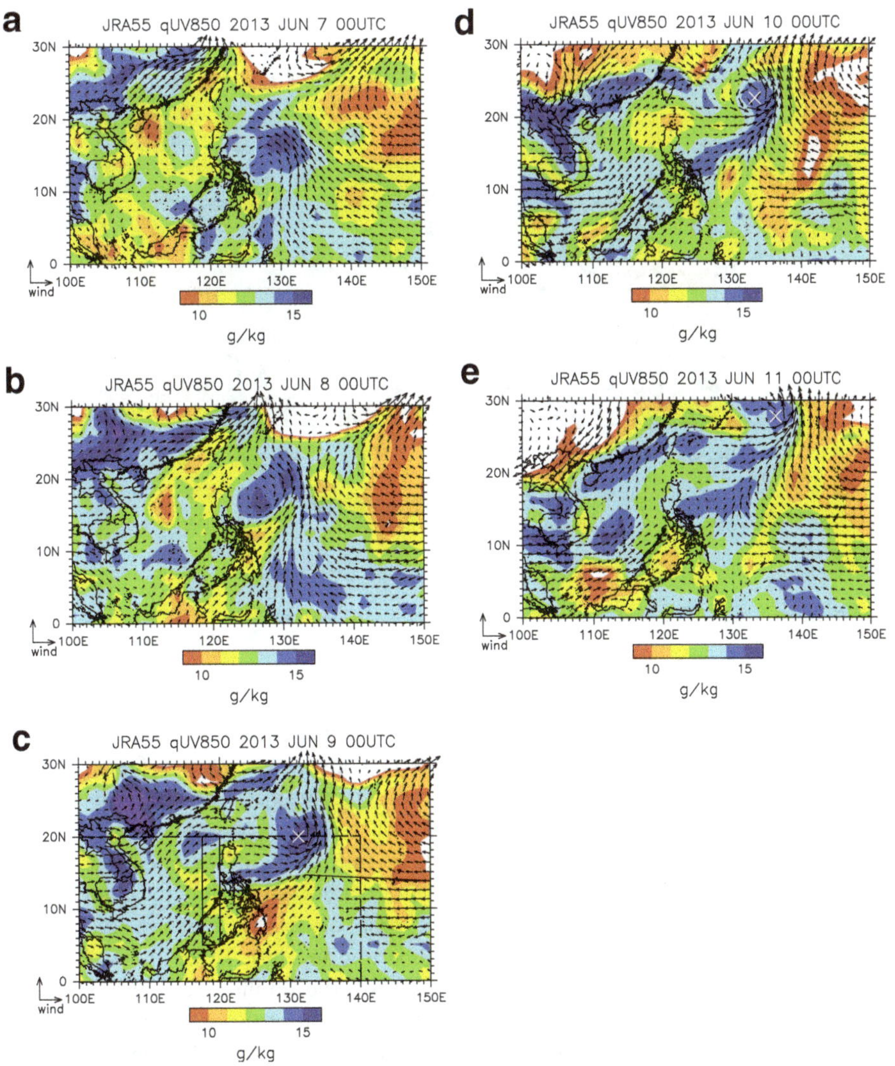

Fig. 10 Horizontal structure of water vapor mixing ratio (shading; g/kg) and wind (vector) at 850 hPa from June 7 (**a**) to 11 (**e**), 2013. Unit vectors represent 25.0 m/s. Crosses indicate the centers of a tropical storm. Solid and dashed boxes in c are the cross-sectional area used in Fig. 9 and the area used to calculate PTCe

$\mathrm{PTCe}(i) = \frac{1}{11}\sum_{i-5}^{i+5}\mathrm{TC}(i)$.

where i is from 1956 to 2008, and $\mathrm{TC}(i) = 1$ if a TC existed in year i or 0 if not.

The above expression has a range from 0 to 1. The former indicates no existence of TC during the Philippine summer monsoon onset, while the latter signifies the occurrence of TC in every year of an 11-year sliding window. From the mid-1990s, PTCe increased during summer monsoon onset, and it exceeded 60% (0.6) in the mid-2000s. The period of high PTCe during summer monsoon onset was concurrent with the period of early monsoon onset (Fig. 12). The correlation coefficient for PTCe and monsoon onset date in the 11-year running mean reached − 0.84 at the 99% confidence level. A recent active TC in the PS and the SCS in May and June contributes to early Philippine summer monsoon onset. This suggests our case study of 2013 captured the typical summer monsoon onset conditions of recent years. Kajikawa and Wang (2012) suggested that TC became active in May after the mid-1990s. Before the mid-1990s, the PTCe was small, and TC did not act as the primary trigger for monsoon onset. When there was no TC near the Philippines during summer monsoon onset, a front was located north of the Philippines, and the southwesterly wind converged into the zonal front as observed for 2010 (Fig. 13). Akasaka (2010) described several conditions of atmospheric circulation during the summer monsoon onset. The interdecadal variability of the

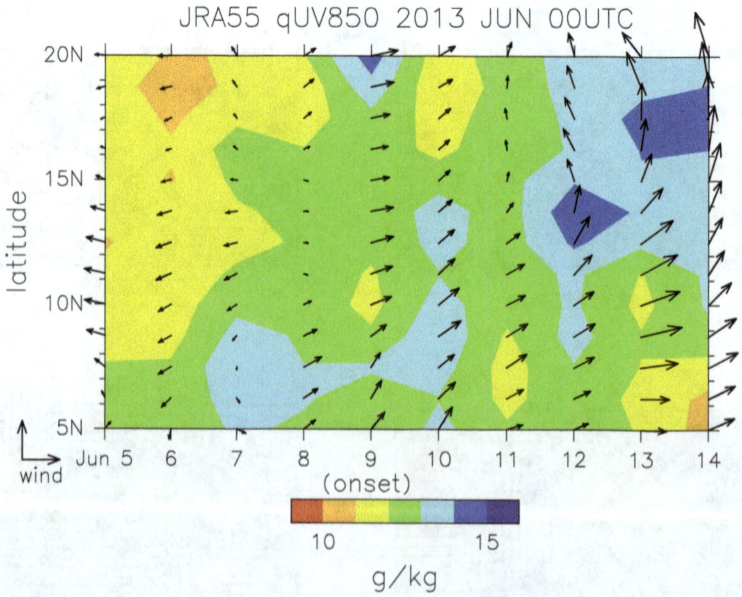

Fig. 11 Cross section (5° N 20° N, 115° E 120° E) of water vapor mixing ratio (shading; g/kg) and wind (vector) at 850 hPa during June 5–14, 2013. Unit vectors represent 10.0 m/s

Philippine summer monsoon onset date is strongly affected by TC activity in the PS and the SCS in May and June. The cause of this interdecadal variability of TC activity will be the subject of future work.

Discussion

Tanaka (1997) noted a delay of summer monsoon onset during El Niño decaying years. However, our longterm dataset demonstrated that the correlation between Philippine summer monsoon onset and the ENSO was significant only during the 1970s to the 1980s. The relationship of SCS SST and the ENSO also showed the highest correlation during the 1970s and the 1980s, as shown in Fig. 4 of Yang et al. (2015). The interdecadal variability of the relationship of SCS SST and the ENSO may influence the interdecadal variability of the relationship of the Philippine summer monsoon onset date and the ENSO.

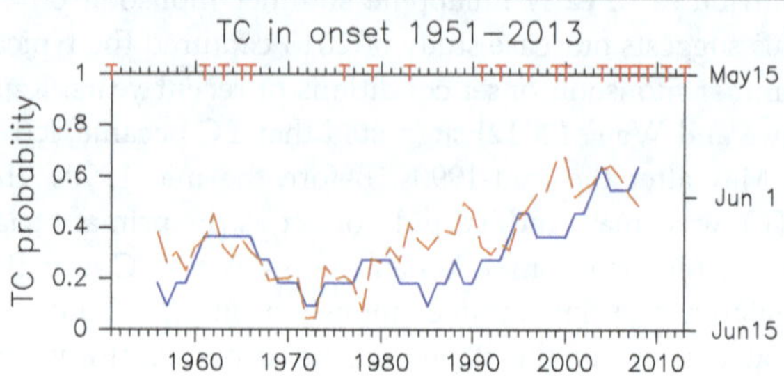

Fig. 12 The probability of TC existence during summer monsoon onset. Existence of TC at monsoon onset (red "T"), 11-year running average of the PTCe (blue line), and Philippine summer monsoon onset date (orange dashed line; same as Fig. 5a) are superimposed

To identify atmospheric structures that trigger summer monsoon onset, we focused on the case of 2013 using a combination of intensive upper-air observations and a reanalysis dataset. The summer of 2013 was the first time that comprehensive upper-air observations were conducted in the Philippines since 1991. The interruption of observation was due to the eruption of Mt. Pinatubo in 1991. In 2013, TC Yagi traveled northward, east of the Philippines, and triggered the Philippine monsoon onset. Moist southwesterly wind was induced in the southwestern Philippines after the monsoon onset, and low-level southerly wind intensified after the monsoon onset with sudden moistening in the northern

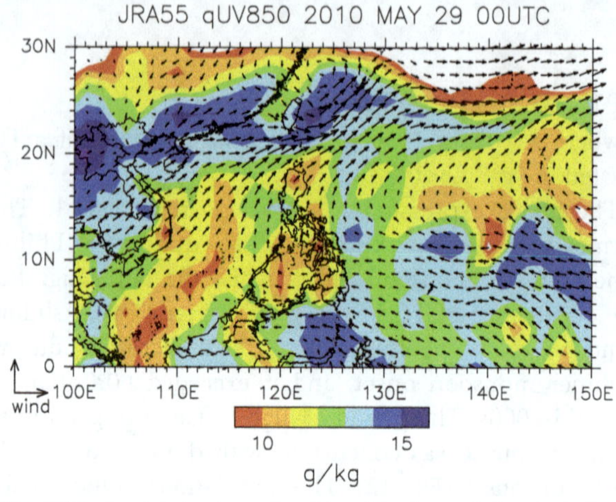

Fig. 13 Horizontal structure of moisture (shading; g/kg) and wind (vector) at 850 hPa on May 29, 2010. Unit vectors represent 25.0 m/s

Philippines. In the climatology analysis, the summer monsoon onset started at the same time in the western Philippines. However, in situ observations demonstrated that when TC triggered the monsoon onset, the moistening process and prevailing wind direction differed between the southern and northern Philippines.

However, the role of TC activity decreased during late summer monsoon onset. Additionally, the historical TC dataset from Kubota (2012) shows high PTCe during the 1910s and the 1920s, which coincide with early summer monsoon onset (not shown). Further investigation is needed to verify the quality of the historical TC dataset. This study demonstrates the importance of historical station data in understanding interdecadal variability during the past 100 years.

Conclusions

The seasonal variation of Philippine rainfall associated with monsoons was investigated. Climatologically, Philippine rainfall increased on the west side of the country during June to October, associated with the summer monsoon, and rainfall increased from October to March on the eastern side, associated with winter monsoons that started in the northern Philippines and propagated southward. Withdrawal of winter monsoon rainfall occurred in December in northern Luzon Island and tended to be delayed moving south.

The long-term variability of summer monsoon onset in the Philippines was investigated using daily rainfall data for 1903–2013. Philippine summer monsoon onset is defined using the rainfall data of eight northwestern stations. The summer monsoon tended to start earlier in May after the mid-1990s. Other early northwestern stations. The summer monsoon tended to start earlier in May after the mid-1990s. Other early of summer monsoon onset was thus identified. Independent surface wind data from ships along the SCS revealed a prevailing westerly wind in May, consistent with early onset before 1940 and after the 1990s.

The influence of TC activity on the variability of summer monsoon onset date was investigated. From the mid-1990s, the probability of TC existence (PTCe) during the summer monsoon onset increased and became more than 60% (0.6), and is correlated with that period having an early monsoon onset. The PTCe was significantly correlated with the monsoon onset date in the Philippines. The recent early monsoon onsets observed after the mid-1990s were influenced by active TC formation in the PS and the SCS. Tropical cyclone activity in the PS and the SCS plays a key role in initiating early Philippine summer monsoon onset.

Abbreviations

ENSO: El Niño-Southern Oscillation; HadISST: Hadley Center Global Sea Ice Sea Surface Temperature; ICOADS: The International Comprehensive Ocean-Atmosphere Dataset; IO: Indian Ocean; JRA55: Japanese 55-year reanalysis; JTWC: Joint Typhoon Warning Center; NAIA: Ninoy Aquino International Airport; OLR: Outgoing longwave radiation; PAGASA: Philippine Atmospheric Geophysical and Astronomical Services Administration; PALAU: Pacific Area Long-term Atmospheric observation for Understanding climate change; PJ: Pacific-Japan; PS: Philippine Sea; PTCe: Probability of TC existence; SCS: South China Sea; SST: Sea surface temperature; TC: Tropical cyclone; WNP: Western North Pacific

Acknowledgements

The PALAU 2013 observation project was conducted under the Implementation Agreement between PAGASA and JAMSTEC. We wish to thank Dr. Ikumi Akasaka of Senshu University and Prof. Masumi Zaiki of Seikei University for collecting the Philippine Weather Bureau station data; Mr. Yoshio Shinohara of the Japan Meteorological Agency for sharing the historical bulletins of the Central Meteorological Observatory of Tokyo, Japan; and Dr. Hiroki Tokinaga of Kyoto University for providing the surface wind dataset of the South China Sea. JM and HK were supported by "Global Environment Research Fund from the Ministry of the Environment Japan" B-061, "Data Integration & Analysis System," "Green Network of Excellence" (GRENE), "Program for Risk Information on Climate Change" (SOUSEI), and Grant-in-Aid for Scientific Research No. (25282085, 15KK0030; PI HK), No. (2024007, 23240122, and 26220202; PI: JM), No. 16H04053, No. 16H03116 and Young Scientific Research No. (21684028; PI HK) was funded by the Japan Society for the Promotion of Science (JSPS). This research was also supported by the Science and Technology Research Partnership for Sustainable Development (SATREPS), Japan Science and Technology Agency (JST)/Japan International Cooperation Agency (JICA).

Funding

JM and HK were supported by the Global Environment Research Fund from the Ministry of the Environment Japan B-061, the Data Integration & Analysis System, Green Network of Excellence (GRENE), the Program for Risk Information on Climate Change (SOUSEI), and Grant-in-Aid for Scientific Research (No. 25282085, 15KK0030; PI HK), (No. 2024007, 23240122, and 26220202; PI: JM), No. 16H04053, No. 16H03116 and Young Scientific Research (No. 21684028; PI HK), funded by the Japanese Society for the Promotion of Science (JSPS). This research was also supported by the Science and Technology Research Partnership for Sustainable Development (SATREPS), the Japanese Science and Technology Agency (JST), and the Japanese International Cooperation Agency (JICA).

Authors' contributions

HK proposed the topic and conceived and designed the study. HK, RS, EC, and FH carried out the intensive observations and obtained the dataset for this study with the cooperation of JAMSTEC and PAGASA. JM, EC, and FH collaborated with the corresponding author and contributed to the interpretation of the monsoon period. All authors read and approved the final manuscript.

Competing interests

The authors declare that they have no competing interests.

Author details

[1]Faculty of Science, Hokkaido University, Kita 10 Nishi 8, Kita-ku, Sapporo 060-0810, Japan. [2]Department of Coupled Ocean-Atmosphere-Land Processes Research, Japanese Agency for Marine-Earth Science and Technology (DCOP, JAMSTEC), Yokosuka, Japan. [3]Department of Geography, Graduate School of Urban Environmental Sciences, Tokyo Metropolitan University, Tokyo, Japan. [4]Philippine Atmospheric, Geophysical and Astronomical Services Administration (PAGASA), Quezon City, Philippines.

References

Akasaka I (2010) Interannual variations in seasonal march of rainfall in the Philippines. Int J Climatol 30:1301–1314

Akasaka I, Morishima W, Mikami T (2007) Seasonal march and spatial difference of rainfall in the Philippines. Int J Climatol 27:715–725

Cayanan EO (2010) A study of heavy rainfall events during the southwest monsoon season in the Philippines. Dissertation, University of the Philippines

Cayanan EO, Chen TC, Argete JC, Yen MC, Nilo PD (2011) The effect of tropical cyclones on southwest monsoon rainfall in the Philippines. J Meteor Soc Japan 89A:123–139

Chang CP, Zhang Y, Li T (2000a) Interannual and interdecadal variations of the east Asian summer monsoon and tropical Pacific SSTs. Part I: roles of the subtropical ridge. J Clim 13:4310–4325

Chang CP, Zhang Y, Li T (2000b) Interannual and interdecadal variations of the east Asian summer monsoon and tropical Pacific SSTs. Part II: meridional structure of the monsoon. J Clim 13:4326–4340

Chen TC, Tsay JD, Yen MC, Cayanan EO (2010) Formation of the Philippine twin tropical cyclones during the 2008 summer monsoon onset. Wea Forecasting 25:1317–1341

Chowdary JS, Xie SP, Tokinaga H, Okumura YM, Kubota H, Johnson N, Zheng XT (2012) Interdecadal variations in ENSO teleconnection to the Indo-Western Pacific for 1870–2007. J Clim 25:1722–1744

Deppermann CE (1939) Some characteristics of Philippine typhoons. Weather Bureau Manila Central Observatory

Durre I, R. Vose RS, Wuertz DB (2006) Overview of the integrated global radiosonde archive. J Clim 19:53–68

Ebita A, Kobayashi S, Ota Y, Moriya M, Kumabe R, Onogi K, Harada Y, Yasui S, Miyaoka K, Takahashi K, Kamahori H, Kobayashi C, Endo H, Soma M, Oikawa Y, Ishimizu T (2011) The Japanese 55-year reanalysis 'JRA-55': an interim report. SOLA 7:149–152

Flores JF, Balagot VF (1969) Climate of the Philippines. Climates of northern and eastern Asia. In: Arakawa H (ed) World survey of climatology, vol 8. Elsevier, Amsterdam, pp 159–213

Hattori M, Matsumoto J, Ogino SY, Enomoto T, Miyoshi T (2016) The impact of additional radiosonde observations on the analysis of disturbances in the South China Sea during VPREX2010. SOLA 12:75–79

Kajikawa Y, Wang B (2012) Interdecadal change of South China Sea summer monsoon onset. J Clim 25:3207–3218

Kobayashi S, Yamamoto H (2013) Wartime meteorological observations of the Japanese military in East Asia and the collection of its neglected data. Hist Geogr 267:82–98

Kubota H (2012) Variability of typhoon tracks and genesis over the Western North Pacific. In: Oouchi K, Fudeyasu H (eds) Cyclones: Formation, Triggers and Control, Nova Science Publishers, Inc., pp 95–114

Kubota H, Shirooka R, Ushiyama T, Chuda T, Iwasaki S, Takeuchi K (2005) Seasonal variations of precipitation properties associated with monsoon over Palau in the western Pacific. J Hydrometeorol 6:518–531

Kubota H, Kosaka Y, Xie SP (2016) A 117-year long index of the Pacific-Japan pattern with application to interdecadal variability. Int J Climatol 36:1575–1589

Kubota H, Wang B (2009) How much do tropical cyclones affect seasonal and interannual rainfall variability over the western North Pacific? J Clim 22:5495–5510

Lau KM, Yang S (1997) Climatology and interannual variability of the southeast Asian summer monsoon. Adv Atmos Sci 14:141–162

Lyon B, Camargo SJ (2008) The seasonally-varying influence of ENSO on rainfall and tropical cyclone activity in the Philippines. Clim Dyn doi https://doi.org/10.1007/s00382-008-0380-z

Lyon B, Cristi H, Verceles ER, Hilario FD, Abastillas R (2006) Seasonal reversal of the ENSO rainfall signal in the Philippines. Geophys Res Lett 33:L24710. https://doi.org/10.1029/2006GL028182

Matsumoto J (1997) Seasonal transition of summer rainy season over Indochina and adjacent monsoon region. Adv Atmos Sci 14:231–245

Murakami T, Matsumoto J (1994) Summer monsoon over the Asian continent and western North Pacific. J Meteor Soc Japan 72:719–745

Nitta T (1987) Convective activities in the tropical western Pacific and their impacts on the Northern Hemisphere summer circulation. J Meteor Soc Japan 65:165–171

Ramage CS (1971) Monsoon Meteorology, Academic Press, New York

Rayner NA, Parker DE, Horton EB, Folland CK, Alexander LV, Rowell DP, Kent EC, Kaplan A (2003) Global analyses of sea surface temperature, sea ice, and night marine air temperature since the late nineteenth century. J Geophys Res 108(D14):4407. https://doi.org/10.1029/2002JD002670

Ropelewski CF, Halpert MS (1987) Global and regional scale precipitation patterns associated with the El Niño/ Southern Oscillation. Mon Wea Rev 115:1606–1626

Saunders MA, Chandler RE, Merchant CJ, Roberts FP (2000) Atlantic hurricanes and NW Pacific typhoons: ENSO spatial impacts on occurrence and landfall. Geophys Res Lett 27:1147–1150

Sola MSJ (1903) Report of the Director of the Philippine Weather Bureau 1902 part II: meteorological service of the Philippine islands, Report of its establishment and development under the Spanish government and its reorganization under the government of the United States (1865–1902). Bureau of Public Printing, Manila

Tanaka M (1992) Intraseasonal oscillation and onset and retreat dates of the summer monsoon over east, southeast Asia and the western Pacific region using GMS high cloud amount. J Meteor Soc Japan 70:613–629

Tanaka M (1997) Interannual and interdecadal variations of the western north Pacific monsoon and Baiu rainfall and their relationship to the ENSO cycles. J Meteor Soc Japan 75:1109–1123

Tokinaga H, Xie SP (2011) Wave- and anemometer-based sea surface wind (WASWind) for climate change analysis. J Clim 24:267–285

Udias A (1996) Jesuits' contribution to meteorology. Bull Am Meteorol Soc 77: 2307–2315

Villafuerte MQ II, Matsumoto J, Akasaka I, Takahashi H, Kubota H, Cinco TA (2014) Long-term trends and variability of rainfall extremes in the Philippines. Atmos Res 137:1–13

Wang B (2006) The Asian Monsoon. Springer Praxis Publishing, New York

Wang B, LinHo (2002) Rainy seasons of the Asian–Pacific monsoon. J Clim 15: 1071–1085

Wang B, Wu R, Fu X (2000) Pacific-East Asian teleconnection: how does ENSO affect East Asian climate? J Clim 13:1517–1536

Wang XL, Feng Y (2013) RHtests_dlyPrcp user manual. Climate Research Division, Atmospheric Science and Technology Directorate, Science and Technology Branch, Environment Canada, http://etccdi.pacificclimate.org/software.shtml. Accessed 7 Aug 2017

Wang XL, Chen H, Wu Y, Feng Y, Pu Q (2010) New techniques for detection and adjustment of shifts in daily precipitation data series. J Appl Meteorol Climatol 49:2416–2436. https://doi.org/10.1175/2010JAMC2376.1

Woodruff SD, Worley SJ, Lubker SJ, Ji Z, Freeman JE, Berry DI, Brohan P, Kent EC, Reynolds RW, Smith SR, Wilkinsond C (2011) ICOADS release 2.5: extensions and enhancements to the surface marine meteorological archive. Int J Climatol 31:951–967

Xie SP, Hu K, Hafner J, Tokinaga H, Du Y, Huang G, Sampe T (2009) Indian Ocean capacitor effect on Indo-western Pacific climate during the summer following El Nino. J Clim 22:730–747

Yang Y, Xie SP, Du Y, Tokinaga H (2015) Interdecadal difference of interannual variability characteristics of South China Sea SSTs associated with ENSO. J Clim 28:7145–7160

5

Seasonal march patterns of the summer rainy season in the Philippines and their long-term variability since the late twentieth century

Ikumi Akasaka[1*], Hisayuki Kubota[2], Jun Matsumoto[3,4], Esperanza O. Cayanan[5], Rosalina G. de Guzman[5] and Flaviana D. Hilario[5]

Abstract

This study investigates the seasonal march patterns of rainfall in the Philippines from 1951 to 2012 and their long-term variability. In order to clarify the dominant patterns in the seasonal march of rainfall, an empirical orthogonal function (EOF) analysis was applied to pentad rainfall data of 30 stations. For the first EOF mode (EOF1), we obtained a pattern related to the summer rainy season. We then applied cluster analysis to the time coefficients of EOF1 in each year to classify the seasonal patterns of the summer rainy season. As a result, the patterns were classified into six clusters. We found a long-term change in the pattern appearances with three anomalous patterns frequently observed since the 1990s: (1) a pattern that has an indistinct dry season and a prolonged peak rainfall, (2) a pattern that has a distinct dry season and an earlier withdrawal of the summer rainy season, resulting in a shortened rainy season, and (3) a pattern with a distinct dry season as well as delayed onset and withdrawal of the summer rainy season. This study also shows the relations between these three patterns and the lower atmospheric circulation at the 850 hPa level around the Philippines. Consequently, large positive and negative anomalies in geopotential height were observed around the Philippines for the distinct and indistinct dry seasons, respectively. The duration and condition of the dry season were greatly affected by the strength and location of the subtropical high especially for February–March. It is also noteworthy that the timing of the onset (withdrawal) of the summer rainy season is clearly related to that of the onset of the westerly (northerly) wind in the zonal (meridional) component around the Philippines. Further, the duration and amount of peak rainfall were directly influenced by the strength of the westerly winds in the zonal component. These three anomalous patterns tended to appear in the years when the warm or cold event of the El Niño–Southern Oscillation (ENSO) occurred. This study suggests that the long-term variability in the seasonal march of rainfall is considerably influenced by the variability in ENSO.

Keywords: Summer rainy season, Seasonal march, Philippines, Long-term variability, El Niño–Southern Oscillation

Introduction

The climate of the Philippines is characterized by a large volume and seasonality of rainfall (Coronas 1920; Flores and Balagot 1969). The seasonality of rainfall is more extreme than that of temperature and is influenced mainly by the Asian monsoon, tropical cyclones, and the orographic effect. Cayanan et al. (2011) argued that heavy rainfall in western Luzon is induced by both the location and the track of the tropical cyclones during June–September and the orographic effect. Kubota and Wang (2009) estimated that the tropical-cyclone-induced rainfall ratio from July to October exceeds 50% in northern Luzon. Moreover, the mean seasonal march of rainfall amount clearly differs between the western and eastern coastal regions of the Philippines because of the Asian monsoon and the orographic effect, as shown by station rainfall data reported by Akasaka et al. (2007) and by the satellite data by Chang et al. (2005). Both the

* Correspondence: akasaka-i@isc.senshu-u.ac.jp
[1]Department of Geography, Senshu University, 2-1-1 Higashi-Mita, Tama-ku, Kawasaki, Kanagawa 214-8580, Japan
Full list of author information is available at the end of the article

rainfall amount and the seasonality of rainfall are important factors for the agriculture and economics of the Philippines. This is because the seasonal march of rainfall is characterized by not only the seasonal rainfall amount but also the onset, peak, withdrawal periods, and duration of the rainy season, which influence planting decisions and the timing of harvesting in the cultivation calendar. Therefore, whether the seasonal march of rainfall will change with future global warming is of great concern to the Philippines. To answer this question with a prognostic analysis, we must first investigate the past interannual and long-term variability in the seasonal march of rainfall and its causes.

Recent studies have explored interannual variability of the onset of the summer rainy season or the summer monsoon in or around the Philippines since the late twentieth century. Moron et al. (2009) used rainfall data collected at 76 stations between 1977 and 2004 and data from the Climate Prediction Center Merged Analysis of Precipitation (CMAP) from 1979 to 2005 to study the spatial and temporal variability of the onset of the summer monsoon and to assess the seasonal predictability of local onset dates. They identified important climatic factors that are essential for the prediction of the onset: the sea surface temperature (SST) over the tropical Pacific and Indian Oceans in March and the wind field at the 850 hPa level in May. Akasaka (2010) studied the interannual variability of the onset of the summer rainy season throughout the Philippines using daily rainfall data collected at 39 stations between 1961 and 2000. The author pointed out that the onset was consistently delayed after the mid-1970s and that long-term changes in the onset might be related to the timing of the shift in the location of the western edge of the subtropical high over the Philippines. Kajikawa and Wang (2012) studied interdecadal changes in the South China Sea (SCS) summer monsoon onset for the period of 1979–2008. They pointed out that the relatively early onset of the SCS appeared during 1994–2008 and that the onset processes were closely connected to the northward seasonal march of the intertropical convergence zone, the enhanced intraseasonal variability, and the number of tropical cyclones over the western North Pacific (WNP) and SCS. However, the interannual variability in the seasonal march of rainfall, including not only the onset but also the withdrawal and duration of the rainy season, has been scarcely discussed in earlier studies. The purpose of this study is to elucidate the interannual variability in the seasonal march patterns of rainfall in the Philippines since the late twentieth century. This study also aims to reveal the relationship between interannual variations in the seasonal march of rainfall and the low-level atmospheric circulation around the Philippines.

In addition, it is well known that the seasonality of rainfall in the Philippines is largely affected by the El Niño–Southern Oscillation (ENSO) (Lyon et al. 2006; Lyon and Camargo 2009; Moron et al. 2009). For example, Ropelewski and Halpert (1987) showed that the El Niño brings about drier conditions around the Philippines between June and November. Using daily rainfall data collected at 35 stations between 1950 and 2010, Villafuerte et al. (2014) also indicated that significantly drier (wetter) conditions appear in the Philippines during the seasons of the El Niño (La Niña) close to the mature ENSO. Cruz et al. (2013) reported a decreasing trend in the southwest monsoon rainfall (June–September) in the Philippines between 1961 and 2010 and suggested a change in climate related to the long-term variability in ENSO. Moreover, Kubota et al. (2016) investigated the longer term variability in the Pacific–Japan (PJ) pattern and its relation to the climates of East Asia and the WNP summer monsoon from 1897 to 2013. They pointed out that the ENSO–PJ relationship had varied on interdecadal time scales and contributed to the interdecadal modulation of climate in the summer of the WNP. However, the variation of the seasonal march pattern of rainfall in relation to ENSO has rarely been studied. This study also discusses the relationship between the interannual and long-term variability in the seasonal march patterns of rainfall and ENSO.

The data and methodology used in this study are described in the next section. The dominant pattern in the spatial and temporal variability of rainfall and its classification are presented. Furthermore, we reveal the characteristics of each seasonal march pattern of the summer rainy season and its relationship with low-level atmospheric circulation. Finally, we discuss the relationship with ENSO and present a summary of the findings of this study.

Methods/Experimental

In this study, we used daily rainfall data between 1951 and 2012 provided by the Philippine Atmospheric, Geophysical and Astronomical Services Administration (PAGASA). Data collected at 35 stations, which included more than 80% of the daily data for the period of 1951 to 2012, were used. Missing rainfall data were not filled in. To investigate the seasonality of rainfall, excluding daily rainfall variations, pentad rainfall data were calculated when missing data of one pentad were less than 1 day.

Firstly, to detect dominant temporal and spatial patterns in the seasonal march of rainfall, an empirical orthogonal function (EOF) analysis was applied to the pentad rainfalls for 30 stations from 1951 to 2012, which had less than 5% missing values. Missing pentad rainfall data were replaced by the 62-year mean values. To consider the statistical distribution, pentad rainfall data were normalized by a cubic root and then standardized for the analysis. Second, to classify the seasonal march patterns of rainfall and reveal the interannual and long-term variability, cluster analysis, using the Euclidean distance and Ward's method, was applied to the time coefficients of the first two EOF modes. In this study, the cluster analysis was stopped at a point where the distances between the merged clusters increased remarkably (Wilks 2006). The characteristics of each EOF mode are described in further detail in the next section. Then, to show the characteristics of seasonality of rainfall exhibited by each cluster, composite analyses were applied to the pentad rainfall for the years classified in each cluster.

The 2.5° × 2.5° grid data of wind fields and geopotential heights at the 850 hPa level from the National Centers for Environmental Prediction/National Center for Atmospheric Research (NCEP/NCAR) reanalysis (Kalnay et al. 1996) from 1951 to 2012 were used. The pentad mean data of wind fields and geopotential heights were calculated using daily data in the same manner as that for rainfall data. To show the key areas and periods in the low-level atmospheric circulation related to the interannual variability in the seasonal march of rainfall, correlation and composite analyses were primarily used. The Oceanic Niño Index (ONI, http://origin.cpc.ncep.noaa.gov/products/analysis_monitoring/ensostuff/ONI_v4.shtml), which is an index used to define ENSO periods shown by the Climate Prediction Center, National Oceanic and Atmospheric Administration (NOAA), was used to describe the relationship with ENSO. The ONI indicates the 3-month running mean of SST anomalies averaged in the Niño 3.4 region (5° N–5° S and 120° W–170° W) calculated from the Extended Reconstructed Sea Surface Temperature (ERSST) version 4 (Huang et al. 2015). In this study, the ONI values for the period of October–December (OND) and January–March (JFM) were utilized because ENSO events that developed and continued for more than 1 year reached a mature phase during these periods (Tomita and Yasunari 1993).

Results and discussion

Temporal and spatial variability of the seasonal march of rainfall

The result of the EOF analysis of pentad rainfall for the period of 1951–2012 showed two dominant EOF modes, as was the case discussed by Akasaka (2010) who applied the EOF analysis of pentad rainfall to data from 39 stations in the Philippines from 1961 to 2000. The combined contribution of the first and second EOF modes (EOF1 and EOF2) was more than 50% since EOF1 and EOF2 accounted for 34.6 and 16.9% of the total variance, respectively.

The spatial pattern of EOF1 indicated a positive sign throughout the Philippines, and high loadings appeared, especially in the western part, at around 10° N–16° N (Fig. 1a). The mean time coefficient of EOF1 averaged for the period between 1951 and 2012 changed sign around the 29th pentad (late May) and the 71st pentad (mid-December), as shown by the red and blue inverse triangles in Fig. 1b. These results indicate that the summer rainy season starts simultaneously in the entire Philippines around late May and withdraws toward mid-December. These characteristics appear particularly in the western part around 10° N–16° N, where the windward side of the southwest monsoon is located. Lower factor loadings of less than 0.5 appeared in the southeastern part of the Philippines, corresponding to the area where it has the rainfall peak for November–January by the northeast monsoon and the orographic effect as shown by Akasaka et al. (2007). That is, EOF1 mainly shows consecutive rainfall variations from late May to mid-December, which are characterized by the onset and withdrawal of the southwest monsoon.

On the other hand, the spatial pattern of EOF2 showed a different signal between the northwestern part of Luzon Island and the eastern part of the Visayas and Mindanao (Fig. 1c). The mean time coefficients of EOF2 changed sign around the 20th pentad (early April) and the 57th pentad (early October) and showed the peak of the positive and negative signs around the 44th–46th pentads (early and mid-August) and from the 65th pentad (mid-November) to the 2nd pentad (early January), respectively (Fig. 1d). These characteristics correspond to regional differences in the period of the peak rainfall. As shown by Akasaka et al. (2007), this is because the peak rainfall in the northwestern part is mainly influenced by the southwest monsoon along with the development of the monsoon trough and because the peak rainfall in the southeastern part is mainly influenced by the northeast monsoon and the easterly wind brought from the subtropical high. In this study, as discussed in the next section, we focus on the interannual variability in the seasonal patterns related to the summer rainy season across the entire Philippines as shown by EOF1.

Seasonal march patterns of the summer rainy season and its long-term variability

To clarify the interannual variability in the seasonal march of rainfall, cluster analysis was applied to the time coefficients of the first two EOF modes. The seasonal march of the summer rainy season stemming from EOF1 showed remarkably distinct differences between each cluster and the long-term changes. Therefore, the characteristics of the patterns of the time coefficients of EOF1 were the primary focus of this study. Figure 2 shows the dendrogram of the cluster analysis for the time coefficients of EOF1 and the corresponding plot of the distance between merged clusters in the cluster

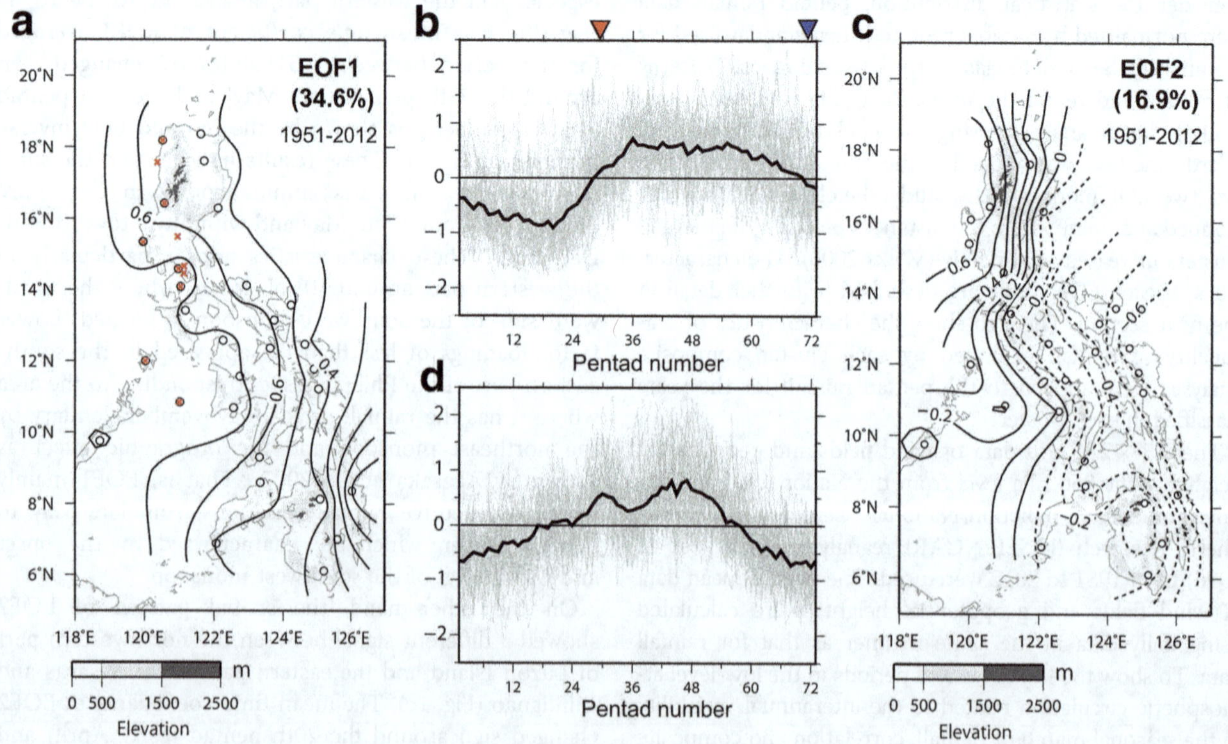

Fig. 1 Factor loadings and the time coefficients of EOF1 and EOF2. The contour interval is 0.1 in **a** and **c**. The open circles in **a** and **c** indicate the observation stations used for the EOF analysis. The red cross marks in **a** indicate the observation stations used in Fig. 4. The thick and thin lines in **b** and **d** indicate the mean time coefficients averaged for 1951–2012 and the time coefficients in each year, respectively. The red and blue inverse triangles in **b** denote the first pentad (the 29th and 71st pentads) when the sign of the mean time coefficients for the EOF1 changed

analysis. We stopped the process in the cluster analysis before the 58th stage because the distance between the merged clusters showed discontinuity after that stage. On the basis of these results, we classified the patterns in the time coefficients of the EOF1 for the period of 1951–2012 into six clusters (Table 1).

The mean time coefficients of EOF1 for each cluster are shown in Fig. 3. In addition, the mean seasonal marches of rainfall in the western coastal region for each cluster are shown in Fig. 4. The observation stations, used in Fig. 4 and shown as red cross marks in Fig. 1a, correspond to those with high loadings in the spatial pattern of EOF1 and those where the windward side of the southwest monsoon is located. Therefore, the seasonal march of the summer rainy season for this area shows more sharp differences in each cluster.

The averaged time coefficients of cluster 1 (C1) displayed a normal pattern similar to that for the averaged 62 years, as denoted by the dotted line in Fig. 3. In the seasonal march of rainfall pattern of C1 (Fig. 4a), the onset of the summer rainy season, indicated by a red inverse triangle, started around the 27th pentad (mid-May), with the peak rainfall appearing between the 44th and 46th pentads (early and mid-August). The summer rainy season gradually began to withdraw from the northern to southern parts of the region toward the 69th pentad (early December). Note that the region north of 16° N only had one peak rainfall amount, and the seasonal march of the rainy season for any other cluster did not display such characteristics.

The mean time coefficients for cluster 2 (C2) changed sign dramatically around the 26th pentad (early May; Fig. 3b). This corresponds to the abrupt increase in rainfall amounts around early May when the rainy season begins earlier than average (Fig. 4b). Furthermore, the peaks in rainfall appeared intermittently to the north of 14° N around mid-May, late June, and late August. The mean time coefficients changed sign from positive to negative at almost the same time compared with the 62-year mean.

In the mean time coefficients for cluster 3 (C3), the change from negative to positive occurred almost simultaneously with the 62-year mean (Fig. 3c). The positive value started to become negative around the 60th pentad (late October), earlier than the average for 62 years. As shown in Fig. 4c, this corresponds to a pattern of earlier withdrawal of the summer rainy season around late October.

Cluster 4 (C4) shows an anomalous pattern in which the mean time coefficients largely vary from the 3rd pentad (mid-January) to the 24th pentad (late April)

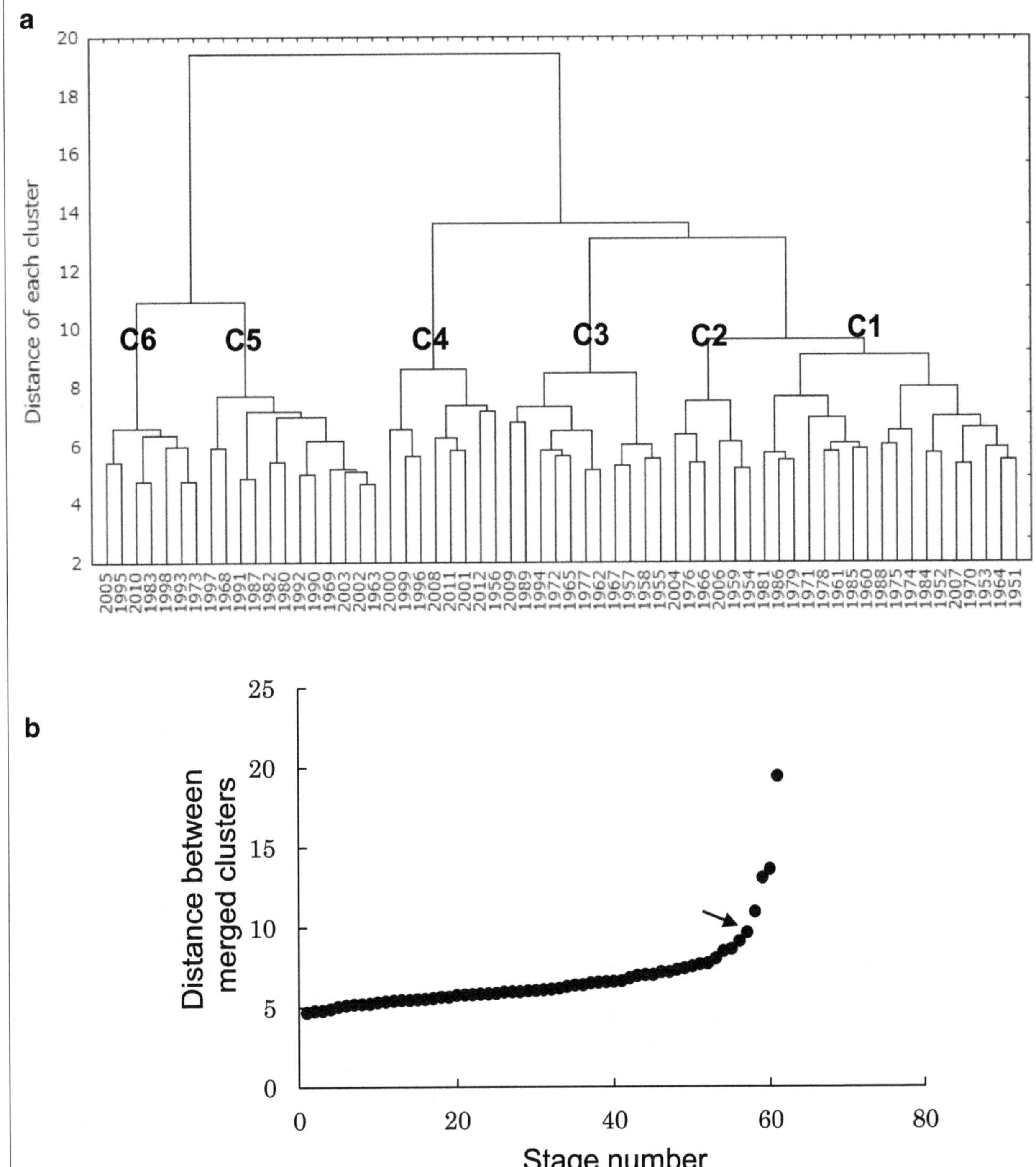

Fig. 2 Dendrogram of cluster analysis to the time coefficients of EOF1 (**a**) and the corresponding plot of the distance between the merged clusters (**b**). The arrow in **b** indicates the stage number just before distances between the merged clusters become large in the process of cluster analysis

toward the positive phase, higher than the 62-year mean time coefficients (Fig. 3d). Such variations were not shown in the mean time coefficients for any other clusters for the period. This indicates that there exists an indistinct dry season between January and April as compared to the 62-year mean (Fig. 4d). The summer rainy season started around the 24th pentad (late April), earlier than the 62-year mean, and withdrew toward the end of the year. Moreover, in Fig. 4d, it is also noteworthy that the period of the peak rainfall was

Table 1 Patterns classification in the time coefficients of EOF1 for 1951–2012 based on cluster analysis

Patterns in the time coefficients of EOF1	Years
C1	1951, 1952, 1953, 1960, 1961, 1964, 1970, 1971, 1974, 1975, 1978, 1979, 1981, 1984, 1985, 1986, 1988, 2007
C2	1954, 1959, 1966, 1976, 2004, 2006
C3	1955, 1957, 1958, 1962, 1965, 1967, 1972, 1977, 1989, 1994, 2009
C4	1956, 1996, 1999, 2000, 2001, 2008, 2011, 2012
C5	1963, 1968, 1969, 1980, 1982, 1987, 1990, 1991, 1992, 1997, 2002, 2003
C6	1973, 1983, 1993, 1995, 1998, 2005, 2010

significantly prolonged, with large rainfall periods from the 36th to the 54th pentads (late June to late September) to the north of 14.5° N compared with any other clusters.

In the mean time coefficients for cluster 5 (C5), a change into a negative sign occurred around the 67th pentad (late November), earlier than the 62-year mean time coefficients (Fig. 3e). The mean time coefficients for C5 fell remarkably below those averaged for 62 years from the 9th pentad to the 26th pentad (early February to early May). These correspond to an earlier withdrawal and a subsequent distinct dry season (Fig. 4e). The summer rainy season started at almost the same time as the 62-year mean. Additionally, the clear double peaks in the rainfall amounts during the 40th–52nd pentads (mid-July to mid-September) are shown to the north of 14.5° N.

The mean time coefficients for cluster 6 (C6) fell toward the negative from the 1st pentad (early January) and continued to be significantly below zero until the 30th pentad (late May; Fig. 3f). That is, both C5 and C6 denote a distinct dry season before the monsoon onset. Additionally, the change in the sign from negative (positive) to positive (negative) is delayed compared to the 62-year average (Fig. 3f). The changes indicate that an increase (decrease) in the amount of rainfall associated with the onset (withdrawal) of the summer rainy season tended to be delayed (Fig. 4f). Moreover, peak rainfall was not clearly shown to the north of 15.5° N.

Figure 5a reveals the notable long-term variability in the seasonal march pattern of the summer rainy season. Before 1990, C1 and C3 were frequently observed, and C4 and C6 appeared only rarely in 1956, 1973, and 1983 before 1990. In particular, C1 was a normal pattern for the seasonal march of the summer rainy season prior to 1990. However, C1 appeared only in 2007 after 1990. Thus, the dominant patterns, which appeared more frequently, have changed from C1 and C3 to C4, C5, and C6 since the early 1990s. It is noteworthy that C4 and C6 have appeared frequently since 1996 and 1993, respectively, whereas C5 has sometimes been observed before the early 1990s. This means that an anomalous pattern, which has an indistinct dry season and a prolonged peak rainfall shown by C4, has appeared frequently since the late 1990s. The opposite pattern, with a distinct dry season notably shown by C5 and C6, has also been frequently observed since the early 1990s. In particular, C6 showed an anomalous pattern, characterized by both delayed onset and withdrawal of the rainy season, and has frequently appeared since the early 1990s. That is, the seasonal march pattern in the summer rainy and dry seasons has changed since the early 1990s.

Characteristics of the low-level atmospheric circulation related to the seasonal march of the summer rainy season

To clarify the relationship between the seasonal march of the summer rainy season and the low-level atmospheric circulation fields, the correlation maps between the time coefficients of EOF1 and the geopotential heights and wind fields at the 850 hPa level are shown for the period between 1951 and 2012 in Fig. 6. The area with a significant negative correlation in the geopotential height widely spreads from India to the Philippines, especially around the center of the SCS and the Philippines (Fig. 6a). Similarly, but slightly to the south, the area with a significant positive correlation in the zonal component of the wind field was broadly distributed from around the south of India to the south of the Philippines (Fig. 6b). In the meridional component of the wind field, significant positive correlations were dispersed around the Bay of Bengal and from the southern part of the SCS to the northeastern part of the Philippines (figure not shown). This suggests that the interannual variations in the time coefficients of EOF1 were closely related to those in South and Southeast Asian summer monsoons because the areas with a significant correlation of the same sign widely spread from around India to the Philippines. This was especially true for the geopotential height and the zonal component of the wind field: the lower (higher) geopotential height and the strengthened (weakened) westerly winds around South and Southeast Asia not only led to an increase (decrease) in the amount of rainfall but also changed the seasonal march of the summer rainy season throughout the Philippines.

Next, in order to discuss the variability in the geopotential heights and wind fields related to the anomalous patterns in the seasonal march of the summer rainy season, time-latitude cross sections of the lower atmospheric circulation averaged between 120° E and 125° E are shown for C4, C5, and C6 in Figs. 7, 8, and 9. This longitudinal zone, including the Philippines, corresponds to that where higher correlations were revealed in the correlation maps shown in Fig. 6. In particular, the seasonality of the lower

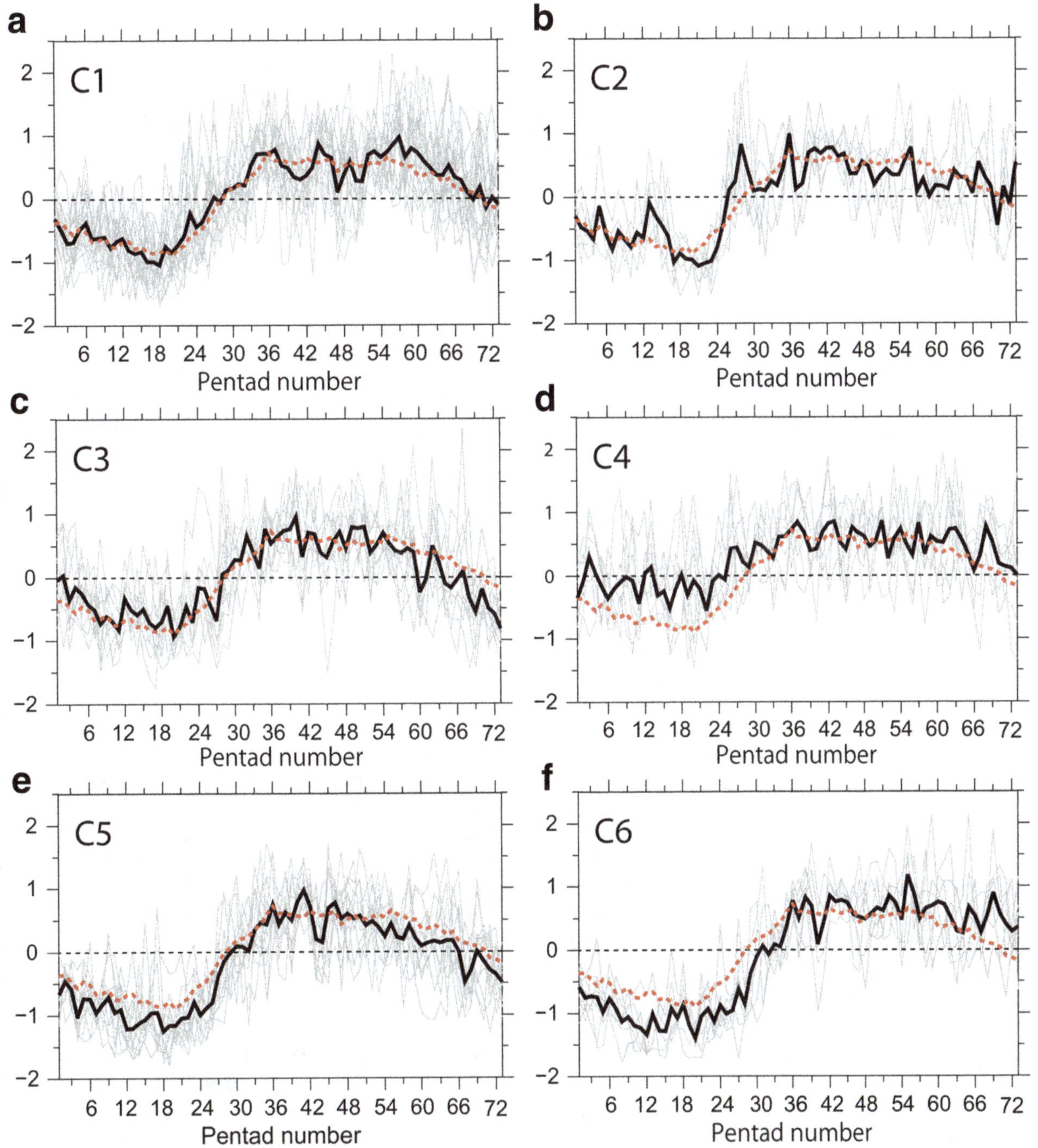

Fig. 3 a–**f** Classification of patterns of the time coefficients of EOF1 using cluster analysis. The dotted red line and thick solid black line indicate the mean time coefficients averaged for 1951–2012 and for the years classified in each cluster (**a** C1, **b** C2, **c** C3, **d** C4, **e** C5 and **f** C6), respectively. The gray thin line indicates the time coefficients in each year

atmospheric circulation years classified in C4, C5, and C6 was compared to that classified in C1, which was the normal pattern prior to 1990. The latitudinal cross sections for C1 are shown in Fig. 7a, d, g.

In the years classified in C4, the geopotential heights were significantly lower around 15° S–15° N from the first to the 18th pentad (January–March; Fig. 7b, c). The easterly and northerly winds were significantly weakened between the equator and 15° N for this period compared to C1 (Fig. 7e, f and h, i). The wind direction of the zonal (meridional) component started to change into a westerly (southerly) wind between the equator and

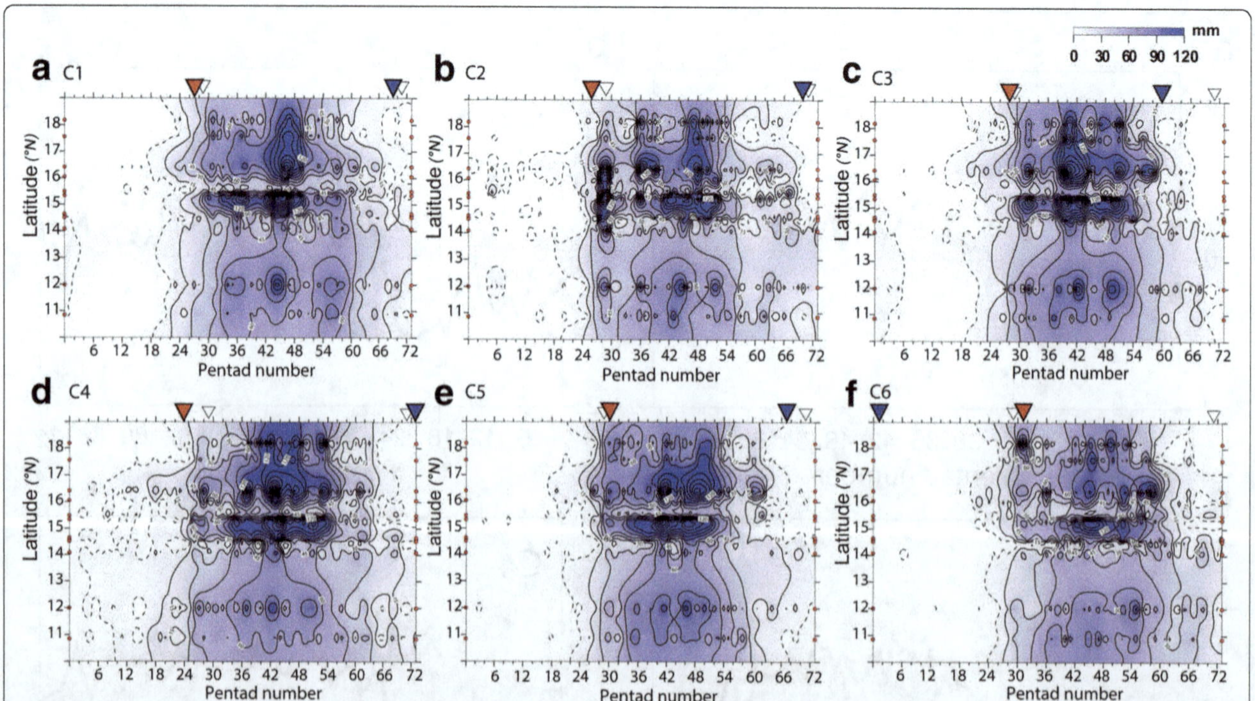

Fig. 4 a–f Time-latitude cross section of the pentad rainfall along the western coastal region in each cluster (**a** C1, **b** C2, **c** C3, **d** C4, **e** C5 and **f** C6). The years used for the composite analyses are listed in Table 1. The red circles on the vertical lines correspond to the latitude of the observation stations shown in Fig. 1a as red cross marks. The red (blue) inverse triangles on the horizontal line indicate the first pentad when the sign of the time coefficients changed after (before) the 18th pentad from negative to positive (from positive to negative) for each cluster. The white inverse triangles indicate the same as Fig. 1b. The contour interval is 15 mm, and the dotted line indicates a contour of 5 mm

20° N around the 26th pentad (early May), although the onset of the southwesterly wind is in mid-May on average, as shown by Akasaka et al. (2007) and Moron et al. (2009). These characteristics in the lower atmospheric circulation anomalies correspond to the weakened subtropical high from January to March and the earlier onset of the southwesterly wind. Thus, it is suggested that these lead to an indistinct dry season during January–March and an earlier onset of the rainy season in the years classified in C4. Additionally, strong westerly winds were seen around 15° N not only for the 41st–44th pentads (late July to early August) but also for the 48th–54th pentads (late August to late September; Fig. 7e, f); the period for strong westerly winds is longer than C1 (Fig. 7d). This corresponds to the prolonged peak rainfall with a large rainfall as shown in Fig. 4d.

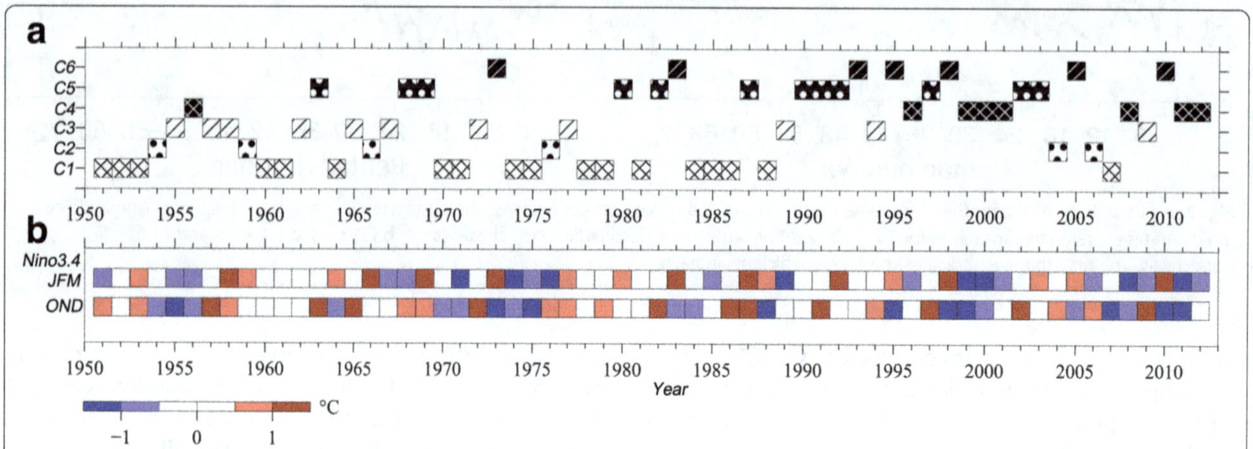

Fig. 5 Long-term variability in the seasonal march pattern of the summer rainy season. **a** A diagram of each cluster for the period of 1951–2012. **b** ONI (the 3-month mean SST anomalies averaged in Niño 3.4) for January–March (JFM) and for October–December (OND)

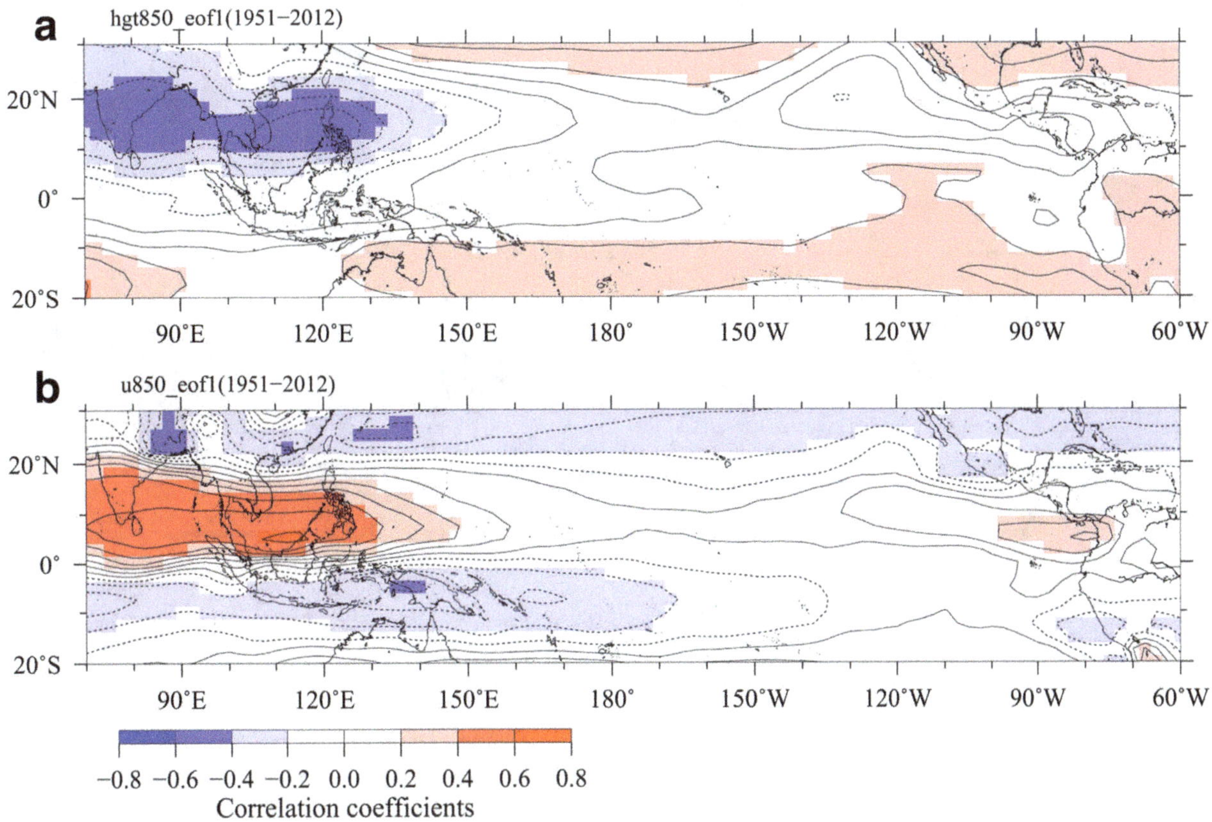

Fig. 6 Correlation maps between the time coefficients of EOF1 and geopotential height (**a**) and zonal component of the wind field (**b**) at the 850 hPa level from 1951 to 2012. The contour interval is 0.1. The solid and dashed lines indicate positive and negative correlation coefficients, respectively. Correlation coefficients above 0.1 or below − 0.1 indicate a significant correlation at the two-sided 99% level according to Student's *t* test

In the years classified in C5, it is noteworthy that the broad and continuous significant positive geopotential height anomalies, indicating a strengthened subtropical high, were observed from 10° S to 25° N especially for the 10th–12th pentads (mid-February to the beginning of March) and the 52nd–56th pentads (mid-September to early October; Fig. 8a, b). It has been suggested that the subtropical high is stronger and the convective activity is suppressed around the Philippines, especially for these periods. The westerly winds started to blow throughout the Philippines around the 32nd pentad (early June; Fig. 8c), although the onset of the westerly wind was the 30th pentad (late May) in C1 (Fig. 7d). For the 40th–42nd pentads and the 48th–52nd pentads (mid-July to late July and late August to mid-September), the westerly wind strongly blew from 5° N to 20° N (Fig. 8c, d). This indicates that peak rainfall clearly appears during this period, as shown in Fig. 4e. Additionally, it is noteworthy that the northerly wind of the meridional component started to blow around the northern part of the Philippines from the 50th pentad (early September), which is earlier than C1 (Fig. 8e, f). The northerly wind then progressed toward the south until the 54th pentad (late September). Significantly high geopotential height anomalies were also shown around 20° S–20° N after the 52nd pentad (mid-September; Fig. 8b). As a result, C5 indicates an earlier withdrawal of the summer rainy season and a distinct dry season.

In the years classified in C6, the higher geopotential height was located around 15° N–20° N, which corresponds to the latitude of the Luzon Island, around the 1st–28th pentads (January to mid-May; Fig. 9a). The significant positive anomalies of the geopotential heights appeared between 20° S and 20° N, especially from the 7th–10th pentads (early and mid-February; Fig. 9b). C5 also showed similar characteristics (Fig. 8b). This means that both C5 and C6 have the distinct dry season before the monsoon onset (Fig. 4e, f). The distinct dry season shown in C5 and C6 is closely related to the strengthened subtropical high around the Philippines, especially for February–March (Additional file 1: Figure S11). In the wind fields, it is noteworthy that the easterly wind continued around the Philippines until the 40th pentad (mid-July), and subsequently a weak westerly wind started to blow (Fig. 9c). The negative anomalies of the zonal wind were significantly and continuously shown

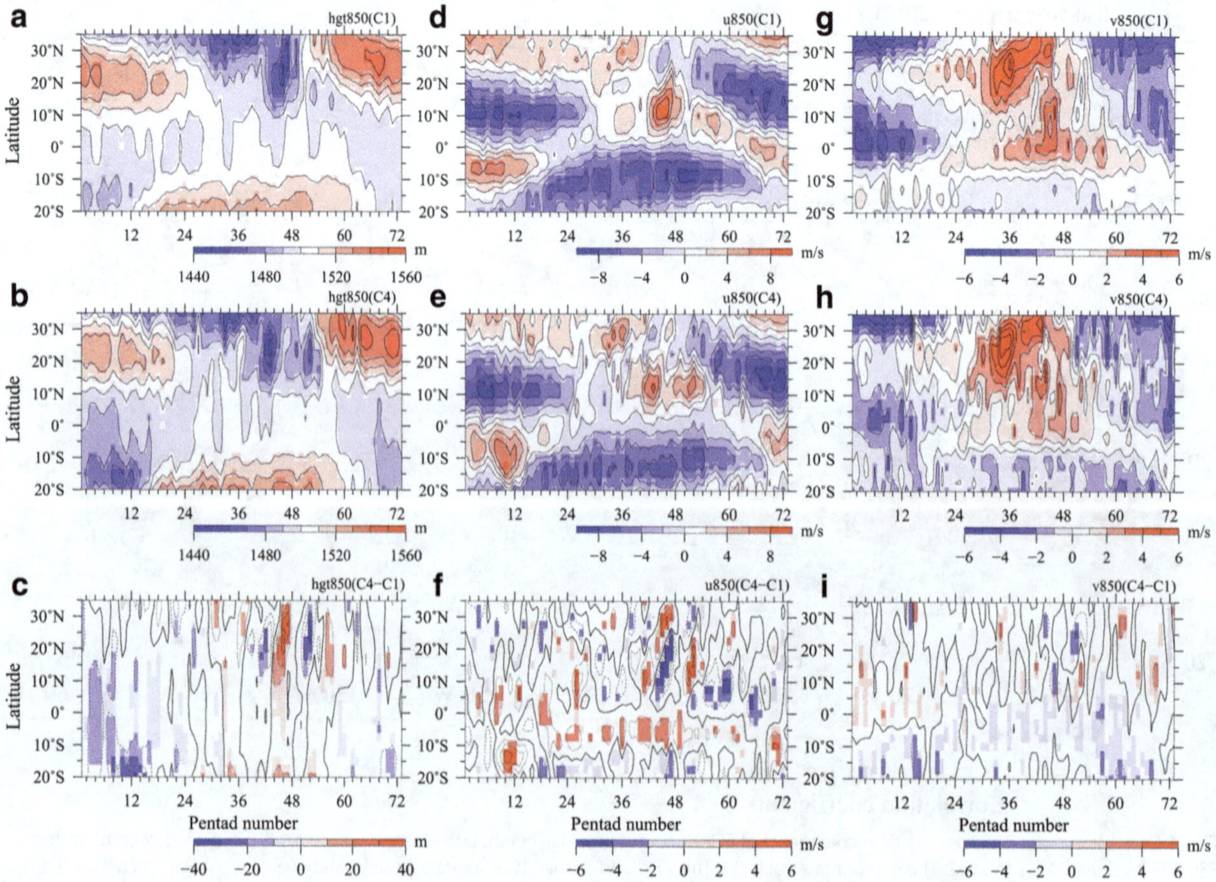

Fig. 7 Time-latitude cross section of the pentad mean geopotential height and zonal and meridional components of the wind fields at the 850 hPa level averaged for the years classified in C1 and C4. The figures in the left column indicate geopotential heights for C1 (**a**) and C4 (**b**) and the anomalies between C4 and C1 (**c**). The figures in the middle and right columns indicate zonal and meridional components of the wind field for C1 (**d**, **g**) and C4 (**e**, **h**) and the anomalies between C4 and C1 (**f**, **i**), respectively. The colored grids in **c**, **f**, and **i** indicate the significant anomalies at the two-sided 90% level according to Student's *t* test

around the Philippines from the 21st to the 47th pentads (mid-April to late August; Fig. 9d). Regarding this, the southerly wind of the meridional component started to blow around the Philippines from the 7th pentad (the beginning of February), earlier than that for C1 (Fig. 9e, f). The negative anomalies of the meridional wind were observed around 5° S–0° N during the 20th–60th pentads (early April to late October; Fig. 9f). Significant positive anomalies of geopotential height were observed around 15° N–30° N from the 42nd to the 48th pentads (late July to late August; Fig. 9b). That is, from the beginning of February to early July, the southeasterly winds related to the strengthened subtropical high continue to blow around the Philippines. In mid-July, the summer rainy season starts with a weakened southwest monsoon around the Philippines. Subsequently, the northerly wind of the meridional component started to blow around the entire Philippines from the 54th to the 60th pentads (late September to late October), slower than that for C1 (Fig. 9e). The onset of the northerly winds led to the withdrawal of the summer rainy season as shown by Akasaka et al. (2007). With such changes in the lower atmospheric circulation, in the year classified as C6, there were a distinct dry season, a delayed onset and withdrawal of the summer rainy season, and indistinct peaks in the summer rainfall amounts.

Relationship between the seasonal march of the summer rainy season and ENSO

In order to discuss the relationship between the interannual and long-term variability in the seasonal march of the summer rainy season and the lower atmospheric circulation, correlation maps between the geopotential heights and the time coefficients of EOF1 for the years classified in each cluster are shown in Fig. 10. The correlation patterns for C1–C3 have similar characteristics to the correlation map for 1951–2012 shown in Fig. 6a (Fig. 10a–c). However, the area with a higher correlation in C4 is limited only around the Philippines as compared to the correlation maps for the other clusters (Fig. 10d).

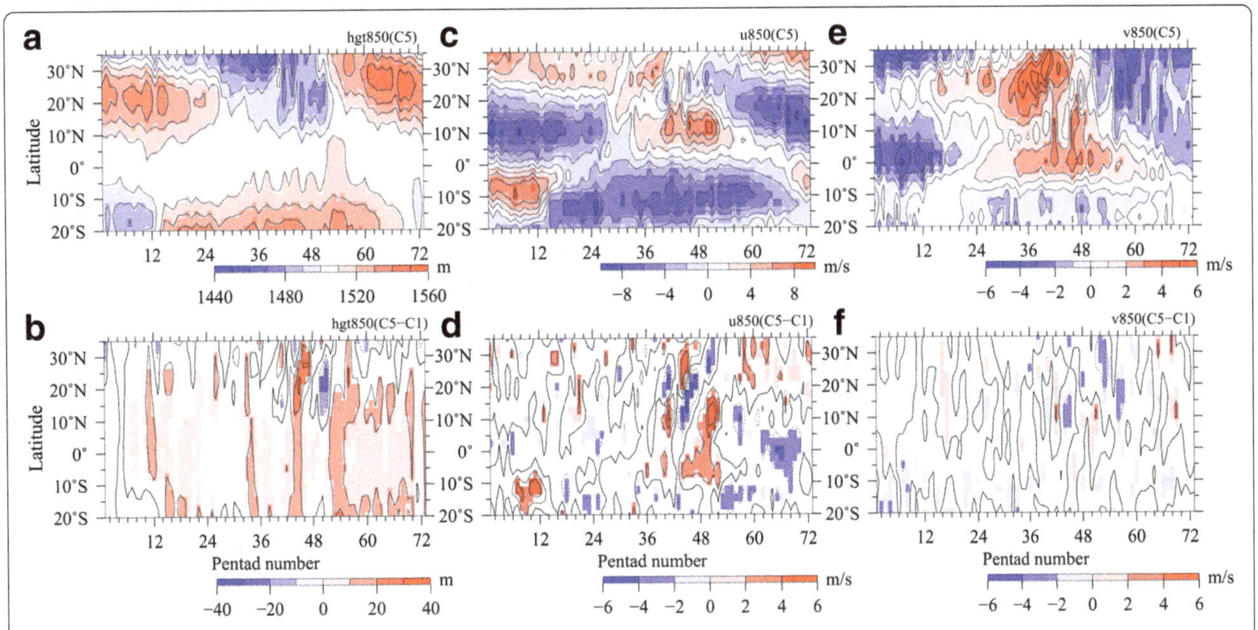

Fig. 8 The same as Fig. 7 except for C5. Figures related to C1 are omitted

Conversely, in the correlation maps for C6, the area with a higher negative correlation mostly appeared from India to the Philippines, and the area with the highest negative correlation coefficient was revealed especially around the Philippines (Fig. 10f). On the other hand, areas with a higher positive correlation were broadly expressed around the tropical eastern Pacific. The east–west contrast of the correlation over the tropical Pacific indicates that C6 is a pattern related to ENSO. The correlation map for C5 revealed a similar pattern to C6, although the positive correlation over the tropical eastern Pacific was lower than that for C6 (Fig. 10e). The correlation patterns between the wind fields and the time coefficients of EOF1 for C5 and C6 also showed such a contrast in correlation (figure not shown). Moreover, Fig. 5b shows that C5 tended to appear in the years when the El Niño condition occurred in JFM and/or OND, except in 1990; C6 tended to appear in the years when the El Niño condition continued from OND in the prior year to the following season (JFM), and La Niña developed for

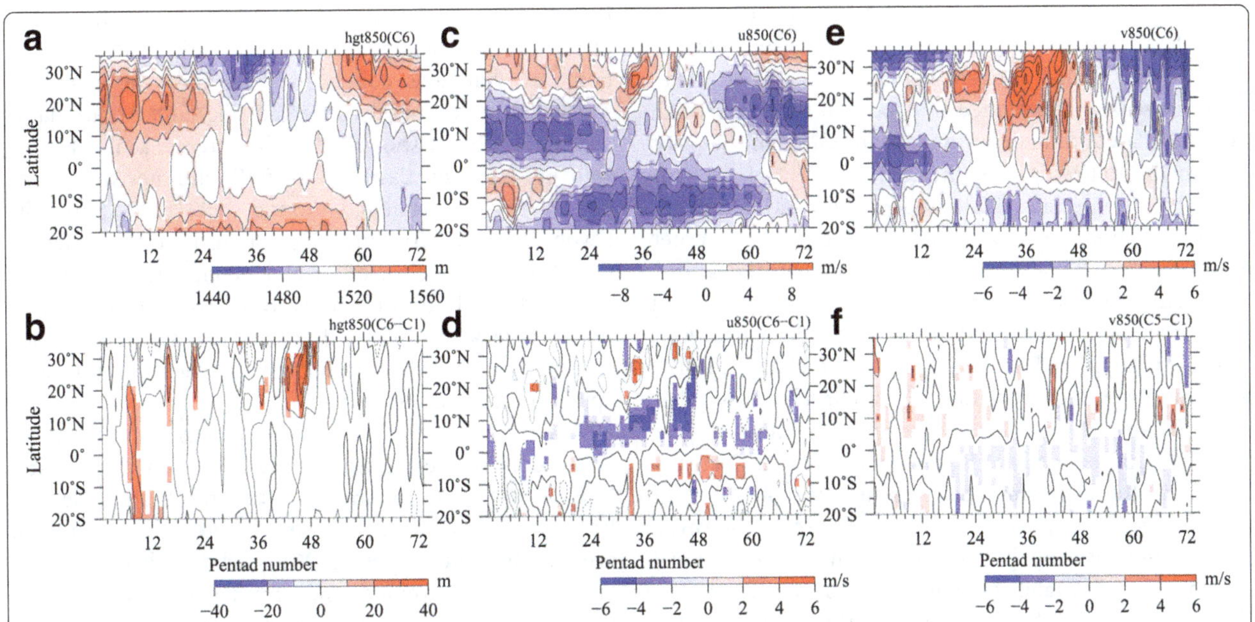

Fig. 9 The same as Fig. 7 except for C6. Figures related to C1 are omitted

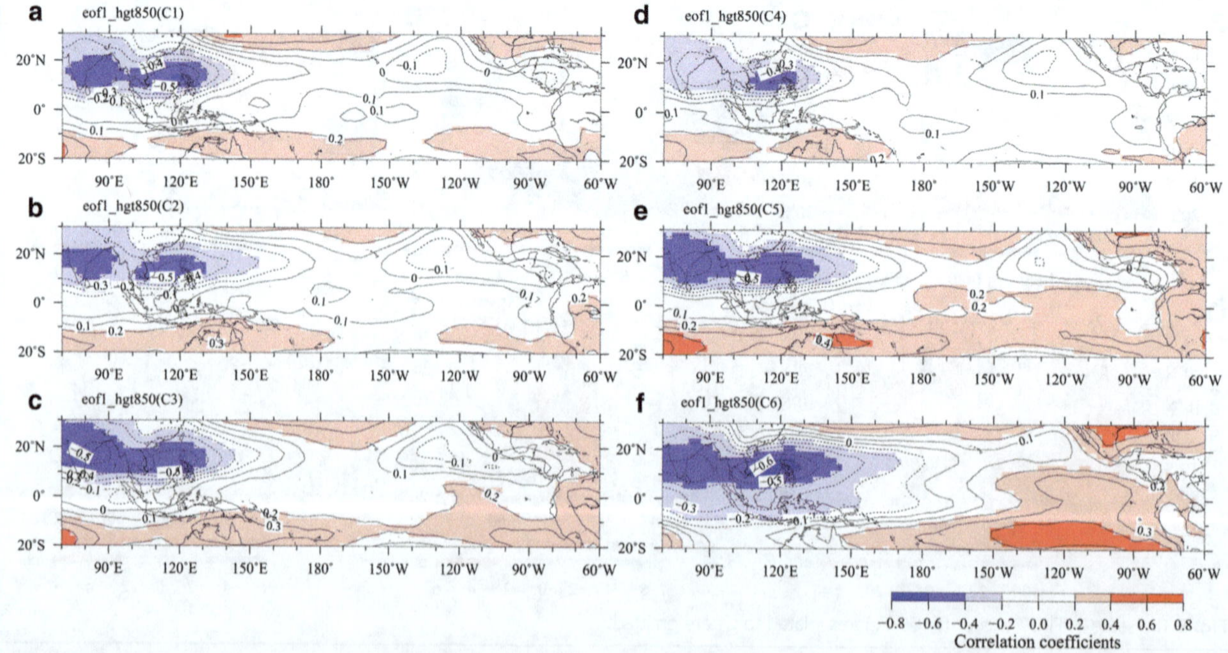

Fig. 10 Correlation maps between the time coefficients of EOF1 and geopotential height at the 850 hPa level during the year classified in **a** C1, **b** C2, **c** C3, **d** C4, **e** C5, and **f** C6. Correlation coefficients above 0.2 or below − 0.2 indicate a significant correlation at the two-sided 99% level according to Student's *t* test

OND, except in 1993. These results suggest that both C5 and C6 are closely related to ENSO.

As pointed out by earlier studies (e.g., Wang et al. 2000; Watanabe and Jin 2002), during the El Niño period and the following season, the Philippine Sea anticyclone developed and suppressed the convective activities over the WNP. Therefore, the seasonal marches of the summer rainy season in C5 and C6 might have revealed similar characteristics as a distinct dry season and the delayed onset of the summer rainy season. On the other hand, C4 had different characteristics as an indistinct dry season and tended to appear in the years when the La Niña condition continued from OND of the prior year to the following season (JFM) or OND of the year (Fig. 5b). Chou et al. (2003) studied the relationship between the interannual variability in the WNP summer monsoon and ENSO and pointed out that the strong WNP summer monsoon tended to occur during an El Niño developing year and a La Niña decaying year. Thus, it is suggested that C4 is related to changes in the lower atmospheric circulation by La Niña because C4 is observed in the La Niña decaying or continuing year (Fig. 5b). Consequently, it is suggested that ENSO greatly influences the anomalous patterns in seasonal march of rainfall like C4, C5, and C6. It is worth noting, however, that the influences of El Niño and La Niña on the seasonal march patterns of rainfall are not always symmetrical.

Moreover, anomalous patterns in the summer rainy season have frequently been observed especially since the early 1990s. One possible reason is an interdecadal or centennial variability in ENSO shown by some earlier studies. For example, Chung and Tim (2013) studied the interdecadal relationship between the trade wind and SST with the change in El Niño types over the tropical eastern Pacific for the period between 1980 and 2010. They pointed out that the mean state over the equatorial Pacific showed a regime change to the La Niña-like interdecadal mean state in 1998/1999 with the transition phase of the Pacific Decadal Oscillation (PDO). The point of change is coincident with the year when C4 started to be frequently observed (Fig. 5b). Additionally, Yim et al. (2008) showed that the interdecadal relationship between the East Asian–WNP summer monsoons and ENSO has changed around 1993/1994. Similarly, Kwon et al. (2005) showed the interdecadal change in the atmospheric circulation and precipitation over East Asia and WNP around the early 1990s. The time when the interdecadal shift in the land–air–sea interactions occurred is coincident with the year when C6 started to appear more frequently (Fig. 5b).

Meanwhile, on the basis of both longer observational and model results of sea surface level over the tropics, Vecchi et al. (2006) pointed out that the Walker circulation weakened from the late nineteenth century to around 2000. Kubota et al. (2017), who studied the influence of tropical cyclones and summer monsoon onset in the Philippines from 1903 to 2013, also pointed out a weak correlation between ENSO and the Philippine summer monsoon onset date since around the 1980s. Similarly, Chowdary et al. (2012), who studied interdecadal variations in ENSO teleconnections to the Indo-western Pacific for the period between 1870 and 2007, pointed out that the centennial modulation of ENSO teleconnections to the Indo-northwest Pacific region was correlated not with the PDO but rather with ENSO variance itself. These results by earlier studies suggest that the interdecadal or slower variability in ENSO is associated with the long-term change in the seasonal march pattern of rainfall for the period of 1951–2012. Thus, to clarify its relation with the interdecadal or slower variability in ENSO, it will be essential to study the relationship between the longer term variability in the seasonal march of rainfall and the slower modulation of air–sea interactions over the tropics.

Conclusions

This study showed the long-term variability in the seasonal march patterns of the summer rainy season in the Philippines from 1951 to 2012 for the first time. On the basis of the EOF analysis of the pentad rainfall, we focused on a dominant pattern in the temporal and spatial structures of rainfall shown by EOF1: a pattern related to the summer rainy season with increases in rainfall from mid-May and decreases in rainfall toward mid-December throughout the Philippines. The patterns obtained from the time coefficients of EOF1 were classified into six clusters by cluster analysis.

The six clusters showed notable long-term variability: three of the six clusters tended to appear more frequently since the early 1990s. Conversely, the normal pattern before 1990 (C1) appeared only in 2007. The three patterns revealed the following characteristics: (1) a pattern which has an indistinct dry season and a prolonged peak rainfall (C4), (2) a pattern which has a distinct dry season and a shortened summer rainy season (C5), and (3) a pattern with a distinct dry season and a delayed onset and withdrawal of the summer rainy season (C6).

In particular, C4 exhibited a significant anomalous pattern, showing an indistinct dry season for January–March. The indistinct dry season was closely related to significantly negative geopotential height anomalies around 15° S–15° N, especially from January to March. Conversely, C5 and C6 are distinguished by a dry season, and they were correlated to significantly positive geopotential height anomalies for February–March. This means that the duration and condition of the dry season are influenced by the strength and location of the subtropical high around the Philippines especially for February–March (Additional file 1: Figure S11). Moreover, C4 has an earlier onset of the summer rainy season related to an earlier onset of the westerly winds and the prolonged peak rainfall related to the strong westerly winds. Consequently, these changes in the lower atmospheric circulation have led to wetter conditions throughout the years classified in C4. This study suggests that changes in the lower atmospheric circulation that brings wetter conditions have increased in frequency since the late 1990s.

On the other hand, C6 shows drier conditions: there were not only a distinct dry season but also a delayed onset and withdrawal of the summer rainy season and indistinct peak rainfall with small rainfall amounts compared with any other patterns. The delayed onset (withdrawal) of the summer rainy season was closely correlated to the delayed onset of the westerly (northerly) wind in the zonal (meridional) component. This resulted in a shorter duration of the westerly winds. The indistinct peak rainfall was significantly connected to the weakened westerly winds around the Philippines during the summer monsoon. This study also revealed that these changes in the lower atmospheric circulation related to C6 have tended to occur since the early 1990s.

This study suggests that ENSO characterizes the seasonal pattern of the summer rainy season. This is because the correlation maps between the time coefficients of EOF1 and geopotential heights for C5 and C6 showed a pattern related to El Niño: they indicated an east–west contrast in the correlation between the WNP and the equatorial eastern Pacific. Moreover, El Niño tended to occur in JFM and/or OND of the years classified in C5 and C6. Conversely, La Niña tended to continue from OND of the prior year to the following season (JFM) or OND of the year classified in C4. This study showed that these three anomalous patterns in seasonal rainfall have appeared frequently since the early 1990s. This suggests that long-term changes in the seasonal pattern of the summer rainy season are greatly influenced by those in ENSO. To clearly understand the cause of the long-term variability in the seasonality of the summer rainy season, clarification regarding its relation to the interdecadal or slower modulation of ENSO in the near future is necessary.

Abbreviations

CMAP: Climate Prediction Center Merged Analysis of Precipitation; ENSO: El Niño–Southern Oscillation; EOF: Empirical orthogonal function; ERSST: Extended Reconstructed Sea Surface Temperature; NCEP/NCAR: National Centers for Environmental Prediction/National Center for Atmospheric Research; NOAA: National Oceanic and Atmospheric Administration; ONI: Oceanic Niño Index; PAGASA: Philippine Atmospheric, Geophysical and Astronomical Services Administration; PDO: Pacific Decadal Oscillation; PJ: Pacific–Japan; SCS: South China Sea; SST: Sea Surface Temperature; WNP: Western North Pacific

Acknowledgements

We would like to thank Professor Wataru Morishima, Nihon University for the helpful suggestions. We sincerely thank two anonymous reviewers for the fruitful and constructive comments to our manuscript.

Funding

This research was supported by the Green Network of Excellence program and the Data Integration and Analysis System through the National Key Technology,

Marine Earth Observation Exploration System of Ministry of Education, Culture, Sports, Science and Technology, Japan. This research was also supported in part by the JSPS KAKENHI Grant-in-Aid for Scientific Research of the MEXT 26220202 (Leader: Jun Matsumoto) and 15K16283 (Leader: Ikumi Akasaka). JM was supported by the Tokyo Human Resources Fund of Tokyo Metropolitan University from the Tokyo Metropolitan Government.

Authors' contributions
IA proposed the topic and designed the study. EC, GR, and FH prepared the rainfall data. IA analyzed the data. HK and JM helped in the interpretation of data and collaborated with the corresponding author in the construction of manuscript. All authors read and approved the final manuscript.

Authors' information
IA is an associate professor at Senshu University. HK is an associate professor at Hokkaido University, and JM is a professor at Tokyo Metropolitan University. EO, GR, and FH are researchers at PAGASA.

Competing interests
The authors declare that they have no competing interests.

Author details
[1]Department of Geography, Senshu University, 2-1-1 Higashi-Mita, Tama-ku, Kawasaki, Kanagawa 214-8580, Japan. [2]Hokkaido University, Kita 10, Nishi 8, Kita-ku, Sapporo, Hokkaido 060-0808, Japan. [3]Department of Geography, Tokyo Metropolitan University, 1-1 Minami-Osawa, Hachioji, Tokyo 192-0397, Japan. [4]Department of Coupled Ocean-Atmosphere-Land Processes Research, Japanese Agency for Marine-Earth Science and Technology,, Yokosuka, Kanagawa 237-0061, Japan. [5]Philippine Atmospheric, Geophysical and Astronomical Services Administration, Science Garden Complex, Diliman, 1101 Quezon City, Philippines.

References
Akasaka I (2010) Interannual variations in seasonal march of rainfall in the Philippines. Int J Climatol 30:1301–1314. https://doi.org/10.1002/joc.1975.

Akasaka I, Morishima W, Mikami T (2007) Seasonal march and spatial difference of rainfall in the Philippines. Int J Climatol 27:715–725. https://doi.org/10.1002/joc.1428

Cayanan EO, Chen T, Argete J, Ten M, Nilo P (2011) The effect of tropical cyclones on southwest monsoon rainfall in the Philippines. J Meteor Soc Jpn 89:123–139. https://doi.org/10.2151/jmsj.2011-A08

Chang CP, Wang Z, Mcbride J, Liu CH (2005) Annual cycle of Southeast Asia—maritime continent rainfall and the asymmetric monsoon transition. J Clim 18:287–301. https://doi.org/10.1175/JCLI-3257.1

Chou C, Tu JY, Yu JY (2003) Interannual variability of the western North Pacific summer monsoon: differences between ENSO and non-ENSO years. J Clim 16:2275–2287. https://doi.org/10.1175/2761.1

Chowdary JS, Xie SP, Tokinaga H, Okumura YM, Kubota H, Johnson N, Zheng XT (2012) Interdecadal variations in ENSO teleconnection to the Indo-western Pacific for 1870–2007. J Clim 25:1722–1744. https://doi.org/10.1175/JCSI-D-11-0070.1.

Chung PH, Tim L (2013) Interdecadal relationship between the mean state and El Niño types. J Clim 26:361–378. https://doi.org/10.1175/JCLI-D-12-00106.1

Coronas J (1920) The climate and weather of the Philippines, 1903–1918. Manila Observatory, Philippines

Cruz FT, Narisma GT, Villafuerte MQ II, Chua KC, Olaguera LM (2013) A climatological analysis of the southwest monsoon rainfall in the Philippines. Atmos Res 122:609–616. https://doi.org/10.1016/j.atmosres.2012.06.010

Flores JF, Balagot VF (1969) Climate of the Philippines. In: Arakawa H (ed) Climate of northern and eastern Asia, World Survey of Climatology, vol 8. Elsevier, Amsterdam

Huang B, Banzon VF, Freeman E, Lawrimore J, Liu W, Peterson TC, Smith TM, Thorne PW, Woodruff SD, Zhang HM (2015) Extended Reconstructed Sea Surface Temperature version 4 (ERSST.v4): part I. Upgrades and intercomparisons. J Clim 28:911–930. https://doi.org/10.1175/JCLI-D-14-00006.1

Kajikawa Y, Wang B (2012) Interdecadal change of the South China Sea summer monsoon onset. J Clim 25:3207–3218. https://doi.org/10.1175/JCLI-D-11-00207.1

Kalnay E, Kanamitsu M, Kistler R, Collins W, Deaven D, Gandin L, Iredell M, Saha S, White G, Woolen J, Zhu Y, Chelliah M, Ebisuzaki W, Higgins W, Janowiak J, Mo KC, Ropelowski C, Wang J, Leetmaa A, Reynolds R, Jenne R, Joseph D (1996) The NCEP/NCAR 40-year reanalysis project. Bull Am Meteorol Soc 77:437–471

Kubota H, Kosaka Y, Xie SP (2016) A 117-year long index of the Pacific-Japan pattern with application to interdecadal variability. Int J Climatol 36:1575–1589. https://doi.org/10.1002/joc.4441

Kubota H, Shirooka R, Matsumoto J, Cayanan EO, Hilario FD (2017) Tropical cyclone influence on the long-term variability of Philippine summer monsoon onset. Progr Earth Planet Sci 4:27. https://doi.org/10.1186/s40645-017-0138-5

Kubota H, Wang B (2009) How much do tropical cyclones affect seasonal and interannual rainfall variability over the western North Pacific? J Clim 22:5495–5510. https://doi.org/10.1175/2009JCLI2646.1

Kwon MH, Jhun JG, Wang B, An SI, Kug JS (2005) Decadal change in relationship between east Asian and WNP summer monsoons. Geophys Res Lett 32:L16709. https://doi.org/10.1029/2005GL023026

Lyon B, Camargo SJ (2009) The seasonally-varying influence of ENSO on rainfall and tropical cyclone activity in the Philippines. Clim Dyn 32:125–141. https://doi.org/10.1007/s00382-008-0380-z

Lyon B, Cristi H, Verceles ER, Hilario FD, Abastillas R (2006) Seasonal reversal of the ENSO rainfall signal in the Philippines. Geophys Res Lett 33:L24710. https://doi.org/10.1029/2006GL028182

Moron V, Lucero A, Hilario F, Lyon B, Robertson AW, DeWitt D (2009) Spatio-temporal variability and predictability of summer monsoon onset over the Philippines. Clim Dyn 33:1159–1177. https://doi.org/10.1007/s00382-008-0520-5

Ropelewski CF, Halpert MS (1987) Global and regional scale precipitation patterns associated with the El Niño/Southern Oscillation. Mon Wea Rev 115:1606–1626

Tomita T, Yasunari T (1993) On the two types of ENSO. J Met Soc Japan 71:273–284

Vecchi GA, Soden BJ, Wittenberg AT, Held IM, Leetmaa A, Harrison MJ (2006) Weakening of tropical Pacific atmospheric circulation due to anthropogenic forcing. Nature 441:73–76. https://doi.org/10.1038/nature04744

Villafuerte MQII, Matsumoto J, Akasaka I, Takahashi HG, Kubota H, Cinco TA (2014) Long-term trends and variability of rainfall extremes in the Philippines. Atmos Res 137:1–13. https://doi.org/10.1016/j.atmosres.2013.09.021.

Wang B, Wu R, Fu X (2000) Pacific-East Asian teleconnection: how does ENSO affect East Asian climate? J Clim 13:1517–1536

Watanabe M, Jin F (2002) Role of Indian Ocean warming in the development of Philippine Sea anticyclone during ENSO. Geophys Res Lett 29:116.1–116.4. https://doi.org/10.1029/2001GL014318

Wilks DS (2006) Statistical methods in the atmospheric sciences, 2nd edn. Academic Press, Elsevier, Amsterdam

Yim SY, Jhun JG, Yeh SW (2008) Decadal change in the relationship between east Asian–western North Pacific summer monsoons and ENSO in the mid-1990s. Geophys Res Lett 35:L20711. https://doi.org/10.1029/2008GL035751

6

A 2700-year record of ENSO and PDO variability from the Californian margin based on coccolithophore assemblages and calcification

Luc Beaufort[1*] and Michaël Grelaud[2]

Abstract

The El Niño Southern Oscillation (ENSO) and the Pacific Decadal Oscillation (PDO) account for a large part of modern climate variability. Over the last decades, understanding of these modes of climate variability has increased but prediction in the context of global warming has proven difficult because of the lack of pertinent and reproducible paleodata. Here, we infer the dynamics of these oscillations from fossil assemblage and calcification state of coccolithophore in the Californian margin because El Niño has a strong impact on phytoplankton ecology and PDO on the upwelling intensity and hence on the ocean chemistry. Intense Californian upwelling brings water rich in CO_2 and poor in carbonate ions and coccolithophores secrete lower calcified coccoliths. Seasonally laminated sediments of the Santa Barbara Basin are used to document ENSO variability and PDO index for the last 2700 years at a temporal resolution of 3 years. The records present the same characteristics as other PDO or ENSO records from the same area spanning the last centuries. We are therefore confident on the value produced here for the last 2.7 millennia. The records show important centennial variability that is equivalent to solar cycles.

Keywords: El Niño Southern Oscillation, Pacific Decadal Oscillation, Past climate variability, Santa Barbara Basin, Solar cycles, Centennial climatic variability

Introduction

El Niño Southern Oscillation (ENSO) and its longer lived cousin, the Pacific Decadal Oscillation (PDO) (Zhang et al. 1997), are the primary sources of global interannual variability. While the ENSO is related to the interconnections between ocean and atmosphere in the tropical eastern Pacific, the PDO is a combination of different physical processes operating on different time scales, including remote tropical forcing (ENSO), oceanic thermal inertia, and atmospheric forcing in response to the Kuroshiyo-Oyashio dynamics (Newman et al. 2016). Yet, their past dynamics remain poorly understood and described. ENSO and PDO act respectively at interannual and decennial scales with variable periods ranging respectively from 2.5 to 7 years and from 20 to 30 years (Mantua et al. 1997; Philander 1983). The impact of ENSO and PDO on the climate of the 20th century, particularly the unusual strong events of 1982–83 and 1997–98 (El Niño) and the climate regime shifts of 1946 and 1977 (PDO), led to a better assessment and understanding of these cycles. However, understanding their changing intensity and frequency during the 21st century is important issues which need to be resolved if we want to predict how these oscillations will be affected by new boundary conditions. By combining recent and past marine sedimentary archives, we aim to monitor the natural variability of these oscillations during the past 2700 years.

The Santa Barbara Basin (SBB) is a sensitive recorder of past climate variability (Behl and Kennett 1996; Biondi et al. 1997; Hendy and Kennett 2000). It is an ideal site to document past variations of ENSO and PDO because of seasonally laminated sediments with excellent preservation during warm periods and high

* Correspondence: beaufort@cerege.fr
[1]Aix Marseille University, CNRS, IRD, Collège de France, CEREGE, Avenue Louis Philibert, BP8013545 Aix-en-Provence, Cedex 04, France
Full list of author information is available at the end of the article

sedimentation rates (Behl and Kennett 1996; Grelaud et al. 2009a). Located in the center of the California Current System (CCS), the SBB is seasonally affected by two opposite currents: the southward California Current (CC) during late spring and summer; and the northward Davidson Current (DC, also known as the California Countercurrent or the California Undercurrent) during late fall and winter. The intensity of these two currents is modulated by the North Pacific High (NPH) and the Aleutian Low (AL). Upwelling activity in the area is related to the strength of the CC, as it brings deeper waters rich in nutriment to the surface. The waters are characterized by low pH and carbonate ion concentrations (Alin et al. 2012). At interannual and decennial scale, ENSO and PDO impact directly the intensity of these atmospheric cells, leading to measurable variations of CC and DC flows (Bograd and Lynn 2001; Chavez et al. 2003). During the cold phases of ENSO or PDO (La Niña or negative PDO), the AL is centered over the northwest Pacific and the NPH intensifies leading to strengthening of the CC, associated upwelling, and a weakening of DC. The opposite occurs during warm phases of ENSO or PDO (El Niño or positive PDO), as the AL strengthens and moves southward over the central Pacific, reducing the intensity of the NPH. ENSO and PDO are modulating the strength of the upwelling system in SBB.

The impact of ENSO and PDO in the SBB is inferred from the interplay of the currents within the CCS whose dynamics can be estimated by carefully analyzing fossil assemblages of microalgae growing in the upper photic zone: the coccolithophores (Calcihaptophycidae, Haptophyta). Their productivity depends directly on the physicochemical characteristics of the CCS. Coccolithophore assemblages are very sensitive to seasonality and can be used to diagnose changes in ENSO as evidenced in 20th century records (De Bernardi et al. 2005; De Bernardi et al. 2008; Grelaud et al. 2008). In some species or group of species, the morphology of the scales constituting their exoskeleton is related to physicochemical parameters of the ocean such as temperature for their size (Bollmann et al. 2002) or concentration in carbonate ion for their mass (Beaufort et al. 2011). We will infer dynamics of the upwelling with this latter parameter. We also present the record of the relative abundances of *Gephyrocapsa oceanica* (%GEO), a subtropical species characteristic of DC intensity. In the SBB, its abundances are strongly dependent of temperature associated with warm El Niño events (De Bernardi et al. 2005; Grelaud et al. 2008).

Methods/Experimental

The data of coccolithophore assemblages and morphology were extracted from two sedimentary archives retrieved in the SBB: a multi-core was used to cover the last 100 years (BASIN04-1MC3, 2004, 34°13.41′N, 120° 01.53′W, 588 m water depth) while a giant piston core was used to cover the last 2.7 kyr (MD02-2503, 2002, 34°17.17′N, 120°02.17 W, 569 m water depth) (Fig. 1). For the chronology of core BASIN04-1MC3, see Grelaud et al. (2009b).

The chronology of the MD02-2503 combines five ^{14}C AMS dating and a tuning of its magnetic susceptibility to

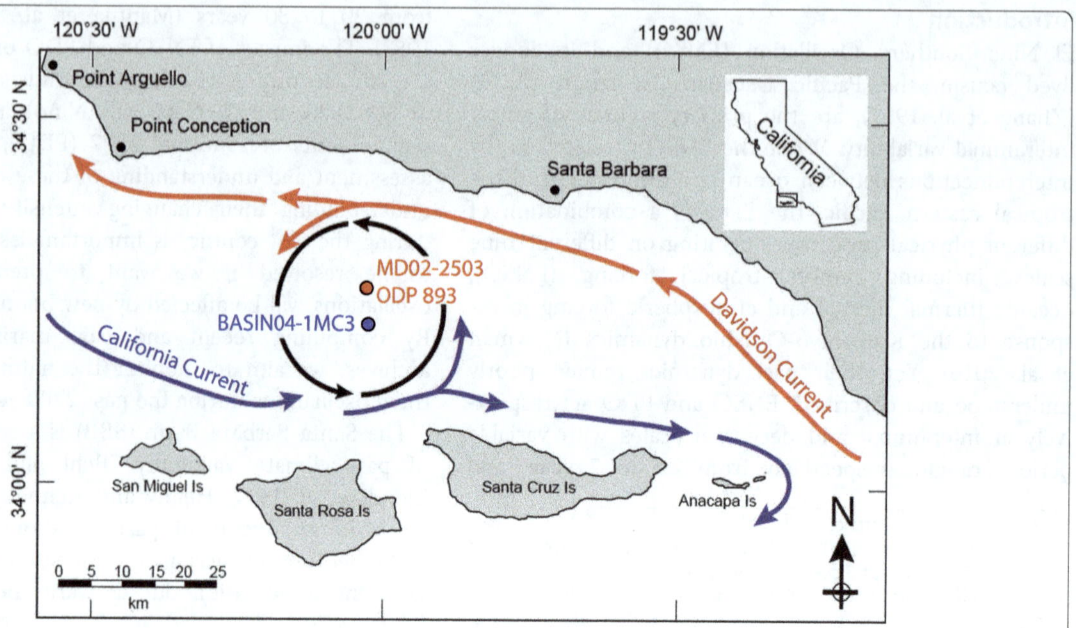

Fig. 1 Location of IMAGES VIII core MD02-2503, BASIN 04 MultiCore 1MC3, and ODP Site 893A. The *arrows* represent the direction of warm (*red*) and cold (*blue*) currents and the counterclockwise circulation (*black*) in the Santa Barbara Basin

that of the nearby Ocean Drilling Program (ODP) Site 893A (Grelaud et al. 2009b) (Fig. 1) which combines 25 ^{14}C dates in the Holocene (Hendy et al. 2002; Roark et al. 2003). The 3.5 uppermost meters of the MD02-2503, representing the last 2700, were sampled every 0.5 cm leading to a resolution of ~3.6 year. The chronology of this section has been slightly reevaluated from Grelaud et al. (2009b) by using the age model of ODP Site 893A (Schimmelmann et al. 2006) and its revised version (Schimmelmann et al. 2013) (Fig. 2). The synchronization between core MD02-2503 and ODP Site 893 imposes a precision in the chronology of the core MD02-2503 estimated to be around 30 years. With this chronology, the age at the core top is around 1916. The presence at about 20 cm of many shell debris can be attributed to the Macoma event which is dated between 1820 and 1850 (Burke et al. 1996).

The sediment of the upper 3.5 m of Core MD02-2503 is composed in majority of hemipelagic laminated sediments (diatom nannofossil clayey silt and diatom nannofossil silty clay). Eighteen millimetric to centimetric gray layers, representing flood deposits (Behl 1995), were observed within this interval of 3.5 m. The levels at which those layers occur are in good agreement with that found in ODP 893A (Schimmelmann et al. 2006). The coccoliths in those layers are at least 10 times less abundant than in the rest of the core. Although there is no shift in the morphology inside and outside, these layers have been withdrawn from this study. The chronology has been established accordingly, as we consider these layers as geological instantaneous events.

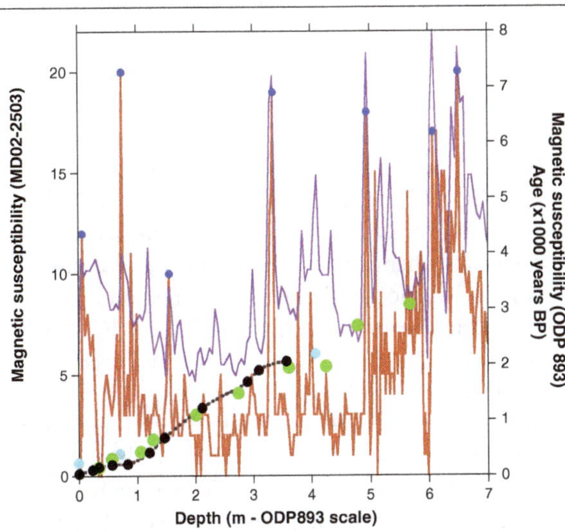

Fig. 2 Chronology of Core MD02-2503: the magnetic susceptibility of core MD02-2503 (*red*) (Beaufort et al. 2002) is compared to that of ODP 989 (*purple*) (Kennett and Ingram 1995) and 7 tie points (*blue*) are chosen to match the two records. Secondary magnetic susceptibility peaks indicate that this match is precise at +/−5 cm between those tie points. Some minor differences appear between the two records when susceptibility is low; those discrepancies are not used for chronology. ^{14}C dates for MD02-2503 (*light blue*) and ODP 893 (*green*) from planktonic foraminifera. In *black* is the revised varved chronology based on ^{14}C date of terrestrial fossils (Schimmelmann et al. 2013)

The sample preparation and analyses for coccolithophore assemblages are the same as for Grelaud et al. (2009a): smear slides were prepared according to standard procedure and were analyzed with an automated system of coccoliths recognition (SYRACO, Beaufort and Dollfus 2004; Dollfus and Beaufort 1999). An average of 600 (BASIN04-1MC3) and 900 (MD02-2503) individuals was counted in each samples and 6 species represented more than 96% of the assemblages: *Emiliania huxleyi, Florisphaera profunda, Gephyrocapsa ericsonii, Gephyrocapsa muellerae, Gephyrocapsa oceanica,* and *Helicosphaera carteri.*

The average mass of calcite of the coccoliths belonging to the Noelaerhabdaceae family was precisely estimated by using the birefringence properties of the calcite when viewed in cross polarized light and by following the method described by Beaufort (2005).

Results and discussion

A Coccolithophore markers of ENSO

The high-resolution record (four samples per year) of the coccolith assemblages present in core BASIN04 and covering the last 100 years (Grelaud et al. 2009b) indicates that higher relative abundances of *G. oceanica* are associated with El Niño years (Fig. 3). This relatively warm water species get into the Santa Barbara Basin in higher relative abundance when the California current relaxes (De Bernardi et al. 2005; De Bernardi et al. 2008; Grelaud et al. 2009b). The relative abundances of this species in the sediment of the SBB can be then used to track the occurrence of past El Niño events. As the North Pacific Ocean is at the end of the ocean thermohaline circulation system (Broecker and Peng 1982), the upwelled waters in the CCS are rich in CO_2 (Feely et al. 2008). During the upwelling season, i.e., in early spring and summer (Pennington and Chavez 2000), surface waters are relatively cold, rich in CO_2 and nutrients, low in pH, and poor in oxygen and carbonate ions (Alin et al. 2012). The upwelling is intensified during La Niña and decreased during El Niño events, and these have strong consequences on the water chemistry of the Santa Barbara Basin. For example, the regional average of carbonate ion concentration was lower during the La Niña event of 2008 ($[CO_3^{2-}]$ = 125 μmol kg^{-1}) and higher during the El Niño event of 2006 ($[CO_3^{2-}]$ = 168 μmol kg^{-1}) as reported by CalCOFI hydrographic cruises (2005–2011) (Alin et al. 2012).

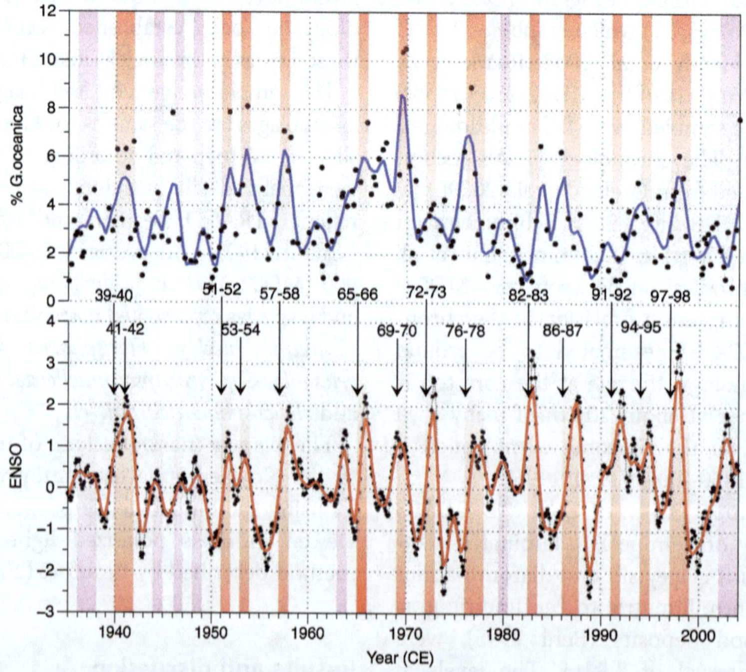

Fig. 3 *Top*: percentage of *G. oceanica* in the coccolith assemblage in *black line* with *dots*. The *blue line* is the record smoothed by a locally weighted regression (LOESS). *Bottom*: NINO3 index (*black line with dots* original—*red* smoothed (LOESS) record.) (Kaplan et al. 1998; Reynolds et al. 2002) (data available at http://iridl.ldeo.columbia.edu/). Major El Niño events are indicated by *numbers*, *arrows*, and *orange bands*. Minor El Niño by *purple bands*. (Adapted from Grelaud et al. (2009b))

When producing their carbonated exoskeleton, coccolithophores are dependent on the carbonate chemistry of the surrounding water (Beaufort et al. 2011; Langer et al. 2006; Langer et al. 2009; Muller et al. 2012; Ridgwell et al. 2009; Riebesell et al. 2000; Rost et al. 2008; Trimborn et al. 2007). We tested here how the seasonal and interannual changes in $[CO_3^{2-}]$ induced by CCS upwelling dynamics impact coccolithophore calcification. Using the data published by Grelaud et al. (2009b), we studied the coccolith mass (CM) of *Gephyrocapsa ericsonii*, *G. muellerae*, and *Emiliania huxleyi*. *Gephyrocapsa oceanica* being one of the heaviest species belonging to the Noelaerhabdaceae family, any change in relative abundance would have a strong impact on the average mass of the group. In consequence, the mass of *G. oceanica* was not included in order to have two independent proxies (coccolith mass and percentage of *G. oceanica*). The differences in coccolith mass are attributed here as the relative abundance of the two *E. huxleyi* morphotypes that present distinct degree of calcification and that are both abundant in this area: Winter (1985) described in the CCS a heavy form of *E. huxleyi* referred as «warm» and a light one referred as «cold». Although they were observed living in the same area together, he found «difficult to relate either temperature or salinity alone to placolith type». This early observation by Winter (1985) fits perfectly with the morphometric data we produced here: results show that the coccolith mass, smoothed with a locally weighted regression (LOESS) (Cleveland and Devlin 1988), generally increases during El Niño events (Fig. 4). A general relation between coccolith mass and $[CO_3^{2-}]$ has been established (Beaufort et al. 2011) to express the importance of the ocean carbonate chemistry on the calcification of the Noelaerhabdaceae (the general regression line plotted in (Beaufort et al. 2011) Fig. 1 is $[CO_3^{2-}] = 20 \times CM + 71.2$). The mass is not considered to produce a reliable proxy for past $[CO_3^{2-}]$ as other factor such as temperature, salinity, and nutriment (e.g., Henderiks et al. 2011) may have also influence on the mean coccolith mass. We can use here this relation to check if the measured coccolith mass is in agreement with some known levels of carbonate concentration measured in SBB. In Fig. 4, the upper right axes represent the concentration of carbonate ions in a scale corresponding to the coccolith mass taken from Beaufort et al. (2011). Associated with these axes are the ranges of $[CO_3^{2-}]$ estimated from the CalCOFI hydrographic measurements from the California Current System between 2005 and 2011 (Alin et al. 2012). The green box represents the

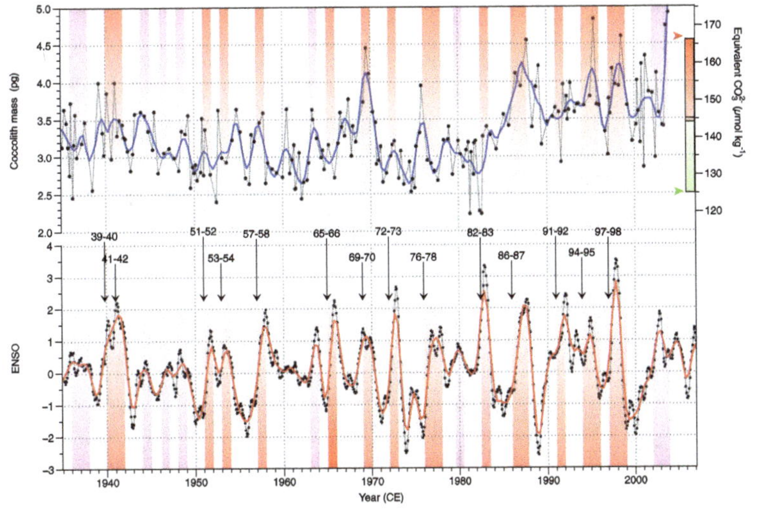

Fig. 4 Twentieth century records of the variability of the coccolithophore nannoflora. *Top*: mean mass in picogram of coccoliths belonging to the species *E. huxleyi*, *G. ericsonii*, and *G. muellerae* in *black line* with *dots*. The *blue line* is smoothed (LOESS) records. The scale on the right represents the concentration in ion bicarbonate equivalent to the coccolith mass from the relation proposed in Beaufort et al. (2011). The *boxes* associated with these *axes* are the ranges of [CO_3^{2-}] between 2005 and 2011 (Alin et al. 2012). *Green* upwelling season, *Red* no-upwelling season. *Red arrowhead* represents the maximal value obtained during an El Niño event, and the *green arrowhead* the minimal value obtained during a La Niña event. *Bottom*: NINO3 index (*black line with dots* original—*red* smoothed (LOESS) record.) (Kaplan et al. 1998; Reynolds et al. 2002) (data available at http://iridl.ldeo.columbia.edu/). Major El Niño events are indicated by *numbers*, *arrows*, and *orange bands*. Minor El Niño by *purple bands*. (Adapted from Grelaud et al. (2009b))

range obtained during the upwelling season and the red during the no-upwelling season. The red arrow represents the maximal value obtained during an El Niño event and the green arrow the minimal value obtained during a La Niña year. Those ranges are in good agreement with the expected coccolith mass–[CO_3^{2-}] equivalent. Increases of coccolith mass in the SBB can be interpreted as an increase of [CO_3^{2-}] related to a decrease in the intensity of the upwelling during an El Niño event.

The anthropogenic input of CO_2 in the atmosphere diffuses into the ocean from its surface. In the California Margin, the last time the upwelled waters were exposed to the atmosphere was 50 years ago (Feely et al. 2008). The ocean acidification is therefore somehow delayed in this area and the pH is more dependent on the upwelling activity than on the recent increase of atmospheric CO_2. The modulation of El Niño's by the Pacific Decadal Oscillation appears in this record (Fig. 5): during high PDO's events (1925–1945 and 1980–2000), the mass of coccolithophores is generally higher than average and the same is also true for secondary events marked by red bands in Fig. 5. The imprint of El Niño on the coccolith mass shades slightly the match between CM and PDO at high frequency. Although the CM and %GEO records result from independent sources, they both increase during El Niño events (Figs. 3 and 4). The two proxies follow slightly differently these climate processes as each responds also differently to other forcing (e.g., nutrient content, seasonal dynamics, cloud cover…). In order to obtain a unique marker of PDO and ENSO in SBB, we propose to combine %GEO and CM because of their visual resemblances. Although they are independent markers, having different units and representing different proxies—one diagnostic of the dynamics of the upwelling (CM), the other of SST (%GEO)—their dynamics has a common forcing that is the PDO-ENSO system. Their combination will allow easier comparison with published variables and pictures in a unique way the dynamics of the ENSO in that region. %GEO and CM were combined by averaging them after standardization (subtracting their average of each variable and by dividing them by their standard deviation. Their sum is divided by two in order to keep the values between −1 and 1:

$$\begin{aligned}\text{COCOC} = {} & ((\%\text{GEO}-\text{mean}(\%\text{GEO}))/\text{std}(\%\text{GEO})) \\ & + ((\text{CM}-\text{mean}(\text{CM})/\text{std}(\text{CM})))/2.\end{aligned}$$

This new combined marker is labeled COCOC that stands for combined coccolith proxies. It features the common variations of the two proxies, for example, COCOC will increase significantly during %GEO and CM common rise.

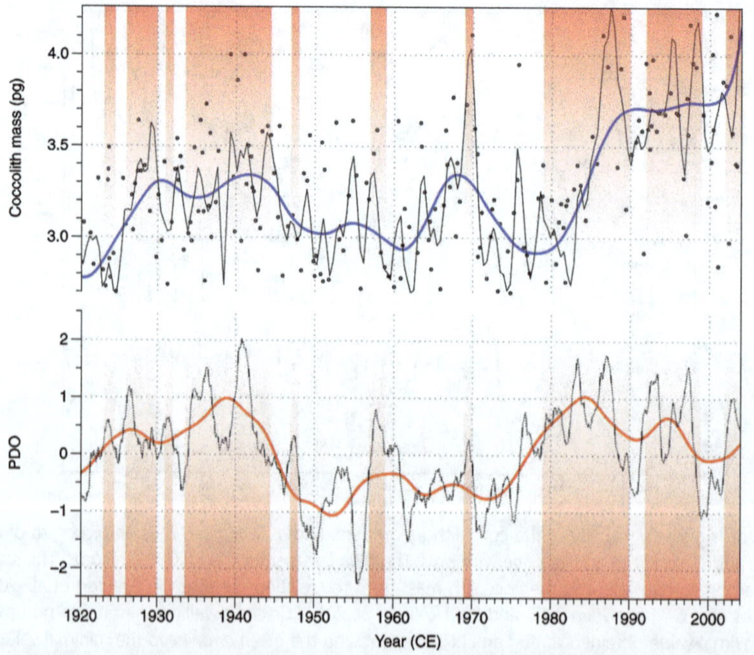

Fig. 5 Comparison between the mass of coccoliths of *E. huxleyi*, *G. ericsonii*, and *G. muellerae* (*Top*) and Pacific Decadal Oscillation (PDO) index (Mantua and Hare 2002). *Red bands* highlight periods of high PDO

B Coccolithophores, assemblages, and morphometry during the last 2700 years

The morphometric (CM) and assemblage (%GEO) records of coccoliths are both showing similar rapid changes as well as a long-term decrease over the last 2700 years (Fig. 6). The polynomial regressions presented in Fig. 6 indicate that the coccolith mass and the relative abundance of *G. oceanica* gradually decreased from about 3.65 pg and 4% respectively (2700 years ago) to about 3.2 pg and 1.5% between 1400 and 1800. This could be attributed to a predominantly negative phase of the PDO; however, as the decrease is more pronounced in %GEO than that in CM, it more certainly results from a decrease in sea surface temperature (SST) during that time. Such decrease has already been documented in many parts of the ocean including the SBB (McGregor et al. 2015). The relation between SST and %GEO is confirmed by the comparison, over the last 700 years, of our results with a high-resolution record of alkenones measured at ODP Site 893 in the SBB (Zhao et al. 2000). Both records are presented in Fig. 7 according to the revised chronology (Schimmelmann et al. 2013). The smoothed (LOESS) records reveal a common regularity of the centennial variability in both records (Fig. 7). The cross spectral analysis (Blackman-Tukey) performed on the two time series shows that the coherency between SST and %GEO for the last 700 years is significant ($P > 0.95$) at three frequencies (1/118, 1/36, and ~1/23 years^{-1}) and relatively in phase (within the chronological constraint) between the two records (Fig. 8). The 23 year period is often reported for the PDO (Biondi et al. 2001; Biondi et al. 1997; Minobe 1999). Although at its lower range, the 36 year period could be assimilated to the pentadecadal (30–70 years) PDO variability (Minobe 1999). The significant relation between SST and %GEO confirms that %GEO can be used to diagnose SST evolution in the SBB. The presence of those frequencies in the time series shows the constant feature of the PDO over the last 700 years. The imprint of the PDO on the upwelling and SST of SBB confirms that it is an important factor of SST variability in this region (Biondi et al. 1997). Although, in some rare parts of the records, a divergence appears: for example, in year 450, %GEO peaks higher than CM. We do not have explanation for those differences that reflect the complex response of an ecological system to climate.

Increases of CM or %*G. oceanica* occur synchronously in both records at a regular centennial pacing (Figs. 6 and 9). The centennial variability is well expressed and highly coherent. The records, smoothed with a LOESS, reveal the regularity of the centennial variability (Fig. 9). The regular peaks of %GEO and CM covering the last 2700 year depict the alternation of sustained phases of ENSO-like and/or PDO conditions at centennial

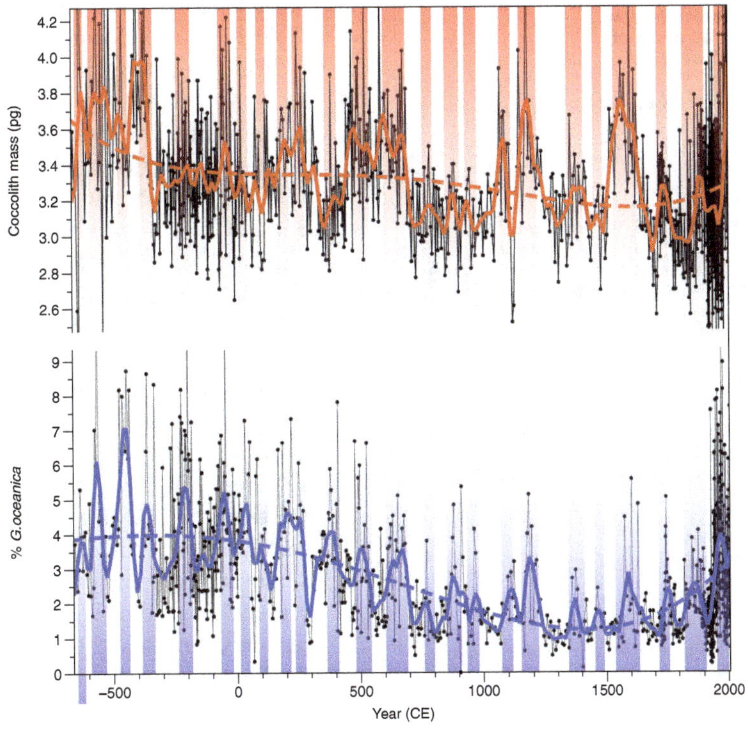

Fig. 6 Two thousand and seven hundred-year records of variability of the coccolithophore nannoflora. *Top—red*: coccolith mass of *E. huxleyi*, *G. ericsonii*, and *G. muellerae*. *Bottom—blue*: relative abundance of *G. oceanica*. *Colored lines* represent the LOESS function and the *dotted lines* represent polynomial of fourth order to depict the long-term tendencies. *Vertical colored bars* are synchronous events common to the two series

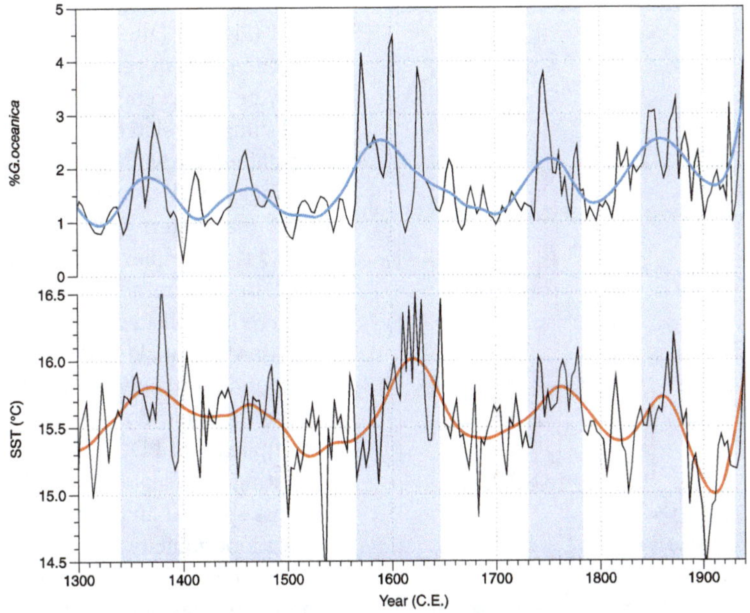

Fig. 7 Comparison of the abundance of *G. oceanica* and SST for the last 700 years. *Top*: Records of %*G. oceanica*. *Bottom*: SST (UK37 from Zhao et al. (2000)). Both records are from the same position in the Santa Barbara Basin. *Colored lines* represent the LOESS smoothing function for the two series. *Vertical bands* are long-term increases of temperature detected by the two proxies

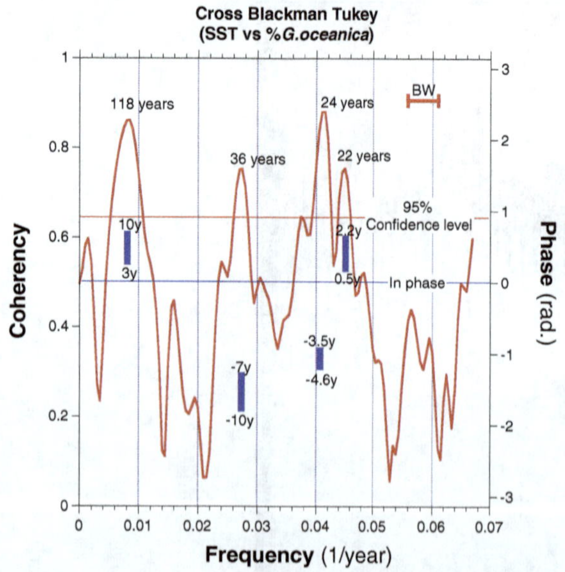

Fig. 8 Cross spectral analysis (Blackman-Tukey) between the 700 year record of SST (Zhao et al. 2000) and %*G. oceanica* in the Santa Barbara Basin. No-Zero coherence at 95% is higher than a confidence level of 0.665 (*red horizontal line*). The bandwidth is 0.0046 (*red horizontal bar marked BW*). Lengths of significant periods are indicated in the *graph*. For those, the phase between the two series is indicated by *vertical blue lines* representing the phase error bar at 80% with the upper and lower limits expressed in years. The *horizontal blue line* is the zero phase. For positive phase value, SST is leading and for negative phase values, %*G. oceanica* is leading

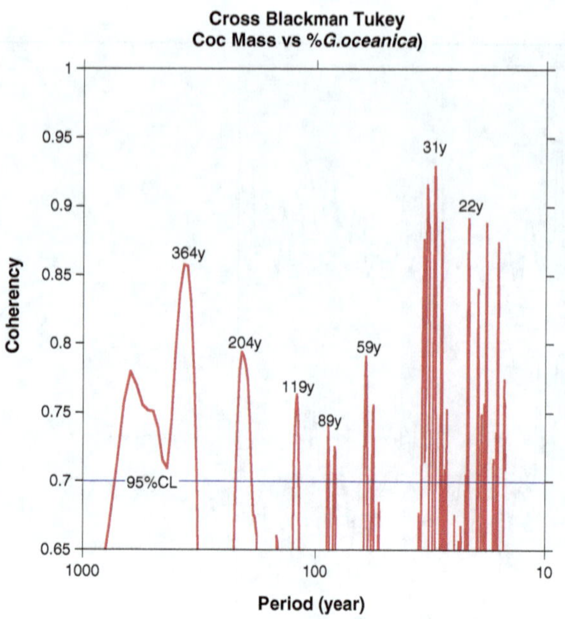

Fig. 9 Cross spectral analysis (Blackman-Tukey) between the 2700 year records of coccolith mass of (*Gephyrocapsa ericsonii, G. muelerae, E. huxleyi*) and %*G. oceanica* in the Santa Barbara Basin. No-Zero coherence at 95% is higher than a confidence level of 0.705 (*blue horizontal line*). The bandwidth is 0.00093. Lengths of significant periods are indicated in the *graph*

scale. Singular spectrum analysis (SSA) (Vautard and Ghill 1989) is performed on the composite record COCOC to determine the relative importance in the total variance of those differences of frequency components. The time series is decomposed in forty independent principal components (PC). The first PC (PC1) contains 46.3% of the total variance; PC2, 8.9%; PC3, 5.2%; and all the other PCs (4 to 40) together, 39.6%. The first three PCs capture more than half of the variance as it is shown in Fig. 12. PC2 appears to fluctuate at about 200 years periodicity and PC3 with pseudoperiods centered on 100 years. Higher important frequencies are centered on period of 30 years (Fig. 10). The strongest PDO+ events and in a lesser extent the strongest PDO− events appear during times of high values of the PC2 (Fig. 11). This indicates that the centennial variability acts also as a modulator of the PDO variability and/or that during an increase of centennial mean of PDO, the negative PDO events remain strong. In other words, the amplitude of PDO cycle increases during centennial "warm" events.

Such a centennial variability is observed in the derivative $\delta^{18}O$ from Pyramid Lake, Nevada, (Benson et al. 2002) and is characteristic of drought occurrences over western USA. The long-term tendency is also similar between the records (see polynomials in Fig. 12). Each of the 17 episodes of drought occurring with a pseudoperiod of 150 year described in Benson et al. (2002) for the last 2700 years corresponds to decreases of %GEO and CM (Fig. 12) (albeit a shift of 50 years probably due to difference in record chronology). This relation may be interpreted as a weakening of the DC, presumably related to La Niña-like conditions. Sustained La Niña-like conditions and associated changes in PDO appear to drive the drought periods over the western USA.

PDO is a complex combination of different dynamical parameters (Newman et al. 2016). The different published PDO proxies (e.g., Benson et al. 2003; MacDonald and Case 2005, COCOC) are responding mainly to one of the dynamical parameter and do not necessarily englobe the entire complexity of PDO. This explains some differences existing between all these PDO reconstructions.

Cross spectral analysis of %GEO versus CM reveals pseudoperiodicity around 100 and 200 years (Fig. 9). Those records are highly coherent with each other and with that of $\delta^{18}O$ from Pyramid Lake, pointing to the highest coherency (>95%) at frequencies of 1/213, 1/193, and 1/96 years^{-1}. These frequencies are also very similar to those commonly cited as de

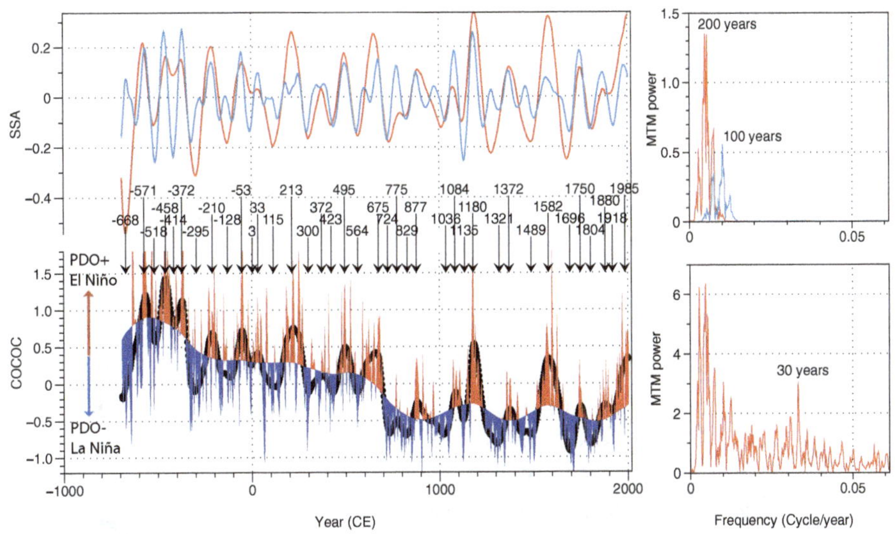

Fig. 10 Variability of the COCOC (composite coccolith proxy) during the last 2700 years in the Santa Barbara Basin. *Bottom left panel* shows the COCOC record in *red* and *blue color*, the limit between *red* and *blue* is a detrended COCOC record by the 1st PC of a SSA (Vautard and Ghill 1989) and the 2nd PC of a SSA (the embedding dimension is 40) in *black dotted line*. *Top left panel* represents the 2nd (*red*) and 3rd (*blue*) PCs of the SSA. The dates of each minima (*blue text*) and maxima (*red text*) of the 2nd PC are given. *Right panels* represent the spectral analysis performed with the Multi-Taper Method (Thomson 1982) of the 2nd PC (*red in top panel*), 3rd PC (*blue*), and the COCOC record detrended with the 1st PC (*red in the bottom panel*). The principal period is given in year

Vries or Suess (~1/200 year^{-1}) and Geisberg (1/83 year^{-1}) cycles found in cosmogenic (^{14}C and ^{10}Be) records (e.g., Emile-Geay et al. 2007; Neftel et al. 1981; Ogurtsov et al. 2002; Suess 1980). Similar cycles are also found in records of the PDO (Benson et al. 2003) and of the Pacific-North American teleconnection pattern (PNA) (Liu et al. 2014; Steinman et al. 2012). This similarity of the variability points to an influence of solar radiation on the ENSO. Mann et al. (2005) studied the importance of volcanic activity and solar irradiance on the ENSO dynamics for the last 1000 years by using Zebiak and Cane (1987) model of the tropical Pacific coupled ocean–atmosphere system. In this model, ENSO reacts to solar irradiance due to a zonal asymmetry in the response of the tropical pacific: a heating due to higher irradiance would cool the East Pacific, inducing a La Niña-like climatic system.

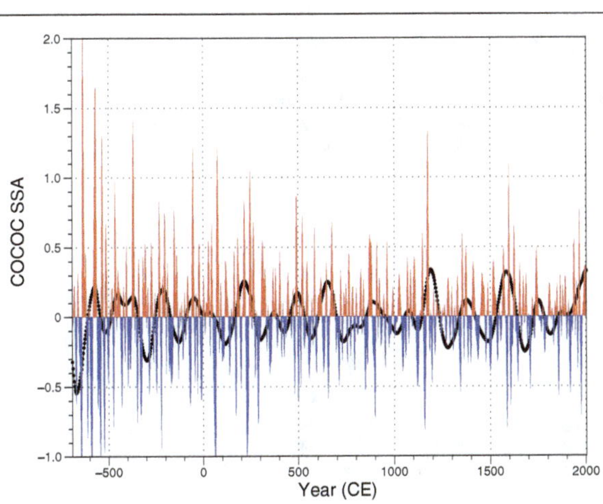

Fig. 11 Modulation of the multi-decadal amplitude of COCOC by its centennial to bicentennial scale variability: The *red* and *blue lines* represent all the PCs higher than 4 (e.g., mutli-decadal amplitude), and the *dotted black line* the two PC (centennial to bicentennial variability)

Such climatic variability appears to be coherent with the amount of heat received by the Earth from the sun: an apparent relation exists between these data and those of the total solar irradiance anomalies (TSIa). Phases of La Niña-like and associated drought conditions over western USA coincide with decrease or negative values of TSIa. Direct solar irradiance including also volcanic activity is controlling ENSO (Liu et al. 2015) by playing a role in the energy received at the ocean level: the influence of both solar irradiance and volcanic activity on ENSO has been tested by numerical simulations (Mann et al. 2005). The results of those simulations have been compared visually to the composite of the coccolithophore records. This indicates that the ENSO simulation with the solar irradiance as sole forcing exhibits synchronous events with coccolithophore record (COCOC) (Fig. 13); the ENSO simulation using both solar irradiance and volcanic activity hardly shows any similarity with coccolithophore records (Fig. 13).

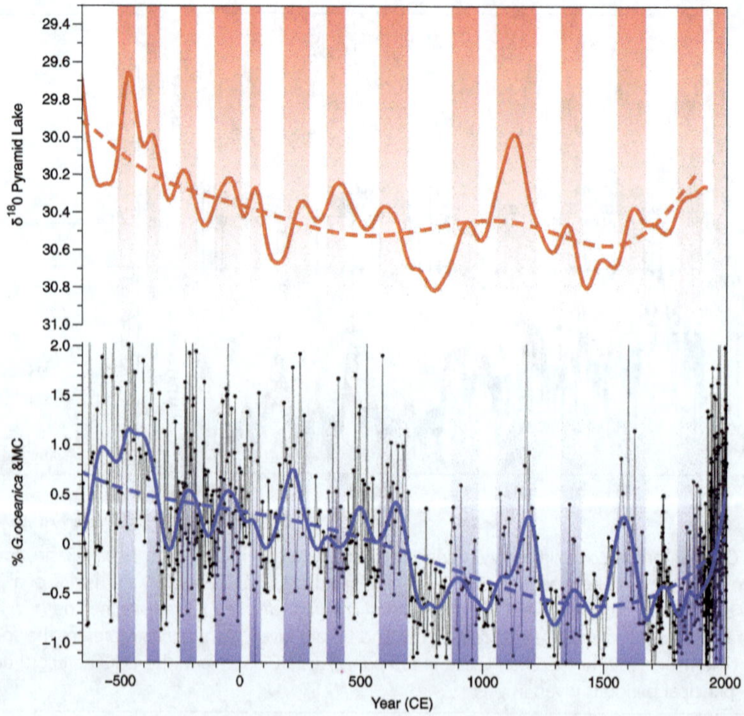

Fig. 12 Two thousand and seven hundred-year records in the *top panel* and in *red*, the smoothed (10 year) record of $\delta^{18}O$ record in core PLC97-1 from Pyramid Lake (Benson et al. 2002) shifted (arbitrarily) by +50 years, and in the *bottom panel* and in *blue*, the COCOC record. The *blue solid line* represents the LOESS function. The *colored dotted lines* represent polynomial of fourth order to depict the long-term tendencies. *Vertical colored bars* are events common to the two series

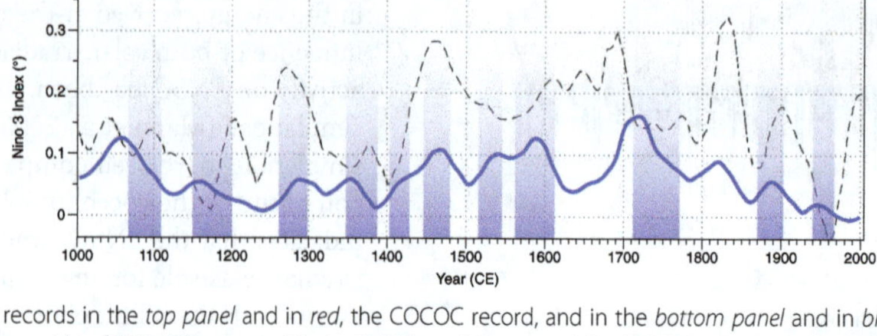

Fig. 13 One thousand-year records in the *top panel* and in *red*, the COCOC record, and in the *bottom panel* and in *blue*, the smoothed (40 year) record of the simulations (ensemble of 100 realizations with the Zebiak and Cane model) of NINO3 to solar radiative forcing in *blue* and to volcanic and solar radiative forcing (*black dotted line*) adapted from Mann et al. (2005). The *colored solid lines* represent the LOESS functions of those records. *Vertical colored bars* are events common to the two series

This could be indicative that solar activity is responsible in a large part of the ENSO and PDO dynamics and that volcanic activity as limited impact on PDO/ENSO dynamics. This is even further demonstrated by the lack of synchronism between the timing of the large PDO events found in the COCOC record and the timing of the major volcanic eruptions occurred during the last 2500 years described by Sigl et al. (2015). This would explain why there is stable centennial variability similar to that of the solar cycle (knowing that volcanic activity should not be cyclic). The persistence of quasi-bicentennial cycle during the entire Holocene in the North America has been recently described elsewhere (Liu et al. 2014) and confirms the permanence of the centennial to bicentennial dynamics in the eastern Pacific and North America.

Conclusions

We presented here two 2700 years' long records from the Santa Barbara Basin of ENSO/PDO proxies based on coccolithophores: one resulting from the relative abundance in the assemblage of one warm species *Gephyrocapsa oceanica* and the other from the degree of calcification of a group of three abundant species (*Emiliania huxleyi, G. ericsonii, G. muellerae*). The first proxy appears to be related primarily to SST changes and to ENSO, and the second proxy is more related to the dynamics of the upwelling and on PDO. Although the two proxies are independent, they show similar dynamics:

1. A general decrease (in abundance and mass) over the last millennia until 300 years ago. Since then, the values increase. This recent increase is particularly pronounced for the mass values that show the highest in the 2700 record in the last 30 years. This general trend is in accordance with the general decrease of SST in the ocean and in particular in the coast of California (McGregor et al. 2015). The impact of ocean acidification in the most recent part of the record may be limited because the relatively old waters upwelled in this area delay the drop of pH by 50 years (Feely et al. 2008).
2. In addition to this long-term variation, there is a strong variability at multi-decadal (periods of 23 and 39 years), centennial, and bicentennial that are typical of ENSO and PDO in that region. The multi-decadal amplitude is modulated by the centennial/bicentennial variability, implying that both phases (negative and positive) of the PDO are increased during high mean COCOC value events at centennial to bicentennial scale.

The presence in those records of de Vries or Suess (~1/200 year^{-1}) and Geisberg (1/83 year^{-1}) cycles that are associated to solar radiation is indicative of the radiative forcing as an important factor of climatic variability in the eastern Pacific regions.

Abbreviations

AL: Aleutian Low; CC: California Current; CCS: California Current System; CM: Coccolith mass; COCOC: Composite coccoliths records; DC: Davidson Current; ENSO: El Niño Southern Oscillation; NPH: North Pacific High; PDO: Pacific Decadal Oscillation; SBB: Santa Barbara Basin

Acknowledgements

We thank the IMAGES program, IPEV, and the crew of the RV Marion-Dufresne for retrieving the core MD02-2503. LB gratefully acknowledges the invitation and support by Pr Masanobu Yamamoto to participate to the INQUA 2015 symposium held in Nagoya, Japan.

Funding

We thank the Agence Nationale de la Recherche for its financial support to LB under projects ANR-12-BS06-0007–CALHIS and JPI Blemont project ANR-15-JCLI-0003-05–PACMEDY. This work is a contribution to the EU FP7 project MedSeA (grant agreement no. 265103).

Authors' contributions

LB proposed the topic. MG carried out the experimental study. LB and MG interpreted the data. LB wrote the first drafts of the manuscript and both approved the final manuscript.

Competing interests

The authors declare that they have no competing interests.

Author details

[1]Aix Marseille University, CNRS, IRD, Collège de France, CEREGE, Avenue Louis Philibert, BP8013545 Aix-en-Provence, Cedex 04, France. [2]Institute of Environmental Science and Technology, Autonomous University of Barcelona (UAB), Bellaterra 08193, Spain.

References

Alin SR, Feely RA, Dickson AG, Martin Hernandez-Ayon J, Juranek LW, Ohman MD, Goericke R (2012) Robust empirical relationships for estimating the carbonate system in the southern California Current System and application to CalCOFI hydrographic cruise data (2005–2011). J. Geophys. Res.-Oceans 117. doi:10.1029/2011JC007511

Beaufort L (2005) Weight estimates of coccoliths using the optical properties (birefringence) of calcite. Micropaleontol 51:289–298

Beaufort L, the members of the scientific party (2002) MD126 MONA (Marges Ouest Nord Americaines) IMAGES VIII. Insitut Polaire Francais Paul Emile Victor (IPEV), Brest, p 453

Beaufort L, Dollfus D (2004) Automatic recognition of coccolith by dynamical neural network. Mar Micropaleont 51/1-2:57–73

Beaufort L, Probert I, de Garidel-Thoron T, Bendif EM, Ruiz-Pino D, Metzl N, Goyet C, Buchet N, Coupel P, Grelaud M, Rost B, Rickaby REM, de Vargas C (2011) Sensitivity of coccolithophores to carbonate chemistry and ocean acidification. Nature 476:80–84

Behl RJ. (1995) Sedimentary Facies and Sedimentology of the Late Quaternary Santa Barbara Basin, Site 893. In: Kennett JP, J.G. B, Lyle M (eds). Proceedings of the Ocean Drilling Program, Scientific Results, vol. 146. Ocean Drilling Program, College Station, pp 295–308

Behl RJ, Kennett JP (1996) Brief interstadial events in the Santa Barbara Basin, NE Pacific, during the past 60 kyr. Nature 379:243–246

Benson L, Kashgarian M, Rye R, Lund S, Paillet F, Smoot J, Kester C, Mensing S, Meko D, Lindstrom S (2002) Holocene multidecadal and multicentennial droughts affecting Northern California and Nevada. Quat Sci Rev 21:659–682

Benson L, Linsley B, Smoot J, Mensing S, Lund S, Stine S, Sarna-Wojcicki A (2003) Influence of the Pacific Decadal Oscillation on the climate of the Sierra Nevada, California and Nevada. Quat Res 59:151–159

Biondi F, Lange CB, Hughes MK, Berger WH (1997) Inter-decadal signals during the last millennium (AD 1117–1992) in the varve record of Santa Barbara basin, California. Geophys Res Lett 24:193–196

Biondi F, Gershunov A, Cayan DR (2001) North Pacific decadal climate variability since 1661. J Clim 14:5–10

Bograd SJ, Lynn RJ (2001) Physical-biological coupling in the California Current during the 1997–99 El Nino-La Nina cycle. Geophys Res Lett 28:275–278

Bollmann J, Henderiks J, Brabec B (2002) Global calibration of Gephyrocapsa coccolith abundance in Holocene sediments for paleotemperature assessment. Paleoceanogr 17:10.1029

Broecker W, Peng TH (1982) Tracer in the sea. Columbia University, New-York

Burke S, Dunbar R, Berger W (1996) Benthic and pelagic Foraminifera of the Macoma layer, Santa Barbara Basin. Oceanogr Lit Rev 1:64–65

Chavez FP, Ryan J, Lluch-Cota SE, Niquen M (2003) From anchovies to sardines and back: multidecadal change in the Pacific Ocean. Science 299:217–221

Cleveland WS, Devlin SJ (1988) Locally weighted regression: an approach to regression analysis by local fitting. J Am Stat Assoc 83:596–610

De Bernardi B, Ziveri P, Erba E, Thunell RC (2005) Coccolithophore export production during the 1997–1998 El Nino event in Santa barbara basin (California). Mar Micropaleontol 55:107–125

De Bernardi B, Ziveri P, Erba E, Thunell RC (2008) Calcareous phytoplankton response to the half century of interannual climatic variability in Santa Barbara Basin (California). Paleoceanography 23. doi:10.1029/2007pa001503

Dollfus D, Beaufort L. (1999) Fat neural network for recognition of position-normalised objects. Neural Networks 12:553–560

Emile-Geay J, Cane M, Seager R, Kaplan A, Almasi P (2007) El Niño as a mediator of the solar influence on climate. Paleoceanography 22. 10.1029/2006PA001304

Feely RA, Sabine CL, Hernandez-Ayon JM, Ianson D, Hales B (2008) Evidence for upwelling of corrosive "acidified" water onto the continental shelf. Science. 320:1490–1492

Grelaud M, Schimmelmann A, Beaufort L (2008) Coccolithophore response to climate and surface hydrography in Santa Barbara Basin, California, AD 1917–2004. Biogeosciences Discuss 5:4129–4159

Grelaud M, Schimmelmann A, Beaufort L (2009a) Coccolithophore response to climate and surface hydrography in Santa Barbara Basin, California, AD 1917–2004. Biogeosciences 6:2025–2039

Grelaud M, Beaufort L, Cuven S, Buchet N (2009b) Glacial to interglacial primary production and El Nino—Southern Oscillation dynamics inferred from coccolithophores of the Santa Barbara Basin. Paleoceanography 24(PA1203):1201–1215

Henderiks J, Winter A, Elbrächter M, Feistel R, Van der Plas A, Nausch G, Barlow R (2011) Environmental controls on Emiliania huxleyi morphotypes in the Benguela coastal upwelling system (SE Atlantic). Mar Ecol Prog Ser 448:51–66

Hendy IL, Kennett JP (2000) Dansgaard-Oeschger cycles and the California Current System: planktonic foraminiferal response to rapid climate change in Santa Barbara Basin, Ocean Drilling Program Hole 893A. Paleoceanography 15:30–42

Hendy IL, Kennett JP, Roarkc EB, Ingramc BL (2002) Apparent synchroneity of submillennial scale climate events between Greenland and Santa Barbara Basin, California from 30–10 ka. Quat Sci Rev 21:1167–1184

Kaplan A, Cane MA, Kushnir Y, Clement AC, Blumenthal MB, Rajagopalan B (1998) Analyses of global sea surface temperature 1856–1991. J Geophys Res Oceans 103:18567–18589

Kennett J, Ingram B (1995) Paleoclimatic evolution of Santa Barbara basin during the last 20 kyr: marine evidence from Hole 893A. Proc Ocean Drill Program Sci Results 146(Part 2):309–325

Langer G, Geisen M, Baumann K-H, Kläs J, Riebesell U, Thoms S, Young JR (2006) Species-specific responses of calcifying algae to changing seawater carbonate chemistry. Geochem Geophys Geosyst 7:Q09006

Langer G, Nehrke G, Probert I, Ly J, Ziveri P (2009) Strain-specific responses of Emiliania huxleyi to changing seawater carbonate chemistry. Biogeosci Discuss 6:4361–4383

Liu Z, Yoshimura K, Bowen GJ, Buenning NH, Risi C, Welker JM, Yuan F (2014) Paired oxygen isotope records reveal modern North American atmospheric dynamics during the Holocene. Nat Commun 5. doi:10.1038/ncomms4701

Liu F, Chai J, Huang G, Liu J, Chen ZY (2015) Modulation of decadal ENSO-like variation by effective solar radiation. Dyn Atmos Oceans 72:52–61

MacDonald GM, Case RA (2005) Variations in the Pacific Decadal Oscillation over the past millennium. Geophys Res Lett 32:4

Mann ME, Cane MA, Zebiak SE, Clement A (2005) Volcanic and solar forcing of the tropical Pacific over the past 1000 years. J Clim 18:447–456

Mantua NJ, Hare SR (2002) The Pacific decadal oscillation. J Oceanogr 58:35–44

Mantua NJ, Hare SR, Zhang Y, Wallace JM, Francis RC (1997) A Pacific interdecadal climate oscillation with impacts on salmon production. Bull Am Meteorol Soc 78:1069–1079

McGregor HV, Evans MN, Goosse H, Leduc G, Martrat B, Addison JA, Mortyn PG, Oppo DW, Seidenkrantz M-S, Sicre M-A, Phipps SJ, Selvaraj K, Thirumalai K, Filipsson HL, Ersek V (2015) Robust global ocean cooling trend for the pre-industrial Common Era. Nat Geosci 8:671–677

Minobe S (1999) Resonance in bidecadal and pentadecadal climate oscillations over the North Pacific: role in climatic regime shifts. Geophys Res Lett 26: 855–858

Muller MN, Beaufort L, Bernard O, Pedrotti ML, Talec A, Sciandra A (2012) Influence of CO2 and nitrogen limitation on the coccolith volume of Emiliania huxleyi (Haptophyta). Biogeosciences 9:4155–4167

Neftel A, Oeschger H, Suess HE (1981) Secular non-random variations of cosmogenic carbon-14 in the terrestrial atmosphere. Earth Planet Sci Lett 56:127–147

Newman M, Alexander MA, Ault TR, Cobb KM, Deser C, Di Lorenzo E, Mantua NJ, Miller AJ, Minobe S, Nakamura H (2016) The Pacific decadal oscillation, revisited. J Clim 29:4399–4427

Ogurtsov M, Nagovitsyn YA, Kocharov G, Jungner H (2002) Long-period cycles of the Sun's activity recorded in direct solar data and proxies. Sol Phys 211:371–394

Pennington JT, Chavez FP (2000) Seasonal fluctuations of temperature, salinity, nitrate, chlorophyll and primary production at station H3/M1 over 1989–1996 in Monterey Bay, California. Deep-Sea Res II Top Stud Oceanogr 47:947–973

Philander SGH (1983) El Niño Southern Oscillation phenomena. Nature 302:295–301

Reynolds RW, Rayner NA, Smith TM, Stokes DC, Wang W (2002) An improved in situ and satellite SST analysis for climate. J Clim 15:1609–1625

Ridgwell A, Schmidt DN, Turley C, Brownlee C, Maldonado MT, Tortell P, Young J (2009) From laboratory manipulations to Earth system models: scaling calcification impacts of ocean acidification. Biogeosciences 6:2611–2623

Riebesell U, Zondervan I, Rost B, Tortell PD, Zeebe R, Morel FM (2000) Reduced calcification of marine plankton in response to increased atmospheric CO2. Nature 407:364–367

Roark EB, Ingram BL, Southon J, Kennett JP (2003) Holocene foraminiferal radiocarbon record of paleocirculation in the Santa Barbara Basin. Geology 31:379–382

Rost B, Zondervan I, Wolf-Gladrow DA (2008) Sensitivity of phytoplankton to future changes in ocean carbonate chemistry: current knowledge, contradictions and research directions. Mar Ecol Prog Ser 373:227–237

Schimmelmann A, Lange CB, Roark EB, Ingram BL (2006) Resources for paleoceanographic and paleoclimatic analysis: a 6,700-year stratigraphy and regional radiocarbon reservoir-age (ΔR) record based on varve counting and 14C-AMS dating for the Santa Barbara Basin, offshore California, USA. J Sediment Res 76:74–80

Schimmelmann A, Hendy IL, Dunn L, Pak DK, Lange CB (2013) Revised ~2000-year chronostratigraphy of partially varved marine sediment in Santa Barbara Basin, California. GFF 135:258–264

Sigl M, Winstrup M, McConnell JR, Welten KC, Plunkett G, Ludlow F, Buntgen U, Caffee M, Chellman N, Dahl-Jensen D, Fischer H, Kipfstuhl S, Kostick C, Maselli OJ, Mekhaldi F, Mulvaney R, Muscheler R, Pasteris DR, Pilcher JR, Salzer M, Schupbach S, Steffensen JP, Vinther BM, Woodruff TE (2015) Timing and climate forcing of volcanic eruptions for the past 2,500 years. Nature 523:543–549

Steinman BA, Abbott MB, Mann ME, Stansell ND, Finney BP (2012) 1500 year quantitative reconstruction of winter precipitation in the Pacific Northwest. Proc Natl Acad Sci 109:11619–11623

Suess HE (1980) The radiocarbon record in tree rings of the last 8000 years. Radiocarbon 22:200–209

Thomson DJ (1982) Spectrum estimation and harmonic analysis. Proc IEEE 70: 1055–1096

Trimborn S, Langer G, Rost B (2007) Effect of varying calcium concentrations and light intensities on calcification and photosynthesis in Emiliania huxleyi. Limnol Oceanogr 52:2285–2293

Vautard R, Ghill M (1989) Singular spectrum analysis in nonlinear dynamics, with applications to paleoclimatic time series. Physica D 35:395–424

Winter A (1985) Distribution of living coccolithophores in the California current system, Southern California borderland. Marine Microplaleontol 9:385–393

Zebiak SE, Cane MA (1987) A model El Niño-Southern Oscillation. Mon Weather Rev 115:2262–2278

Zhang Y, Wallace JM, Battisti DS (1997) ENSO-like interdecadal variability: 1900–93. J Clim 10:1004–1020

7

Toward reduction of the uncertainties in climate sensitivity due to cloud processes using a global non-hydrostatic atmospheric model

Masaki Satoh[1*], Akira T. Noda[2], Tatsuya Seiki[2], Ying-Wen Chen[1], Chihiro Kodama[2], Yohei Yamada[2], Naomi Kuba[1] and Yousuke Sato[3,4]

Abstract

In estimates of climate sensitivity obtained from global models, the need to represent clouds introduces a great deal of uncertainty. To address this issue, approaches using a high-resolution global non-hydrostatic model are promising: the model captures cloud structure by explicitly simulating meso-scale convective systems, and the results compare reasonably well with satellite observations. We review the outcomes of a 5-year project aimed at reducing the uncertainty in climate models due to cloud processes using a global non-hydrostatic model. In our project, which was conducted as a subgroup of the Program for Risk Information on Climate Change, or SOUSEI, we use the non-hydrostatic icosahedral atmospheric model (NICAM) to study cloud processes related to climate change. NICAM performs numerical simulations with much higher resolution (about 7 km or 14 km mesh) than conventional global climate models (GCMs) using cloud microphysics schemes without a cumulus parameterization scheme, which causes uncertainties in climate projection.

The subgroup had three research targets: analyzing cloud changes in global warming simulations with NICAM with the time-slice approach, sensitivity of the results to the cloud microphysics scheme employed, and evaluating circulation changes due to global warming. The research project also implemented a double-moment bulk cloud microphysics scheme and evaluated its results using satellite observation, as well as comparing it with a bin cloud microphysics scheme. The future projection simulations show in general increase in high cloud coverage, contrary to results with other GCMs. Changes in cloud horizontal-size distribution size and structures of tropical/extratropical cyclones can be discussed with high resolution simulations. At the conclusion of our review, we also describe the future prospects of research for global warming using NICAM in the program that followed SOUSEI, known as TOUGOU.

Keywords: NICAM, Cloud microphysics scheme, Cloud changes, Cloud feedback, Global non-hydrostatic model, Global warming

* Correspondence: satoh@aori.u-tokyo.ac.jp
[1]Atmosphere and Ocean Research Institute, The University of Tokyo, 5-1-5 Kashiwanoha, Kashiwa, Chiba 277-8568, Japan
Full list of author information is available at the end of the article

Introduction

The representation of clouds is widely regarded as the largest source of uncertainty in estimates of climate sensitivity obtained by global climate models (GCMs) (Boucher et al. 2013; Soden and Held 2006; Dufresne and Bony 2008; Vial et al. 2013; Schneider et al. 2017; Zelinka et al. 2017). A useful approach to reduce the uncertainties associated with the simulation of clouds is to develop appropriate cloud schemes for very fine resolution global models and evaluate them with observed data particularly from satellite remote sensing (Bony et al. 2015; Klein and Hall 2015).

The Program for Risk Information on Climate Change (referred to as the SOUSEI program[1]) was promoted by the Japanese Ministry of Education, Culture, Sports, Science and Technology (MEXT) from October 2012 to March 2017 with the aim of generating the basic information required to manage the various risks resulting from climate change. Using world-class supercomputers, such as Earth Simulator[2], the SOUSEI program conducted research and development across four linked themes: (A) prediction and diagnosis of imminent global climate change expected to occur within a few years or decades, (B) research on greenhouse gas emission scenarios and associated long-term climate change projections, (C) development of probabilistic climate change projection techniques, and (D) development of technology for precise impact assessment of climate change. Under theme A, a subgroup focused on reducing the uncertainty in climate models due to cloud processes using a global non-hydrostatic icosahedral atmospheric model (NICAM; Tomita and Satoh 2004; Satoh et al. 2008, 2014); we refer to it as the NICAM subgroup.

NICAM has been used for more than a decade to assess atmospheric changes due to global warming (Satoh et al. 2014, 2017). Because of its high resolution, NICAM is expected to provide more information about multi-scale convective systems such as tropical cyclones (TCs) and Madden-Julian oscillations, compared to currently used coarser resolution GCMs. NICAM can also be used with an explicit cloud microphysics scheme and without a convective parameterization scheme to realistically reproduce the structure of meso-scale convective systems which compare reasonably well with satellite observations (Hashino et al. 2013, 2016).

However, the use of NICAM for climate studies has been limited because computational resources were not sufficiently powerful to simulate longer integrations on a climate scale. Yamada et al. (2010) and Yamada and Satoh (2013) examined NICAM projections of the response of simulated TCs to imposed future global warming conditions. Thanks to the growth in computing power in the past decade (Satoh et al. 2017), multi-decadal simulations using NICAM have been achieved with horizontal resolutions of about 14 km (Kodama et al. 2015; Satoh et al. 2015; Yamada et al. 2017).

The aims of the NICAM subgroup, which comprised the authors of this review, were aligned with the demand for a more comprehensive understanding of cloud processes proposed by Bony et al. (2015). The research was focused on three themes. First, as a basis for global non-hydrostatic climate simulations, we implemented a newly developed double-moment bulk cloud microphysics scheme with six water categories (NDW6; originally described by Seiki and Nakajima 2014), which is suitable for the analysis of cloud radiation feedback. We evaluated the results of the new scheme using satellite observations and compared it with other cloud microphysics schemes, including a bin scheme, to understand the uncertainties related to cloud microphysics schemes. Second, we analyzed the cloud changes due to global warming in simulations using NICAM with the new version of NDW6 and compared the results to those of other microphysics schemes. Third, we analyzed changes in circulation due to global warming, particularly features as cloud system organization, storm tracks, and TCs.

Review

Here, we review the development of a new cloud microphysics scheme and compare it with other cloud microphysics schemes, particularly using a large-eddy simulation model and a kinematic driver. We analyze the NICAM simulation results in terms of clouds and circulation changes by comparing the simulations under present-day and global warming scenarios. We also present an analysis of cloud changes, sensitivities to cloud microphysics schemes, and changes in storm tracks and TCs.

Development of a double-moment cloud microphysics scheme

Efforts to improve a cloud microphysics scheme for NICAM face particular challenges. Here, we focus on improving global simulations of high clouds by elaborating ice cloud microphysics and refining model horizontal and vertical resolution. Most GCMs in the Coupled Model Intercomparison Project Phase 5 (CMIP5[3]) had large uncertainty in ice water content (IWC) in the upper troposphere (Waliser et al. 2009; Jiang et al. 2012; Li et al. 2012). This uncertainty is expected to cause large biases in longwave cloud radiative forcing (LWCRF), because high clouds are known to make a large positive contribution to the total LWCRF (Jensen et al. 1994). Flato et al. (2013) showed that LWCRF in the tropics and mid-latitude varied considerably among individual GCMs. Common biases in LWCRF are apparent in simulated results from GCMs of both the Coupled Model Intercomparison Project Phase 3 (CMIP3) and CMIP5 (Li et al. 2013), so the cloud processes related to the LWCRF over such regions have hindered the accuracy of GCMs. In the SOUSEI program, we improved the NICAM double-moment bulk cloud microphysics scheme with six water categories (NDW6). The formula of the cloud microphysics processes in NDW6 mainly follows that described by Seifert and Beheng (2006) (hereafter referred to as the SB scheme) and was implemented into NICAM by Seiki and Nakajima (2014) with minor modifications. The key characteristics of the SB scheme are consistency between particle size distributions and growth equations: tendencies of the number and mass concentration are derived by integrating theoretical particle growth equations weighted by particle size distributions. In addition to the SB scheme, Seiki and Nakajima (2014) designed consistency between the particle shape, terminal velocity, and growth equation. The self-consistency results in a decrease in uncertainties in NDW6, and hence, is important for global climate simulations. The cloud microphysical processes solved in NDW6 are schematically illustrated in Fig. 1, and details of the development of NDW6 are described in the "Appendix" section.

Cloud microphysics schemes used in NICAM

NICAM has been used to understand the roles of convective clouds, particularly over the tropics, in climate systems and the response of such clouds to global warming. For this purpose, single-moment bulk cloud microphysics schemes (referred to here as SMs) have been used in NICAM simulations because changes in mass concentration of precipitating hydrometeors and associated latent heat release, which are explicitly calculated by SMs, involve dynamical feedback to cloud development such as cold pool and precipitation drag.

NICAM can be used with various SMs (Kessler 1969; Grabowski 1998; Lin et al. 1983; Hong et al. 2004; Tomita 2008). In particular, the NICAM single-moment bulk cloud microphysics scheme with six water categories (NSW6) developed by Tomita (2008), which predicts the mass concentration of vapor, cloud water, rain, cloud ice, snow, and graupel, has been evaluated by comparing with multi-satellite observations (Satoh et al. 2010; Kodama et al. 2012; Hashino et al. 2013, 2016). Roh and Satoh (2014) improved NSW6 by examining the diagnosis of particle size distribution and ice cloud microphysics related to snow and graupel. Using their scheme, the simulated vertical profiles of precipitating hydrometeors were in good agreement with those of the Tropical Rainfall Measuring Mission (TRMM) and CloudSat satellite observations (Roh et al. 2017).

Cloud microphysics schemes in NICAM have been improved to simulate persistent clouds, such as cirrus clouds, or light snow over the tropics and mid-latitude, because such high clouds have large impacts on LWCRF (Chen et al. 2000). For this purpose, during the SOUSEI program, we improved the ice cloud microphysics in the original NDW6 developed by Seiki and Nakajima (2014). In general, double-moment bulk cloud microphysics schemes (referred to here as DMs) predict the number concentration of hydrometeors as well as their mass concentration, and so, more precisely represented the evolution of particle size distribution in cloud growth over time compared with SMs. As a result, DMs can capture the transition of cloud systems from weak or moderate organization to a strong convective phase,

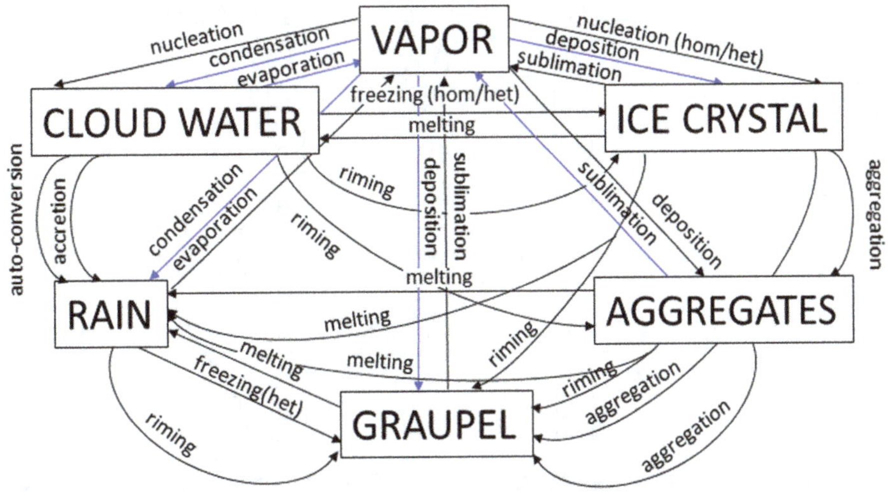

Fig. 1 Schematic representation of cloud microphysical processes solved in NDW6. Black lines show the processes that change the number and mass concentrations of hydrometeors, while blue lines show the processes that change only the mass concentrations of hydrometeors. Hom: homogenous, het: heterogenous

whereas SMs are often optimized for capturing the mature stage of convective systems. Other differences between SMs and DMs are discussed by Igel et al. (2015).

Global simulations using NDW6

During the SOUSEI program, global simulations using NICAM with NDW6 were conducted and evaluated. Figure 2 shows the zonal mean values of simulated outgoing longwave radiation (OLR) at the top of the atmosphere using NSW6 and different versions of NDW6 averaged over the last 5 days of the simulation period. The simulation was conducted for 10 days initiated from 15 June 2008; the 10-day simulation period is sufficient for spin-up of the response of cloud distribution to changes in cloud microphysical settings (e.g., Kodama et al. 2012). The development of various version of NDW6 is described in the Appendix and summarized in Table 1. The OLR is a good proxy for evaluating the global distribution of cirrus clouds because of their strong LWCRF. For reference, satellite observations from the Clouds and the Earth's Radiant Energy System (CERES) level3 SYN1-deg_Ed3A product (Doelling et al. 2013) are superimposed (Fig. 2). The global average OLR of the observation was 241.72 W m^{-2} in this period. The deviation of the global averaged OLR of each simulation from the observation was 3.08, − 18.44, − 2.50, − 1.13, 4.62, and 3.76 for NSW6, SN14, S14, S15, S15 V, and S15VH, respectively (refer to the Appendix and Table 1 for the meaning of these symbols).

The NICAM simulation with NSW6 originally had a large bias in OLR and was optimized by Kodama et al. (2012) to reduce the global averaged bias by changing the aggregation efficiency and terminal velocity of cloud ice (see Fig. 2a). The bias in OLR was globally significant in SN14 (see Fig. 2b) because NDW6 had not been used for global simulations at that time. From SN14 to S14, this severe bias was reduced by changing the heterogeneous ice nucleation scheme (Seiki et al. 2014). These studies demonstrated that optimization of OLR using in situ observation was also effective for global simulations (Figs. 2b, c). From S14 to S15, minor improvements in OLR are shown (Figs. 2c, d). In NSW6, S14, and S15, a positive bias over the tropics and a negative bias over the sub-tropics were commonly observed and compensated for each other, so the global averaged bias was relatively small. Seiki et al. (2015b) found that the negative bias over the sub-tropics originated from thin cirrus clouds near the tropopause and decreased as the vertical resolution increased up to 400 m (Fig. 2e). Furthermore, Seiki et al. (2015b) suggested that the positive bias over the tropics was sensitive to horizontal resolution (Fig. 2f) and could therefore be resolved by increasing the horizontal resolution to finer than 7 km to represent convective clouds over the tropics. In terms of the geographical distribution of OLR values, S15VH reduced the bias over both the tropics and sub-tropics. On the basis of these results, we currently believe that S15VH is the best NDW6 for reproducing OLR.

The global simulation of cirrus clouds was also evaluated by comparing to CALIPSO and CloudSat satellite observations with a satellite simulator named the Joint Simulator for Satellite Sensors (Hashino et al. 2013). Within the simulation period, S15VH closely reproduced the vertical profiles of IWC, ice number concentration (NI), and ice effective radius (REI) (Seiki et al. 2015b).

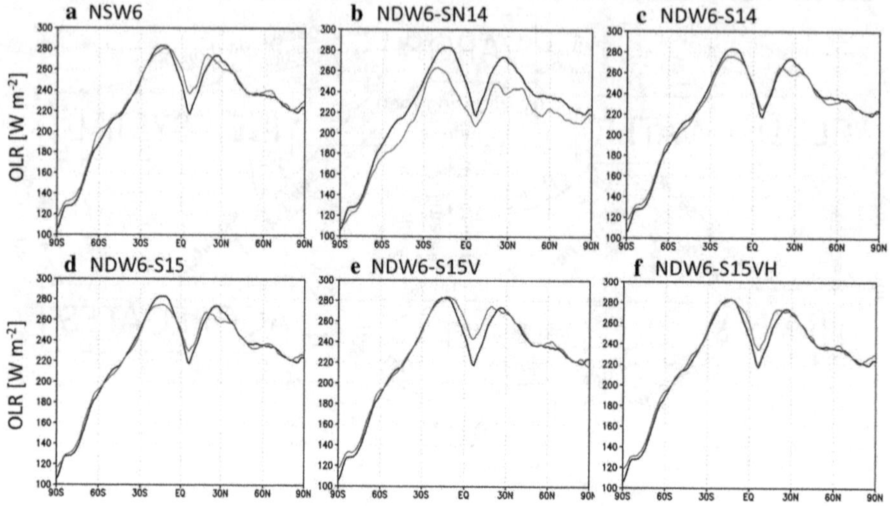

Fig. 2 Comparison of zonal mean values of outgoing longwave radiation. The simulations (blue lines) and the satellite observation (black lines) are averaged over the last 5 days of the 10-days simulation from June 15, 2008, for **a** NSW6, **b** NDW6-SN14, **c** NDW6-S14, **d** NDW6-S15, **e** NDW6-S15V, and **f** NDW6-S15VH. Refer to Table 1 for descriptions of the NDW6 versions

Table 1 Summary of different versions of NDW6

Abbreviated name (reference paper)	Main features
SN14 (Seifert and Beheng 2006; Seiki and Nakajima 2014)	Minor changes from Seifert and Beheng (2006): the power law relationships between x and D and between x and A are assumed and the terminal velocity is diagnosed using x, D, and A.
S14 (Seiki et al. 2014)	Cloud optical properties are updated from Mie scattering to non-spherical scattering. The heterogeneous ice nucleation scheme proposed by Meyers et al. (1992) is replaced by the scheme proposed by Phillips et al. (2007).
S15 (Seiki et al. 2015b)	Homogeneous ice nucleation is newly implemented by referring to Ren and McKenzie (2005) and Kärcher et al. (2006). Values of the sticking efficiency at temperature lower than 253 K is limited to its value at 253 K in aggregation between ice hydrometeors.
S15V (Seiki et al. 2015b)	The maximum vertical grid spacing is limited to 400 m below an altitude of 20 km. Accordingly, the number of vertical layers increases from 40 to 74.
S15VH (Seiki et al. 2015b)	Horizontal grid size is refined from 14 to 7 km. Accordingly, the model time step is changed from 30 to 15 s.

Seiki et al. (2015a) performed seasonal global simulations using S14 and compared them with the simulations produced using NSW6 by Kodama et al. (2012) to examine improvements in atmospheric state via cirrus cloud radiative forcing after feedback of dynamics to cloud distribution. The simulations were initiated from 1 June 2004 and were integrated for 3 months. Seiki et al. (2015a) found that NSW6 had severe warm bias over the tropics, which was more than 2 K in the entire free troposphere and up to 7 K near the tropopause. This bias was also found in 20-year simulations (Kodama et al. 2015) and hence was robust when using NSW6. Using NDW6, this bias was significantly reduced by 4 K on average in the free troposphere because NDW6 accurately represent the spatial distribution of cirrus clouds, particularly near the tropopause (200 to 100 hPa), and the LWCRF of cirrus cloud. Kodama et al. (2015) found that the auto-conversion of cloud ice to snow was the key to reproducing cirrus clouds and their LWCRF. The auto-conversion was generally modeled by a decay equation of cloud ice with an empirical time-scale for ease of tuning. The threshold of the mixing ratio of cloud ice to initiate the conversion and the time-scale are necessarily assumed as global constants in SMs. Seiki et al. (2015b) noted that the conversion rate strongly depends on the cross section, terminal velocity, and number concentration of ice particles based on the theory of cloud microphysics. NDW6 realistically simulated the auto-conversion of cloud ice to snow by solving a theoretical collection equation, with predicted mass concentration and number concentration of hydrometeors. Since the auto-conversion process was sensitive to the simulated amount of cloud ice, which could persist near the tropopause, such an empirical formulation used for SMs should be avoided for more precise calculations of cloud and radiation feedback.

In the SOUSEI program, NDW6 was shown to more realistically simulate cirrus clouds, making it suitable for global long-term simulations of climate studies. Cloud radiative properties should be refined by improving ice cloud microphysics. We mainly used OLR as a proxy of bulk structure of ice clouds, such as the spatial distribution of cirrus clouds. The shortwave (SW) radiation and backscattered radar/lidar signals were sensitive to cloud microphysical structures such as IWC, NI, REI, and the shape of ice particles. Improvement in ice cloud microphysics and the refinement of the model's spatial resolution were important for reducing bias in the global radiation budget and atmospheric temperature (Seiki et al. 2015a, b). Despite developing a new and improved cloud microphysics scheme, the continuous development of cloud microphysics schemes is still necessary. For example, our cloud microphysics schemes assumed at most five categories of hydrometeors (Fig. 1) although various ice species were modeled for atmospheric simulation (Chen and Lamb 1994a, b, 1999; Hashino and Tripoli 2007, 2008, 2011a, b; Harrington et al. 2013a, b). Uncertainties in ice species could be partially constrained by comparing simulated results with satellite observations (e.g., Li et al. 2010; Lang et al. 2014). Efforts are underway to couple NDW6 with aerosol transport models to focus on the effects of aerosols on the ice nucleation process, as suggested by Seiki et al. (2014), but we do not cover that work here.

Sensitivity of cloud microphysics schemes

A cloud microphysics scheme is at the heart of cloud-resolving models. Cloud-resolving simulations that represent clouds using a cloud microphysics scheme aid in reducing the uncertainties in climate predictions related to clouds. SMs have been implemented in NICAM and have been used in global-scale cloud-resolving simulations; an SM by Grabowski (1998) and NSW6 by Tomita (2008) are mostly commonly used, such as by Miura et al. (2007) and Kodama et al. (2015), respectively. However, DMs or more sophisticated schemes are required to appropriately calculate cloud and radiation processes and the effects of aerosols on clouds (Grabowski and Morrison 2016). NDW6 was implemented in NICAM following Seiki and Nakajima (2014) and has been improved further, as discussed earlier in this review. In addition to NDW6, a spectral bin scheme by Suzuki et al. (2010a) has been implemented in NICAM and is now being tested. However, it is not easy to conduct global cloud-resolving simulations coupled with a spectral bin scheme due to limitations in the currently available computing power.

Simple models, such as a single-column model (e.g., Guo et al. 2011; Suzuki et al. 2015) and a kinematic model (e.g., the Kinetic Driver model or KiD; Shipway and Hill 2012), have been used to investigate the sensitivity of cloud microphysics schemes and to interpret their differences. Although these simple models enable us to interpret the differences of cloud microphysics schemes easily, they ignore atmospheric cloud feedback processes. To help us understand the sensitivity of cloud microphysics schemes including feedback processes, idealized fine-resolution simulations have been conducted using large-eddy simulation (LES) models.

Sensitivity using a large-eddy simulation model

LES models provide an additional approach to examine the differences between cloud microphysics schemes. However, because LES models have a large computational cost due to their fine spatial grid resolution, they have been used mainly for numerical simulations targeting shallow clouds in a limited domain. A number of intercomparison studies using LES models have been conducted (e.g., Bretherton et al. 1999a, b; Siebesma et al. 2003; Stevens et al. 2001, 2005; Ackerman et al. 2009; van Zanten et al. 2011; Blossey et al. 2013). These intercomparison studies showed that the results of the LES models have large variability in general. Among them, van Zanten et al. (2011) claimed that the cloud microphysical scheme is one of the main sources of this variability. However, their study is based on the results of several models with their own dynamical cores and physical schemes, which include different turbulent schemes and cloud microphysical schemes. Thus, it is difficult to identify the cause of the variability.

To overcome this difficulty, Sato et al. (2015) investigated the sensitivity of cloud microphysics schemes to trade wind cumulus simulated by an LES model with interchanging cloud microphysics schemes. They used an LES model included in Scalable Computing for Advanced Library and Environment (SCALE; Nishizawa et al. 2015; Sato et al. 2015). The configuration of the simulations was the same as that used in the intercomparison study, Rain in Cumulus over the Ocean (RICO; van Zanten et al. 2011). The cloud microphysics schemes used were SM, DM, and a spectral bin microphysics scheme (Suzuki et al. 2010a). SM and DM were based on NSW6 and NDW6, respectively, with slight modifications when implementing them in SCALE.

Sato et al. (2015) found that the dependence of the cloud microphysics scheme on vertically integrated liquid water (liquid water path) and cloud cover was small (Figs. 3a, b). By contrast, the vertical profiles of liquid water mixing ratios were very variable (Fig. 3c), which resulted in differences in the turbulent structure of the boundary layer (e.g., Fig. 3d). Their analyses showed that these differences derived mainly from differences in the schemes with respect to the generation of cloud droplets (i.e., activation of aerosols, nucleation of cloud droplets) and the rate of conversion of small cloud droplets to raindrops. A saturation adjustment, which was only applied to the cloud droplet formation process for SMs, generated a large amount of cloud water, which resulted in rapid conversion of cloud droplets to raindrops. As a result of differences in the new-particle generation process, the conversion speeds of SMs and DMs were fast and slow, respectively, and the conversion speed of the spectral bin scheme was somewhere in between. The fast conversion promoted the generation of raindrops and precipitation, which led to the evaporation of raindrops below clouds (sub-cloud layer) and stabilized the boundary layer.

In contrast, when the conversion speed was slow, small cloud droplets remained in the atmosphere and reached upper layers, as small droplets are more easily carried by updrafts compared to large raindrops. With such small droplets, the latent heat of the condensation process was released at higher layers. As a result, in the slow conversion scheme (i.e., DMs), the development of the boundary layer was active, and the boundary layer height was high (Fig. 3c).

The results from these sensitivity simulations suggest potential modifications to improve the simple microphysics scheme (i.e., SMs). In addition, they aid in the interpretation of differences between the results of NICAM coupled with SMs, DMs, and the spectral bin scheme.

Comparison with a bin microphysics scheme

Another approach to examine sensitivities to cloud microphysics schemes is the use of simplified models without dynamical feedback of cloud microphysics. A kinematic model is widely used to separately examine sensitivities to microphysics processes (Shipway and Hill 2012; Hill et al. 2015). Grabowski (2016) recently proposed a "piggybacking" approach that separates dynamical and microphysical impacts on deep convection.

Kuba et al. (2014, 2015) used a two-dimensional kinematic driver model (Shipway and Hill 2012) to study shallow cloud properties simulated by a spectral bin cloud microphysics scheme and examined the differences between NDW6 and the bin scheme; the bin scheme examined here is a cloud microphysical model

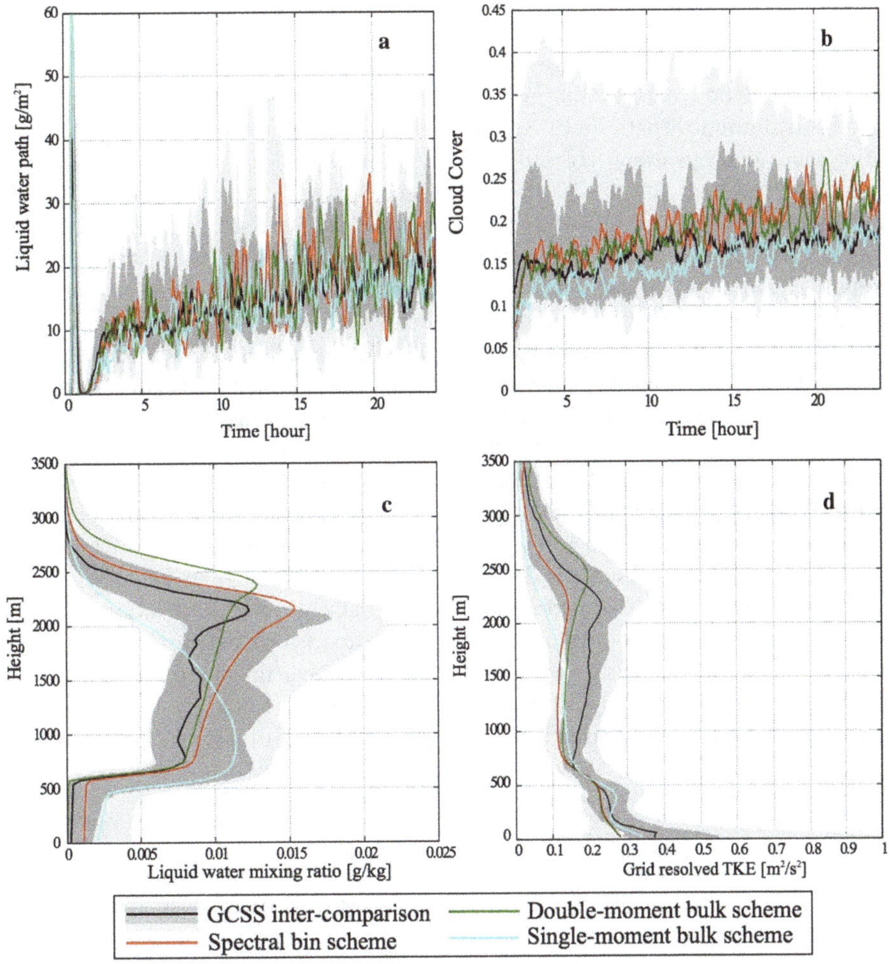

Fig. 3 Comparison of the bulk cloud microphysics schemes and spectral bin scheme. Time evolution of the **a** liquid water path and **b** cloud cover, and 4-h averaged vertical profile of **c** liquid water mixing ratio and **d** grid-resolved turbulent kinetic energy averaged over the entire calculation domain. The red, green, and cyan lines indicate the results of the spectral bin, double-moment bulk cloud microphysics scheme, and single-moment bulk cloud microphysics scheme, respectively. The black line, dark gray shading, and light gray shading indicate the median, range between the first and third quartiles, and range between maximum and minimum values of an intercomparison study of Global Energy and Water EXchange project (GEWEX) Cloud System Study (GCSS), respectively (van Zanten et al. 2011). Adapted from Sato et al. (2015)

with a two-moment bin method based on Kuba and Fujiyoshi (2006) and modified by Kuba and Murakami (2010), and will be referred to as the KF-bin scheme. To investigate the relationships between cloud microphysical processes and satellite observational data, Suzuki and Stephens (2008) examined the relationships between mean radar reflectivity, Z_e, and mean particle effective radius, R_e, values for shallow clouds retrieved from satellite observations and used this information to identify warm cloud microphysical processes. Suzuki et al. (2010b) further analyzed the contoured frequency by optical depth (CFODD) for vertical profiles of radar reflectivity as a function of cloud optical depth from cloud top to investigate cloud microphysical processes based on a methodology adapted from Nakajima et al. (2010).

Kuba et al. (2014) reproduced a diagram proposed by Suzuki and Stephens (2008) using a kinematic driver model (Shipway and Hill 2012) with the KF-bin scheme and a forward simulator of satellite measurements. Changes in the size distribution of cloud droplets and raindrops were traced in the two-dimensional kinematic driver model (Shipway and Hill 2012) for a warm cloud. Kuba et al. (2014) calculated radar reflectivity and optical thickness using the numerical data output from this model as the input for the radar simulator developed by Masunaga and Kummerow (2005) in the Joint Simulator for Satellite Sensors (Hashino et al. 2013).

In the Z_e-R_e diagram reproduced by Kuba et al. (2014), the relationships between Z_e and R_e follow four paths during the lifetime of warm clouds. In the first path, both Z_e and R_e increase with an approximate sixth-power dependency, indicating a stage of condensational growth of droplets without raindrops in the clouds. In the second path, Z_e increases sharply while R_e

stays almost constant, indicating that few raindrops emerge in the clouds. In the third path, R_e increases while Z_e stays almost constant, indicating a stage of coalescence of droplets; precipitation reaches the ground at the end of this path. In the fourth path, both Z_e and R_e decrease, indicating that raindrops evaporate and precipitation weakens.

Kuba et al. (2015) followed Kuba et al. (2014) and investigated the relationships between horizontally averaged radar reflectivity, Z_m, and cloud optical depth from cloud top, τ_d, to examine cloud microphysical processes inferred from the CFODD by Suzuki et al. (2010b) and Sato et al. (2012). Kuba et al. (2015) used the same kinematic model with the KF-bin scheme and the satellite simulator to depict vertical profiles of radar reflectivity of warm clouds in the same way as Kuba et al. (2014). Figure 4 shows the relationship between Z_m and τ_d at 2.5-min intervals in cases of continental cloud condensation nuclei (Figs. 4a–d) and maritime cloud condensation nuclei (Figs. 4e–h). This indicates three patterns in the relationships related to cloud microphysical processes. In the first path, Z_m increases slightly with a decrease in τ_d or with an increase in altitude (A arrows in Figs. 4a, b, e). In the second path, Z_m increases with an increase in τ_d in the upper part of cloud (B arrows in Figs. 4c, d, f–h). Then, in the third path, Z_m decreases with an increase in τ_d in the lower part of cloud and under the cloud base (C arrows in Figs. 4f, g). The cloud microphysical processes affect the relationship between Z_m and τ_d (Fig. 4). When Z_m increases linearly with a decrease in τ_d (A arrows in Fig. 4), the number of small raindrops increases. This is caused by coalescence between cloud droplets (i.e., the auto-conversion process). When Z_m increases with τ_d in the upper part of cloud (B arrows in Fig. 4), the mode radius of raindrops increases. This is caused by coalescence between falling raindrops and cloud droplets (i.e., the accretion process). In the third path, Z_m decreases with τ_d in the lower part of cloud and under the cloud base (C arrows in Fig. 4). This is caused by a decrease in the number of raindrops with τ_d in the lower part of cloud and under the cloud base (i.e., sedimentation and evaporation). Kuba et al. (2015) also found that the bulk collection efficiency has a partially positive correlation with the slope factor of the relationships between Z_m and τ_d (changing rate of Z_m to τ_d) near the cloud top in a later stage of cloud lifetime. Overall, Kuba et al. (2015) reproduced the basic characteristics of the vertical structures of the CFODD

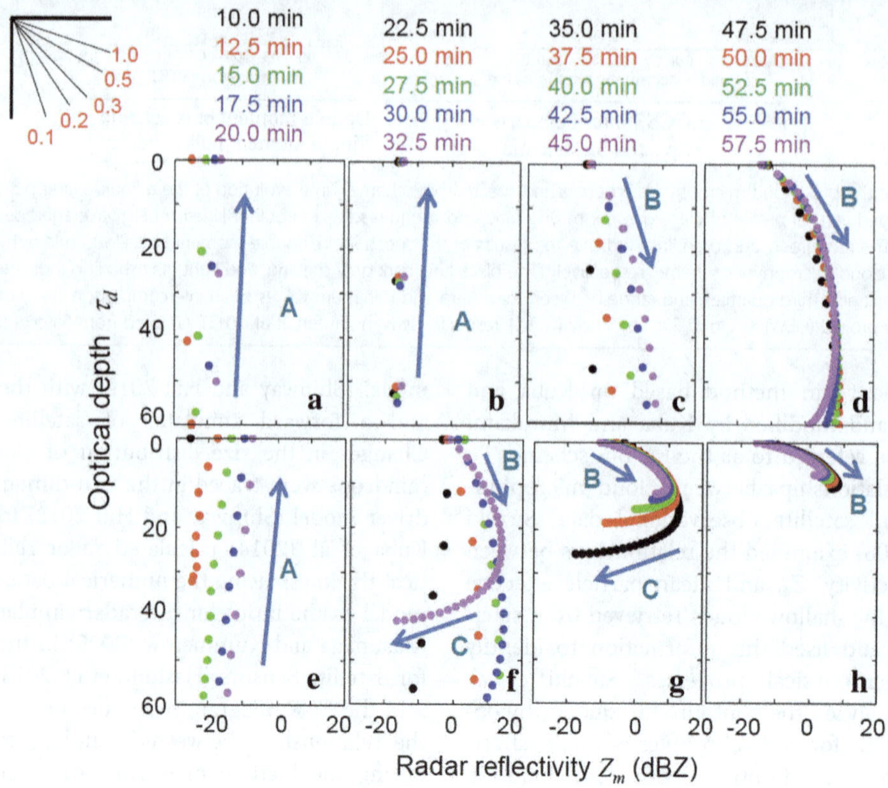

Fig. 4 Relationships between horizontally averaged radar reflectivity Z_m and optical depth from cloud top, τ_d, for cloud condensation nuclei with weak updraft. **a–d** Continental cloud condensation nuclei. **e–h** Maritime cloud condensation nuclei. Arrays A, B, and C indicate the first, second, and third path, respectively (refer to the text). The diagram in the upper left shows the slope factor defined in Suzuki et al. (2010). Adapted from Fig. 5 of Kuba et al. (2015)

shown by Suzuki et al. (2010b) and Sato et al. (2012) and suggest that the methodology using the kinematic driver is useful for studying cloud microphysical processes.

Kuba et al. (2018, personal communication) further compare different cloud microphysics schemes and propose a method for improving cloud microphysics schemes. They compared NDW6 with the KF-bin scheme using the two-dimensional kinematic driver model and a forward simulator of satellite measurements, following the approach by Kuba et al. (2014, 2015). They found that accumulated surface rainfall amounts are relatively similar but that the vertical distributions of rain-water mass are different. They determined that this difference is caused by overestimations of falling velocity of raindrops in NDW6. They also found that the slope factor of the relationships between Z_m and τ_d near the cloud top in a later stage of cloud lifetime is smaller in NDW6 than in the KF-bin scheme, which is also caused by the overestimation of raindrop falling velocity in NDW6. Furthermore, Kuba et al. (2018, personal communication) found that suitable choices for a shape parameter, ν_r, of a gamma distribution assumed as raindrop size distribution in NDW6 can improve the estimation of raindrop falling velocity. Observational data on raindrop number at ground level and from satellites can be used to adjust ν_r in NDW6 if the data can be classified by the environmental conditions of clouds.

Cloud and circulation changes

Climate projection simulations by NICAM

Climate projection simulations using NICAM have been conducted in a time slice approach by focusing on the changes in clouds and circulation systems, such as tropical and extra-tropical cyclones, between two different climates. That is, a set of experiments are conducted under the current climate condition and under the future climate condition by using the specified sea surface temperature (SST) of the respective situations, rather than an atmosphere-ocean coupled model. To be precise, the recent Atmospheric Model Intercomparison Project (AMIP) type and future climate experiments using NICAM by Kodama et al. (2015) and Yamada et al. (2017) were conducted with a slab ocean model nudged toward the specified SST.

In earlier attempts, Miura et al. (2005) examined simulated cloud changes by comparing an aqua planet experiment under a control condition with that under a warmer condition with SST uniformly increased by 4 K. Their projected cloud changes were the first result from a global non-hydrostatic model with cloud microphysics schemes and without cumulus parameterization. The projected changes were found to be very different from those obtained by coarser-resolution GCMs with cumulus parameterization. Around the same time, Wyant et al. (2006) simulated a cloud response using a GCM with super parameterization (superparameterized Community Atmosphere Model; SP-CAM). Wyant et al. (2006) compared their result with the NICAM results of Miura et al. (2005) and found similarities and differences between the two models. In the high latitudes poleward of 50°, both models have strong cloud fraction increases despite their different surface boundary conditions. In the subtropics and tropics, the models showed opposite response of clouds: SP-CAM cloud fraction increased in the Intertropical Convergence Zone (ITCZ), while the NICAM cloud fraction decreased in the ITCZ. Such differences in cloud feedback in low latitudes by high-resolution models are reviewed by Bretherton (2015).

Collins and Satoh (2009), Satoh et al. (2012), and Tsushima et al. (2014) further analyzed a series of the perpetual July experiments between the control and the warming conditions based on the NICAM simulation results by Iga et al. (2007, 2011). In the tropical to sub-tropical area, they all showed that upper clouds increase as climate becomes warmer but ice water path (IWP) decreases. Such cloud responses are different from other GCMs (Tsushima et al. 2014), and the difference have been found to be robust. Satoh et al. (2012) examined the reason for the decrease in IWP and argued that the future weakening of circulation leads to a reduction of the ice production rate. This mechanism was recently used by Bony et al. (2016) to explain the "shrinking" of tropical upper clouds simulated by their GCMs. However, as described above, the NICAM results show an increase in upper clouds, so more detailed discussions are required to resolve this contradiction.

Chen et al. (2016) conducted similar experiment as Collins and Satoh (2009), Satoh et al. (2012), and Tsushima et al. (2014), but used two different cloud microphysics schemes: NSW6 and NDW6. The results of high-cloud changes and IWP were very similar in the two schemes, suggesting that these responses are robust, at least when using NICAM.

These studies raise the question of why such high-cloud responses occur and why the high-cloud responses of NICAM are different from those of the other GCMs. Noda et al. (2014) partly answered the latter question by analyzing the areal size distribution and radiative effects of high clouds. Changes in cloud clusters are a main focus of the NICAM subgroup of the SOUSEI program. Because cloud distributions simulated by NICAM look similar to those observed in satellite images (Miura et al. 2007), the area coverage of high clouds is analyzed and compared with observations (Inoue et al. 2008). Using this method, Noda et al. (2014) compared the distribution of cloud size in the two climates and found that the number of smaller high clouds is more drastically increased in the warmer climate and that the cloud radiative forcing by smaller clouds less than about 50 km in radius contributed to the positive feedback by LWCRF. Such small-scale clouds are not generally explicitly represented by ordinary GCMs and parameterized by cloud macro-physics schemes such as cloud fraction, but they are explicitly represented in NICAM. These findings suggest that assumptions introduced in the cloud macro-physics schemes should be examined more closely and compared with explicit calculations using cloud microphysics schemes to understand the different responses of high clouds between models. Noda et al. (2016) extended this approach to examine the fixed anvil temperature (FAT) hypothesis (Hartmann and Larson 2002) as described below.

Projected changes in circulation systems are also examined using NICAM simulations. Since deep convective circulations are explicitly represented by NICAM, future changes in cloud systems can be examined together with changes in meso-scale properties and their embedding synoptic structures. TC changes were examined by Yamada et al. (2010), who used NICAM to show decreases in the number of TCs and more frequent intensive cyclones, consistent with other GCMs (Knutson et al. 2010). They further examined the eye wall structures and showed that eye wall height tends to increase with TC intensity. Yamada and Satoh (2013) examined projected changes in ice and liquid clouds associated with TCs. These studies are based on the results of 5-month simulations, which are too short to explore basin-scale differences or inter-annual variabilities. Recently, however, Yamada et al. (2017) used the K computer to extend these experiments to a multi-decadal time scale: the AMIP-type experiment under the present-day climate condition and the corresponding experiment under the future climate condition (Kodama et al. 2015; Satoh et al. 2017). Yamada et al. (2017) showed basin-scale changes in TCs and seasonal variations. In general, the global TC statistics from the NICAM simulations were consistent with those reported by Knutson et al. (2010). Yamada et al. (2017) further analyzed the structural changes in TCs and showed that the horizontal scale of TCs increased with warming when compared between TCs of the same categories in the two climates. Using these simulation results, Satoh et al. (2015) analyzed convective mass flux associated with TCs and proposed a constraint for a future decrease in the number of TCs.

Future changes in extra-tropical cyclones are examined by Kodama et al. (2014) using the aqua planet experiments conducted by Yoshizaki et al. (2012). Their simulation shows that, similar to the TC response, the frequency of extra-tropical cyclones decreased but the cyclones slightly intensified on average. The storm track moved poleward, consistent with other GCMs.

In the subsequent subsections, we review the outcomes of the SOUSEI program, the analysis of cloud size statistics, upper cloud changes, and extra-tropical cyclone changes in more detail.

Changes in the size distribution of high clouds

As reviewed in the previous section, increased high-cloud coverage is consistently observed in NICAM simulations (Collins and Satoh 2009; Satoh et al. 2012; Tsushima et al. 2014; Chen et al. 2016). To explore the characteristic change in high clouds, we show the size distribution of high cloud in Fig. 5, along with satellite data. Here, we show additional cloud size analysis of the 1-year present-day control simulations and global warming simulations at mesh size of 7 km and 14 km used by Noda et al. (2016). We defined a high-cloud region as one where the modeled OLR was less than 210 W m^{-2}. Cloud sizes are shown as an equivalent radius of a circle, and thus, the minimum radius is approximately half of the model grid size (Noda et al. 2012, 2014).

The NICAM result shows good agreement with the satellite data except for clouds with a radius smaller than 30 km, which show negative bias in the frequency. However, this bias is improved by reducing the grid size from 14 to 7 km, consistent with Inoue et al. (2008). Under the warmer condition, the number of clouds increases in all radius bins. In particular, smaller clouds increase in frequency more than larger clouds. The smaller clouds contribute more to the increase in the total high-cloud coverage under the warmer condition.

Changes in OLR due to changes in high clouds directly affect projected changes to the global radiative budget. We define a diagnostic equation for the changes in OLR which decomposes relative contributions of various cloud properties: cloud emissivity (ε), cloud-top temperature (T_{CT}), and clear-sky OLR (F^{CLR}). The response of OLR, F, to global warming Δ, can be approximated as.

$$\Delta\overline{F}^{(i)} \cong \left(\frac{\partial F}{\partial \varepsilon}\right)_{T_{\mathrm{CT}},F^{\mathrm{CLR}}} \Delta\overline{\varepsilon}^{(i)} + \left(\frac{\partial F}{\partial T_{\mathrm{TC}}}\right)_{\varepsilon,F^{\mathrm{CLR}}} \Delta\overline{T_{\mathrm{CT}}}^{(i)} + \left(\frac{\partial F}{\partial F^{\mathrm{CLR}}}\right)_{\varepsilon,T_{\mathrm{CT}}} \Delta\overline{F^{\mathrm{CLR}}}^{(i)} \equiv F_\varepsilon \Delta\overline{\varepsilon}^{(i)} + F_T \Delta\overline{T_{\mathrm{CT}}}^{(i)} + F_F \Delta\overline{F^{\mathrm{CLR}}}^{(i)}, \tag{1}$$

where

$$F_\varepsilon = \sigma\left(\overline{T_{\mathrm{CT}}}^{(i)}\right)^4 - \overline{F^{\mathrm{CLR}}}^{(i)}, \tag{2}$$

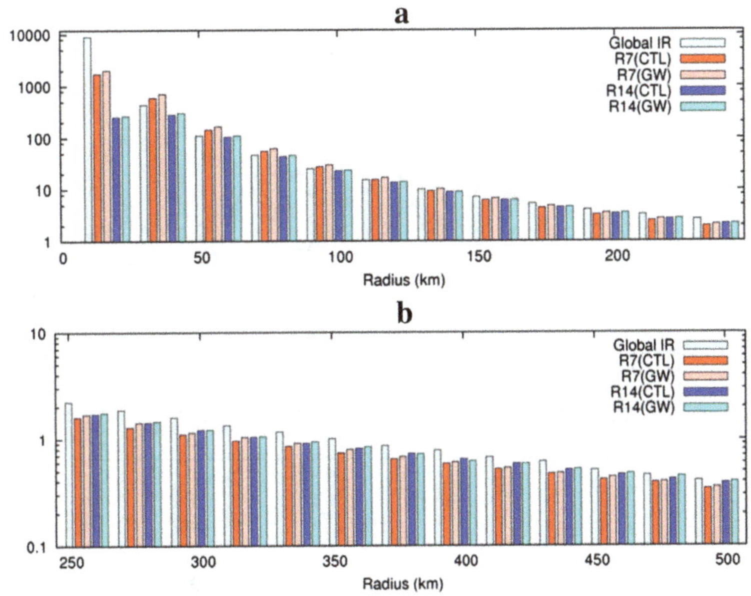

Fig. 5 Frequency distribution of high-cloud size. The radius of high clouds is derived from global infrared data (Global IR) and from present-day control simulations (CTL) and global warming simulations (GW) at mesh resolutions of 7 km (R7) and 14 km (R14). Cloud size (radius) is binned every 20 km. The top panel **a** shows for the size range between 0 and 250 km, while the bottom panel **b** shows the size range between 250 and 500 km. Adapted from Noda et al. (2014a)

$$F_T = 4\overline{\varepsilon}^{(i)}\sigma\left(\overline{T_{CT}}^{(i)}\right)^3, \tag{3}$$

$$F_F = 1-\overline{\varepsilon}^{(i)}. \tag{4}$$

An overbar with a prefix i denotes an average over ith cloud area. Those values are hereafter shown by being binned according to the cloud radius. Using the above formulae, one can evaluate the contributions of ε, T_{CT}, and F^{CLR} to changes in F.

Figure 6 shows the contribution of each term of (1); the vertical axis is the accumulated contribution of the cloud size whose radius is less than the value in the horizontal axis. The contribution of each term clearly differs depending on cloud size; the change in T_{CT} contributes more to clouds smaller than 400 km in radius. In contrast, changes in both T_{CT} and ε contribute almost equally to clouds larger than 400 km. The contribution of the change in F^{CLR} is minor relative to the other two elements, and it acts to reduce OLR. The contribution of ε increases rapidly for clouds up to 300 km in radius, followed by an almost linear increase, and then asymptotically approaches a constant value (Fig. 6b). In contrast, the contribution of T_{CT} increases almost linearly, independent of cloud size, and eventually approaches a constant value. The net contributions of the two elements are comparable. The contribution of F^{CLR} is one order smaller than that of the other elements.

Hartmann and Larson (2002) claimed that T_{CT} is kept almost constant, irrespective of surface temperature, which is referred to as the FAT hypothesis. Our analysis reveals that the FAT hypothesis is valid for smaller high clouds for which the contribution of T_{CT} change is smaller than that of ε change. However, the validity of the hypothesis can be uncertain for larger clouds because changes to both T_{CT} and ε contribute almost equally to OLR change; the former contribution is not negligible. In other words, whether the FAT hypothesis holds or not depends on which size of cloud primarily increases or decreases due to global warming. This suggests that the size distribution of high clouds should be examined for more reliable projections of radiation budget.

High cloud changes in two different microphysics scheme experiments

Clouds play an important role in the uncertainty of the climate projection under the global warming condition because of their influence on the energy balance of the atmosphere (Hartmann et al. 1992). The uncertainty of GCM-simulated climate projection comes from the large spread in cloud responses to global warming as found in CMIP3 and CMIP5 (Vial et al. 2013). In particular, we focus on cloud ice simulations (Waliser et al. 2009; Li et al. 2013).

Chen et al. (2016) examined how cloud responce changes by using NICAM with two cloud microphysics schemes, NSW6 and NDW6, and compared these results with those of the GCM simulations from CMIP3 and CMIP5. The aim of their study was to reveal

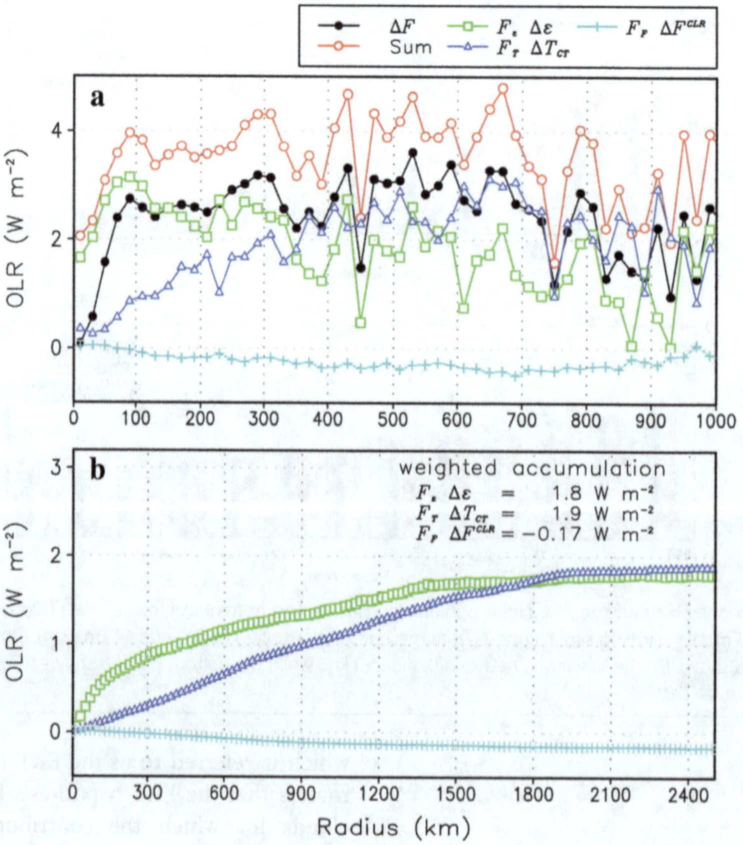

Fig. 6 Budget analysis of outgoing longwave radiation changes due to global warming. Cloud size is binned by radius (20 km bins), showing **a** mean values of outgoing longwave radiation (OLR) for each cloud size and **b** the OLR area-ratio-and-frequency-of-occurrence-weighted accumulation in the *x* (radius) direction (i.e., $\sum_{ir=1}^{\infty} \varphi_{ir} \frac{\pi r_{ir}^2}{S} P_{ir}$, where $S = \sum_{ir=1}^{\infty} \pi r_{ir}^2$, φ_{ir}, and r_{ir} denote values of each term shown in **a** and cloud radius at the *ir*th bin, respectively. *P* denotes the probability distribution function of high cloud number). The original OLR values (black), diagnosed values from Eq. (2) (red), and the contributions of each term on the right hind side of Eq. (2) (green, blue, and cyan) are plotted. Adapted from Noda et al. (2016)

whether the responses of clouds, particularly high clouds, to global warming in NICAM simulations are robust irrespective of the choice of cloud microphysics schemes. Chen et al. (2016) further calculated the cloud radiative feedback of the results from NSW6 and NDW6 using the cloud radiative kernel proposed by Zelinka et al. (2012). Because the available integration time was limited to 1 year for NSW6 and to the boreal summer experiment from June to August for NDW6, Chen et al. (2016) focused on the longwave (LW) cloud feedback, which exhibit a smaller seasonal dependency compared with SW cloud feedback. The 1-year experiment with NSW6 is referred to as NSW6-ANNUAL, the boreal summer experiment with NSW6 as NSW6-JJA, and the experiment with NDW6 as NDW6-JJA.

Figure 7 plots the global mean SW, LW, and net cloud radiative feedback corresponding to all clouds and each type of clouds, with clouds classified by cloud top pressure (CTP) and cloud optical depths (COD). The results from CMIP5 and the Cloud Feedback Model Intercomparison Project Phase 2 (CFMIP2[4]) for each model and their average are also plotted for comparison. The comparison between NSW6-ANNUAL and NSW6-JJA indicates that the seasonal variation in the net cloud feedback comes mainly from the seasonal variation in the SW cloud feedback and that the LW cloud feedback are almost stable irrespective of season. The differences in the LW cloud feedback between NSW6-JJA and NDW6-JJA are then examined; the total LW cloud feedback and those of almost all the cloud categories for both NSW6 and NDW6 are within the uncertainty of the CMIP5/CFMIP2 model ensemble (Fig. 7). However, there is an exception; the LW cloud feedback attributed to the thin-cloud response to global warming are considerably larger than those in the CMIP5/CFMIP2 model ensembles.

Figure 8 decomposes the LW cloud feedback shown in Fig. 7 into each cloud category for NSW6-JJA (Fig. 8a) and NDW6-JJA (Fig. 8b). The results demonstrate that the large positive LW cloud feedback are attributed mostly to cirrus clouds (high and thin clouds) in both NSW6-JJA and NDW6-JJA. This large cirrus cloud

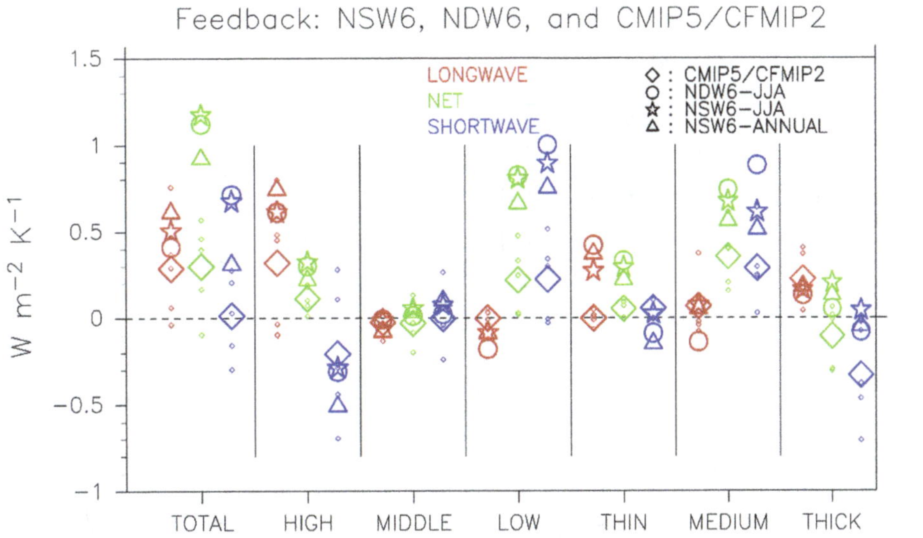

Fig. 7 The global mean cloud feedback. Clouds are classified by cloud top pressure (CTP) and cloud optical depth (COD): high clouds, CTP ≤ 440 hPa; middle clouds, 440 hPa ≤ CTP < 680 hPa; low clouds, 680 hPa ≤ CTP < 1000 hPa; thin clouds, 0.3 ≤ COD < 3.6; medium clouds, 3.6 ≤ COD < 23; thick clouds, 23 ≤ COD). Longwave, shortwave, and net cloud feedback are shown in red, blue, and green, respectively, for NSW6-ANNUAL (triangles), NSW6-JJA (stars), NDW6-JJA (circles), and CMIP5/CFMIP2 average (diamonds). The small dots indicate each member of the CMIP5/CFMIP2 models (courtesy of Dr. Mark Zelinka). Adapted from Chen et al. (2016)

contribution to the LW cloud feedback is dominated by clouds with CTP < 180 hPa for both NSW6-JJA and NDW6-JJA. For NSW6-JJA, clouds with CTP < 180 hPa and COD < 3.6 contribute to the positive LW cloud feedback, whereas clouds with CTP > 180 hPa contribute to negative LW cloud feedback. For NDW6-JJA, among high clouds, clouds with CTP < 310 hPa contribute to positive LW cloud feedback, whereas LW cloud feedback of clouds with CTP < 180 hPa and 1.3 < COD < 3.6 are slightly negative. This difference is consistent with the difference in cloud fraction response to global warming (not shown).

Figure 9 shows the zonal mean distributions of cirrus cloud (high and thin cloud) change and the LW cloud feedback attributed to cirrus cloud changes for NSW6-JJA (Fig. 9a) and NDW6-JJA (Fig. 9b). For NSW6-JJA, the cirrus cloud fraction significantly increases in the tropics and, as a result, the LW cloud feedback attributed to cirrus clouds are mostly dominant in the tropics. In contrast, although the NDW6-JJA cirrus cloud fraction changes mainly in the tropics, appreciable changes occur in the mid-latitudes. This means that the LW cloud feedback from cirrus clouds dominate in the tropics, although they are non-negligible in the mid-latitudes.

As described above, Chen et al. (2016) indicated that projected changes of high clouds simulated by NICAM with the two different cloud microphysics schemes show similar increases in high and thin clouds over the tropics and the subtropics. Chen et al. (2016) further analyzed the difference between the two schemes and found that the change in the effective radius of ice particles has a non-negligible impact on LWCRF in NDW6 through emissivity of clouds. This result demonstrates the importance of ice cloud microphysics in making more accurate assessments of cloud radiative forcing.

Clouds, precipitation, and storm changes in the multi-decadal experiments

As introduced earlier in this review, a long-term multi-decadal simulation using NICAM with a horizontal 14 km mesh has been achieved (Kodama et al. 2015; Satoh et al. 2015, 2017; Yamada et al. 2017). The NICAM 14 km AMIP-type simulations were performed for the present climate of 1979–2008 and the future climate of 2074–2100. The present simulation was analyzed by Kodama et al. (2015) and Satoh et al. (2017), while Satoh et al. (2015) and Yamada et al. (2017) discussed projection of TCs using the present and the future simulations. Kodama et al. (2015) and Satoh et al. (2015) only used the first 20 years simulation from June 1979 to May 1998, because of the simulation was not finished at the time these papers were published. In this and the next subsections, we review future changes in cloud and precipitation and associated circulation changes, such as TCs and storm tracks, due to global warming using outputs of long-term NICAM simulations.

NSW6 was utilized to explicitly calculate cloud microphysics without a cumulus parameterization scheme. The atmospheric radiation transfer scheme MSTRNX (Sekiguchi and Nakajima 2008), a second-order

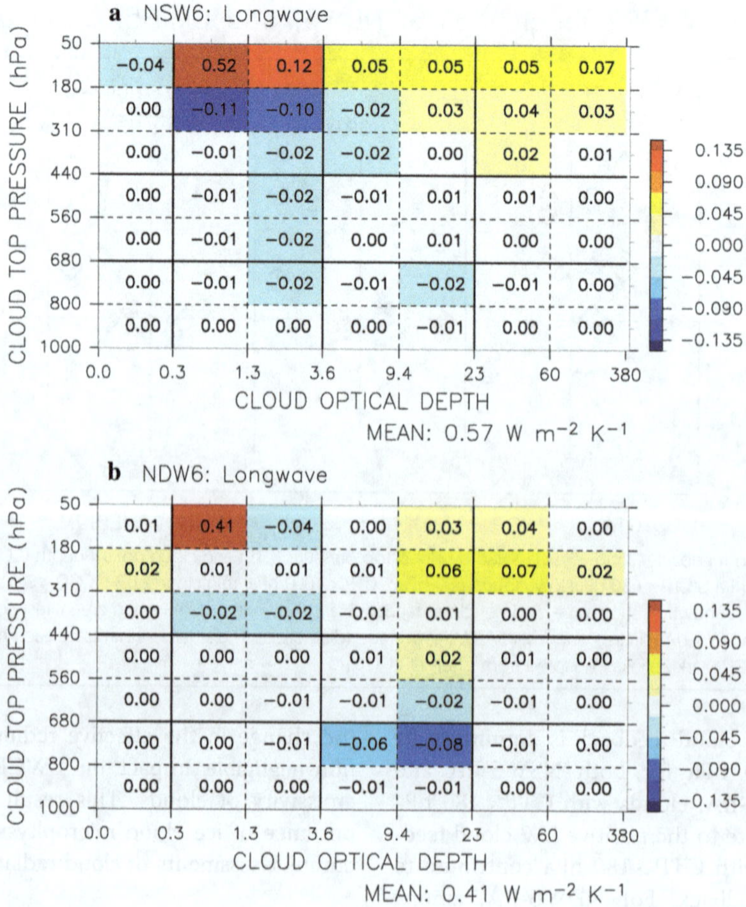

Fig. 8 Longwave cloud feedback [W m^{-2} K^{-1}] on International Satellite Cloud Climatology Project (ISCCP) diagrams estimated from the cloud radiative kernel. Longwave cloud feedback in **a** NSW6-JJA and **b** NDW6-JJA are plotted in the ISCCP diagram with cloud top pressure (vertical axis) and cloud optical depth (lateral axis). Adapted from Chen et al. (2016)

turbulence closure scheme (Nakanishi and Niino 2006; Noda et al. 2010) and the land surface model MATSIRO (Takata et al. 2003) have been used. A slab ocean model was utilized to simulate SSTs, which were nudged toward Taylor-corrected (Taylor et al. 2000) HadISST1 values in the present-day climate simulation. In the future climate simulation, the simulated SSTs were nudged toward future SSTs, which were estimated by adding the SST differences between the present-day (1979–2003) and future (2075–2099) climate periods to the HadISST1 values. The SST differences were obtained from outputs of the CMIP3 dataset of the World Climate Research Program (WCRP) (Mizuta et al. 2008).

Figure 10 shows the climatological daytime cloud amount and its future change for high, middle, and low clouds. The cloud amount was estimated using the International Satellite Cloud Climatology Project (ISCCP) simulator (Klein and Jakob 1999; Webb et al. 2001). The high-cloud amount is increased, as shown by the upper panels of Fig. 10, consistent with the results of previous studies using NICAM (Satoh et al. 2012; Chen et al. 2016). The low-cloud amount is decreased between 60°S and 60°N, which is prominent near the west coast of continents, and is increased at high latitudes (polar side of about 60°S and 60°N). The decrease in the low-cloud amount is also related to an increase in cloud-top height due to warming SST. The reduction of the low-cloud amount is similarly seen in many conventional GCMs (Zelinka et al. 2013, 2016, 2017).

Figure 11 shows the seasonal climatological means of the simulated precipitation rate and its future change during June–July–August (JJA), September–October–November (SON), December–January–February (DJF), and March–April–May (MAM). Under the future climate condition, the global averaged precipitation rate increased by approximately 1.0 mm day^{-1} for all seasons. The distribution of zonal mean precipitation rates is shown in the right panels of Fig. 11. The model projects that the precipitation rate is increased in the tropics, slightly decreased over the subtropics, and increased at the higher latitudes. An increase is prominent near ITCZ throughout all seasons. Clear decreases are seen over the

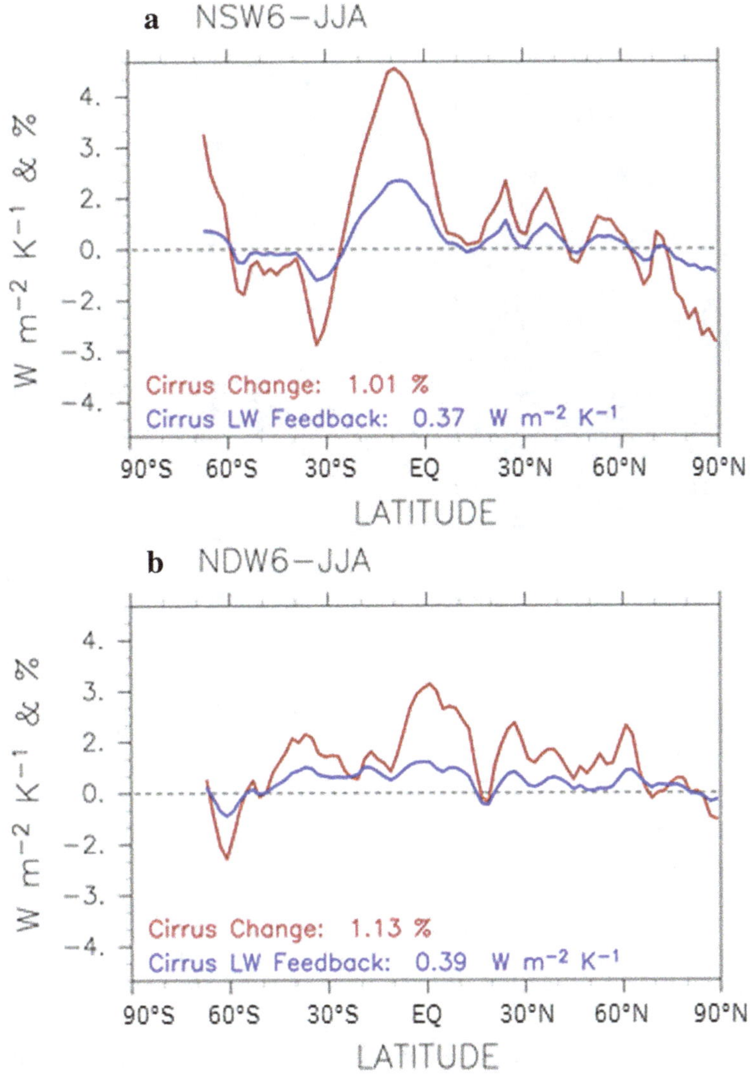

Fig. 9 The zonal mean distribution of cloud changes. Cirrus cloud changes (red lines) and longwave cloud feedback contributed by cirrus clouds changes (blue lines) are shown for **a** NSW6-JJA and **b** NDW6-JJA. Note that the longwave cloud radiative feedback are calculated using the cloud radiative kernel proposed by Zelinka et al. (2012). Adapted from Chen et al. (2016)

subtropical eastern North Pacific during JJA, and the tropical Indian Ocean during JJA and SON.

The change in TC activities is documented in Satoh et al. (2015) and Yamada et al. (2017). Precipitation related to TCs has been investigated in previous studies (Rodgers et al. 2000; Scoccimarro et al. 2014), which used an observational dataset to show that TCs contributed considerably to rainfall over the subtropical eastern North Pacific, where the contribution was more than 40% at maximum. Figure 12 shows the contributions of rainfall associated with TCs to the total rainfall in the present-day and future climate simulations for JJA, SON, DJF, and MAM. Here, rainfall associated with TCs is defined as rainfall within 500 km from the center of the TC (hereafter, referred to as TC-rainfall). The TC-rainfall rate is high over the subtropics during all seasons in both climate conditions (right panels of Fig. 12). In the present-day climate simulation, the contribution of TC-rainfall to total rainfall is more than 60% during JJA and SON over the subtropical eastern North Pacific, where TCs are frequently generated during these seasons. This result is qualitatively consistent with observational studies (Rodgers et al. 2000; Scoccimarro et al. 2014). Under the future climate condition, the TC-rainfall rate during JJA is decreased over the subtropical eastern North Pacific (center panels of Fig. 12), where the total precipitation rate is also decreased (center panels of Fig. 11). Therefore, future change in TCs appears to contribute to the decrease in precipitation over the subtropical eastern North Pacific in the future climate simulation.

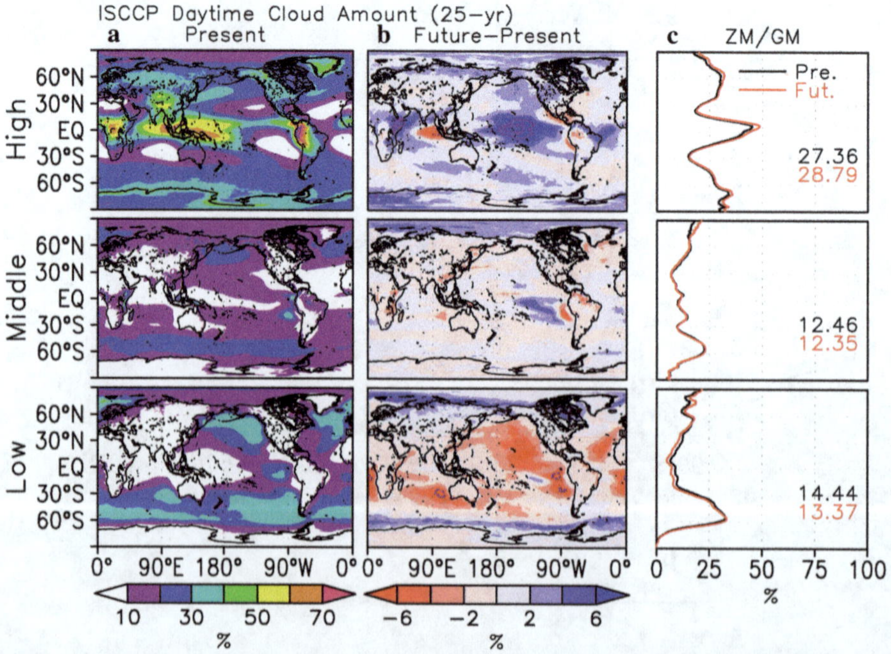

Fig. 10 Annual mean International Satellite Cloud Climatology Project (ISCCP) daytime cloud amount (%). Panels represent high, middle, and low clouds from top to bottom. **a** Results of the present-day climate simulation (1979–2003) with ISCCP simulator. **b** Differences between the future climate (2075–2099) and present-day climate simulations with ISCCP simulator. **c** The zonal mean (ZM) and global mean (GM) values for the present-day climate simulation (black) and the future climate simulation (red). Adapted from Chen et al. (2016)

Storm track and its cloud radiative effect

It is still unclear how clouds interact with large-scale circulation in the extratropics. Many studies using GCMs project a poleward shift of the storm track, a reduction in the total number of extratropical cyclones, and an increase in the number of intense extratropical cyclones due to global warming (Geng and Sugi 2003; Yin 2005; Mizuta 2012). Most GCMs also project a negative cloud feedback by SW radiation in the higher latitudes (Zelinka et al. 2012) in association with an increase in liquid water optical depth (Ceppi et al. 2015; Bodas-Salcedo et al. 2016). It has also been suggested that enhanced extratropical cyclone intensities due to global warming induce more SW cooling and more LW warming (Tselioudis and Rossow 2006). In contrast, a poleward shift of the storm track does not seem to induce a significant positive SW cloud radiative effect in the southern hemisphere (Grise and Polvani 2014). These previous studies employed conventional GCMs with a mesh size of O (10^2 km). Much higher horizontal resolution is necessary to reproduce fine-scale structures of extratropical cyclones such as cold and warm fronts.

Bengtsson et al. (2009) and Champion et al. (2011) used global models with 40–60 km horizontal resolutions and projected an increase in precipitation around extratropical cyclones due to global warming. They also suggested a need for non-hydrostatic model simulations with finer horizontal resolution to study the responses of precipitation and wind fields. Miura et al. (2005) analyzed a set of short-term aqua planet experiments and found that high-latitude total cloud amount responded differently to an increase in SST between 7 km mesh NICAM and T42 (2.8°) GCM. Kodama et al. (2014) further analyzed the output of aqua planet experiments using 14 km mesh NICAM and showed an increase in liquid water optical depth and enhancement of SW cooling, consistent with the GCM results.

In this subsection, we show additional analysis of the cloud radiative effect in the extratropics using the output data of the present-day and future climate simulations described in the previous subsections. We focus on the southern hemisphere for simplicity. Mean sea level pressure, temperature, and wind fields from the JRA-55 reanalysis dataset (Kobayashi et al. 2015; Harada et al. 2016) and cloud radiative effect from CERES SYN1deg computed product (Loeb et al. 2009) are used to evaluate the performance of the present-day climate simulation.

We detected and tracked extratropical cyclones in the reanalysis and simulations based on the mean sea level pressure. The mean structure of a field variable around extratropical cyclones was diagnosed by compositing it against cyclone centers following the methodology described by Kodama et al. (2014) with slight modifications. In addition, the relationship between the location of the mid-latitude low-level jet and the magnitude of

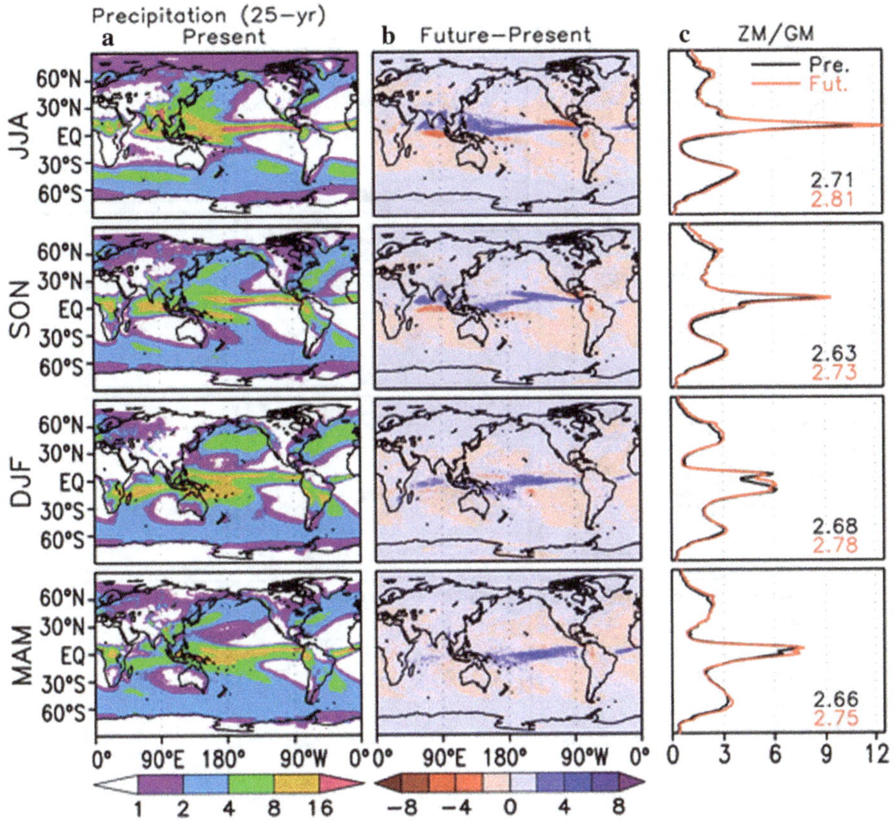

Fig. 11 Seasonal mean climatological precipitation rates (mm day^{-1}) throughout the year. Panels show June-July-August (JJA), September-October-November (SON), December-January-February (DJF), and March-April-May (MAM). **a** Results of the present-day climate simulation (1979–2003). **b** Differences between future climate (2075–2099) and the present-day climate simulation. **c** The zonal mean (ZM) and global mean (GM) values for the present-day climate simulation (black) and the future climate simulation (red)

the cloud radiative effect was analyzed following Grise and Polvani (2014). First, anomalies from monthly mean climatology were taken for LW and SW cloud radiative effects and peak latitude of the zonal mean zonal wind at 1500 m in altitude (hereafter, jet latitude). Then, regressions on the jet latitude were calculated to obtain changes in the LW and SW radiative effects by jet shift of 1° in latitude.

Figure 13a, d shows the mean horizontal structures of the observed LW and SW cloud radiative effects around extratropical cyclones in the southern hemisphere. Net cloud radiative effect is negative as a result of a compensation between SW cooling and LW warming. A characteristic comma-like pattern is evident around the poleward and eastward sides of the cyclone center. Overall, results from the present-day climate simulation (Figs. 13b, e) capture the observed horizontal structure of the LW and SW cloud radiative effects around extratropical cyclones. The simulated LW cloud radiative effect agrees well with observation, with slightly warmer bias. The model underestimates SW cloud radiative cooling by roughly 10 W m^{-2} or more, which is consistent with an underestimation of the zonal mean low- and middle-level cloud fractions (Kodama et al. 2015).

Figure 13c, f shows the responses of LW and SW cloud radiative effects to global warming. SW cloud radiative cooling is enhanced in the entire extratropics and peaks near the cyclone center. LW cloud radiative warming is slightly increased around the poleward and eastward sides of the cyclone center. These changes in SW and LW cloud radiative effects resemble changes in the liquid and ice water paths, respectively (not shown), which are consistent with the results of the aqua planet experiments (Kodama et al. 2014). The magnitudes of the changes in SW and LW cloud radiative effects (scaled by changes in surface air temperature) are far greater in the aqua planet experiments than in this study.

Figure 14 shows changes in the LW and SW cloud radiative effects. This reflects effects of the 1° poleward shift of the jet latitude. The simulated jet, on average, is located at 50°S. As the jet shifts poleward, the anomaly of the LW cloud radiative effect becomes positive in the

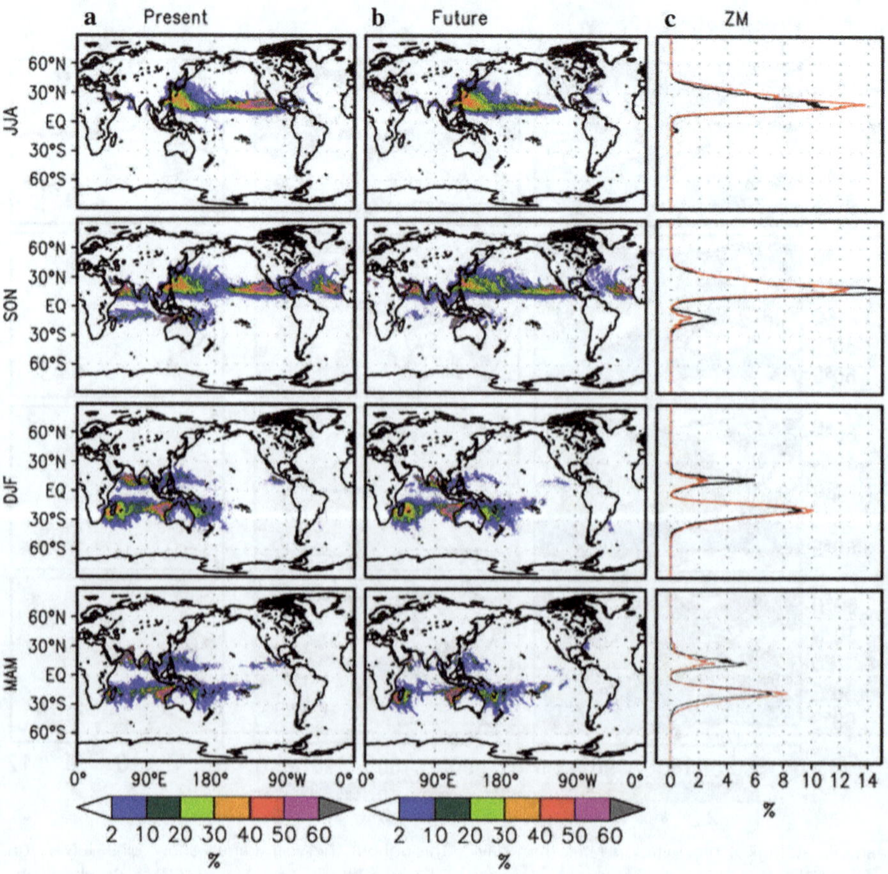

Fig. 12 Seasonal mean climatological contributions (%) of rainfall associated with tropical cyclones to total rainfall. Results of the **a** present-day climate simulation (1979–2003) and **b** the future climate simulation (2075–2099). **c** The zonal mean (ZM) values for the present-day climate simulation (black) and for the future climate simulation (red). The rainfall associated with tropical cyclones is defined as rainfall within 500 km from the center of tropical cyclones. The panels follow the same seasons as defined in Fig. 11

poleward flank of the jet latitude and negative in the equator flank. The response of SW cloud radiative effects to the jet shift is small in the extratropics and negative in the subtropics. Grise and Polvani (2014) showed that, as the jet shifts poleward, half of the CMIP5 models (type I models in their study) reproduce the positive anomaly of SW cloud radiative effect induced by the poleward jet shift, whereas the other half (type II models) show no significant changes in SW cloud radiative effect. Analysis of the CERES product supports type II models (Grise and Polvani 2014). Our model is also consistent with the type II models, supporting the discovery that the position of the dynamic storm track does not influence the cloud radiative effect in the mid-latitudes.

Discussion

Cloud feedback

For the past decade, seasonal- to annual-scale simulations by NICAM have been used to investigate clouds, circulation, and climate sensitivity. Using outputs of 14 km mesh present-day and future climate simulations by NICAM (Kodama et al. 2015), we estimated the cloud feedback by analyzing the dataset for the climate-scale output (Fig. 15). As a result, the cloud feedback parameters derived from the NICAM climate simulations (Fig. A2 in Chen et al. 2016; Fig. 15) are quantitatively similar to those from seasonal- to annual-scale simulations (Fig. 7). Specifically, positive LW radiative feedback from thin cloud (mainly cirrus) changes, which has been suggested to be different from other GCMs by Chen et al. (2016), is still found in NICAM climate simulations. In addition, low-level cloud is decreased more due to global warming compared with the CMIP3 and CMIP5 models, leading to a greater positive SW feedback. Note that the low-level cloud fraction simulated by NICAM is largely underestimated compared with satellite observations (Kodama et al. 2015). Comparison of Figs. 15 and 7 also indicates that the long-term climate-scale simulation is necessary to quantitatively assess cloud feedback and therefore climate sensitivity. For example, the total cloud feedback obtained from the short-term NICAM

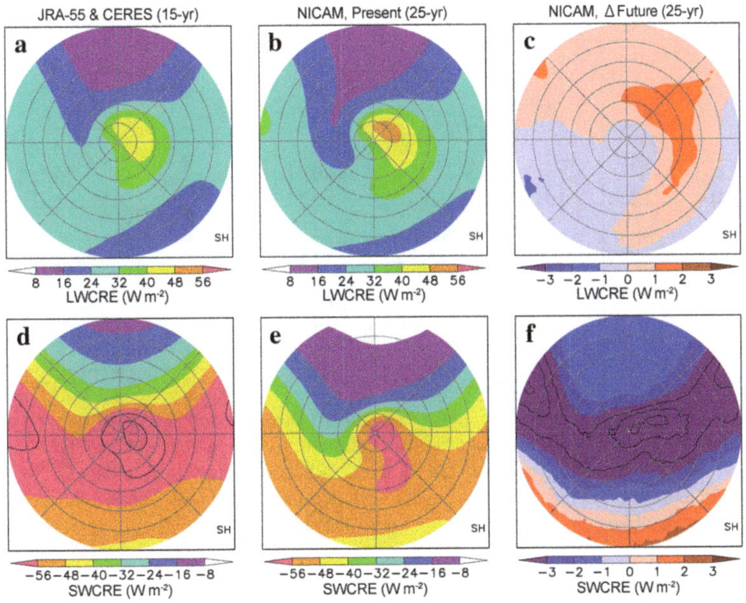

Fig. 13 Composite of longwave and shortwave cloud radiative effects averaged over extratropical cyclones in the southern hemisphere. **a**, **d** JRA-55 reanalysis and CERES computed product for March 2000 to February 2015. **b**, **e** NICAM for 1979–2003. **c**, **f** NICAM for 2075–2099 minus that for 1979–2003. In addition to color scales, additional contours are added in **d** and **f**; contours show − 64 and − 72 W m^{-2} in **d** and − 4, − 5 and − 6 W m^{-2} in **f**. The concentric gray circles represent increments of 500 km from the center of extratropical cyclones. The composites are meridionally reversed, and gray straight lines indicate (clockwise from top) poleward, pole-eastward, eastward, equator-eastward, equatorward, equator-westward, westward, and pole-westward directions from the center of extratropical cyclones

simulation (Fig. 7) is greater than that obtained in almost all the CMIP3 and CMIP5 models, whereas the total cloud feedback from the long-term NICAM climate simulation (Fig. 15) is comparable with that of the CMIP3 and CMIP5 model ensembles.

In the near future, historical simulations using NICAM with 56, 28, and 14 km meshes will be performed under the protocol of the Coupled Model Intercomparison Project Phase 6 (CMIP6) High Resolution Model Intercomparison Project (HighResMIP; Haarsma et al. 2016). The use of multiple horizontal resolutions may provide us with an explanation of the different cloud responses among GCMs and NICAM (and global cloud-system resolving models). The influence of vertical resolution on cloud feedback should also be investigated, as a vertical resolution of 400 m

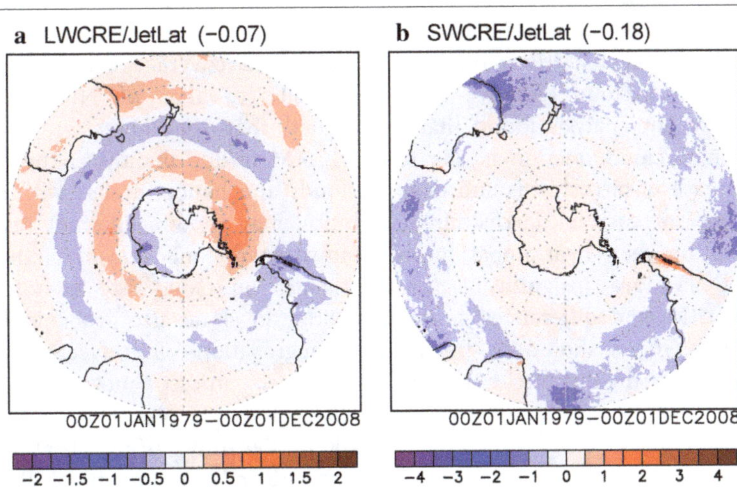

Fig. 14 Response of cloud radiative effects to poleward jet shift in a present-day climate simulation (W m^{-2} degree^{-1}). **a** Longwave cloud radiative effects (LWCRE); **b** shortwave cloud radiative effects (SWCRE). Gray dotted circles represent latitudes in 10° increments between 20°S and 80°S. Values at the top of each panel are spatial mean regressions averaged over 30°S–60°S. See main text for analysis method

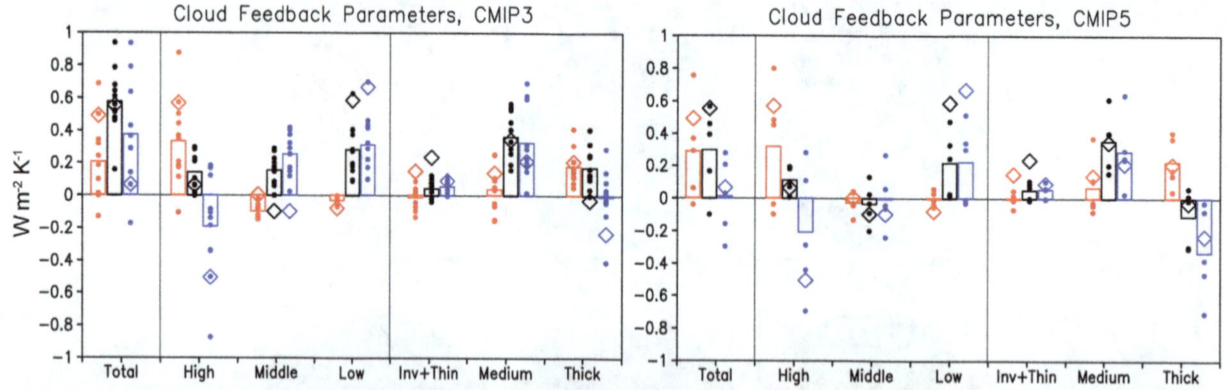

Fig. 15 Cloud feedback parameters (W m^{-2} K^{-1}) contributed from cloud radiative effects. Cloud categories follow those described in Fig. 7. "Inv" means invisible clouds. Longwave (red), shortwave (blue), and net cloud radiations (black) are derived from NICAM (rhombus), CMIP3 (left panel, dot), and CMIP5 models (right panel, dots). Ensemble means of CMIP3 and CMIP5 models are shown as bars. For NICAM, outputs of present-day (1979–2003) and future (2075–2099) climate simulations are used. Cloud feedback parameters of CMIP3 and CMIP5 models were provided by Dr. Mark Zelinka and are the same as in Fig. 8 of Zelinka et al. (2012) except that a small bug has been fixed

is necessary to resolve cirrus clouds (Seiki et al. 2015b). The HighResMIP simulation, which covers 1950–2014, will also provide us with the first opportunity to discuss decadal variability using a global cloud-system resolving model. For example, Zhou et al. (2016) analyzed their own historical and future climate simulations by using a GCM and found that the cloud feedback parameter derived from the recent 30-year simulation is smaller than that from the recent 100-year and future climate simulations, seemingly due to a different spatial pattern of SST rise.

Constraining climate sensitivity using observation is also of great interest to reduce uncertainty in climate sensitivity. It has been suggested from observed interannual variability that high-level cloud fraction tends to decrease as SST increases (Zelinka and Hartmann 2011; Su et al. 2017). By breaking down high-cloud change into optical depth, however, cirrus cloud can cause a significant increase in cloud radiative feedback (Zhou et al. 2014). Recent analysis found that a net decrease in tropical high-cloud fraction is due to a compensation of a decrease in cirrostratus and deep convective clouds and an increase in cirrus cloud, and that the relationship between near-global mean high-cloud fraction and SST are insignificant (Liu et al. 2017). These similarities and differences among observations, NICAM, and GCMs should be further investigated to offer clues to reducing uncertainty in climate sensitivity.

Future directions

Despite the outcomes of the SOUSEI project, we acknowledge the need for further studies in some areas. One of these areas concerns an increased high-cloud coverage due to global warming, as analyzed by Chen et al. (2016). The NICAM simulations show an increase in high-cloud coverage together with a reduction in IWP in the tropics. We need to understand the relationship between changes in IWP and cloud coverage in the tropics, how this response is consistent with observations, and whether it is robust or different among other climate models. For example, Satoh et al. (2012) used a simple model to find that reduced convective mass fluxes led to less IWP. Bony et al. (2016) compared the results of GCMs that were run in a radiative-convective equilibrium (RCE) configuration and found that GCMs generally showed reduction of high-cloud coverage. They attribute this partly to a reduced divergent flow near the tropopause, which is caused by reduced convective mass flux in a stabilized atmosphere, and partly to aggregation of convective clouds. The response of high-cloud coverage shown by Bony et al. (2016) is opposite to that of NICAM. Noda et al. (2014) argued that the increase in high-cloud coverage in NICAM is due mainly to the contribution of smaller clouds and hence models with higher horizontal resolution may show a different response for high clouds. However, the mechanism of the dependency of cloud changes on cloud size is not well understood. The change in cloud size probability distribution is related to aggregation of convective clouds, and this issue has been studied most actively since a study on self-aggregation of convective cloud (Bretherton et al. 2005). Recently, Wing et al. (2018) proposed intercomparison experiments of RCE to comprehensively understand the response of convective clouds. RCE simulations with NICAM suggest that the response of high clouds in RCE are related to cloud microphysics processes and turbulence (Ohno and Satoh 2018).

In terms of important research topics that should be addressed by the climate research community in the next decade, Bony et al. (2015) proposed a study on "Cloud, Circulation and Climate Sensitivity" as one of the WCRP Grand Challenges. In particular, they asked four questions: What role does convection play in cloud feedback? What controls the position, strength, and variability of storm tracks? What controls the position, strength, and variability of tropical rain belts? What role does convective aggregation play in climate? High-resolution global nonhydrostatic models, including NICAM, are expected to play increasingly important roles in answering these questions. In particular, remarkable progress in computer technology now enables global cloud-system-resolving simulations on a multi-decadal scale (e.g., Kodama et al. 2015). We now need to understand the robustness of cloud feedback processes by investigating differences between such high-resolution climate simulations and conventional GCM simulations.

To contribute to the forthcoming CMIP6 project (Eyring et al. 2016), we plan to conduct climate projection studies using NICAM in cooperation with the new call for climate projection studies in Japan, called the Integrated Research Program for Advancing Climate Models (referred to as the TOUGOU program[5]), which began in May 2017. As part of CMIP6, HighResMIP places more importance on higher resolution simulations to assess more detailed changes of extreme events (Haarsma et al. 2016), and we will be expected to contribute to it. We plan to perform a series of AMIP-type climate simulations using NICAM with 14, 28, and 56 km meshes under a protocol of HighResMIP in order to analyze cloud and precipitation changes and their relationships to large-scale circulations, as was analyzed using a previous data set (Noda and Satoh 2014; Noda et al. 2015).

Using a hierarchy of numerical models is also an effective way to accelerate improvements in our understanding of cloud responses to a warming world. Idealized frameworks such as aqua planet simulations would provide us with an essential mechanism to understand cloud response due to global warming (Miura et al. 2005; Stevens and Bony 2013; Kodama et al. 2014). In particular, a new intermodal comparison project using RCE simulations was recently initiated (Wing et al. 2018), and RCE results with NICAM (Ohno and Satoh 2018) show different high-cloud sensitivity to SST from those of GCMs (Bony et al. 2016). Climate research has suffered from the uncertainty of modeled cloud behavior and associated large-scale circulation fields even in the idealized aqua planet assumption, due to uncertainty arising from cumulus parameterization (Williamson et al. 2013; Stevens and Bony 2013). The use of high-resolution modeling that directly calculate cloud processes using cloud microphysics schemes without cumulus parameterization is expected to improve our understanding of cloud responses to global warming, further reducing the uncertainties derived from different model results.

Interaction between clouds and circulation in the storm-track region is one of the key challenges in the WCRP Grand Challenge, "Clouds, Circulation and Climate Sensitivity" (Bony et al. 2015). It is of great interest to investigate interaction among radiation, latent heat, and the development and track of extratropical cyclones using NICAM, which explicitly treats cloud microphysical processes without the aid of convection schemes. Latent heat may not only accelerate the development of extratropical cyclones but also affect their track and size. A baroclinic lifecycle experiment suggested a decrease in the horizontal scale of extratropical cyclones in a moister world (Booth et al. 2013). Li et al. (2015) showed suppression of the storm-track activity and a significant change in the zonal mean basic state by switching off cloud-radiation interactions in a GCM.

It is now possible to observe fine-scale horizontal and vertical structures of extratropical cyclones using high-resolution satellite data with active sensors. Some studies have evaluated cloud and precipitation processes around extratropical cyclone simulated by GCMs using CloudSat (Posselt et al. 2008; Naud et al. 2010; Field et al. 2011) and CALIPSO (Naud et al. 2010; Bodas-Salcedo et al. 2014). Now, the Global Precipitation Measurement (GPM) (Hou et al. 2014) radar products, which cover both the tropics and extratropics, are available, and the vertical structure of frontal precipitation in the extratropics simulated by 3.5 km mesh NICAM was evaluated as a case study (Kotsuki et al. 2014).

In the near future, the HighResMIP dataset will help us to improve our understanding of model bias including resolution dependency and confidence in future changes of cloud and precipitation associated with extratropical cyclones through more comprehensive analysis of the forthcoming high-resolution climate simulations.

Conclusions

This article reviews the outcomes of the NICAM subgroup of the 5-year SOUSEI program, which aimed to reduce the uncertainty in climate models due to cloud processes using a global non-hydrostatic model. NICAM was used to conduct numerical simulations with much higher resolution than conventionally used GCMs by explicitly calculating cloud processes using cloud microphysics schemes without a cumulus parameterization scheme. In NICAM, clouds are more realistically calculated with resolving meso-scale convective systems, and it is thought that uncertainties of cloud simulations derived from cumulus parameterizations are reduced.

In this review, we describe a newly developed double-moment bulk cloud microphysics scheme and stress importance of evaluation using satellite observation. We explore the uncertainties of the cloud microphysics scheme using a large-eddy simulation model and a kinematic model and compare it with a bin microphysics scheme. We discuss clouds and changes in weather systems, such as tropical and extratropical cyclones, using climate projection simulations by NICAM with the time-slice approach. In particular, we focus on changes in high clouds. Based on this work, we propose analysis using the size distribution of high clouds. We also discuss dependencies of high-cloud changes in experiments using two different microphysics schemes. Clouds, precipitation, and storm changes are also analyzed using the multi-decadal experiments by NICAM.

We estimated the cloud feedback parameters simulated by NICAM and discussed their uncertainties. Finally, we explore the prospects for research into global warming using NICAM, which are currently being pursued in the TOUGOU program.

It is thought that high-resolution global non-hydrostatic models without cumulus parameterization will reduce uncertainties due to clouds in estimates of cloud sensitivity. Although the ambiguities related to cumulus parameterization are absent, another type of uncertainty is introduced by the cloud microphysics schemes. However, clouds simulated by these models are more feasible to direct comparison and evaluation with satellite observational data, which cover the entire globe with high resolution. Comprehensive evaluations of simulated clouds are achievable using satellite simulators (Hashino et al. 2013). Cloud microphysics schemes also offer various choices for constructing schemes: classification of hydrometeors, number of moments of prognostic variables or treatment of size distributions of hydrometeors, and cloud habits such as shape and density. While double-moment bulk cloud microphysics schemes are expected to show good performance as shown in this study, single-moment bulk cloud microphysics schemes are still useful (Roh and Satoh 2014). Although bin cloud microphysics schemes used in high-resolution global models require computer power that is an order of magnitude larger than bulk cloud microphysics schemes, they are used as a reference for other simpler cloud microphysics schemes.

High-resolution global non-hydrostatic models capture meso-scale convective systems over the globe and then reproduce large-scale organized convective systems, including TCs, extratropical cyclones, and cloud clusters in the tropics. Evaluations of these meso-scale convective systems add another dimension to model evaluations, which were not possible for GCMs. Cloud size analysis is an example of the evaluation of cloud systems (Fig. 5). Through these efforts, the use of this type of high-resolution models will lead to reduction in the uncertainty in climate sensitivity related to cloud processes.

Even though high-resolution global non-hydrostatic models reproduce realistic cloud structure, they still have biases in climatology, such as distributions of precipitation (Kinter et al. 2013; Kodama et al. 2015). They still have other ambiguities of GCMs including shallow clouds and planetary boundary layer schemes. Coupling to ocean and land models also introduces sources of model bias. Although uncertainties related to cloud processes may be reduced, continuous efforts are required to improve model biases and total reproducibility of climatology. The new DYAMOND project[6] involves the intercomparison of global non-hydrostatic models with mesh size less than 5 km. Studies such as this one will help to elucidate the advantages and limitations of high-resolution global non-hydrostatic models.

Endnotes

[1]SOUSEI Program https://www.jamstec.go.jp/sousei/eng/. Accessed 15 Oct 2018.

[2]Earth Simulator http://www.jamstec.go.jp/es/en/. Accessed 15 Oct 2018.

[3]CMIP5 https://cmip.llnl.gov/cmip5/. Accessed 15 Oct 2018.

[4]CFMIP2 https://www.earthsystemcog.org/projects/cfmip/cfmip2-cmip5. Accessed 15 Oct 2018.

[5]TOUGOU program http://www.jamstec.go.jp/tougou/eng/. Accessed 15 Oct 2018.

[6]DYAMOND https://www.esiwace.eu/services/dyamond. Accessed 15 Oct 2018.

Appendix

Description of NDW6

In the SOUSEI program, Seiki et al. (2014, 2015a, b) evaluated and improved the ice cloud microphysics in NDW6. Seiki et al. (2014) proposed a simple but efficient method to model non-sphericity of ice particles. Uncertainties in ice cloud modeling originate partly from the uncertainties in the interrelationships among the mass x, maximum dimension D, projected area to the flow (cross section) A, and terminal velocity of ice particles, which are generally determined according to in situ measurements (Ono 1969; Auer and Veal 1970; Locatelli and Hobbs 1974; Heymsfield and Kajikawa 1987; Mitchell 1996). In NDW6, the power law relationships between x and D and between x and A are assumed as follows:

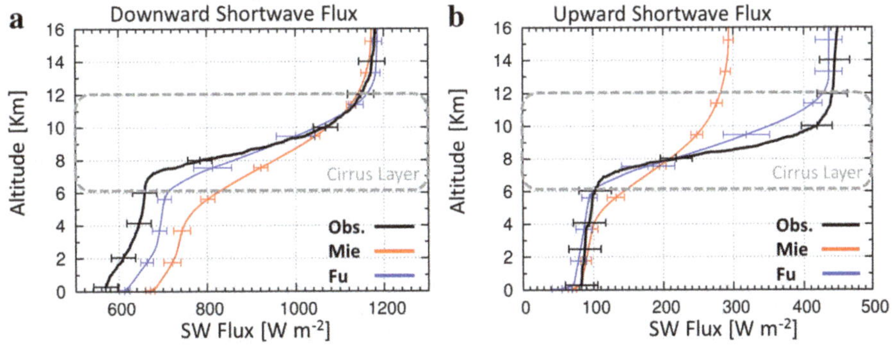

Fig. 16 The vertical profiles of shortwave fluxes. Downward shortwave flux **a** and upward shortwave flux **b** are shown for observation (black line) and for simulations using the radiation database calculated according to the Mie theory (red line) and the Fu database (blue line), assuming non-spherical particles. The error bars in the observed profiles indicate the estimated possible range, and in the simulated profiles indicate the standard deviation within the 63 sampled profiles. The gray dashed boxes indicate the cirrus layer. The simulated profiles are from the M92PLT experiment described in Seiki et al. (2014). The target cirrus clouds had an optical thickness of 6.1 ± 0.5 over the land around Japan (30°N to 40°N and 130°E to 142°E) on 8 June 1995. Adapted from Seiki et al. (2014)

$$D = a_D x^{b_D},$$

$$A = a_A x^{b_A},$$

where a_D, b_D, a_A, and b_A are constants determined experimentally from in situ observations by Mitchell (1996). The terminal velocities of ice particles are approximated using values from a pre-calculated database according to the theoretical methodology proposed by Bohm (1989) and Mitchell (1996). Seiki et al. (2014) chose the assemblages of planer polycrystals in cirrus clouds for snow, lump graupel for graupel, and hexagonal columns for cloud ice (with hexagonal plate as an optional extra for cloud ice) from Mitchell's database, and the abovementioned power law relationships were defined for each hydrometeor category. Vapor deposition/sublimation, melting, collisional growth, and gravitational sedimentation were calculated using the power law relationships. In Seiki et al. (2015b), homogeneous ice nucleation was newly implemented by referring to Ren and MacKenzie (2005) and Kärcher et al. (2006). In addition, Seiki et al. (2015b) reduced the sticking efficiency to its lower limit at 253 K in ice crystal aggregation, because the parameterized sticking efficiency was originally determined by laboratory experiments within temperature ranges mostly between 253 and 273 K (see Figs. 14–18 in Pruppacher and Klett 1997).

The differences between the versions of NDW6 referred to in this review are summarized in Table 1. We used different versions of NDW6 from Seiki and Nakajima (2014), referred to as SN14; Seiki et al. (2014), referred to as S14; and Seiki et al. (2015b), referred to as S15. S14 used the same settings as P07COL in Seiki et al. (2014), and S15 used the same settings as G9CTL in Seiki et al. (2015b). These simulations were performed using NICAM with 14 km horizontal resolution and 40 vertical layers. Seiki et al. (2015b) also performed simulations with finer spatial resolution using the same settings as S15, except for the resolution. NICAM had 74 vertical layers in S15 V (the same as G9Z4 in Seiki et al. 2015b) and had 74 vertical layers with 7 km horizontal resolution in S15VH (the same as G10Z4 in Seiki et al. 2015b).

Seiki et al. (2014) calculated cloud optical properties (the volume extinction coefficient, the volume absorption coefficient, the asymmetry factor, and the second moment of phase function) and prepared a look-up table of the effective radius in the range from 1 μm to 1 mm to cover precipitating hydrometeors for a broadband radiative transfer model MSTRNX (Sekiguchi and Nakajima 2008). The optical properties of cloud water and rain were calculated according to the Mie theory, and the optical properties of cloud ice, snow, and graupel (assuming non-spherical ice particles) were provided by Fu (1996) and Fu et al. (1998). This database of optical properties was first implemented in NDW6 and NSW6 (in 2014 and 2017, respectively; the range of the effective radii was limited to less than 80 μm and all hydrometeors were assumed to be spherical).

It is significant that the asymmetry factor decreases as the aspect ratio departs from unity for non-spherical particles (e.g., Fu 2007), because the cloud albedo is known to increase as the asymmetry factor decreases (Liou 2002). Seiki et al. (2014) evaluated the improvement in the simulated radiative fluxes by using the Fu database for ice particles and comparing them to in situ balloon-borne sensor observations of cirrus clouds along a mid-latitude frontal system. They found that the simulated radiative fluxes using the Fu database, particularly in the SW range, were in good agreement with observations, in contrast to the simulated fluxes using the database calculated according to the Mie theory (see Fig. 16).

When using the Fu database, the downward (upward) SW flux significantly decreases (increases) through the cirrus layer. The remaining differences between the simulated and observed SW fluxes originated from both model and observation errors and are discussed in Seiki et al. (2014).

Abbreviations

AMIP: Atmospheric Model Intercomparison Project; CERES: Clouds and the Earth's Radiant Energy System; CFMIP2: The Cloud Feedback Model Intercomparison Project Phase 2; CFODD: Contoured frequency by optical depth; CMIP3: Coupled Model Intercomparison Project Phase 3; CMIP5: Coupled Model Intercomparison Project Phase 5; CMIP6: Coupled Model Intercomparison Project Phase 6; COD: Cloud optical depths; CTL: Control simulation; CTP: Cloud top pressure; DJF: December–January–February; DM: Double-moment bulk cloud microphysics scheme; DYAMOND: DYnamics of the Atmospheric general circulation Modeled On Non-hydrostatic Domains; ERA-Interim: European Centre for Medium-Range Weather Forecasts interim reanalysis; FAT: Fixed anvil temperature; GCM: General circulation model; GPM: Global Precipitation Measurement; GW: Global warming simulation; HighResMIP: High Resolution Model Intercomparison Project; ISCCP: International Satellite Cloud Climatology Project; ISO: Intra-seasonal oscillations; ITCZ: Intertropical Convergence Zone; IWC: Ice water content; IWP: Ice water path; JAMSTEC : Japan Agency for Marine-Earth Science and Technology; JJA: June–July–August; KiD: The Kinetic Driver; LES: Large-eddy simulation; LW: Longwave; LWCRF: Longwave cloud radiative forcing; MAM: March–April–May; NDW6: NICAM double-moment water 6 cloud microphysics scheme; NI: Ice number concentration; NICAM: Non-hydrostatic icosahedral atmospheric model; NOAA: National Oceanic and Atmospheric Administration; NSW6: NICAM single-moment water 6 cloud microphysics scheme; OLR: Outgoing longwave radiation; PD: Present-day simulation; RCE: Radiative-convective equilibrium; REI: Ice effective radius; RICO: Rain in Cumulus over the Ocean; SCALE: Scalable Computing for Advanced Library and Environment; SM: Single-moment bulk cloud microphysics scheme; SON: September–October–November; SP-CAM: Superparameterized Community Atmosphere Model; SST: Sea surface temperature; SW: Shortwave; TC: Tropical cyclone; TRMM: Tropical Rainfall Measuring Mission; WCRP: World Climate Research Program

Acknowledgements

This work is supported by the Program for Risk Information on Climate Change (SOUSEI) and the Integrated Research Program for Advancing Climate Models (TOUGOU) funded by the Ministry of Education, Culture, Sports, Science, and Technology (MEXT), Japan. The NICAM simulations of NDW6 were performed on the Earth Simulator at Japan Agency for Marine-Earth Science and Technology (JAMSTEC). The development of NICAM used in this study and its experiments were conducted in part by the Strategic Programs for Innovative Research (SPIRE) Field 3 (Projection of Planet Earth Variations for Mitigating Natural Disasters), which is promoted by MEXT. The contributions of all the members of the NICAM development team, particularly for Dr. Hisashi Yashiro, are greatly appreciated. The numerical simulation by Kodama et al. (2015) and the NICAM simulations of NSW6 in Chen et al. (2016) were performed on the K computer at the RIKEN Advanced Institute for Computational Science (Proposal numbers hp120279, hp120313, hp130010, hp140219, and hp150213). This review was also supported by SCALE library, which was developed by Team-SCALE of RIKEN Advanced Institute for Computational Sciences (http://scale.aics.riken.jp). The data from the Global Energy and Water Exchange Project (GEWEX) Cloud System Study (GCSS) intercomparison studies, which we used in several figures, were downloaded from http://projects.knmi.nl/rico/.

Funding

This study was supported by the Program for Risk Information on Climate Change (SOUSEI) and Integrated Research Program for Advancing Climate Model (TOUGOU), which are promoted by the Ministry of Education, Culture, Sports, Science and Technology (MEXT), Japan.

Authors' contributions

MS led the studies using NICAM in the SOUSEI program and coordinated this review. AN is leading the NICAM studies in the follow-up TOUGOU program. MS contributed to the sections "Introduction," subsection "Future directions" of the section "Discussion," and the section "Summary." TS is responsible for the subsection "Development of a double-moment cloud microphysics scheme" in the section "Review." In the subsection "Sensitivity of cloud microphysics schemes," "A large-eddy simulation model" was contributed by YS and "Comparison with a bin microphysics scheme" by NK. In the subsection "Cloud and circulation changes," "Climate projection simulations by NICAM" was contributed by MS and AN, "Changes of size distribution of high clouds" by AN, "High cloud changes in two different microphysics scheme experiments" by YC, "Clouds, precipitation, and storm changes in the multi-decadal experiments" by CK and YY, and "Storm track and its cloud radiative effect" by CK, who also contributed to "Cloud feedback" in the subsection "Discussion." The numerical simulations of the multi-decadal NICAM experiments were conducted mainly by YY, CK, and AN. All authors read and approved the final manuscript.

Authors' information

MS started development of NICAM around the year 2000, and since then, the NICAM studies have been conducted mainly in three institutes: Atmosphere and Ocean Research Institute of the University of Tokyo, JAMSTEC, and the Advanced Institute for Computational Science, RIKEN (currently, RIKEN Center for Computational Science).

Competing interests

The authors declare that they have no competing interests.

Author details

[1]Atmosphere and Ocean Research Institute, The University of Tokyo, 5-1-5 Kashiwanoha, Kashiwa, Chiba 277-8568, Japan. [2]Japan Agency for Marine-Earth Science and Technology, 3173-15, Showa-machi, Kanazawa-ku, Yokohama, Kanagawa 236-0001, Japan. [3]Department of Applied Energy, Graduate School of Engineering, Nagoya University, Furo cho, Chikusa-ku, Nagoya, Aichi 464-8603, Japan. [4]RIKEN Center for Computational Science, 7-1-26, Minatojima-minami-machi, Chuo-ku, Kobe, Hyogo 650-0047, Japan.

References

Ackerman AS, van Zanten MC, Stevens B, Savic-Jovcic V, Bretherton CS, Chlond A, Golaz J-C, Jiang H, Khairoutdinov M, Krueger S, Lewellen DC, Lock A, Moeng C-H, Nakamura K, Petters MD, Snider JR, Weinbrecht S, Zulauf M (2009) Large-eddy simulations of a drizzling, stratocumulus-topped marine boundary layer. Mon Weather Rev 137:1083–1110. https://doi.org/10.1175/2008MWR2582.1

Auer AH, Veal DL (1970) The dimension of ice crystals in natural clouds. J Atmos Sci 27:919–926

Bengtsson L, Hodges KI, Keenlyside N (2009) Will extratropical storms intensify in a warmer climate? J Clim 22:2276–2301. https://doi.org/10.1175/2008JCLI2678.1

Blossey PN, Bretherton CS, Zhang M, Cheng A, Endo S, Heus T, Liu Y, Lock AP, De Roode SP, Xu K-M (2013) Marine low cloud sensitivity to an idealized climate change: the CGILS LES intercomparison. J Adv Model Earth Syst 5:234–258. https://doi.org/10.1002/jame.20025

Bodas-Salcedo A, Andrews T, Karmalkar AV, Ringer MA (2016) Cloud liquid water path and radiative feedbacks over the Southern Ocean. Geophys Res Lett 43: 10938–10946. https://doi.org/10.1002/2016GL070770

Bodas-Salcedo A, Williams KD, Ringer MA, Beau I, Cole JNS, Dufresne J-L, Koshiro T, Stevens B, Wang Z, Yokohata T (2014) Origins of the solar radiation biases over the Southern Ocean in CFMIP2 models. J Clim 27:41–56. https://doi.org/10.1175/JCLI-D-13-00169.1

Bohm HP (1989) A general equation for the terminal fall speed of solid hydrometeors. J Atmos Sci 45:2419–2427. https://doi.org/10.1175/1520-0469(1989)046<2419:AGEFTT>2.0.CO;2

Bony S, Stevens B, Coppin D, Becker T, Reed KA, Voigt A, Medeiros B (2016) Thermodynamic control of anvil cloud amount. Proc Natl Acad Sci U S A 113:8927–8932. https://doi.org/10.1073/pnas.1601472113

Bony S, Stevens B, Frierson DMW, Jakob C, Kageyama M, Pincus R, Shepherd TG, Sherwood SC, Siebesma AP, Sobel AH, Watanabe M, Webb MJ (2015) Clouds, circulation and climate sensitivity. Nat Geosci 8:261–268. https://doi.org/10.1038/ngeo2398

Booth JF, Wang S, Polvani L (2013) Midlatitude storms in a moister world: lessons from idealized baroclinic life cycle experiments. Clim Dyn 41:787–802. https://doi.org/10.1007/s00382-012-1472-3

Boucher O, Randall D, Artaxo P, Bretherton C, Feingold F, Forster P, Kerminen VM, Kondo Y, Liao H, Lohmann U, Rasch P, Satheesh SK, Sherwood S, Stevens B, Zhang XY (2013) Clouds and aerosols. In: Stocker TF, Qin D, Plattner GK, Tignor M, Allen SK, Boschung J, Nauels A, Xia Y, Bex V, Midgley PM (eds) Climate change 2013: the physical science basis. Contribution of Working Group I to the Fifth Assessment Report of the Intergovernmental Panel on Climate Change. Cambridge University Press, Cambridge

Bretherton CS (2015) Insights into low-latitude cloud feedbacks from high-resolution models. Phil Trans R Soc A 373:20140415. https://doi.org/10.1098/rsta.2014.0415

Bretherton CS, Blossey P, Khairoutdinov M (2005) An energy-balance analysis of deep convective self-aggregation above uniform SST. J Atmos Sci 62:4273–4292. https://doi.org/10.1175/JAS3614.1

Bretherton CS, Krueger SK, Wyant MC, Bechtold P, Van Meijgaard E, Stevens TJ (1999b) A GCSS boundary-layer cloud model intercomparison study of the first ASTEX Lagrangian experiment. Bound-Layer Meteorol 93:341–380. https://doi.org/10.1023/A:1002005429969

Bretherton CS, Macvean M, Bechtold P, Chlond A, Cotton WR, Cuxart J, Cuijpers H, Mhairoutdinov M, Kosovic B, Lewellen D, Moeng C-H, Siebesma P, Stevens B, Stevens DE, Sykes I, Wyant MC (1999a) An intercomparison of radiatively driven entrainment and turbulence in a smoke cloud, as simulated by different numerical models. Q J R Meteorol Soc 125:391–423. https://doi.org/10.1002/qj.49712555402

Ceppi P, Hartmann DL, Webb MJ (2015) Mechanisms of the negative shortwave cloud feedback in high latitudes. J Clim 29:139–157. https://doi.org/10.1175/JCLI-D-15-0327.1

Champion AJ, Hodges KI, Bengtsson LO, Keenlyside NS, Esch M (2011) Impact of increasing resolution and a warmer climate on extreme weather from northern hemisphere extratropical cyclones. Tellus A Dyn Meteorol Oceanogr 63:893–906. https://doi.org/10.1111/j.1600-0870.2011.00538.x

Chen J-P, Lamb D (1994a) The theoretical basis for the parameterization of ice crystal habits: growth by vapor deposition. J Atmos Sci 51:1206–1221. https://doi.org/10.1175/1520-0469(1994)051<1206:TTBFTP>2.0.CO;2

Chen J-P, Lamb D (1994b) Simulation of cloud microphysical and chemical processes using a multicomponent framework. Part I: description of the microphysical model. J Atmos Sci 51:2613–2630. https://doi.org/10.1175/1520-0469(1994)051<2613:SOCMAC>2.0.CO;2

Chen J-P, Lamb D (1999) Simulation of cloud microphysical and chemical processes using a multicomponent framework. Part II: microphysical evolution of a wintertime orographic cloud. J Atmos Sci 56:2293–2312. https://doi.org/10.1175/1520-0469(1999)056<2293:SOCMAC>2.0.CO;2

Chen T, Rossow WB, Zhang Y (2000) Radiative effects of cloud-type variations. J Clim 13:264–286. https://doi.org/10.1175/1520-0442(2000)013<0264:REOCTV>2.0.CO;2

Chen Y-W, Seiki T, Kodama C, Satoh M, Noda AT, Yamada Y (2016) High cloud responses to global warming simulated by two different cloud microphysics schemes implemented in the nonhydrostatic icosahedral atmospheric model (NICAM). J Clim 29:5949–5964. https://doi.org/10.1175/JCLI-D-15-0668.1

Collins WD, Satoh M (2009) Simulating global clouds, past, present, and future. In: Heintzenberg J, Charlson RJ (eds) Clouds in the perturbed climate system: their relationship to energy balance, atmospheric dynamics, and precipitation, Struengmann forum report, vol 2. MIT Press, Cambridge

Doelling DR, Loeb NG, Keyes DF, Nordeen ML, Morstad D, Nguyen C, Wielicki BA, Young DF, Sun M (2013) Geostationary enhanced temporal interpolation for CERES flux products. J Atmos Ocean Technol 30:1072–1090. https://doi.org/10.1175/JTECH-D-12-00136.1

Dufresne J-L, Bony S (2008) An assessment of the primary sources of spread of global warming estimates from coupled atmosphere–ocean models. J Clim 21:5135–5144. https://doi.org/10.1175/2008JCLI2239.1

Eyring V, Bony S, Meehl GA, Senior CA, Stevens B, Stouffer RJ, Taylor KE (2016) Overview of the coupled model intercomparison project phase 6 (CMIP6) experimental design and organization. Geosci Model Dev 9:1937–1958. https://doi.org/10.5194/gmd-9-1937-2016

Field PR, Bodas-Salcedo A, Brooks ME (2011) Using model analysis and satellite data to assess cloud and precipitation in midlatitude cyclones. Q J R Meteorol Soc 137:1501–1515. https://doi.org/10.1002/qj.858

Flato G, Marotzke J, Abiodun B, Braconnot P, Chou SC, Collins W, Cox P, Driouech F, Emori S, Eyring V, Forest C, Gleckler P, Guilyardi E, Jakob C, Kattsov V, Reason C, Rummukainen M (2013) Evaluation of climate models. In: Climate change 2013: the physical science basis. Contribution of working group I to the fifth assessment report of the Intergovernmental Panel on Climate Change. Cambridge University Press, Cambridge

Fu Q (1996) An accurate parameterization of the solar radiative properties of cirrus clouds for climate models. J Atmos Sci 9:2058–2082. https://doi.org/10.1175/1520-0442(1996)009<2058:AAPOTS>2.0.CO;2

Fu Q (2007) A new parameterization of an asymmetry factor of cirrus clouds for climate models. J Atmos Sci 64:4140–4150. https://doi.org/10.1175/2007JAS2289.1

Fu Q, Yang P, Sun WB (1998) An accurate parameterization of the infrared radiative properties of cirrus clouds for climate models. J Clim 11:2223–2237. https://doi.org/10.1175/1520-0442(1998)011<2223:AAPOTI>2.0.CO;2

Geng Q, Sugi M (2003) Possible change of extratropical cyclone activity due to enhanced greenhouse gases and sulfate aerosols – study with a high-resolution AGCM. J Clim 16:2262–2274. https://doi.org/10.1175/1520-0442(2003)16<2262:PCOECA>2.0.CO;2

Grabowski WW (1998) Toward cloud resolving modeling of large-scale tropical circulation: a simple cloud microphysics parameterization. J Atmos Sci 55:3283–3298. https://doi.org/10.1175/1520-0469(1998)055<3283:TCRMOL>2.0.CO;2

Grabowski WW (2016) Separating dynamical and microphysical impacts of aerosols on deep convection applying piggybacking methodology. In: The European Geosciences Union General Assembly, Vienna, Austria, 17–22 April 2016

Grabowski WW, Morrison H (2016) Untangling microphysical impacts on deep convection applying a novel modeling methodology. Part II: double-moment microphysics. J Atmos Sci 73:3749–3770. https://doi.org/10.1175/JAS-D-15-0367.1

Grise KM, Polvani LM (2014) Southern hemisphere cloud-dynamics biases in CMIP5 models and their implications for climate projections. J Clim 27:6074–6092. https://doi.org/10.1175/JCLI-D-14-00113.1

Guo H, Golaz J-C, Donner LJ (2011) Aerosol effects on stratocumulus water paths in a PDF-based parameterization. Geophys Res Lett 38:L17808. https://doi.org/10.1029/2011GL048611

Haarsma RJ, Roberts MJ, Vidale PL, Senior CA, Bellucci A, Bao Q, Chang P, Corti S, Fučkar NS, Guemas V, von Hardenberg J, Hazeleger W, Kodama C, Koenigk T, Leung LR, Lu J, Luo J-J, Mao J, Mizielinski MS, Mizuta R, Nobre P, Satoh M, Scoccimarro E, Semmler T, Small J, von Storch J-S (2016) High resolution model Intercomparison project (HighResMIP v1.0) for CMIP6. Geosci Model Dev 9:4185–4208. https://doi.org/10.5194/gmd-9-4185-2016

Harada Y, Kamahori H, Kobayashi C, Endo H, Kobayashi S, Ota Y (2016) The JRA-55 reanalysis: representation of atmospheric circulation and climate variability. J Meteorol Soc Jpn 94:269–302. https://doi.org/10.2151/jmsj.2016-015

Harrington JY, Sulia K, Morrison H (2013a) A method for adaptive habit prediction in bulk microphysical models. Part I: theoretical development. J Atmos Sci 70:349–364. https://doi.org/10.1175/JAS-D-12-040.1

Harrington JY, Sulia K, Morrison H (2013b) A method for adaptive habit prediction in bulk microphysical models. Part II: parcel model corroboration. J Atmos Sci 70:365–376. https://doi.org/10.1175/JAS-D-12-0152.1

Hartmann DL, Larson K (2002) An important constraint on tropical cloud-climate feedback. Geophys Res Lett 29:12-1–12-4. https://doi.org/10.1029/2002GL015835

Hartmann DL, Ockert-Bell ME, Michelsen ML (1992) The effect of cloud type on Earth's energy balance: global analysis. J Clim 5:1281–1304. https://doi.org/10.1175/1520-0442(1992)005<1281:TEOCTO>2.0.CO;2

Hashino T, Satoh M, Hagihara Y, Kato S, Kubota T, Matsui T, Nasuno T, Okamoto H, Sekiguchi M (2016) Evaluating arctic cloud radiative effects simulated by NICAM with A-train. J Geophys Res Atmos 121:7041–7063. https://doi.org/10.1002/2016JD024775

Hashino T, Satoh M, Hagihara Y, Kubota T, Matsui T, Nasuno T, Okamoto H (2013) Evaluating cloud microphysics from NICAM against CloudSat and CALIPSO. J Geophys Res Atmos 118:7273–7292. https://doi.org/10.1002/jgrd.50564

Hashino T, Tripoli GJ (2007) The spectral ice habit prediction system (SHIPS). Part I: model description and simulation of the vapor deposition process. J Atmos Sci 64:2210–2237. https://doi.org/10.1175/JAS3963.1

Hashino T, Tripoli GJ (2008) The spectral ice habit prediction system (SHIPS). Part II: simulation of nucleation and depositional growth of polycrystals. J Atmos Sci 65:3071–3094. https://doi.org/10.1175/2008JAS2615.1

Hashino T, Tripoli GJ (2011a) The spectral ice habit prediction system (SHIPS). Part III: description of the ice particle model and the habit-dependent aggregation model. J Atmos Sci 68:1125–1141. https://doi.org/10.1175/2011JAS3666.1

Hashino T, Tripoli GJ (2011b) The spectral ice habit prediction system (SHIPS). Part IV: box model simulations of the habit-dependent aggregation process. J Atmos Sci 68:1142–1161. https://doi.org/10.1175/2011JAS3667.1

Heymsfield AJ, Kajikawa M (1987) An improved approach to calculating terminal velocities of plate-like crystals and graupel. J Atmos Sci 44:1088–1099. https://doi.org/10.1175/1520-0469(1987)044<1088:AIATCT>2.0.CO;2

Hill AA, Shipway BJ, Boutle IA (2015) How sensitive are aerosol-precipitation interactions to the warm rain representation? J Adv Model Earth Syst 7:987–1004. https://doi.org/10.1002/2014MS000422

Hong S-Y, Dudhia J, Chen S-H (2004) A revised approach to ice microphysical processes for the bulk parameterization of clouds and precipitation. Mon Weather Rev 132:103–120. https://doi.org/10.1175/1520-0493(2004)132,0103:ARATIM.2.0.CO;2

Hou AY, Kakar RK, Neeck S, Azarbarzin AA, Kummerow CD, Kojima M, Oki R, Nakamura K, Iguchi T (2014) The global precipitation measurement mission. Bull Am Meteorol Soc 95:701–722. https://doi.org/10.1175/BAMS-D-13-00164.1

Iga S, Tomita H, Tsushima Y, Satoh M (2007) Climatology of a nonhydrostatic global model with explicit cloud processes. Geophys Res Lett 34:L22814. https://doi.org/10.1029/2007GL031048

Iga S, Tomita H, Tsushima Y, Satoh M (2011) Sensitivity of Hadley circulation to physical parameters and resolution through changing upper-tropospheric ice clouds using a global cloud-system resolving model. J Clim 24:2666–2679. https://doi.org/10.1175/2010JCLI3472.1

Igel AL, Igel MR, Van den Heever SC (2015) Make it a double? Sobering results from simulations using single-moment microphysics schemes. J Atmos Sci 72:910–925. https://doi.org/10.1175/JAS-D-14-0107.1

Inoue T, Satoh M, Miura H, Mapes B (2008) Characteristics of cloud size of deep convection simulated by a global cloud resolving model over the western tropical Pacific. J Meteorol Soc Jpn 86A:1–15. https://doi.org/10.2151/jmsj.86A.1

Jensen EJ, Kinne S, Toon OB (1994) Tropical cirrus cloud radiative forcing: sensitivity studies. Geophys Res Lett 21:2023–2026. https://doi.org/10.1029/94GL01358

Jiang JH, Su H, Zhai C, Perun VS, Genio AD, Nazarenko LS, Donner LJ, Horowitz L, Seman C, Cole J, Gettelman A, Ringer MA, Rotstayn L, Jeffrey S, Wu T, Brient F, Dufresne J-L, Kawai H, Koshiro T, Watanabe M, LÉcuyer TS, Volodin EM, Iversen T, Drange H, Mesquita MDS, Read WG, Waters JW, Tian B, Teixeira J, Stephens GL (2012) Evaluation of cloud and water vapor simulations in CMIP5 climate models using NASA A-train satellite observations. J Geophys Res 117:D14105. https://doi.org/10.1029/2011JD017237

Kärcher B, Hendricks J, Lohmann U (2006) Physically based parameterization of cirrus cloud formation for use in global atmospheric models. J Geophys Res 111:D01205. https://doi.org/10.1029/2005JD006219

Kessler E (1969) On the distribution and continuity of water substance in atmospheric circulations. In: Meteorological monographs, no. 32. American Meteorological Society, Boston

Kinter JL, Cash B, Achuthavarier D, Adams J, Altshuler E, Dirmeyer P, Doty B, Huang B, Jin EJ, Marx L, Manganello J, Stan C, Wakefield T, Palmer T, Hamrud M, Jung T, Miller M, Towers P, Wedi N, Satoh M, Tomita H, Kodama C, Nasuno T, Oouchi K, Yamada Y, Taniguchi H, Andrews P, Baer T, Ezell M, Halloy C, John D, Loftis B, Mohr R, Wong K (2013) Revolutionizing climate modeling with project Athena: A multi-institutional, international collaboration. Bull Amer Meteorol Soc 94:231-245. https://doi.org/10.1175/BAMS-D-11-00043.1.

Klein SA, Hall A (2015) Emergent constraints for cloud feedbacks. Current Climate Change Reports 1:276–287. https://doi.org/10.1007/s40641-015-0027-1

Klein SA, Jakob C (1999) Validation and sensitivities of frontal clouds simulated by ECMWF model. Mon Weather Rev 127:2514–2531. https://doi.org/10.1175/1520-0493(1999)127<2514:VASOFC>2.0.CO;2

Knutson TR, McBride JL, Chan J, Emanuel K, Holland G, Landsea C, Held I, Kossin JP, Srivastava K, Sugi M (2010) Tropical cyclones and climate change. Nat Geosci 3:157–163. https://doi.org/10.1038/ngeo779

Kobayashi S, Ota Y, Harada Y, Ebita A, Moriya M, Onoda H, Onogi K, Kamahori H, Kobayashi C, Endo H, Miyaoka K, Takahashi K (2015) The JRA-55 reanalysis: general specifications and basic characteristics. J Meteorol Soc Japan 93:5–48. https://doi.org/10.2151/jmsj.2015-001

Kodama C, Iga S, Satoh M (2014) Impact of the sea surface temperature rise on storm-track clouds in global non-hydrostatic aqua-planet simulations. Geophys Res Lett 41:3545–3552. https://doi.org/10.1002/2014GL059972

Kodama C, Noda AT, Satoh M (2012) An assessment of the cloud signals simulated by NICAM using ISCCP, CALIPSO, and CloudSat satellite simulators. J Geophys Res 117:D12210. https://doi.org/10.1029/2011JD017317

Kodama C, Yamada Y, Noda AT, Kikuchi K, Kajikawa Y, Nasuno T, Tomita T, Yamaura T, Takahashi HG, Hara M, Kawatani Y, Satoh M, Sugi M (2015) A 20-year climatology of a NICAM AMIP-type simulation. J Meteorol Soc Japan 93: 393–424. https://doi.org/10.2151/jmsj.2015-024

Kotsuki S, Terasaki K, Miyoshi T (2014) GPM/DPR precipitation compared with a 3.5-km-resolution NICAM simulation. SOLA 10:204–209. https://doi.org/10.2151/sola.2014-043

Kuba N, Fujiyoshi Y (2006) Development of a cloud microphysical model and parameterizations to describe the effect of CCN on warm cloud. Atmos Chem Phys 6:2793–2810. https://doi.org/10.5194/acp-6-2793-2006

Kuba N, Hashino T, Satoh M, Suzuki K (2014) Relationships between layer-mean radar reflectivity and columnar effective radius of warm cloud: numerical study using a cloud microphysical bin model. J Geophys Res Atmos 119: 3281–3294. https://doi.org/10.1002/2013JD020276

Kuba N, Murakami M (2010) Effect of hygroscopic seeding on warm rain clouds – numerical study using a hybrid cloud microphysical model. Atmos Chem Phys 10:3335–3351. https://doi.org/10.5194/acp-10-3335-2010

Kuba N, Suzuki K, Hashino T, Seiki T, Satoh M (2015) Numerical experiments to analyze cloud microphysical processes depicted in vertical profiles of radar reflectivity of warm clouds. J Atmos Sci 72:4509–4528. https://doi.org/10.1175/JAS-D-15-0053.1

Lang SE, Tao W-K, Chern J-D, Wu D, Li X (2014) Benefits of a fourth ice class in the simulated radar reflectivities of convective systems using a bulk microphysics scheme. J Atmos Sci 71:3583–3612. https://doi.org/10.1175/JAS-D-13-0330.1

Li J-LF, Waliser DE, Chen W-T, Ghan B, Kubar T, Stephens G, Ma H-Y, Deng M, Donner L, Seman C, Horowitz L (2012) An observationally based evaluation of cloud ice water in CMIP3 and CMIP5 GCMs and contemporary reanalyses using contemporary satellite data. J Geophys Res 117:D16105. https://doi.org/10.1029/2012JD017640

Li J-LF, Waliser DE, Stephens G, Lee S, L'Ecuyer T, Kato S, Loeb N, Ma H-Y (2013) Characterizing and understanding radiation budget biases in CMIP3/CMIP5 GCMs, contemporary GCM, and reanalysis. J Geophys Res Atmos 118:8166–8184. https://doi.org/10.1002/jgrd.50378

Li X, Tao W-K, Matsui T, Liu C, Masunaga H (2010) Improving a spectral bin microphysical scheme using TRMM satellite observations. Q J R Meteorol Soc 136:382–399. https://doi.org/10.1002/qj.569

Li Y, Thompson DWJ, Bony S (2015) The influence of atmospheric cloud radiative effects on the large-scale atmospheric circulation. J Clim 28:7263–7278. https://doi.org/10.1175/JCLI-D-14-00825.1

Lin YL, Farley RD, Orville HD (1983) Bulk parameterization of the snow field in a cloud model. J Appl Meteorol 22:1065–1092. https://doi.org/10.1175/1520-0450(1983)022<1065:BPOTSF>2.0.CO;2

Liou K-N (2002) An introduction to atmospheric radiation. Academic Press, San Diego

Liu R, Liou K, Su H, Gu Y, Zhao B (2017) High cloud variations with surface temperature from 2002 to 2015: contributions to atmospheric radiative cooling rate and precipitation changes. J Geophys Res Atmos 122:5457–5471. https://doi.org/10.1002/2016JD026303

Locatelli JD, Hobbs PV (1974) Fall speeds and masses of solid precipitation particles. J Geophys Res 79:2185–2197. https://doi.org/10.1029/JC079i015p02185

Loeb NG, Wielicki BA, Doelling DR, Smith GL, Keyes DF, Kato S, Manalo-Smith N, Wong T (2009) Toward optimal closure of the Earth's top-of-atmosphere radiation budget. J Clim 22:748–766. https://doi.org/10.1175/2008JCLI2637.1

Masunaga H, Kummerow CD (2005) Combined radar and radiometer analysis of precipitation profiles for a parametric retrieval algorithm. J Atmos Ocean Technol 22:909–929. https://doi.org/10.1175/JTECH1751.1

Meyers MP, DeMott PJ, Cotton WR (1992) New primary ice-nucleation parameterizations in an explicit cloud model. J Appl Meteorol 31:708–720. https://doi.org/10.1175/1520-0450(1992)031<0708:NPINPI>2.0.CO;2

Mitchell DL (1996) Use of mass and area-dimensional power laws for determining precipitation particle terminal velocities. J Atmos Sci 53:1710–1723. https://doi.org/10.1175/1520-0469(1996)053<1710:UOMAAD>2.0.CO;2

Miura H, Satoh M, Nasuno T, Noda AT, Oouchi K (2007) A madden-Julian oscillation event realistically simulated by a global cloud-resolving model. Science 318:1763–1765. https://doi.org/10.1126/science.1148443

Miura H, Tomita H, Nasuno T, Iga S, Satoh M, Matsuno T (2005) A climate sensitivity test using a global cloud resolving model under an aqua planet condition. Geophys Res Lett 32:L19717. https://doi.org/10.1029/2005GL023672

Mizuta R (2012) Intensification of extratropical cyclones associated with the polar jet change in the CMIP5 global warming projections. Geophys Res Lett 39:L19707. https://doi.org/10.1029/2012GL053032

Mizuta R, Adachi Y, Yukimoto S, Kusunoki S (2008) Estimation of the future distribution of sea surface temperature and sea ice using the CMIP3 multi-model ensemble mean. In: Technical reports of the meteorological research institute, vol. 56. Meteorological Research Institute, Tsukuba. https://doi.org/10.11483/mritechrepo.56

Nakajima TY, Suzuki K, Stephens GL (2010) Droplet growth in warm water clouds observed by the A-train. Part II: a multisensor view. J Atmos Sci 67:1897–1907. https://doi.org/10.1175/2010JAS3276.1

Nakanishi M, Niino H (2006) An improved Mellor-Yamada level-3 model: its numerical stability and application to a regional prediction of advection fog. Bound-Layer Meteorol 119:397–407. https://doi.org/10.1007/s10546-005-9030-8

Naud CM, Del Genio AD, Bauer M, Kovari W (2010) Cloud vertical distribution across warm and cold fronts in cloudsat-CALIPSO data and a general circulation model. J Clim 23:3397–3415. https://doi.org/10.1175/2010JCLI3282.1

Nishizawa S, Yashiro H, Sato Y, Miyamoto Y, Tomita H (2015) Influence of grid aspect ratio on planetary boundary layer turbulence in large-eddy simulations. Geosci Model Dev 8:3393–3419. https://doi.org/10.5194/gmd-8-3393-2015

Noda AT, Oouchi K, Satoh M, Tomita H (2012) Quantitative assessment of diurnal variation of tropical convection simulated by a global nonhydrostatic model without cumulus parameterization. J Clim 25:5119–5134. https://doi.org/10.1175/JCLI-D-11-00295.1

Noda AT, Oouchi K, Satoh M, Tomita H, Iga S, Tsushima Y (2010) Importance of the subgrid-scale turbulent moist process: cloud distribution in global cloud-resolving simulations. Atmos Res 96:208–217. https://doi.org/10.1016/j.atmosres.2009.05.007

Noda AT, Satoh M (2014) Intermodel variances of subtropical stratocumulus environments simulated in CMIP5 models. Geophys Res Lett 41:7754–7761. https://doi.org/10.1002/2014GL061812

Noda AT, Satoh M, Yamada Y, Kodama C, Miyakawa T, Seiki T (2015) Cold and warm rain simulated using a global nonhydrostatic model without cumulus parameterization, and its responses to global warming. J Meteorol Soc Japan 93:181–197. https://doi.org/10.2151/jmsj.2015-010

Noda AT, Satoh M, Yamada Y, Kodama C, Seiki T (2014) Responses of tropical and subtropical high-cloud statistics to global warming. J Clim 27:7753–7768. https://doi.org/10.1175/JCLI-D-14-00179.1

Noda AT, Seiki T, Satoh M, Yamada Y (2016) High cloud size dependency in the applicability of the fixed anvil temperature hypothesis using global non-hydrostatic simulations. Geophys Res Lett 43:2307–2314. https://doi.org/10.1002/2016GL067742

Ohno T, Satoh M (2018) Roles of cloud microphysics on cloud responses to sea surface temperatures in radiative-convective equilibrium experiments using a high-resolution global nonhydrostatic model. J Adv Model Earth Syst 10. https://doi.org/10.1029/2018MS001386

Ono A (1969) The shape and riming properties of ice crystals in natural clouds. J Atmos Sci 26:138–147. https://doi.org/10.1175/1520-0469(1969)026<0138:TSARPO>2.0.CO;2

Phillips VTJ, Donner LJ, Garner ST (2007) Nucleation processes in deep convection simulated by a cloud-system-resolving model with double-moment bulk microphysics. J Atmos Sci 64:738–761. https://doi.org/10.1175/JAS3869.1

Posselt DJ, Stephens GL, Miller M (2008) CLOUDSAT: adding a new dimension to a classical view of extratropical cyclones. Bull Am Meteorol Soc 89:599–609. https://doi.org/10.1175/BAMS-89-5-599

Pruppacher HR, Klett JD (1997) Microphysics of clouds and precipitation. Kluwer Academic Publishers, Dordrecht

Ren C, MacKenzie AR (2005) Cirrus parameterization and the role of ice nuclei. Q J R Meteorol Soc 131:1585–1605. https://doi.org/10.1256/qj.04.126

Rodgers EB, Adler RF, Pierce HF (2000) Contribution of tropical cyclones to the North Pacific climatological rainfall as observed from satellites. J Appl Meteorol 39:1658–1678. https://doi.org/10.1175/1520-0450(2000)039<1658:COTCTT>2.0.CO;2

Roh W, Satoh M (2014) Evaluation of precipitating hydrometeor parameterizations in a single-moment bulk microphysics scheme for deep convective systems over the tropical Central Pacific. J Atmos Sci 71:2654–2673. https://doi.org/10.1175/JAS-D-13-0252.1

Roh W, Satoh M, Nasuno T (2017) Improvement of a cloud microphysics scheme for a global nonhydrostatic model using TRMM and a satellite simulator. J Atmos Sci 74:167–184. https://doi.org/10.1175/JAS-D-16-0027.1

Sato Y, Nakajima TY, Nakajima T (2012) Investigation of the vertical structure of warm-cloud microphysical properties using the cloud evolution diagram, CFODD, simulated by a three-dimensional spectral bin microphysical model. J Atmos Sci 69:2012–2030. https://doi.org/10.1175/JAS-D-11-0244.1

Sato Y, Nishizawa S, Yashiro H, Miyamoto Y, Kajikawa Y, Tomita H (2015) Impacts of cloud microphysics on trade wind cumulus: which cloud microphysics processes contribute to the diversity in a large eddy simulation? Prog Earth Planet Sci 2(1):23. https://doi.org/10.1186/s40645-015-0053-6

Satoh M, Iga S, Tomita H, Tsushima Y, Noda AT (2012) Response of upper clouds due to global warming tested by a global atmospheric model with explicit cloud processes. J Clim 25:2178–2191. https://doi.org/10.1175/JCLI-D-11-00152.1

Satoh M, Inoue T, Miura H (2010) Evaluation of cloud properties of global and local cloud system resolving model using CALIPSO and CloudSat simulators. J Geophys Res 115:D00H14. https://doi.org/10.1029/2009JD012247

Satoh M, Matsuno T, Tomita H, Miura H, Nasuno T, Iga S (2008) Nonhydrostatic icosahedral atmospheric model (NICAM) for global cloud resolving simulations. J Comput Phys 227:3486–3514. https://doi.org/10.1016/j.jcp.2007.02.006

Satoh M, Tomita H, Yashiro H, Kajikawa Y, Miyamoto Y, Yamaura T, Miyakawa T, Nakano M, Kodama C, Noda AT, Nasuno T, Yamada Y, Fukutomi Y (2017) Outcomes and challenges of global high-resolution non-hydrostatic atmospheric simulations using K computer. Prog Earth Planet Sci 4:13. https://doi.org/10.1186/s40645-017-0127-8

Satoh M, Tomita H, Yashiro H, Miura H, Kodama C, Seiki T, Noda AT, Yamada Y, Goto D, Sawada M, Miyoshi T, Niwa Y, Hara M, Ohno Y, Iga S, Arakawa T, Inoue T, Kubokawa H (2014) The non-hydrostatic icosahedral atmospheric model: description and development. Prog Earth Planet Sci 1:18. https://doi.org/10.1186/s40645-014-0018-1

Satoh M, Yamada Y, Sugi M, Kodama C, Noda AT (2015) Constraint on future change in global frequency of tropical cyclones due to global warming. J Meteorol Soc Japan 93:489–500. https://doi.org/10.2151/jmsj.2015-025

Schneider T, Teixeira J, Bretherton CS, Brient F, Pressel KG, Schär C, Siebesma AP (2017) Climate goals and computing the future of clouds. Nat Clim Chang 7:3–5. https://doi.org/10.1038/nclimate3190

Scoccimarro E, Gualdi S, Villarini G, Vecchi GA, Zhao M, Walsh K, Navarra A (2014) Intense precipitation events associated with landfalling tropical cyclones in response to a warmer climate and increased CO_2. J Clim 27:4642–4654. https://doi.org/10.1175/JCLI-D-14-00065.1

Seifert A, Beheng KD (2006) A two-moment cloud microphysics parameterization for mixed-phase clouds. Meteorog Atmos Phys 92:45–66. https://doi.org/10.1007/s00703-005-0112-4

Seiki T, Kodama C, Noda AT, Satoh M (2015a) Improvement in global cloud-system resolving simulations by using a double-moment bulk cloud microphysics scheme. J Clim 28:2405–2419. https://doi.org/10.1175/JCLI-D-14-00241.1

Seiki T, Kodama C, Satoh M, Hashino T, Hagihara Y, Okamoto H (2015b) Vertical grid spacing necessary for simulating tropical cirrus clouds with a high-resolution atmospheric general circulation model. Geophys Res Lett 42:4150–4157. https://doi.org/10.1002/2015GL064282

Seiki T, Nakajima T (2014) Aerosol effects of the condensation process on a convective cloud simulation. J Atmos Sci 71:833–853. https://doi.org/10.1175/JAS-D-12-0195.1

Seiki T, Satoh M, Tomita H, Nakajima T (2014) Simultaneous evaluation of ice cloud microphysics and nonsphericity of the cloud optical properties using hydrometeor video sonde and radiometer sonde in situ observations. J Geophys Res Atmos 119:6681–6701. https://doi.org/10.1002/2013JD021086

Sekiguchi M, Nakajima T (2008) A k-distribution-based radiation code and its computational optimization for an atmospheric general circulation model.

J Quant Spectrosc Radiat Transf 109:2779–2793. https://doi.org/10.1016/j.jqsrt.2008.07.013

Shipway BJ, Hill AA (2012) Diagnosis of systematic differences between multiple parametrizations of warm rain microphysics using a kinematic framework. Q J R Meteorol Soc 138:2196–2211. https://doi.org/10.1002/qj.1913

Siebesma AP, Bretherton CS, Brown A, Chlond A, Cuxart J, Duynkerke PG, Jiang H, Khairoutdinov M, Lewellen D, Moeng C-H, Sanchez E, Stevens B, Stevens DE (2003) A large eddy simulation intercomparison study of shallow cumulus convection. J Atmos Sci 60:1201–1219

Soden BJ, Held IM (2006) An assessment of climate feedbacks in coupled ocean atmosphere models. J Clim 19:3354–3360. https://doi.org/10.1175/JCLI3799.1

Stevens B, Ackerman AS, Albrecht BA, Brown AR, Chlond A, Cuxart J, Duynkerke PG, Lewellen DC, Macvean MK, Neggers RAJ, Sánchez E, Siebesma AP, Stevens DE (2001) Simulations of trade wind cumuli under a strong inversion. J Atmos Sci 58:1870–1891. https://doi.org/10.1175/1520-0469(2001)058<1870:SOTWCU>2.0.CO;2

Stevens B, Bony S (2013) What are climate models missing? Science 340:1053–1054. https://doi.org/10.1126/science.1237554

Stevens B, Moeng C-H, Ackerman SD, Bretherton CS, Chlon A, de Roode S, Edwards J, Golaz J, Jiang H, Khairoutdinov M, Kirkpatrick MP, Lewellen DC, Lock A, Müller F, Stevens DE, Whelan E, Zhu P (2005) Evaluation of large-eddy simulations via observations of nocturnal marine stratocumulus. Mon Weather Rev 133:1443–1462. https://doi.org/10.1175/MWR2930.1

Su H, Jiang JH, Neelin JD, Shen TJ, Zhai C, Yue Q, Wang Z, Huang L, Choi Y-S, Stephens GL, Yung YL (2017) Tightening of tropical ascent and high clouds key to precipitation change in a warmer climate. Nat Commun 8:15771. https://doi.org/10.1038/ncomms15771

Suzuki K, Nakajima T, Nakajima TY, Khain AP (2010a) A study of microphysical mechanisms for correlation patterns between droplet radius and optical thickness of warm clouds with a spectral bin microphysics cloud model. J Atmos Sci 67:1126–1141. https://doi.org/10.1175/2009JAS3283.1

Suzuki K, Nakajima TY, Stephens GL (2010b) Particle growth and drop collection efficiency of warm clouds as inferred from joint CloudSat and MODIS observations. J Atmos Sci 67:3019–3032. https://doi.org/10.1175/2010JAS3463.1

Suzuki K, Stephens G (2008) Global identification of warm cloud microphysical processes with combined use of A-train observations. Geophys Res Lett 35:L08805. https://doi.org/10.1029/2008GL033590

Suzuki K, Stephens G, Bodas-Salcedo A, Wang M, Golaz J-C, Yokohata T, Koshiro T (2015) Evaluation of the warm rain formation process in global models with satellite observations. J Atmos Sci 72:3996–4014. https://doi.org/10.1175/JAS-D-14-0265.1

Takata K, Emori S, Watanabe T (2003) Development of the minimal advanced treatments of surface interaction and runoff. Glob Planet Chang 38:209–222. https://doi.org/10.1016/S0921-8181(03)00030-4

Taylor KE, Williamson D, Zwiers F (2000) The sea surface temperature and sea-ice concentration boundary conditions for AMIP II simulations. In: Program for climate model diagnosis and intercomparison report 60. Lawrence Livermore National Laboratory, Livermore

Tomita H (2008) New microphysical schemes with five and six categories by diagnostic generation of cloud ice. J Meteorol Soc Japan 86A:121–142. https://doi.org/10.2151/jmsj.86A.121

Tomita H, Satoh M (2004) A new dynamical framework of nonhydrostatic global model using the icosahedral grid. Fluid Dyn Res 34:357–400. https://doi.org/10.1016/j.fluiddyn.2004.03.003

Tselioudis G, Rossow WB (2006) Climate feedback implied by observed radiation and precipitation changes with midlatitude storm strength and frequency. Geophys Res Lett 33:L02704. https://doi.org/10.1029/2005GL024513

Tsushima Y, Iga S, Tomita H, Satoh M, Noda AT, Webb M (2014) High cloud increase in a perturbed SST experiment with a global nonhydrostatic model including explicit convective processes. J Adv Model Earth Syst 6:571–585. https://doi.org/10.1002/2013MS000301

van Zanten MC, Stevens B, Nuijens L, Siebesma AP, Ackerman AS, Burnet F, Cheng A, Couvreux F, Jiang H, Khairoutdinov M, Kogan Y, Lewellen DC, Mechem D, Nakamura K, Noda A, Shipway BJ, Slawinska J, Wang S, Wyszogrodzki A (2011) Controls on precipitation and cloudiness in simulations of trade-wind cumulus as observed during RICO. J Adv Model Earth Syst 3:M06001. https://doi.org/10.1029/2011MS000056

Vial J, Dufresne J-L, Bony S (2013) On the interpretation of inter-model spread in CMIP5 climate sensitivity estimates. Clim Dyn 41:3339–3362. https://doi.org/10.1007/s00382-013-1725-9

Waliser DE, Li J-LF, Woods CP, Austin RT, Bacmeister J, Chern J, Genio AD, Jiang JH, Kuang Z, Minnis P, Platnick S, Rossow WB, Stephens GL, Sun-Mack S, Tao W-K, Tompkins AM, Vane DG, Walker C, Wu D (2009) Cloud ice: a climate model challenge with signs and expectations of progress. J Geophys Res 114:D00A2. https://doi.org/10.1029/2008JD010015

Webb M, Senior C, Bony S, Morcrette JJ (2001) Combining and ERBE and ISCCP data to assess clouds in the Hadley Centre, ECMWF and LMD atmospheric climate models. Clim Dyn 17:905–922. https://doi.org/10.1007/s003820100157

Williamson DL, Blackburn M, Nakajim K, Ohfuchi W, Takahashi YO, Hayashi Y-Y, Nakamura H, Ishiwatari M, McGregor JL, Borth H, Wirth V, Frank H, Bechtold P, Wedi NP, Tomita H, Satoh S, Zhao M, Held IM, Suarez MJ, Lee M-I, Watanabe M, Kimoto M, Liu Y, Wang Z, Molod A, Rajendran K, Kitoh A, Stratton R (2013) The aqua planet experiment (APE): response to changed SST profile. J Meteorol Soc Japan 91A:57–89. https://doi.org/10.2151/jmsj.2013-A03

Wing AA, Reed KA, Satoh M, Stevens B, Bony S, Ohno T (2018) Radiative-convective equilibrium model intercomparison project. Geosci Model Dev 11:793–813. https://doi.org/10.5194/gmd-2017-213

Wyant MC, Khairoutdinov M, Bretherton CS (2006) Climate sensitivity and cloud response of a GCM with a superparameterization. Geophys Res Lett 33:L06714. https://doi.org/10.1029/2005GL025464

Yamada Y, Oouchi K, Satoh M, Tomita H, Yanase W (2010) Projection of changes in tropical cyclone activity and cloud height due to greenhouse warming: global cloud-system-resolving approach. Geophys Res Lett 37:L07709. https://doi.org/10.1029/2010GL042518

Yamada Y, Satoh M (2013) Response of ice and liquid water paths of tropical cyclones to global warming simulated by a global nonhydrostatic model with explicit cloud microphysics. J Clim 26:9931–9945. https://doi.org/10.1175/JCLI-D-13-00182.1

Yamada Y, Satoh M, Sugi M, Kodama C, Noda AT, Nakano M, Nasuno T (2017) Response of tropical cyclone activity and structure to global warming in a high-resolution global nonhydrostatic model. J Clim 30:9703–9724. https://doi.org/10.1175/JCLI-D-17-0068.1

Yin JH (2005) A consistent poleward shift of the storm tracks in simulations of 21st century climate. Geophys Res Lett 32:L18701. https://doi.org/10.1029/2005GL023684

Yoshizaki M, Iga S, Satoh M (2012) Eastward-propagating property of large-scale precipitation systems simulated in the coarse-resolution NICAM and an explanation of its formation. SOLA 8:21–24. https://doi.org/10.2151/sola.2012-006

Zelinka MD, Hartmann DL (2011) The observed sensitivity of high clouds to mean surface temperature anomalies in the tropics. J Geophys Res Atmos 116:D23103. https://doi.org/10.1029/2011JD016459

Zelinka MD, Klein SA, Hartmann DL (2012) Computing and partitioning cloud feedbacks using cloud property histograms. Part I: cloud radiative kernels. J Clim 25:3715–3735. https://doi.org/10.1175/JCLI-D-11-00248.1

Zelinka MD, Klein SA, Taylor KE, Andrews T, Webb MJ, Gregory JM, Forster PM (2013) Contributions of different cloud types to feedbacks and rapid adjustments in CMIP5. J Clim 26:5007–5027. https://doi.org/10.1175/JCLI-D-12-00555.1

Zelinka MD, Randall DA, Webb MJ, Klein SA (2017) Clearing clouds of uncertainty. Nat Clim Chang 7:674–678. https://doi.org/10.1038/nclimate3402

Zelinka MD, Zhou C, Klein SA (2016) Insights from a refined decomposition of cloud feedbacks. Geophys Res Lett 43:9259–9269. https://doi.org/10.1002/2016GL069917

Zhou C, Dessler AE, Zelinka MD, Yang P, Wang T (2014) Cirrus feedback on interannual climate fluctuations. Geophys Res Lett 41:9166–9173. https://doi.org/10.1002/2014GL062095

Zhou C, Zelinka MD, Klein SA (2016) Impact of decadal cloud variations on the Earth's energy budget. Nat Geosci 9:871–874. https://doi.org/10.1038/ngeo2828

8

Maritime continent coastlines controlling Earth's climate

Manabu D. Yamanaka[1,2*], Shin-Ya Ogino[1], Pei-Ming Wu[1], Hamada Jun-Ichi[1,3], Shuichi Mori[1], Jun Matsumoto[1,3] and Fadli Syamsudin[4]

Abstract

During the Monsoon Asian Hydro–Atmosphere Scientific Research and Prediction Initiative (MAHASRI; 2006–16), we carried out two projects over the Indonesian maritime continent (IMC), constructing the Hydrometeorological Array for Intraseasonal Variation–Monsoon Automonitoring (HARIMAU; 2005–10) radar network and setting up a prototype institute for climate studies, the Maritime Continent Center of Excellence (MCCOE; 2009–14). Here, we review the climatological features of the world's largest "regional" rainfall over the IMC studied in these projects. The fundamental mode of atmospheric variability over the IMC is the diurnal cycle generated along coastlines by land–sea temperature contrast: afternoon land becomes hotter than sea by clear-sky insolation before noon, with the opposite contrast before sunrise caused by evening rainfall-induced "sprinkler"-like land cooling (different from the extratropical infrared cooling on clear nights). Thus, unlike the extratropics, the diurnal cycle over the IMC is more important in the rainy season. The intraseasonal, seasonal to annual, and interannual climate variabilities appear as amplitude modulations of the diurnal cycle. For example, in Jawa and Bali the rainy season is the southern hemispheric summer, because land heating in the clear morning and water vapor transport by afternoon sea breeze is strongest in the season of maximum insolation. During El Niño, cooler sea water surrounding the IMC makes morning maritime convection and rainfall weaker than normal. Because the diurnal cycle is almost the only mechanism generating convective clouds systematically near the equator with little cyclone activity, the local annual rainfall amount in the tropics is a steeply decreasing function of coastal distance (*e*-folding scale 100–300 km), and regional annual rainfall is an increasing function of "coastline density" (coastal length/land area) measured at a horizontal resolution of 100 km. The coastline density effect explains why rainfall and latent heating over the IMC are twice the global mean for an area that makes up only 4% of the Earth's surface. The diurnal cycles appearing almost synchronously over the whole IMC generate teleconnections between the IMC convection and the global climate. Thus, high-resolution (<< 100 km; << 1 day) observations and models over the IMC are essential to improve both local disaster prevention and global climate prediction.

Keywords: Indonesian maritime continent, Atmosphere–ocean–land interaction, Convective clouds, Multiple scales, Diurnal cycle, Sea–land breeze circulation

Introduction

Tropical rainfall in the global climate

Recognition of rainfall as one of the most important quantities for classifying climate and its global quantitative description was established by Köppen (1918), who divided the tropical climate (*A*; classified by temperature in his first world climate map in 1884) into rainforest (*Af*) and the savanna (*Aw*), according to vegetation. This is also physically reasonable, because in essence, the temperature and rainfall represent the sensible (radiative) and latent heating processes, respectively, that maintain the climate (Fig. 1a). Note that rainfall is a flux quantity (amount per area per time) dependent on horizontal and temporal scales (such as the hourly, daily, monthly, and annual rainfall measured by a single-station rain gauge, in a radar-observed area, and over an integrated region) and that it fluctuates greatly because individually short-lived small

* Correspondence: mdy@jamstec.go.jp
[1]Japan Agency for Marine–Earth Science and Technology (JAMSTEC), Natsushima–cho, Yokosuka 237-0061, Japan
[2]Professor Emeritus, Kobe University, Rokkodai–cho, Nada, Kobe 657-8501, Japan
Full list of author information is available at the end of the article

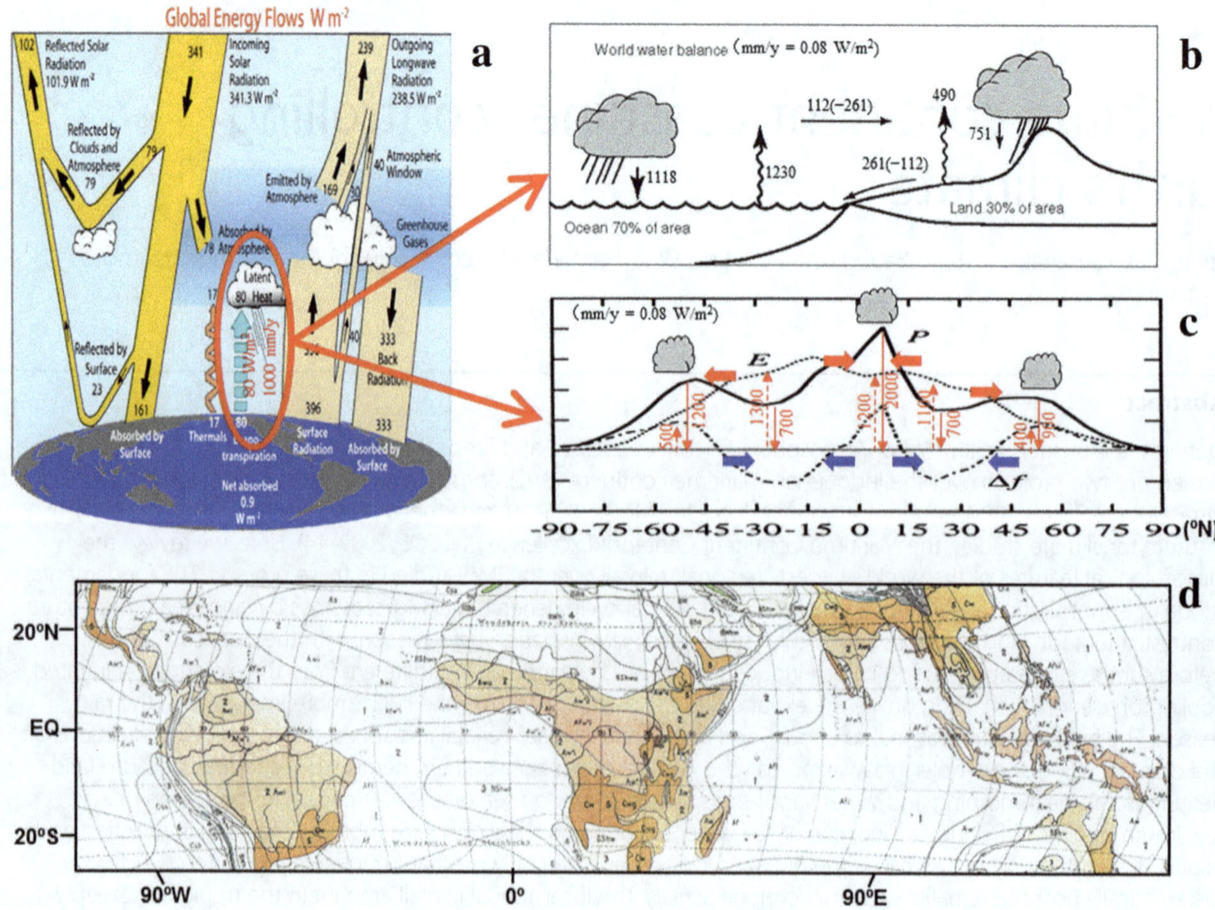

Fig. 1 **a** The global energy and water cycle (modified from Trenberth et al. 2009) and its alternative distributions: **b** in sea and land (modified from Hartmann 1994, with values according to Trenberth et al. 2009) and **c** in latitude (modified from Hartmann 1994, which was based on Baumgartner and Reichel 1975; originally much older, at latest up to Meinardus' analysis published in 1934; see Jaeger 1983). **d** Low-latitude part of Köppen's (1918) original climate classification map showing already both latitudinal and coastal dependencies

precipitating clouds are organized into multiple scale structures. In this article, we consider mainly a "regional" rainfall with a scale somewhat larger than that supposed (seasonal–annual and 10^2–10^3 km) by Köppen on the basis of vegetation.

Köppen classified low latitude regions of South America and Africa and most of the "East Indies" as *Af* climate regions defined by a minimum monthly rainfall ≥ 60 mm/month (total annual rainfall >> 720 mm/year), but carefully excluded the inland areas of Sumatera, Kalimantan and Papua,[1] as well as those of South America and Africa (except for the Amazon and Congo River Basins) so that his "idealized" land had *Af* along the equator including inland areas (see Fig. 1d). This view of the rainfall concentration in the equatorial *coastal* areas was still accepted after some revisions up to his updated version (Köppen and Geiger 1954),[2] but is no longer consistent with recent descriptions (Peel et al. 2007).[3] The smaller (Lesser Sunda) islands between Jawa and Papua were also excluded from *Af*, which was recognized as a west–east difference in the Indonesian maritime continent (IMC). At present, a detailed gridded rainfall distribution based on rain gage data (APHRODITE; Yatagai et al. 2012) shows differences between coastal and inland regions, although this is less clear than in the space-borne radar data to be mentioned later.

Köppen's son-in-law Wegener (1929, after revisions of the first edition published in 1915) tried to use his father-in-law's idealized climate distribution (1918) to estimate the geographical location of his proposed early-Mesozoic supercontinent before separation and drift, assuming the *Af* climate near the equator and the coastline. He also speculated on the origin of the East Indies by the association of northward movement of the Indian subcontinent followed by the Australian continent, which is essentially consistent with present-day plate tectonics after a long period of controversy (e.g., Frankel 2012). Köppen and Wegener (1924) evaluated the importance of theories of climate variability associated with insolation changes, developed by Milanković, a Serbian applied mathematician, who published a compilation of these theories (Milanković 1941), which concerned mainly long-term historical climate change (ice ages) but were based on the variation of insolation with the axis of the Earth's rotation that will be described later in this article. The East Indies, that is the present-day Indonesian Archipelago, and surroundings were designated the maritime continent by Ramage (1968), based on the occurrence frequency of thunderstorms, which was the only database covering the whole equatorial region at that time. In the tropics, thunderstorms as well as rainfall are dependent on convective cloud activity, and the Indonesian maritime continent[4] (hereafter referred to as IMC) is known as the world's convection center or the region of maximum rainfall on Earth.

The IMC was one of the most important regions for hydroclimate studies during the Monsoon Asian Hydro–Atmosphere Scientific Research and Prediction Initiative (MAHASRI; 2006–2016) and the Asian Monsoon Years (AMY; 2007–2012)[5] in the World Climate Research Programme (WCRP) (Matsumoto et al. 2016, 2017; Yamanaka et al. 2016). This is simply because the IMC is the region with the world's largest regional rainfall but has no sufficiently reliable observation network. In the subsequent two subsections we summarize the scientific background and implementation of the two projects carried out over the IMC related to MAHASRI and AMY.

The global energy/water cycle and the maritime continent

The Earth's climate system includes latent heat input from the ground to the atmosphere through the cloud–precipitation process (Fig. 1a), which is also a part of the water cycle between the land and ocean through the atmosphere (Fig. 1b). The meridional peak of the cloud–precipitation activity (corresponding to rainfall amounts of about 2000 mm/year ≈ 5.5 mm/day) is located in the tropics along the equator (Fig. 1c). The equatorial rainfall peak is consistent with the intertropical convergence zone (ITCZ) in the general circulation (Fig. 2a) known since Halley's (1686) observations and Hadley's (1735) theory.[6] It is surprising that the maximum value was known rather correctly more than a century ago (cf. Jaeger 1983), when stations were restricted to major cities over land, in particular *along the coastlines*.

Geographical rainfall distributions with the meridional peak around 2000 mm/year in the tropics are estimated also from satellite cloud observations[7] and simulated by numerical climate models, for example, as shown in a report of the Intergovernmental Panel on Climate Change (IPCC) (Randall et al. 2007) (Fig. 3), although peak values are somewhat larger than the direct rainfall observations from rain gages and radars. An IPCC-related study (Yoshida et al. 2005) has compared these estimates and their simulations (the Community Climate System Model version 3 (CCSM3) model; Collins et al. 2006) of the mean annual rainfall amounts in various regions, of which 10 are inside the tropics (see the two columns labeled IPCC07 in Table 1). The IMC is the region of the most active convective clouds producing the largest regional rainfall on Earth (around 2700 mm/year ≈ 7 mm/day). The second highest regional rainfall is found in Central America.

The world's largest regional rainfall over the IMC has often been explained in terms of the warmest seawater surrounding the IMC (Fig. 4b), which looks just like a "dam" for the global "river" of Indonesian throughflow from the Pacific to the Indian Ocean (e.g., Lukas et al. 1996; Gordon 2005). However, rainfall over the open ocean including the ITCZ is less than that over the islands in the IMC. The tropical atmosphere is *conditionally* unstable; convection is developed only when cloud appears, whereas a cloud is generated when convection develops. This paradoxical situation may be

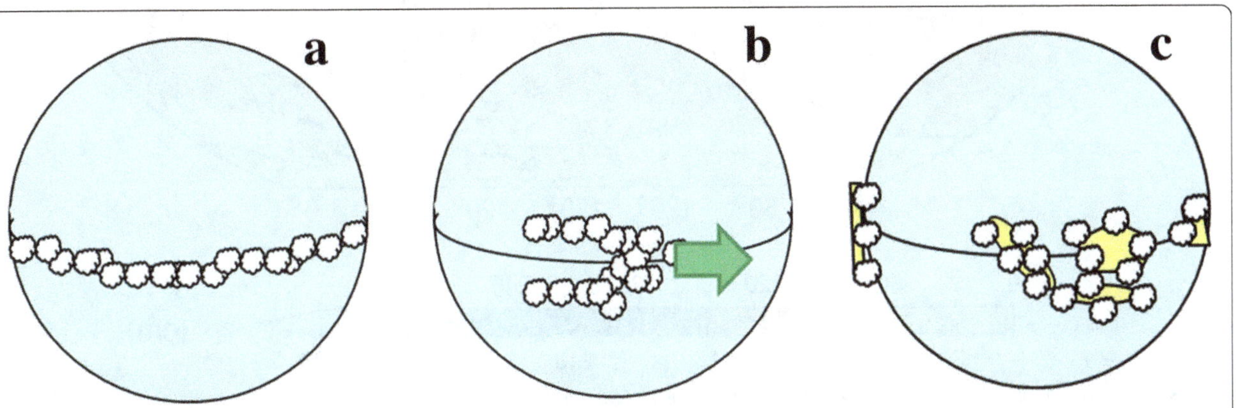

Fig. 2 Schematic diagrams of distributions of tropical precipitation: **a** simple ITCZ as considered by Hadley in 18th century, **b** MJOs propagating eastward as simulated in an "aqua planet" model (e.g., Hayashi and Sumi 1986), and **c** concentration along coastlines as described in this article

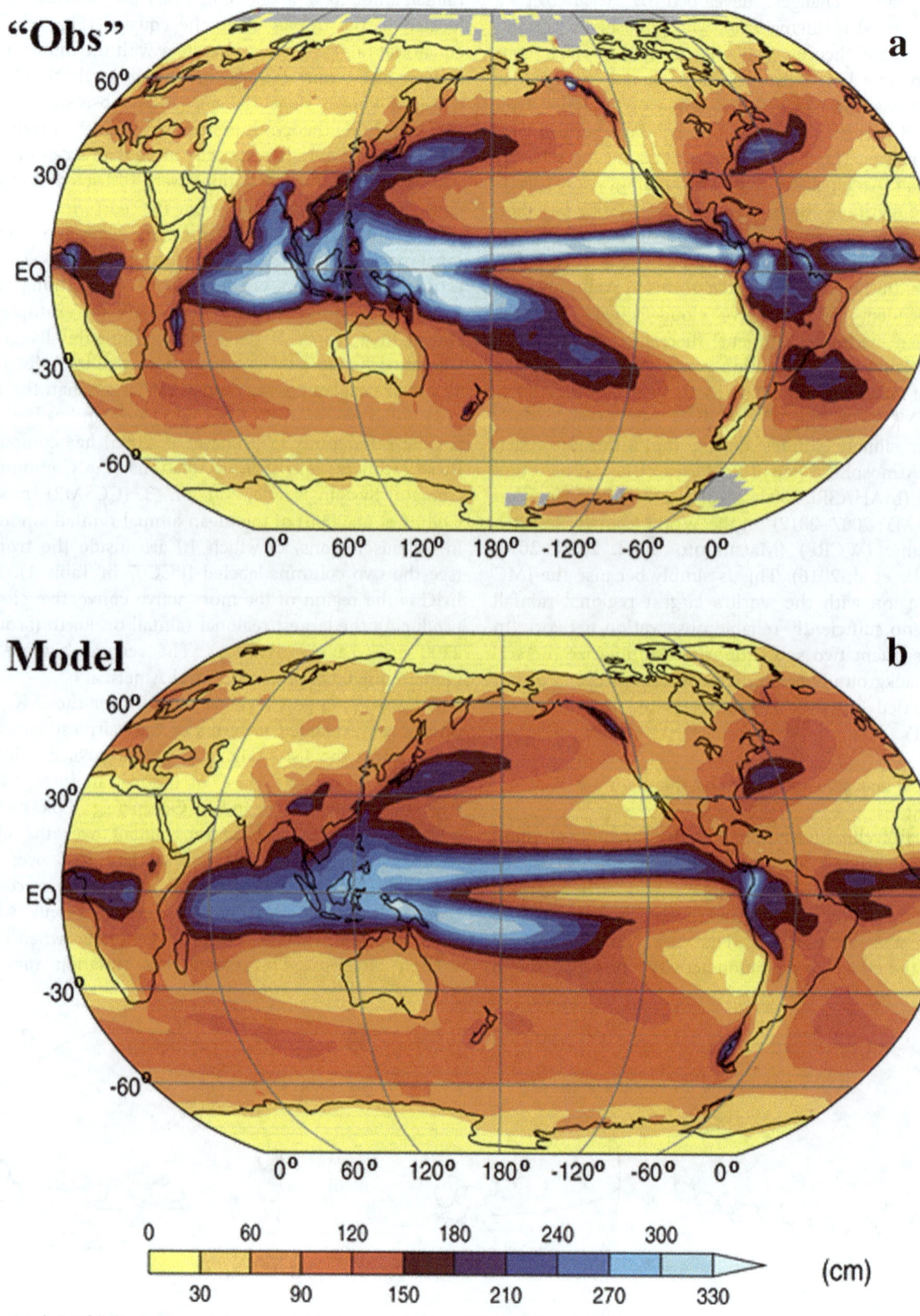

Fig. 3 a, b Rainfall distributions **a** "observed" and **b** simulated for IPCC AR4 (Randall et al. 2007)

Table 1 Ten tropical regions specified in a basic study by the CRIEPI group (Maruyama et al. 2005) to check the reliability of a numerical climate model (CCSM3; Collins et al. 2006). Land areas and coastline lengths were estimated by calculating numbers of 1° longitude × 1° latitude (corresponding to 111 km × 111 km on the equator) grid points close to coastlines (within 0.5°). "IPCC07 Obs/Model" rainfall amounts are obtained from figures of the original paper. "TRMM Dec97–Jan11" is from Ogino et al. (2016)

Area	Latitude (deg N)	Longitude (deg E)	Land area (10^4 km^2)	Coastline (10^2 km)	CRIEPI 1990 – 1999 (mm/year)	CRIEPI CCSM3 (mm/year)	TRMM Dec 97 – Jan 11 (mm/year)
India	10 to 24	71 to 91	144.5	39	1289	1158	1101
Indochina	9 to 21	91 to 121	145.0	79	1237	1816	1789
Sahel	6 to 16	− 5 to 13	180.0	5	1132	921	1319
Central America	6 to 15	− 93 to − 74	47.5	46	1105	2474	2942
Central Africa	− 9 to − 4	11 to 29	227.0	6	2158	1684	1276
Indonesia	− 10 to 9	96 to 153	242.5	333	2632	2737	1718
Amazonia	− 10 to − 1	− 69 to − 49	180.0	0	1974	2342	1657
Southern Africa	− 23 to − 7	11 to 40	410.5	33	1132	947	479
Northern Australia	− 20 to − 10	120 to 145	136.5	65	947	789	483
Brazil	− 22 to − 10	− 65 to − 34	311.5	15	1211	1382	1042

More than 80% of earth's circumference (400 × 10^2 km)

overcome over the open ocean by conditional instability of the second kind (CISK), which generates tropical cyclones in the subtropics and intraseasonal variations (ISVs) along the equator (Ooyama 1971; Lindzen 1974; Hayashi and Sumi 1986). Tropical cyclones generally appear outside the IMC. The ISVs (shown in Fig. 2b), which often have the so-called Matsuno–Gill pattern (Matsuno 1966; Gill 1980), are actually dominant and known as the Madden–Julian oscillations (MJOs) over the open ocean (Madden and Julian 1971, 1994; Nakazawa 1988; Zhang 2005, 2013; Yoneyama et al. 2008, 2013) on both the western and the eastern sides of the IMC (Fig. 5). However, such transient or traveling disturbances are modified or weakened on reaching the IMC (Nitta et al. 1992; Hashiguchi et al. 1995; Widiyatmi et al. 2001; Sakurai et al. 2005; Shibagaki et al. 2006; Hidayat and Kizu 2010; Marzuki et al. 2013) and cannot explain the convective clouds and rainfall over the IMC throughout the year.

On the other hand, the land is heated more easily than the sea (Fig. 4a), but convective clouds generated over the true continents (Africa and South America) are less active than those over the IMC. Therefore, another systematic mechanism is necessary to kick-off an initial updraft in the conditionally unstable tropical atmosphere, and it must work most actively over the IMC.

We have also noticed climatic variabilities of the IMC (cf. Yamanaka 2016): two systematic periodical (annual and diurnal) cycles that are forced astronomically and other components without fixed periodicity (ISVs and interannual variations). The annual cycles of rainy (monsoon) and dry seasons are less clear in the IMC than in adjacent regions such as Indochina (Murakami and Matsumoto 1994) and vary spatially: the seasonal march in the southern hemispheric part (Jawa and Bali) has a southern-hemispheric annual cycle (a clear rainfall maximum in southern summer) (e.g., Haylock and McBride 2001; Hamada et al. 2002; Aldrian and Susanto 2003; McBride et al. 2003; Chang et al. 2004, 2016). The equatorial and northern-hemispheric parts have either a semi-annual cycle (with double peaks near equinox seasons), a northern-hemispheric annual cycle (one peak in northern summer), or the cycle is unclear. The semi-annual cycle is reasonable if we consider that the sun passes the zenith, when the insolation becomes maximum, twice a year. So the rainfall over the IMC is sensitive to the insolation. However, why is the insolation stronger in the rainy and cloudy season? Why is the southern-hemispheric rainy season so dominant in low-latitude

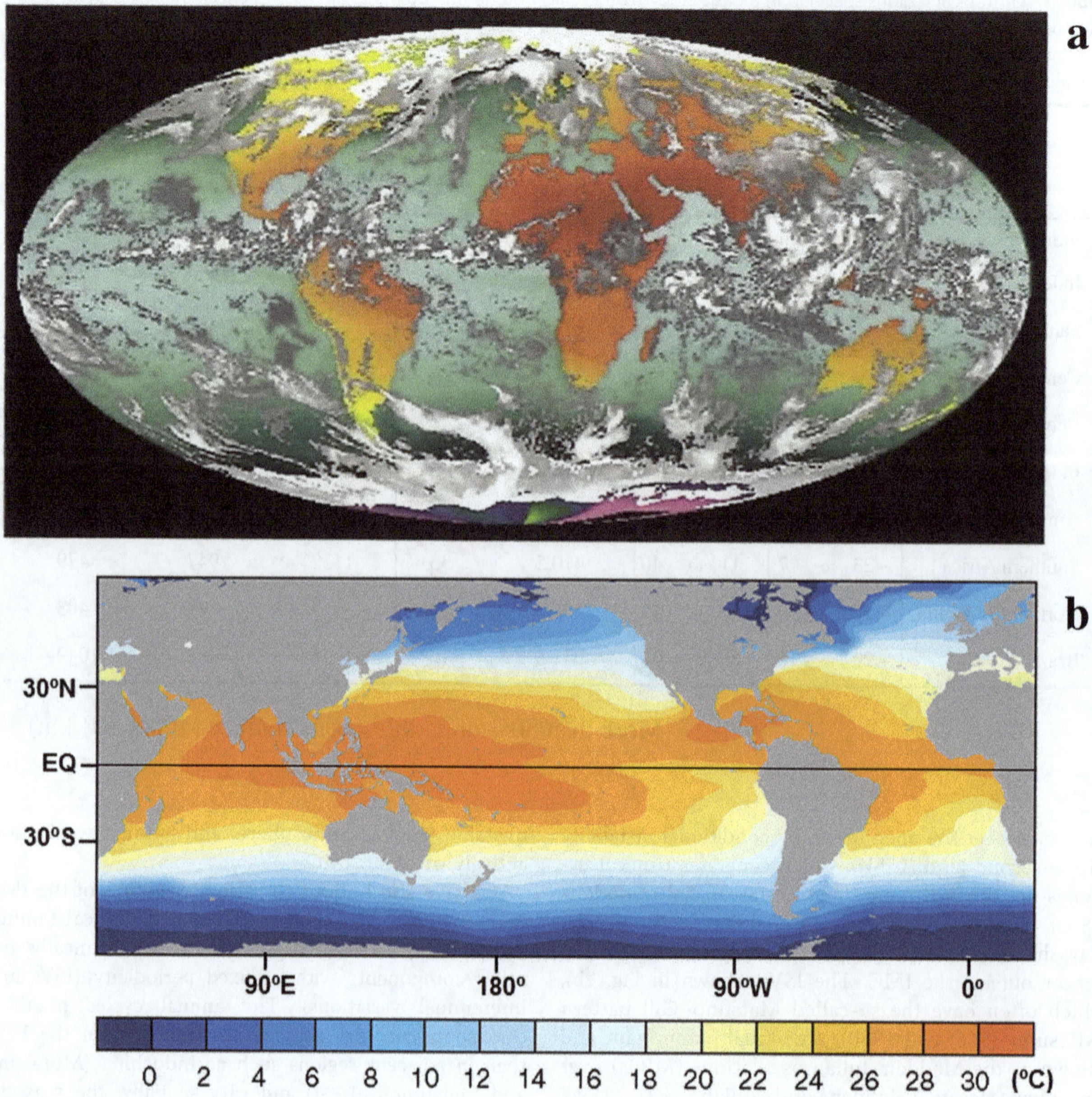

Fig. 4 Distributions of **a** OLR (sea, land, and cloud top) temperature (snapshot, originally from the University of Wisconsin) and **b** SST (annual mean, originally from UK Meteorological Office). Copied from figs. 1.27 and 2.11 of Wallace and Hobbs (2006)

Jawa and Bali? We must answer these questions adequately, as well as establishing the mechanism to explain why the world's largest regional rainfall falls over the IMC.

JEPP–HARIMAU and SATREPS–MCCOE projects

Our observations have been outlined in Yamanaka et al. (2008, 2016) and Yamanaka (2016), and here, we give some details of the background (see also the map in Fig. 5b). The most important aspect in promoting a hydroclimate project over Southeast Asia in the present century is that both infrastructure and the need for observations have grown rapidly with economic growth. More complete understanding and more accurate predictability are requested strongly from industry and the public sectors that have graduated quickly from day-to-day business and a hand-to-mouth existence. The numbers of universities and their students are also increasing rapidly. These circumstances have encouraged governments to promote hydroclimatology with resolution better than annual or continental scale monsoons, as well as with interannual and global scope to maintain sustainable development by mitigating the effects of climate change. Therefore, scientific studies have been conducted in parallel with projects for societal applications (e.g., disaster prevention) and capacity building.

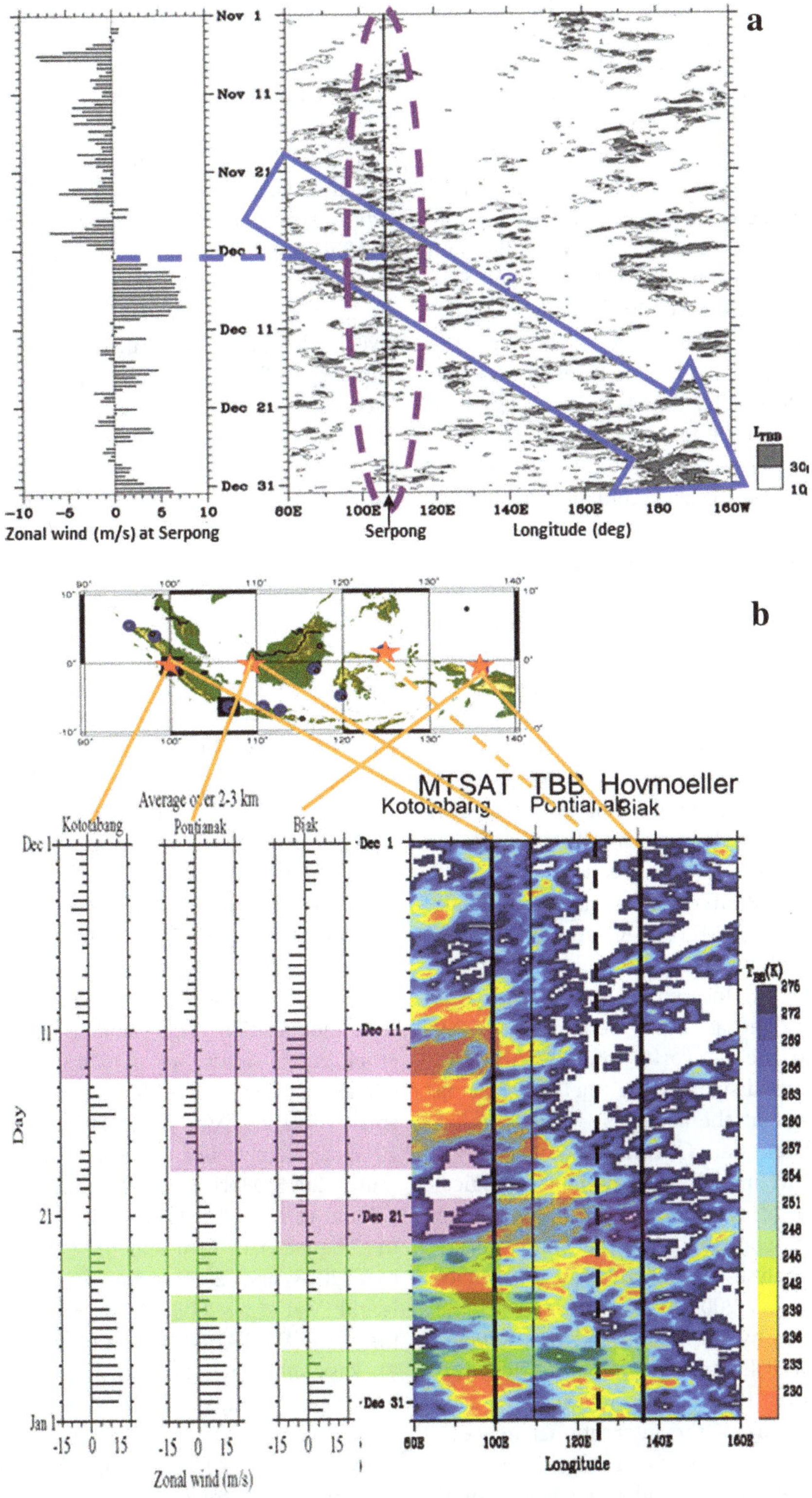

Fig. 5 ISVs arriving at the IMC **a** around the time of rainy season onset (November–December 1992) observed by a wind profiler (12 h averaged) at Serpong (6° S, 107° E) (left) and a GMS (hourly) Hovmöller diagram along 6° S (right) (Hashiguchi et al. 1995) and **b** during a rainy season (December 2007) by a wind profiler network at Kototabang (0° S, 100° E), Pontianak (0° S, 110° E), and Biak (1° S, 136° E) (map and left three panels) and GMS along the equator (right) (Yamanaka et al. 2008), showing modification/decay (smaller hierarchies associated with reversals in profiler zonal winds) mainly due to the diurnal cycle (appearing in GMS clouds) over the IMC

Projects related to MAHASRI and AMY over the IMC are most typical in view of their relevance to the criteria mentioned above. In particular, high-resolution radar observations were strongly requested by both hydroclimatological interest in the world's strongest convective activity and the societal impacts such as frequent flood and landslide disasters. When we proposed a project related to MAHASRI over the IMC, the Japanese Government established the Japan Earth-Observation System Promotion Program (JEPP) contributing to the Global Earth Observation System of Systems (GEOSS; see GEO 2007), which emphasized societal benefits. Our proposal for the Hydrometeorological ARray for Intraseasonal variation –Monsoon AUtomonitoring (HARIMAU) (Yamanaka et al. 2008)[8] was approved as a JEPP project and was conducted from August 2005 to March 2010 (see: http://www.jamstec.go.jp/iorgc/harimau/HARIMAU.html), in collaboration with the Japan Agency for Marine–Earth Science and Technology (JAMSTEC), Kyoto University, Hokkaido University, the Agency for the Assessment and Application of Technology (BPPT), the Indonesian National Institute of Aeronautics and Space (LAPAN), and the Indonesian Agency for Meteorology, Climatology and Geophysics (BMKG). HARIMAU consisted of a radar-profiler network with meteorological Doppler radars of X-band (at Padang; e.g., Kawashima et al. 2011; Mori et al. 2011; Sakurai et al. 2011; Kamimera et al. 2012) and C-band (at Serpong near Jakarta; e.g., Wu et al. 2013; Mori et al., 2018, personal communication), and wind profilers of L-band (at Pontianak, Manado, and Biak; e.g., Tabata et al. 2011a, 2011b). A large very high frequency (VHF) wind profiler called the Equatorial Atmosphere Radar (EAR) located at Kototabang (e.g., Fukao 2006; Shibagaki et al. 2006; Seto et al. 2009) was also included in the network. Rainfall and wind distribution data were displayed via the Internet in near real-time throughout the project.

Since 2005, the Indonesian Government has funded the BMKG's efforts to install over 40 meteorological radars.[9] The number of radiosonde stations was also increased to 22, which contributed to the improved accuracy of objective analyses at meteorological agencies worldwide (cf. Seto et al. 2009).[10] In parallel with the measurements obtained by the radars and radiosondes, JAMSTEC and BPPT collaborated on buoy and shipboard observations over the Indonesian exclusive economic zone (EEZ) as part of the international network of the Tropical Ocean Climate Study (TOCS; Ando et al. 2010). Another JEPP project, the Indian Ocean Moored Network Initiative for Climate Studies (IOMICS), collected data during 2005–2010 as part of the Research Moored Array for African–Asian–Australian Monsoon Analysis and Prediction (RAMA; McPhaden et al. 2009). Ship-based sea water, surface weather, radar, and radiosonde observations on atmosphere–ocean interactions over the Indian Ocean were conducted by the Mirai Indian Ocean Cruise for the Study of the MJO–Convection Onset in 2006 (MISMO2006; Yoneyama et al. 2008).

In 2009, a follow-on project to HARIMAU and IOMICS was approved under the Science and Technology Research Partnership for Sustainable Development (SATREPS). This project established an international center of tropical climatology, meteorology, and oceanography, called the Maritime Continent Center of Excellence (MCCOE) under BPPT in November 2013 and included capacity building for radar operations and buoy manufacturing during February 2010–March 2014 (see http://www.jamstec.go.jp/rigc/tcvrp/satreps_id/ in Japanese). Some observations were conducted also in collaboration with the Cooperative Indian Ocean Experiment on Intraseasonal Variability in the Year 2011 (CINDY2011; Yoneyama et al. 2013). After MCCOE was established, the HARIMAU radars and profilers came under the responsibility of that center. Observational data have contributed to global reanalysis (e.g., Onogi et al. 2007; Kobayashi et al. 2015), as well as high-resolution numerical studies at both regional (e.g., Wu et al. 2008, 2009) and global (e.g., Sato et al. 2009) scales.

The HARIMAU campaign observations were carried out during periods of intense observation almost every year from 2006 to 2011, under a collaboration with MAHASRI and other international projects (Table 2). In particular, HARIMAU2006, 2007 and 2011 were in the westernmost area of the IMC (Propinsi Sumatera Barat or West Sumatra Province), targeting the MJOs from the Indian Ocean and their modification (Yoneyama et al. 2008, 2013; Kamimera et al. 2012; Kubota et al. 2015, with the MISMO and CINDY projects), the diurnal cycles of the major and smaller islands (Wu et al. 2008, 2009; Kamimera et al. 2012), and convective cloud structure and vertical coupling (Fukao 2006; Mori et al. 2011; Kawashima et al. 2006, 2011; Sakurai et al. 2009, 2011, with the Coupling Processes in the Equatorial Atmosphere (CPEA) project). HARIMAU2008 and 2009 were performed over Kalimantan Island (Wu et al. 2008, 2009). HARIMAU2010 focused on the JABODETABEK (Jakarta–Bogor–Depok–Tangerang –Bekasi, or greater Jakarta) area, in collaboration with a hydrological study of the Ciliwung River (Sulistyowati et al. 2014; Mori et al., 2018, personal communication; Katsumata et al., 2018, personal communication).

Many other projects over this important region have been carried out in collaboration with various countries and agencies, for example, ocean–atmosphere interaction studies with China (e.g., Yang et al. 2015), aerosol studies with the USA and others (e.g., Reid et al. 2016), a collaboration within Southeast Asia (e.g., Koh and Teo

Table 2 Intense observation periods (IOPs) of the HARIMAU project under MAHASRI

Campaign name	Intense observation area/period	Stations of major instruments	Collaborated projects	Major results
HARIMAU2006	Sumatera Barat (incl. Siberut Island) 26 Oct–27 Nov 2006	MIA (XDR, AWS, SVC) Tiku (XDR, AWS) Tabing (RS, AWS, RIS) Siberut (RS, AWS, GPS, RIS)	MAHASRI, CPEA, MISMO2006	Vertical coupling Diurnal cycle at small island
HARIMAU2007	Sumatera Barat 14 Apr–25 May 2007	MIA (XDR, AWS, SVC) Tiku (XDR, AWS) Tabing (RS, AWS)	MAHASRI, AMY, PRIMO/Maldives	Details of diurnal cycles
HARIMAU-AMOY	Kalimantan (+Sumatera, Sulawesi) 21–27 Aug 2008, 20 Jan–7 Feb 2009, 20–26 Aug 2009,	Pontianak (WPR, RS, AWS, GPS) Putussibau (AWS, GPS) Balikpapan (AWS, GPS) (Manado (WPR, RS, AWS), Biak (WPR, RS, AWS))	MAHASRI, AMY,	Geographical differences of diurnal cycles Local water cycles
HARIMAU2010 (co-funded with SATREPS)	JABODETABEK/Jawa 16 Jan–14 Feb 2010	Serpong (CDR, WPR, AWS, RSD) Bogor (RS, AWS, RSD, RIS) Pramuka (RS, AWS, RIS) Serang (RS, AWS) Karawang (RS) Cengkareng (RS, AWS)*	MAHASRI, AMY,	Details of diurnal cycles Local circulations River diurnal cycle
HARIMAU2011 (funded by SATREPS)	Sumatera Barat (incl. Sipora Island) 1–31 Dec 2011	MIA (XDR, AWS), Pontiank (RSD) Muara Putus (MPR, AWS) Tabing (RS, AWS) Tuapejat/Sipora (RS, AWS), Pariaman (AWS, RSD, ARG), Sicincin (AWS, RSD), Bawan, (AWS) Kapar, Maninjau, Kandang Empat, Limau Purut, Sungai Sarik, Sintuk, Buayan, Lubuk Minturun, UNAND (ARG)	MAHASRI, AMY, CINDY2011	Organization processes of coastal heavy rainbands Examination for quantitative precipitation estimation Experiment for quantitative precipitation forecasting

Operational observations only

MIA Minangkabau International Airport/Padang, *XDR* X-band Doppler radar, *CDR* C-band Doppler radar, *MPR* X-band multi-parameter radar, *WPR* UHF (L-band) wind profiler radar, *EAR* Equatorial Atmosphere Radar (VHF), *RS* radiosondes, *AWS* automatic weather stations, *ARG* automatic raingauge, *GPS* Global Positioning System (precipitable water), *RSD* disdrometer (raindrop size distribution), *RIS* raindrop isotope sampling, *SVC* Sky-view camera, *UAV* unmanned aerial vehicle

2009), and a collaboration/capacity building for meteorological operations and database construction with Australia, the Netherlands, France, Germany, and others. Collaborations between projects are occurring, and contributions from Indonesian scientists are increasing (see the last paragraph of the "Conclusions" section of this article).

In this article, we review mainly our own observational studies on the mechanisms generating the world's largest regional rainfall over the IMC under the JEPP–HARIMAU and SATREPS–MCCOE projects. The "Review" section is separated into five subsections. The "Diurnal cycle observed around the IMC coastlines" section concerns observational evidence of the diurnal cycle near coastlines, which is the most fundamental and dominant mode over the IMC. The "Local rainfall as a function of coastline length" section describes the general feature of tropical rainfall concentration near coastlines. The "Regional rainfall as a function of 'coastline density'" section explains the largest regional rainfall over the IMC in terms of the concentration of coastal diurnal cycle rainfall. The "Consistency with the global water budget" section estimates the contribution of the IMC rainfall to the global water/energy cycle. The "Control of the global climate" section looks toward connections between the IMC rainfall and the global climate.

Review

Diurnal cycle observed around the IMC coastlines

The dominant periodicity of insolation is diurnal, and sea–land heat capacity contrasts enforce the local diurnal variations in wind, clouds, and rainfall that were known in Batavia (now Jakarta) a century ago (van Bemmelen 1922). These diurnal variations have been confirmed everywhere along the coastal zones of larger islands (Sumatera, Kalimantan, Jawa, Sulawesi, and Papua) of the IMC, based on wind profilers (including raindrop measurements) (e.g., Hashiguchi et al. 1995; Renggono et al. 2001; Hadi et al. 2002; Murata et al. 2002; Araki et al. 2006; Marzuki et al. 2009; Tabata et al. 2011b), precipitation radars (e.g., Kawashima et al. 2006, 2011; Wu et al. 2007, 2013; Sakurai et al. 2009, 2011; Mori et al. 2011; Kamimera et al. 2012), a space-borne precipitation radar (e.g., Sorooshian et al. 2002; Mori et al. 2004; Kikuchi and Wang 2008; Wu et al. 2008, 2009; Peatman et al. 2014), GPS radio occultation (precipitable water or moisture) sounding (e.g., Wu et al. 2003, 2008, 2009), and satellite observations of outgoing longwave (infrared) radiation (OLR) (cloud-top temperature or height) (e.g., Hendon and Woodberry 1993; Nitta and Sekine 1994; Ohsawa et al. 2001; Yang and Slingo 2001; Sakurai et al. 2005; Hamada et al. 2008). As categorized by Kikuchi and Wang

(2008), the diurnal cycle over the IMC is different from that over the open ocean (e.g., Sui et al. 1997).

Figure 6 shows examples of monthly averages of hourly OLR (cloud top temperature) observations by a Geostationary Meteorological Satellite (GMS) over the IMC and surrounding oceans. Individual mesoscale diurnal cycles are generated almost synchronously along coastlines of the archipelago, so that the whole (synoptic scale) IMC oscillates with an almost uniform diurnal cycle, which is consistent with the coastal heavy rainband organization previously reported (e.g., Mori et al. 2004, 2011, (2018, personal communication). We have carried out spectral analyses of the GMS data accumulated for 14 years over the IMC–Pacific region (Fig. 7a, the same region as Fig. 6). Geographical distributions of the frequency power-spectral density for selected components are plotted in Fig. 7b. This clearly shows the primary dominance of diurnal cycles and the secondary importance of annual cycles around major islands of the IMC and surrounding tropical regions. The diurnal cycle is particularly clear just on the landward side of the coastline, whereas the annual cycle appears somewhat vaguely and mainly inland. Figure 7c, d, are examples of frequency and zonal wavenumber spectral analyses. In the frequency domain, a striking spectral line corresponding to the diurnal cycle and the shorter period red noise-like continuum spectrum (to its right) are both stronger around the IMC than over the ocean. The wavenumber domain is only continuum-spectral (except for a flection around 300 km in wavelength, which will be mentioned later). These panels confirm that spatially broad-band (meso-to-synoptic scale) convection is the end result of the diurnal cycle energy input, although intermediate nonlinear processes may play a role and should be investigated further.

Figure 8b, e show the mean annual rainfall (over a period of about 13 years; Ogino et al. 2016) and the local AM–PM rainfall difference (for 3 years; Mori et al. 2004) over the IMC, analyzed from space-borne precipitation radar (PR) data from the Tropical Rainfall Measurement Mission (TRMM) satellite. Rainfall amount is large near the coastlines, and morning (evening) rainfall is concentrated over the sea (land) side of a coastline. In central Africa (Fig. 8a, d) and South America (Fig. 8c, f), there are major inland rainfall regions, but circular distributions of rainfall amount and AM–PM difference are found, albeit faintly, around the Congo and Amazon river basins (which will be discussed later), in addition to remarkable coastal diurnal cycle rainfall areas over the Gulf of Guinea and Lake Victoria (as well as the Gulf of Panama, although this lies outside Fig. 8f). Hirose et al. (2017) show much higher resolution analyses including smaller island features and even a land–large river contrast over the Amazon River.

Figure 9a shows an example of frequency spectra of temperature at the surface and near the tropopause observed by radiosondes launched three hourly from both sides of a coastline near Bengkulu, southwestern Sumatera. Figure 9b, c show mean diurnal variations of the land and marine atmosphere temperature for the lowest 1 km obtained by the radiosonde observations. The diurnal cycle of temperature is striking near the land surface below about 300 m but is weak near the tropopause in spite of the cloud top height having a clear diurnal cycle. In particular, for the lowest 100 m on land, the lapse rate is large in the daytime (below 40 m, each temperature profile is often partly super-adiabatic, hence absolutely unstable around 06 UT = 13 LT), whereas it is negative (that is an inversion, i.e., strongly stable) in the nighttime 15–24 UT (22–07 LT).

Actual surface temperature is determined by complex boundary layer processes of radiation, convection (with eddy motion on smaller scales than the cloud convection) and conduction (molecular properties, less important than the other two in the atmosphere, except just above the surface) interacting with each other (e.g., section 14.1 in Stull 1988). For example, if clouds are generated by surface heating in daytime, the heating becomes immediately reduced by the albedo (parasol effect) of the clouds. Rainfall produced by the clouds on land in nighttime cools down the land surface temperature like a sprinkler (Fig. 10b; Wu et al. 2009), as shown observationally in Fig. 9b and numerically in a global high-resolution model (Sato et al. 2009). As a result, there is almost no "tropical night"[11] in the tropics (Fig. 10b). This sprinkler-like self-cooling process is different from the extratropical diurnal cycle that needs radiative cooling in clear weather through day and night. Thus, over the IMC, the latent heat and the sprinkler-like effect should have comparable magnitudes with the other (sensible or radiative) heating or cooling mechanisms such as insolation, infrared radiation, and the greenhouse effect. Hadi et al. (2002) and Araki et al. (2006) presented climatological analyses of diurnal cycles near Jakarta in northwestern Jawa and discussed differences between dry and rainy seasons and between wind and rainfall.

Usually the land surface, which has a smaller effective heat capacity, is heated or cooled by any thermal forcing more quickly and more largely than the sea surface with larger effective heat capacity, and the horizontal gradient (sea–land difference) of surface temperature (precisely, atmospheric bottom temperature) directly governs the amplitude of diurnal cycles (e.g., Kimura 1975; Rotunno 1983; Niino 1987; Niino et al. 2006). Figure 9d shows mean diurnal variations of the land–sea atmosphere temperature difference for the lowest 1 km obtained by radiosonde observations. Warmer land than sea is observed through the analyzed altitude range (especially

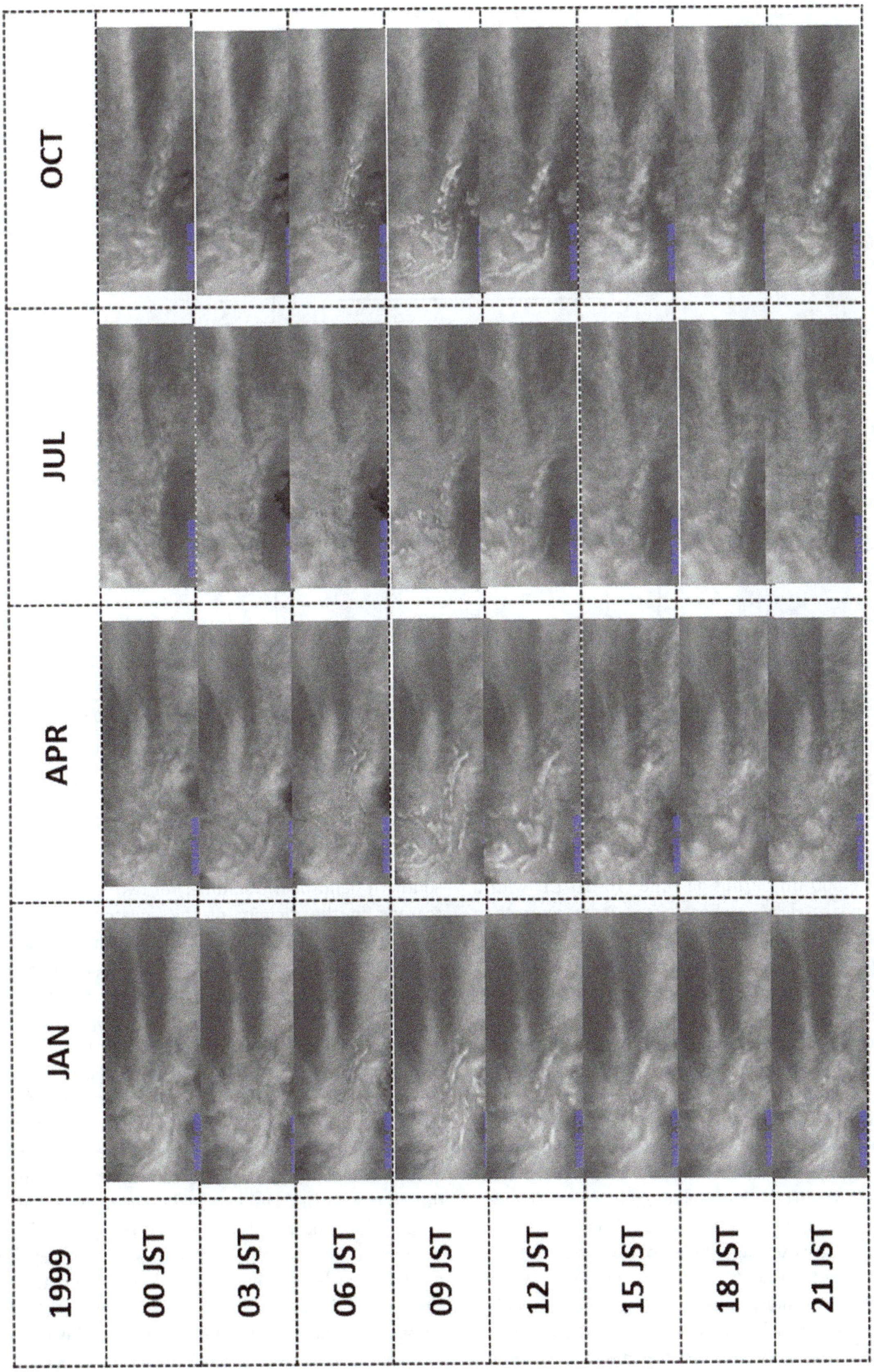

Fig. 6 Monthly-mean 3-hourly GMS cloud-top height (temperature) distributions in 1999. The analyzed region is shown in a map in Fig. 7a. Time is in JST (= LT +2 h)

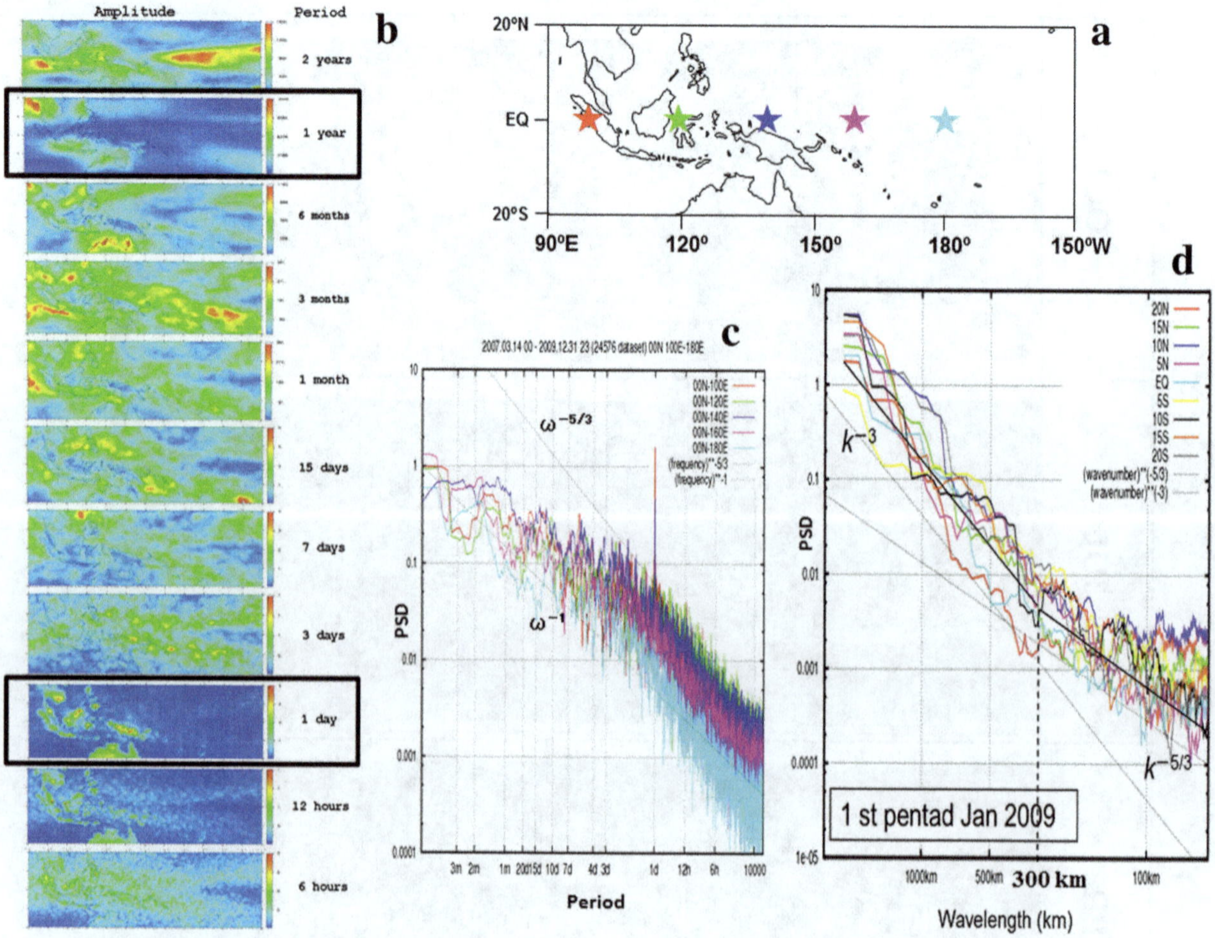

Fig. 7 Spectral analysis of 14-year (1996–2009) hourly GMS–OLR cloud top temperature for the region **a** (Yamanaka 2016). Frequency spectra are **b** plotted geographically for 6 h to 2-year components and are shown **c** for five locations at every 20° longitude from 100° to 180° E along the equator (indicated by stars in **a**). **d** Zonal wavenumber spectra for the 90° E–150° W longitude range at every 5° latitude around the equator

strong below 500 m) during daytime (10–16 LT, with a downward phase delay), whereas sea is warmer than land by a greater amount and for longer (19–07 LT) but only below 200-m altitude. This asymmetric temperature cycle seems consistent with the diurnal cycle rainfall being heavier in the morning on the sea-side than in the evening on the land-side (contrary to the cloud top height shown in Fig. 6) as observed in various coastal areas (Sumatera, Jawa, and Kalimantan) over the IMC (Mori et al. 2004, 2011; Wu et al. 2008, 2009). For the case shown in Fig. 9, detailed features including those associated with an MJO landing are also shown elsewhere (Wu et al. 2017, 2018; Yokoi et al. 2017).

On land, these processes lead to rapid land/hydrosphere–atmosphere water exchange (Sulistyowati et al. 2014) and local air pollutant washout. The resulting clear sky over land from sunrise until around noon provides insolation dependent upon the solar angle, which gives more active convection or a rainy season in each hemispheric summer in spite of the very low latitudes (e.g., Jawa and Bali in southern hemispheric summer) or two (semi-annual) rainy seasons in the vicinity of the equator (Murakami and Matsumoto 1994; Hamada et al. 2002; Aldrian and Susanto 2003). Detailed observations and investigations are needed in future. Air pollution in a mega city such as Jakarta is becoming more serious with rapid development.

Over the open oceans, in the CISK mechanism (e.g., Ooyama 1971; Lindzen 1974; Hayashi and Sumi 1986), cyclonic or wavelike disturbances are generated as feedback that further enhances upward motion, that is the cumulus cloud. Energetically speaking, the kinetic energy is always dissipated in the boundary layer and supplied by the conversion of latent heat obtained by ocean surface evaporation, through the potential energy producing the meridional and zonal circulation. Several other feedback mechanisms have been proposed to explain tropical cyclones and ISVs/MJOs (see, e.g., Wang 2012). For example, in the wind-induced surface heat exchange (WISHE) mechanism (Emanuel 1987; Neelin et al. 1987), surface wind generates waves that enhance evaporation

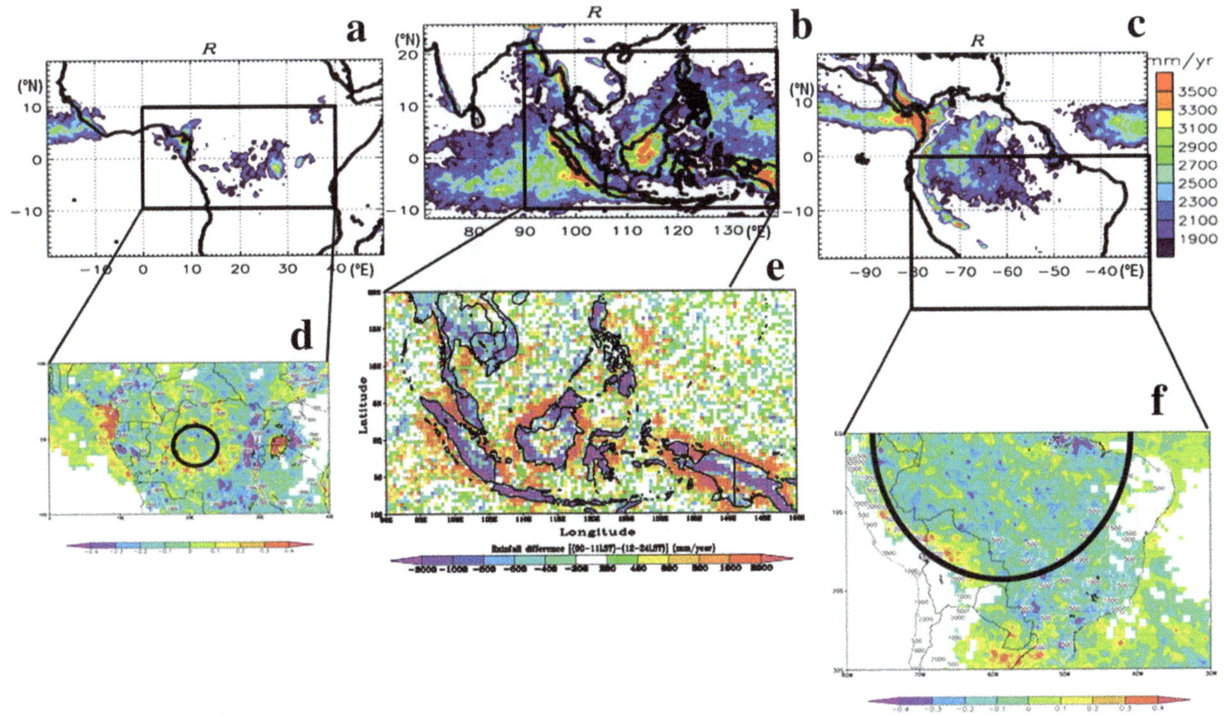

Fig. 8 **a**–**c** annual amount (analyzed for Dec 1997–Jan 2011; Ogino et al. 2016) and **d**–**f** AM–PM difference (Jan 1998–Dec 2000; Mori et al. 2004) of rainfall for **a**, **d** central Africa, **b**, **e** South/Southeast Asia, and **c**, **f** South America observed by the TRMM-PR. Circles in **d**, **f** indicate rough locations of the contour line around which the mountain valley breeze circulation occurs

and latent heat supply. These feedback mechanisms may be suppressed, because the hydrologic and energy cycles close approximately on a diurnal time scale (including clear/dry land in the morning) over the IMC, and the standard features of ISVs/MJOs along the ITCZ over the Indian and Pacific Oceans (e.g., Madden and Julian 1971, 1994; Wheeler and Hendon 2004; Zhang 2005, 2013; Yoneyama et al. 2008, 2013) are greatly modified over the IMC. For example, passage of ISVs/MJOs over the IMC is shifted to the southern side of the equator, which gives a decrease in rainfall over the IMC from west to east (e.g., Nitta et al. 1992; Shibagaki et al. 2006; Hidayat and Kizu 2010; Fujita et al. 2011; Kamimera et al. 2012; Marzuki et al. 2013; Peatman et al. 2014; Zhang and Ling 2017).

Simultaneously, the ISVs/MJOs influence the amplitudes of local diurnal cycles in moisture, stratification stability, and wind over the IMC. The rainfall (particularly over the sea rather than the land) at various locations over the IMC increases during the active/wet phase of ISVs/MJOs (e.g., Hashiguchi et al. 1995; Widiyatmi et al. 2001; Murata et al. 2002; Sakurai et al. 2005, 2009, 2011; Shibagaki et al. 2006; Wu et al. 2007, 2013; Hidayat and Kizu 2010; Mori et al. 2011, (2018, personal communication); Kamimera et al. 2012; Marzuki et al. 2013; Peatman et al. 2014). As a result, the onset and total rainfall amount of the rainy season over the IMC (in particular over the southern hemispheric part) are governed by the passage of active phases of ISVs/MJOs (see Fig. 5a) (Hashiguchi et al. 1995; Renggono et al. 2001; Hamada et al. 2002; Sakurai et al. 2005; Araki et al. 2006; Xie et al. 2006). Interactions with monsoon surges over complex topography (e.g., Chan and Li 2004; Chang et al. 2004, 2016) may also affect the amplitude variations of diurnal cycle rainfall (Wu et al. 2007, 2013; Hattori et al. 2011; Matsumoto et al. 2017), and sometimes generate sub-synoptic scale Borneo (or cold surge) vortices (e.g., Yokoi and Matsumoto 2008; Chen et al. 2013, 2015). These features of rainfall distribution and variation are in general consistent with sea-side dominance of mainly diurnal cycle coastal rainfall (e.g., Mori et al. 2004; Wu et al. 2009; Ogino et al. 2016, 2017).

Correlations between the IMC rainfall and the sea surface temperature (SST) that vary with interannual atmosphere–ocean interactions such as El Niño Southern Oscillation (ENSO) and the Indian Ocean Dipole mode (IOD) have been noted in particular in the dry season, but are notably spatially incoherent in the rainy season (e.g., Haylock and McBride 2001; Hamada et al. 2002; Aldrian and Susanto 2003; McBride et al. 2003). More detailed correlation analyses of recent improved rainfall observations for West Jawa with SST over the Central Pacific and Western Indian Oceans (Hamada et al. 2012) and for the eastern IMC (Kubota et al. 2011; Lestari et al. 2016) are now being applied to the whole of Indonesia. SST variations may reduce or amplify the nighttime sea-to-land temperature gradient and control the morning coastal-sea

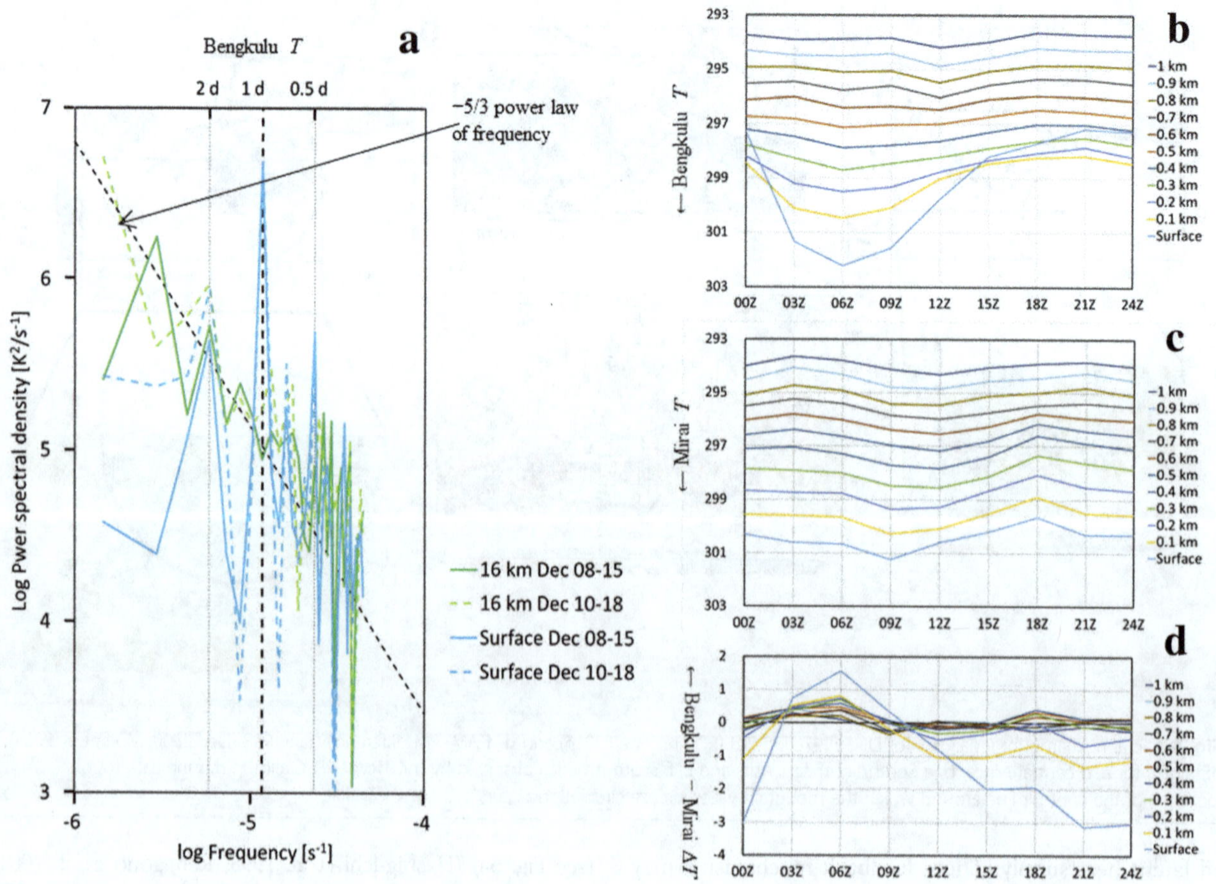

Fig. 9 Examples of **a** frequency power spectra (at a coastal land station in Bengkulu (04°S, 102°E), southwestern coast of Sumatera); **b, c** 10 day-mean land (Bengkulu)/marine (RV Mirai, approximately 50 km SW from Bengkulu) atmospheric temperature diurnal cycles below 1 km altitude; and **d** their difference, obtained during 8–18 December 2015. The abscissas of **b–d** are in UT (= LT−7 h). Details of the observations are described in Yokoi et al. (2017) and Wu et al. (2018)

rainfall in particular in the dry season. In major El Niño and positive IOD years, rainfall is often reduced, even in the rainy season, leading to serious drought and smog disasters over the IMC (e.g., Hashiguchi et al. 2006; Assyakur et al. 2016; Hidayat et al. 2016). In La Niña and negative IOD years, larger moisture transport from warmer sea water enhances diurnal cycle rainfall, which leads to a less clear dry season and flood disasters during cold surges and/or MJO active phases in the rainy season (e.g., Wu et al. 2007, 2013; Matsumoto et al. 2017). These results are essentially consistent with independent similar studies by Qian (2008) and Qian et al. (2013). Decadal or longer-scale rainfall trends (e.g., Endo et al. 2009; Villafuerte and Matsumoto 2015) should also appear through amplitude modulations of the diurnal cycles associated with variations of SST and moisture transport.

Local rainfall as a function of coastal distance

Based on radar observations in Sumatera and Jawa, the rainfall intensity in a typical precipitating cloud system is stronger than 10 mm/h only in a small migrating area of $(10\ \text{km})^2$, which has an echo top of about 10 km and a life time of roughly an hour (e.g., Kawashima et al. 2006, 2011; Wu et al. 2007, 2013; Sakurai et al. 2009, 2011; Mori et al. 2011 (2018, personal communication); Kamimera et al. 2012; Katsumata et al., 2018, personal communication), although Marzuki et al. (2013) noticed variabilities of raindrop size distributions over the IMC. Thus, the precipitation (liquid plus solid) water content of such a developed cloud system is estimated very roughly as

$$1\ \text{t/m}^3 \times \left[10\ \text{mm} \times (10\ \text{km})^2\right]/(10\ \text{km})^3 \approx 1 \times 10^{-6}\ \text{t/m}^3 \approx 1\ \text{g/m}^3, \quad (1)$$

which is somewhat larger than typical values (0.5 g/m³) for oceanic convective clouds. For a broader area around the region of intense rainfall, however, rainfall intensities are around 1 mm/h, and such weaker rainfall may be observed at a station for about 10 h from the evening until the next morning, giving a daily rainfall amount of about 10 mm/day, which is close to the regional-mean annual-

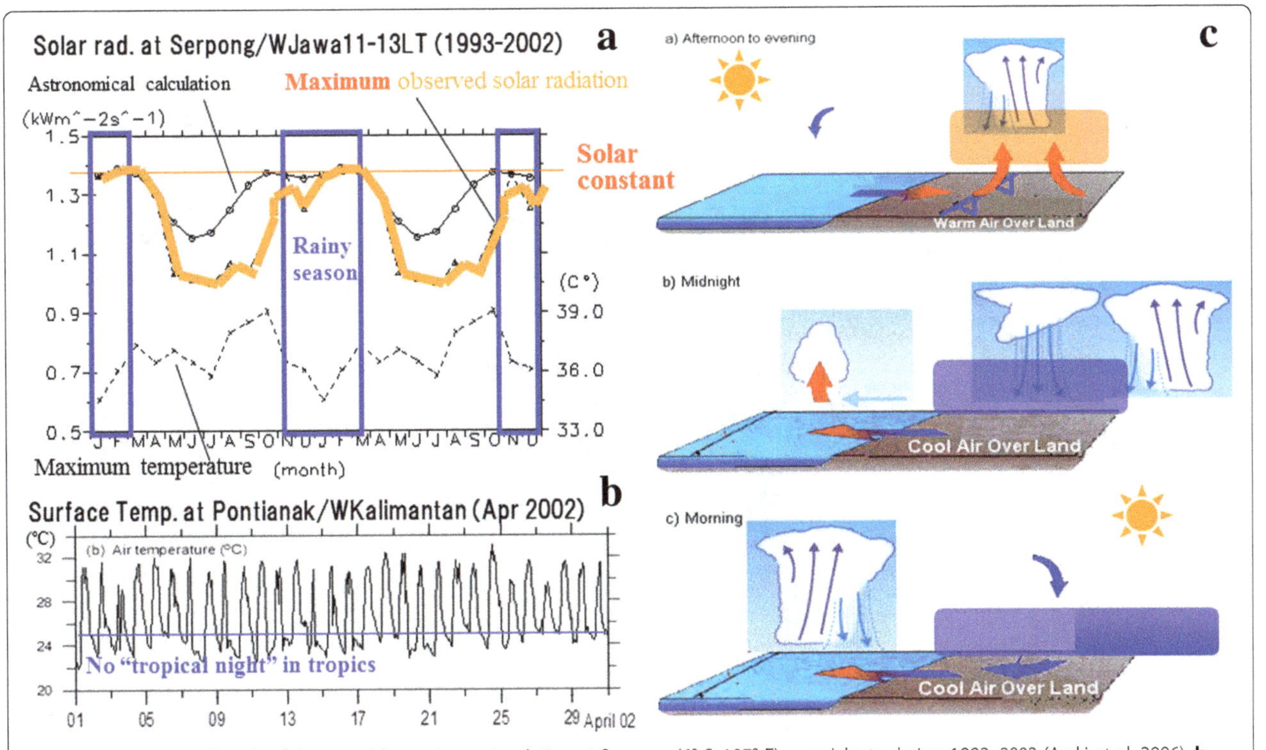

Fig. 10 a Mean annual cycle of the monthly maximum insolation at Serpong (6° S, 107° E) near Jakarta during 1993–2002 (Araki et al. 2006), **b** monthly temperature variation at Pontianak (0°, 109° E), west Kalimantan in April 2002 (Wu et al. 2008), and **c** mechanism (Wu et al. 2009) of the diurnal cycle over the IMC

mean value of 2700 mm/year ≈ 7 mm/day (see IPCC report values in Table 1). Therefore, the annual rainfall amount over the IMC on average is almost explained by such typical diurnal cycle clouds generated along the coastlines almost every day.

As mentioned already, we may consider a very rapid hydrologic cycle near the IMC coastline, unlike in the extratropics where the hydrologic cycle closes on an annual time scale. If so, liquid water transport by river from land to sea should be approximately balanced by water vapor transport by sea breeze from sea to land. We have confirmed that a major river in West Jawa has a diurnal cycle (Sulistyowati et al. 2014). The hydrologic cycle over the IMC has been revealed using isotopic analyses of sampled rain and land water (Kurita et al. 2009; Fudeyasu et al. 2011; Suwarman et al. 2013, 2017; Belgaman et al. 2016; Permana et al. 2016). Furthermore, we know diurnal cycles occur also in the biosphere and humanosphere, which are also concentrated near the coastline.

Radar and satellite observations have shown that the characteristic horizontal scale of the diurnal cycle migrating areas of cloud is of the order of some hundreds of kilometers around a coastline (Fig. 11), which looks somewhat broader than typical values in the extratropics.[12] Clouds and precipitation are concentrated in a narrower updraft/convergence zone of the sea–land breeze circulation cell, but their daylong migration provides rainfall in each place over the whole diurnal cycle area at different local time: mainly on land in the evening and on the sea in the morning. Figure 12 shows the global precipitation and evaporation (30-year average for 1981–2010) as functions of coastline length (Ogino et al. 2017) calculated from an objective analysis dataset called the Japanese 55-year Reanalysis (JRA–55) for 1981–2010 (Kobayashi et al. 2015). Because the amounts and areas of rainfall are always larger in the tropics (in particular the IMC) than in the extratropics, the globally averaged rainfall distribution along the coastal distance is dependent mainly on the tropical distribution (Ogino et al. 2016). As shown in Fig. 13, the local rainfall in the tropics (< 37° in this figure, but not essentially changed even for < 20° as will be discussed later) is a function of coastal distance: 34% of the total tropical rainfall and all areas with > 3000 mm/year are distributed within 300 km of a coastline (Ogino et al. 2016). The characteristic scale 300 km is also found in the spectral analysis of cloud activity[13] (see Fig. 7d), suggesting a sink of energy or enstrophy (cf. Gage and Nastrom 1986) if it is related to the horizontal flow field (e.g., convergence below the cloud base).

For an island smaller than this horizontal scale, the diurnal cycle of sea–land circulation along the coastline is suppressed, which is consistent with theoretical and numerical

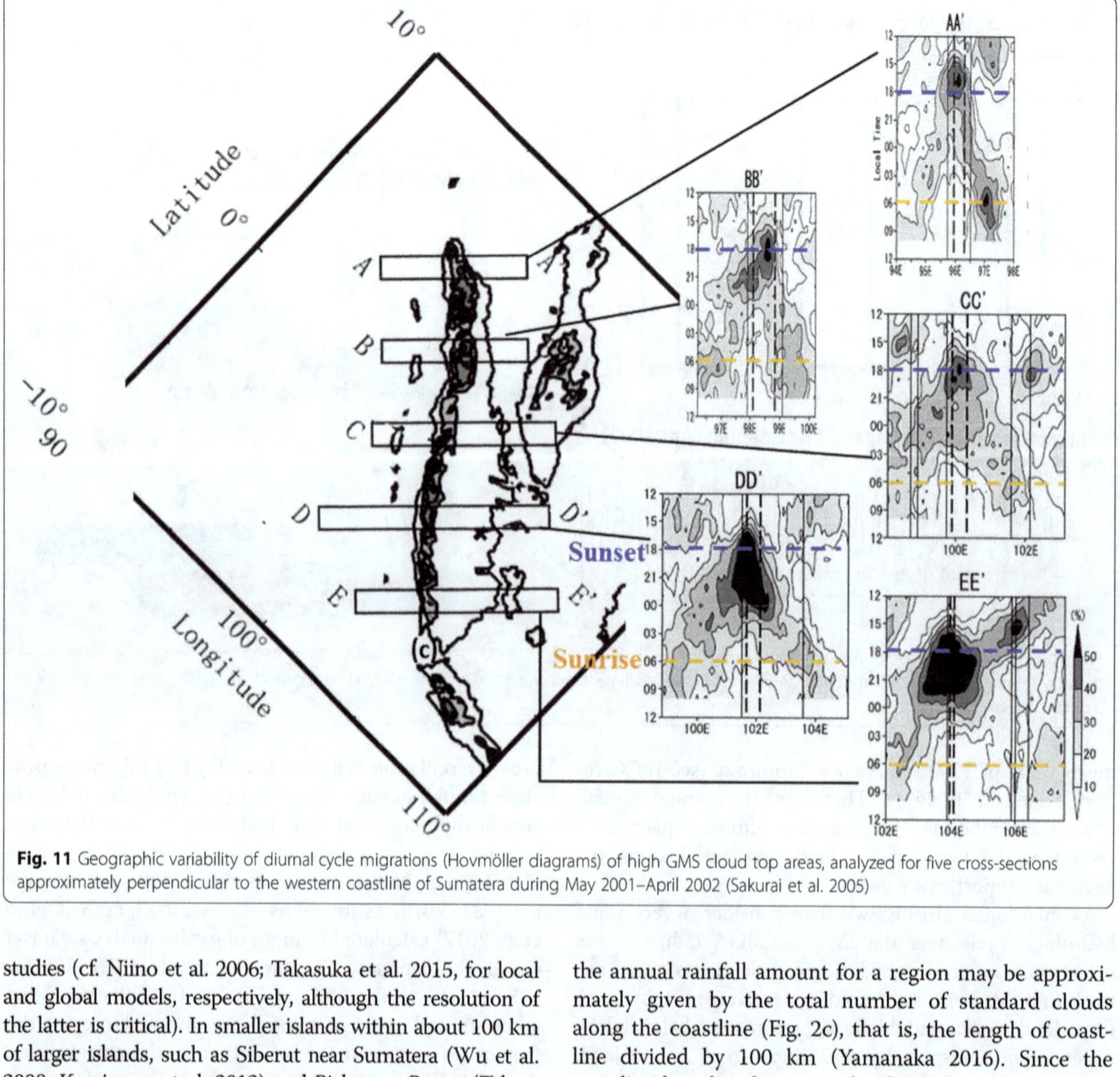

Fig. 11 Geographic variability of diurnal cycle migrations (Hovmöller diagrams) of high GMS cloud top areas, analyzed for five cross-sections approximately perpendicular to the western coastline of Sumatera during May 2001–April 2002 (Sakurai et al. 2005)

studies (cf. Niino et al. 2006; Takasuka et al. 2015, for local and global models, respectively, although the resolution of the latter is critical). In smaller islands within about 100 km of larger islands, such as Siberut near Sumatera (Wu et al. 2008; Kamimera et al. 2012) and Biak near Papua (Tabata et al. 2011b), there are two peaks due to the diurnal cycles of both islands (night and morning by small and large islands, respectively).

The predominance of the coastal diurnal cycle is characteristic in the equatorial tropics, and it disappears with the traveling cyclone activity in the subtropics and extratropics (appearing only on clear days in anticyclones). However, because of its perennial, stationary, and somewhat broader features, the tropical diurnal cycle generates much more rainfall in the annual or longer-period total than (sub)tropical and extratropical cyclones.

Regional rainfall as a function of "coastline density"

Based on an estimate of precipitable water from a standard convective cloud as in Eq. (1), we hypothesize that the annual rainfall amount for a region may be approximately given by the total number of standard clouds along the coastline (Fig. 2c), that is, the length of coastline divided by 100 km (Yamanaka 2016). Since the coastline length is known to be dependent on the measurement resolution (Mandelbrot 1967), we have measured the lengths of coastlines for the ten regions listed in Table 1 with a resolution of 100 km or approximately 1° in latitude (and also in longitude near the equator), and the land area surrounded by such smoothed coastlines. The results are shown in the center two columns of Table 1. Plotting the data in a diagram of annual rainfall amount vs. *coastline density* defined as coastline length divided by land area (Fig. 14),[14] we obtain a roughly linear relationship:

$$\text{Annual regional rainfall } [\text{mm/year}] \approx 2{,}000 \; [\text{mm/year}{\cdot}10^2 \text{ km}] \times \text{Coastline length } [10^2 \text{ km}] / \text{Land area } [10^4 \text{ km}^2] \quad (2)$$

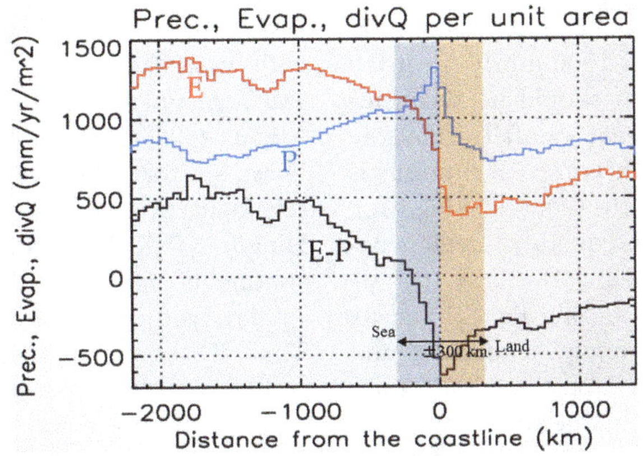

Fig. 12 The 30-year (1981–2010) averaged precipitation (*P*) and evaporation (*E*) calculated from the JRA–55 objective analysis data and their difference (*E–P*, corresponding to the horizontal water vapor flux divergence) as functions of the coastal distance for the entire globe (Ogino et al. 2017)

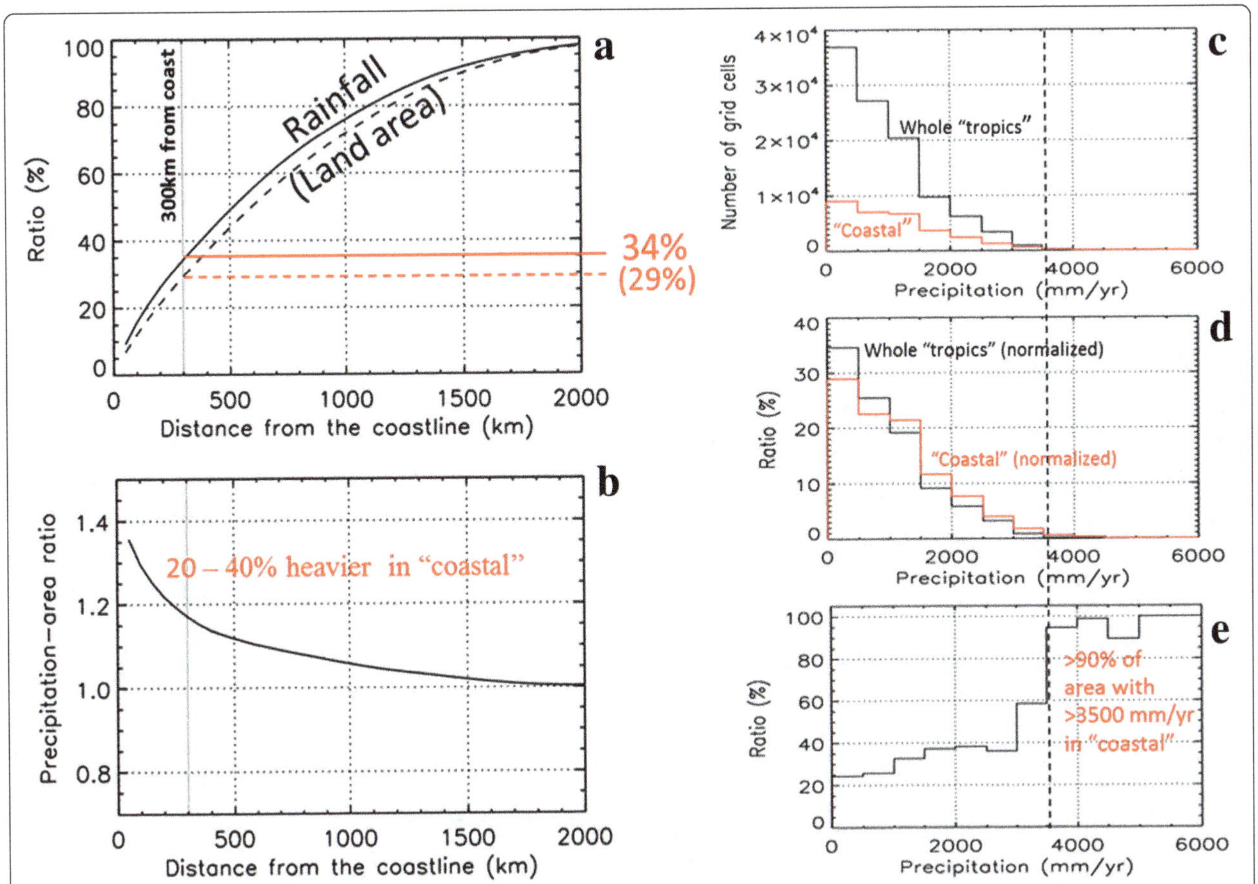

Fig. 13 a Share of rainfall observed by TRMM–PR (latitude < 37°; December 1997–January 2011) and land area and **b** rainfall "heaviness" relative to the whole tropical mean, as functions of the coastal distance. **c** Histogram of grid cell numbers and **d** the normalized histogram, and **e** histogram of coastal share, all as functions of rainfall intensity. After Ogino et al. (2016)

for eight regions except Amazonia and Central Africa. The factor 2000 mm/year in Eq. (2) looks somewhat larger than the coastline peak of 1300 mm/year in the TRMM analysis (Fig. 12), which should be reconsidered after analyzing the regional rainfall from the TRMM data. Note that 2000 mm/year ≈ 5.5 mm/day is the zonal-mean value around the equator (Fig. 1c), suggesting major rainfall in the equatorial region occurs near coastlines.

The approximately linear relationship Eq. (2) and its variance as shown in Fig. 14 imply similarities and variations of precipitating convective clouds and their clusters. Deviations may be partly due to the arbitrary selection and boundary of each region. A systematic difference between the western and eastern coasts of meridionally elongated land areas such as Sumatera, Kalimatan, the Malay Peninsula, and the Indian Subcontinent (e.g., Ogura and Yoshizaki 1988; Murata et al. 2002; Chang et al. 2004, 2016; Xie et al. 2006; Wu et al. 2009) should possibly be included in the linearity factor 2000 mm/year/100 km.

Steep slopes between high mountain ranges and low swampy basins may also produce diurnal variations with mountain–valley breeze-like circulation. In the IMC, mountains of major islands reach significant heights (3000 m or higher) but are close to the coastlines, so that the mountain–valley breezes are not separated from the sea–land breezes (e.g., Yang and Slingo 2001). However, inland regions of the true continents may have diurnal cycles of the mountain–valley type completely separated from the coastlines. Thus, if we take such steep slope effects into account and replace the coastline length by the length of the contour line of an appropriate altitude (probably given by a midpoint between the lifting condensation levels in the morning and evening), we may obtain a corrected result in which the data plots for Central Africa (Fig. 8a, d) and Amazonia (Fig. 8c, f) also approach the linear relationship. This is only a secondary effect because, as shown in the previous subsection and Figs. 12 and 13, the local rainfall distribution is primarily a function of the coastal distance (Ogino et al. 2016, 2017).

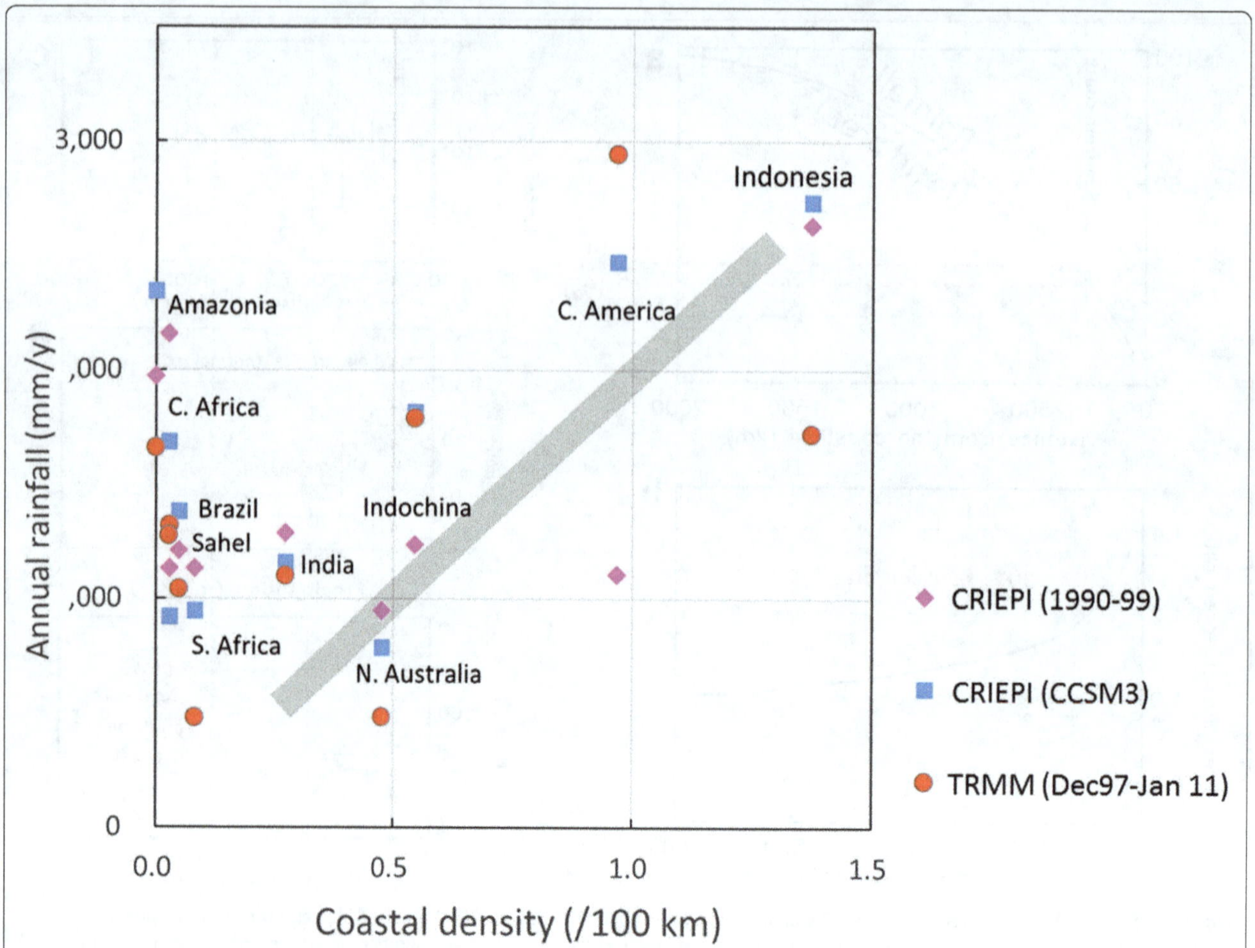

Fig. 14 Regional rainfall listed in Table 1, as a function of "coastal density" (modified from Yamanaka 2016)

Detailed studies of the local rainfall variability due to orography over the IMC have started with a dense mesoscale rain gage network of daily-pentad (so the diurnal cycle is not resolvable) sampling (Hamada et al. 2008), and with raindrop size disdrometers (Marzuki et al. 2013) and wind profilers (Tabata et al. 2011a, 2011b) with high time resolution at almost isolated stations. For local orographic effects on the diurnal cycle rainfall (and wind), we need to analyze meteorological Doppler radar data more precisely in the next step.

In particular, west–east differences of the IMC have been found in climatological (e.g., Köppen 1918; Murakami and Matsumoto 1994; Kubota et al. 2011; Hidayat et al. 2016; Lestari et al. 2016), and cloud-physical (e.g., Marzuki et al. 2013; As-syakur et al. 2016) and water isotope (Suwarman et al. 2013, 2017; Belgaman et al. 2016; Permana et al. 2016) observations. The western part has relatively greater rainfall, a more continental-like raindrop size distribution, and more Indian Ocean monsoon-like origin than the eastern part. The border is roughly the Makassar and Lombok Straits, and a similar separation was noted much earlier in zoology/ecology (e.g., the famous line discovered first by Wallace 1869), geology/geophysics (e.g., Wegener 1929; Frankel 2012), and oceanography (e.g., Lukas et al. 1996; Gordon 2005). The western part consists of three large islands (Kalimantan, Sumatera, and Jawa; the world's 3rd, 6th, and 11th largest) and the Malay Peninsula, which have longer coastlines than the eastern side with two large (the 2nd and 9th; Papua and Sulawesi) and many much smaller islands. This is consistent with the greater rainfall in the western part.

The results described here suggest that a numerical model needs to resolve the equatorial coastlines with sufficiently high resolution (<< 100 km). The importance of the diurnal cycle over the IMC not only in this region but also for the global climate has been suggested by modelers (e.g., Neale and Slingo 2003). Results from a high-resolution model (20 km mesh) indicate that the diurnal cycles over the IMC will be weakened in future mainly due to nighttime land temperature increase associated with global warming (Kitoh and Arakawa 2005), although at present, even megacities (such as Jakarta) in the IMC do not have notable local heat island effects. Much higher resolution models (7 and 3.5 km grids) have simulated local clouds activated by an MJO causing a flood in the Malay Peninsula (Miura et al. 2007) and rainfall coupled with convergence of sea–land breeze circulations over the IMC (Sato et al. 2009). However, the effects of the resolution improvement on the accuracy of the simulation do not seem to be so simple (e.g., Wang et al. 2012). Numerical studies for different coastline features will also be useful for the study of paleoclimate; these will be mentioned later.

Consistency with the global water budget

We have mentioned four expressions of the rainfall distribution. Early studies gave

(i) Meridional distribution as shown in Fig. 1c; and
(ii) Land and marine breakdown as shown in Fig. 1b.

Recent studies reviewed in this article gave distributions that were a function of:

(iii) Coastal distance as shown in Fig. 12; and
(iv) *Coastal density* as shown in Fig. 14.

All these distributions must be consistent with the global total rainfall or latent heating shown in Fig. 1a. Concerning (ii), it is well known (e.g., Jaeger 1983; Hartmann 1994; Trenberth et al. 2009) that, multiplied with the area ratio of land and sea,

$$\text{Global mean rainfall} \approx 1.1\,[\text{m/year}] \times 0.7 + 0.7\,[\text{m/year}] \times 0.3 \approx 1\,[\text{m/year}], \tag{3}$$

which gives, multiplied by the density of water $\approx 1 \times 10^3$ kg/m^3, the annual mass of water falling per unit area $\approx 1 \times 10^3$ kg/year/m^2. Further multiplying by the latent heat of evaporation of water $\approx 2.5 \times 10^6$ J/kg and using 1 year $\approx 3.15 \times 10^7$ s, we confirm the latent heating per unit time per unit area ≈ 80 W/m^2, as shown in Fig. 1a (Trenberth et al. 2009). If we multiply Eq. (3) by the Earth's surface area $\approx 4\pi \times (6.37 \times 10^3\ \text{km})^2 \approx 5 \times 10^{14}$ m^2, we have

$$\text{Global total rainwater volume} \approx 1[\text{m/year}] \times 5 \times 10^{14}[\text{m}^2] \approx 5 \times 10^{14}[\text{m}^3/\text{year}]. \tag{4}$$

If we integrate the meridional distribution (i) over the latitude range $< 37° \approx 8 \times 10^3$ km which is used in Ogino et al. (2016) for calculation of (iii) in the tropics, the mean rainfall estimated from Fig. 1c is about 1.5 m/year; therefore,

$$\text{Tropical rainwater volume} \approx 1.5\,[\text{m/year}] \times 8 \times 10^3\,[\text{km}] \times 4 \times 10^4\,[\text{km}] \times 0.9 \approx 4 \times 10^{14}\,\left[\text{m}^3/\text{year}\right], \tag{5}$$

where a factor 0.9 is used under the trapezoidal approximation. The tropical rainwater Eq. (5) is as much as 80% of the global total Eq. (4). If one third of Eq. (5) is falling in the coastal region as shown in Fig. 13a, we estimate very roughly

$$\text{Tropical coastal rainwater volume} \sim 1 \times 10^{14}\,\left[\text{m}^3/\text{year}\right], \tag{6}$$

which is 20% of the global total Eq. (4). Even if we take a narrower latitude width $< 20° \approx 4.4 \times 10^3$ km, then the mean rainfall is about 1.7 m/year, and the tropical total and coastal rainwater volumes are very roughly (within one significant digit) the same as in Eqs. (5) and (6). This is reasonable, because the latitude range 20° – 37° is the subtropical high-pressure belt with less rainfall.

If we integrate (iii), the precipitation distribution in terms of coastal distance, for the coastal region of width 600 km and length 150,000 km (Ogino et al. 2016), then the mean rainfall estimated from Fig. 12 is about 1.1 m/year; therefore,

$$\text{Tropical coastal rainwater volume} \approx 1.1\,[\text{m/year}] \times 0.6 \times 10^3\,[\text{km}] \times 15 \times 10^4\,[\text{km}] \approx 1 \times 10^{14}\,\left[\text{m}^3/\text{year}\right], \tag{7}$$

which is consistent with Eq. (6). More estimates including the breakdown of (ii) for land and sea rainfall over the tropics as well as other latitudes are given in Ogino et al. (2017).

Finally, concerning the distribution in terms of coastal density (iv), multiplying both sides of Eq. (2) by the regional (land) area and taking the sum over all the regions (with total coastline length $\approx 1.5 \times 10^5$ km from Table 1) we have the estimate

$$\text{Tropical coastal rainwater volume} \approx 2\,\left[\text{m/year}{\cdot}10^2\,\text{km}\right] \times 1.5 \times 10^3\,\left[10^2\,\text{km}\right] \approx 3 \times 10^{13}\,\left[\text{m}^3/\text{year}\right], \tag{8}$$

which is about 1/3 of Eqs. (6) or (7), due to differences of the resolution measuring the coastal length and the width of the coastal region.

As shown in Table 1, the IMC coastline is longer than half the 10-region total, and longer than 20% of the total coastline length of 150,000 km within latitudes $< 37°$. This is why the IMC (both land and sea) occupying only 4% of the Earth's surface generates such a large fraction of the global latent heating. We may imagine that we replace clouds distributed along the ITCZ (Fig. 2a) by clouds along the IMC coastline, which has length comparable to the equatorial circumference (Fig. 2c). The IMC produced tectonically as suggested by Wegener (1929) is now experiencing equatorial climate that will be recorded as fossils of tropical organisms in future, and here we may say in addition that the present climate itself is also dependent on the geomorphology of the IMC! A new climatological classification including the diurnal cycle amplitude may be necessary to replace Köppen's that only considers the annual cycle.

Control of the global climate

The interannual variations of rainfall over the IMC and their relation with ENSO and IOD have been analyzed in several studies (e.g., Hamada et al. 2002, 2008, 2012; Aldrian and Susanto 2003; Kubota et al. 2011; Siswanto et al. 2016; Yanto et al. 2016; Lestari et al. 2016; As-syakur et al. 2016) (Fig. 15). In general, the correlation is better in a dry season (much drier in El Niño and/or positive IOD years). For example, during the recent half century, eight relatively strong El Niño events (large positive values (red) in Fig. 15f, g) in 1957/58, 1965/66, 1972/73, 1982/83 (the 2nd strongest), 1986–88, 1991/92, 1997/98 (the strongest), and 2014–16 (too recent to be shown here) produced less rainfall (negative anomalies in Fig. 15a–e) over the IMC, whereas La Niña events (negative (blue) in Fig. 15f, g) in 1954–56, 1975/76, 1988/89, 1998–2000, 2007/08, and 2010/11 correspond to periods of more rain (maxima in Fig. 15a–e). Furthermore, in the recent two decades after the 1997/98 El Niño, a longer time-scale tendency for increasing rainfall is found in Fig. 15b, d, e, as well as the SST warming. Major features can be explained by modification of the amplitude of the diurnal cycle through SST variations, as mentioned in the first subsection. The evidence of the synchronous diurnal cycle and spatially broad-band convection over the whole IMC (as shown in Figs. 6 and 7) suggests a planetary-wave response is generated, which may induce teleconnections to global climate variations (e.g., Nitta 1987; Chen 2002; for northern- and southern hemispheric summer seasons, respectively) that have been studied much more than the detailed observations inside the tropics. Another possibility is that ENSO changes the monsoon circulation through variations of the IMC convection (e.g., Zhang et al. 1996) which are mainly associated with the diurnal cycles along the coastlines. Neale and Slingo (2003) proposed such impacts of the IMC convection on the global climate, and Slingo et al. (2014) summarized the mechanisms and predictability in the context of recent abnormal weather and climate in the UK.

Another process controlling the global climate may be through the modification of MJOs (cf. Zhang 2005, 2013), which are regarded as an important factor governing the medium-range variability of global weather and climate. A recent study by Zhang and Ling (2017) shows a "barrier effect" of the IMC on more than 75% of MJOs. More detailed observational and theoretical studies may be necessary to identify the mechanisms weakening MJOs, but the synchronous diurnal cycle noted above may play an important role.

In addition, a convection cell such as the sea or land breeze circulation consists dynamically of a pair of internal gravity waves propagating upward and downward, and their diurnal cycle migrations both seaward and landward imply

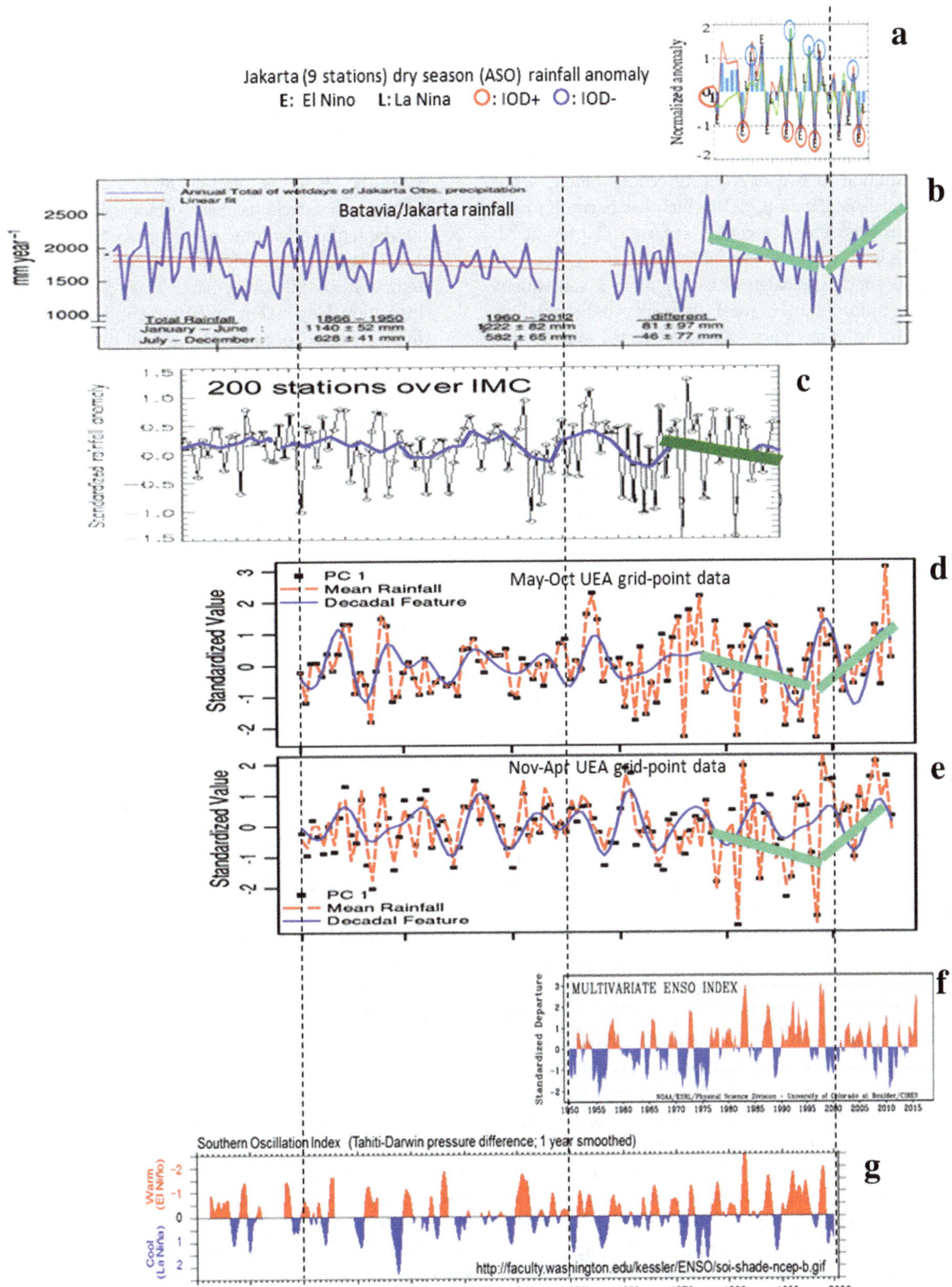

Fig. 15 Interannual rainfall variations of **a** Aug–Oct anomaly (Hamada et al. 2012) and **b** annual total (Siswanto et al. 2016) in Jakarta. **c** Annual (Hamada et al. 2002), **d** May–Oct and **e** Nov – Apr (Yanto et al. 2016) anomalies over the IMC, in comparison with **f** multivariate ENSO (https://www.esrl.noaa.gov/psd/enso/mei/) and **g** original SO (http://faculty.washington.edu/kessler/ENSO/soi-shade-ncep-b.gif) indices

generation of waves with phase velocities and momenta of two opposite directions (e.g., Rotunno 1983; Niino 1987). This feature is important for considering the horizontal propagation of the diurnal cycle generated at a coastline, as well as interaction with other diurnal cycles (e.g., Niino et al. 2006), and to study upward propagation (e.g., Ogino et al. 1995; Tsuda et al. 1995, 2000) and stratospheric mean-flow acceleration in two opposite directions (such as the quasi-biennial oscillation (QBO), which has become somewhat controversial very recently, see e.g., Osprey et al. 2016). This issue will be discussed as a case study based on recent radiosonde campaign observations in a subsequent paper. The contributions of the 11-year solar-cycle variation of insolation to climate are still controversial and do not look so clear in recent studies (e.g., Lean et al. 2005).

For longer time scales (10^2–10^6 years) several studies (such as those cited in Randall et al. 2007) suggest increasing occurrence of El Niño (La Niña) in a globally warmer (colder) climate. If so, the rainy (dry) season is dominant over the IMC in glacial (interglacial) periods (with time scales of a few 10^5 years called the Milanković (1941) cycle). However, in a glacial period the sea level falls, and the IMC became part of real continents (Sunda and Sahul subcontinents) connected to Eurasia and Australia (e.g., Voris 2000), and the IMC "dam" was almost closed, although the Makassar Strait transporting warm water from the Pacific to Indian Ocean (e.g., Lukas et al. 1996; Gordon 2005) still existed. The coastlines were much shorter around the Indonesia throughflow region and around the whole subcontinents, which means broader drier areas inland and less regional rainfall. Therefore, the active water cycle occurs in a much more limited coastal area, and a smaller amount of water is circulated (consistent with a larger amount of water frozen over land). Results from a model run with changing sea level (Braconnot et al. 2007a, 2007b) gave less tropical rainfall in particular over the IMC than from a run without the change (Kitoh and Murakami 2002), although the authors considered that the sea level change in the former is model-dependent and the rainfall decrease is due to atmospheric water vapor decrease with cooling. On much longer scales ($\gtrsim 10^7$ years), the continents and oceans may be aggregated or separated (with a period of a few 10^8 years called the Wilson (1963) cycle), and the results are not so simple, because other conditions (such as land orography, ocean circulation, and atmospheric composition) are not fixed. In general, the inland of a supercontinent is very dry, and a stronger monsoon is expected to or from the surrounding superocean (e.g., Parrish 1993), which is not inconsistent with our idea on the diurnal cycle coastal rainfall in the tropics.

Conclusions

In this article, we review the concentration of tropical rainfall around coastlines and why the world's largest regional rainfall is distributed over the IMC that has the longest total length of coastline. The major contents are summarized as follows:

(i) The cloud–precipitation observation system over the IMC has been improved by the JEPP–HARIMAU and SATREPS–MCCOE projects that have produced scientific results as listed below.
(ii) The diurnal cycle has been established by (i) as the fundamental phenomenon in particular near the coastlines over the IMC. It is forced by the land–sea temperature difference mainly due to the daytime insolation (dependent on latitude and season) and the nighttime sprinkler-like rainfall itself and partly by interannual SST variations. It is synchronized over almost the whole IMC and forces cloud convection systems from meso to synoptic scales.
(iii) Analyses of the TRMM–PR observation data and the JRA-55 objective reanalysis data have shown that the local rainfall in the tropics is found to be a steeply decreasing function of the coastline length, which results mainly from rainfall concentrated near the IMC coastline (ii).
(iv) The regional rainfall as summarized by the IPCC can be expressed by a roughly linear function of the "coastal density" defined as the coastline length divided by the enclosed land area. The world's largest regional rainfall over the IMC is explained by the long coastlines, which is consistent with (ii) and (iii).
(v) The IMC coastline rainfall as mentioned in (ii)–(iv) produces the largest part (roughly 20% or more) of the global latent heating, which is about 22% of the magnitude of the global infrared radiation (see Fig. 1a). Thus, its variations may affect the global climate considerably. This is consistent with the rainfall itself enhancing the diurnal cycle as mentioned in (ii).
(vi)The larger-scale forcing as mentioned in (ii) by the IMC coastline diurnal cycle convection of the intensity mentioned in (v) may generate teleconnections controlling the global climate.

There are several issues which must be clarified. For example, the regional rainfall–coastline density relation (iv) must be improved, using the local rainfall–coastal distance relation (iii). Understanding tropical and global characteristics including the impacts of IMC-like complex land distributions is also quite important for considering the future climate. We now understand that high-resolution observations and models that can detect or reproduce the diurnal cycle and cloud convection over the IMC are necessary in order to improve both local disaster prevention and global climate prediction.

In the next step called Years of the Maritime Continent (YMC; 2017–9), Indonesia as a G20 country should take observations under a more completely multinational framework (see YMC (Years of the Maritime Continent) 2017). As a preliminary activity, pre-YMC campaign observations have been carried out in and off Bengkulu (Wu et al. 2017, 2018; Yokoi et al. 2017), which is located about 400 km south of Padang (Sumatera Barat area in Table 2). The Bengkulu area has not yet been observed well and is almost free from small islands distributed off Padang. Also, the long historical meteorological records of Indonesia are starting to be analyzed by Indonesian scientists (e.g., Siswanto et al. 2016; Yanto et al. 2016; Supari et al. 2016), which will clarify features and mechanisms of the multi-scale interactions. Agricultural applications of climatological studies (Hidayat et al. 2016; Marjuki et al. 2016) are requested strongly by the Indonesian government and society. Interdisciplinary studies with hydrology and oceanography are also developing, although they are out of the scope of this article.

Endnotes

[1]We use Indonesian names for islands of the IMC: Sumatera, Kalimantan, Jawa, Sulawesi and Papua for English names of Sumatra, Borneo, Java, Celebes, and New Guinea, respectively.

[2]A major revision separated *Am* (meaning originally medium and later monsoonal) from *Aw*, dependent on whether the minimum monthly rainfall (< 60 mm/month) is greater than 4% of the annual rainfall or not.

[3]A spline interpolation might smooth variations within the scale of coastal regions (10^2 km).

[4]In the Paleogene (66–23 Ma) the Americas were separated like the present-day Asia–Australia, which was another example of maritime continent.

[5]The AMY program is a cross-cutting coordinated observation and modeling initiative under the International Monsoon Study (IMS) of WCRP, which nearly overlapped with and was promoted together with MAHASRI (see Matsumoto et al. (2017)).

[6]They and Köppen (1918) recognized only the azimuthal distribution of surface wind. After the development of fluid dynamics and its application to atmospheric motion, various definitions were gradually unified (Simpson 1947, who used the term "intertropical convergence").

[7]Rainfall estimates from cloud observations differ from direct measurements, but have merits such as temporal and horizontal continuity (see, e.g., Xie and Arkin 1997).

[8]At this initial stage, ISVs and monsoons were specified as the most important targets; although, the diurnal cycle became the most important target by the end of the period.

[9]As of March 2017, 36 C-band and 4 X-band radars are in operation, and one more radar is being installed in 2017.

[10]Radiosonde observations (originally at 12 stations) reported to the global telecommunication system (GTS) were increased to more than once per day since mid-2005 (approaching twice a day during 2007); their impact on objective analysis has been shown by Seto et al. (2009), and the number of stations has been increased to 22 (in addition to 2 wind-profiler and 73 pilot balloon stations) as of March 2017.

[11]A "tropical night" is defined by the Japan Meteorological Agency as having the daily minimum temperature higher than 25 °C, although another criterion, for example, in Germany (20 °C), is somewhat lower than the Japanese definition and also lower than the usual daily minimum temperature in the IMC.

[12]The typical values are observationally less than 10^2 km in the extratropics and theoretically given as the half horizontal wavelength using the gravity-wave dispersion relation, substituting the height of the sea–land breeze circulation cell as the half vertical wavelength (internal case, latitude < 30°) (e.g., Rotunno 1983; Niino 1987).

[13]We have confirmed that, at least for the coastal regions in the IMC, the diurnal-cycle migration feature of rainfall (based on TRMM; Mori et al. 2004) and that of cloud (based on GMS; Sakurai et al. 2005; shown in Fig. 11) are similar.

[14]The coastal density is 2/radius for a circular land area, and 4/side length for a square land area.

Abbreviations

AM: Ante meridiem (before noon); AMY: Asian Monsoon Years; APHRODITE: Asian Precipitation–Highly Resolved Observational Data Integration Towards Evaluation of Water Resources; BMKG: Indonesian Agency for Meteorology, Climatology and Geophysics; BPPT: Agency for the Assessment and Application of Technology; CCSM3: Community Climate System Model version 3; CINDY2011: Cooperative Indian Ocean Experiment on Intraseasonal Variability in the Year 2011; CISK: Conditional instability of the second kind; CPEA: Coupling Processes in the Equatorial Atmosphere; CRIEPI: Central Research Institute of Electric Power Industry; EAR: Equatorial Atmosphere Radar; EEZ: Exclusive economic zone; ENSO: El Niño southern oscillation; EOS: Earth observation system; G20: Group of Twenty; GEO: Group on Earth Observations; GEOSS: Global Earth Observation System of Systems; GEWEX: Global Energy and Water Cycle Exchanges; GMS: Geostationary Meteorological Satellite; HARIMAU: Hydrometeorological Array for ISV –Monsoon Automonitoring; IMC: Indonesian maritime continent; IOD: Indian Ocean Dipole mode; IOMICS: Indian Ocean Moored Network Initiative for Climate Studies; ISV: Intraseasonal variation; ITCZ: Intertropical convergence zone; JAMSTEC: Japan Agency for Marine–Earth Science and Technology; JEPP: Japan EOS Promotion Program; JICA: Japan International Cooperation Agency; JST (in Acknowledgement): Japan Science and Technology Agency; JST (in Fig. 6): Japanese standard time (= Indonesian east standard time = UT + 9 h); LAPAN: Indonesian National Institute of Aeronautics and Space; LT: Local time (Sumatera, west/central Kalimantan and Jawa: West Indonesian standard time (WIB) = UT + 7 h; Bali, east Kalimantan and Sulawesi: Central Indonesian standard time (WIP) = UT + 8 h; Papua = East Indonesian standard time = UT + 9 h = JST); MAHASRI: Monsoon Asian Hydro –Atmosphere Scientific Research and Prediction Initiative; MCCOE: Maritime Continent Center of Excellence; MEXT: Japanese Ministry of Education, Culture, Sports, Science and Technology; MISMO2006: Mirai Indian Ocean

Cruise for the Study of the MJO–Convection Onset in 2006; MJO: Madden –Julian Oscillation; MOFA: Japanese Ministry of Foreign Affairs; PM: Post meridiem (after noontime); QBO: Quasi-biennial oscillation; RAMA: Research Moored Array for African–Asian–Australian Monsoon Analysis and Prediction; SATREPS: Science and Technology Research Partnership for Sustainable Development; SST: Sea surface temperature; TOCS: Tropical Ocean Climate Study; TRMM: Tropical Rainfall Measuring Mission; UT: Universal time (see LT for further explanation); VHF: Very high frequency; WCRP: World Climate Research Programme; WISHE: Wind-induced surface heat exchange; YMC: Years of the Maritime Continent

Acknowledgements
As mentioned in the text, the contents overviewed in this article were obtained under the two projects, JEPP–HARIMAU and SATREPS–MCCOE. We would like to acknowledge the Japanese and Indonesian Governments, JST, JICA, JAMSTEC, Kyoto University, Hokkaido University, BPPT, LAPAN, and BMKG for their support and collaboration. In completing this article, we thank Dr. Kunio Yoneyama, Director of the Department of Coupled Ocean –Atmosphere–Land Processes Research of JAMSTEC for his supervision and encouragement concerning research toward the YMC and on our submission of an earlier version of this paper to the international seminar at Bengkulu on October 19–20, 2016. The YMC websites at BMKG and JAMSTEC are maintained/updated by Ms. Nurhayati and Dr. Yoneyama, respectively. We also thank Dr. Dawn P. Erlich and other editors of GEWEX NEWS for their comments on a summary of MAHASRI, JEPP–HARIMAU, and SATREPS –MCCOE projects related partly to this article. Comments from Profs. Mike Chen, Masahito Ishihara and Shigeo Yoden, on the authors' presentations related to this article are also greatly acknowledged. The authors wish to thank sincerely Prof. Susumu Kato and the late Prof. Shoichiro Fukao for much important advice and encouragement since the beginning of observational studies over the IMC. Lastly, the authors would like to acknowledge the journal General Chief Editor (Prof. Yasufumi Iryu), an Editor (Prof. Jonny Chan), and two anonymous reviewers for their constructive comments and positive evaluations.

Funding
This study was mainly supported by JEPP and SATREPS.
JEPP was a Japanese Government program operated directly by Japanese Ministry of Education, Culture, Sports, Science and Technology (MEXT).
SATREPS is a Japanese Government program operated jointly by JST and JICA under supervisions of MEXT and MOFA.

Authors' contributions
All authors read and approved the final manuscript. MDY proposed and summarized this article, based on the contents first-authored by the other authors, and contributed mainly to the first subsection of the "Introduction" section, the "Regional rainfall as a function of 'coastline density'" section of the "Review" section, and the "Conclusions" section. SM and HJI carried out the radar and radiosonde observations and operational data collections in Indonesia, as well as data analysis and paper publication in Japan and contributed mainly to the "JEPP–HARIMAU and SATREPS–MCCOE projects" section of the "Introduction" and the "Tropical rainfall in the global climate" section of the "Review". PMW and SYO carried out observations including outside the Indonesian territory, regional numerical modeling, and global data analysis and contributed mainly to the "Local rainfall as a function of coastline length" and "Regional rainfall as a function of 'coastline density'" sections of the "Review" section. JM and FS combined our results with the whole international collaborative studies and contributed mainly to the "The global energy/water cycle and the maritime continent" of the "Introduction" section and the "Control of the global climate" section of the "Review" section.

Authors' information
MDY was the Principal Investigator of the JEPP-HARIMAU (2005–2010) and STREPS-MCCOE (2009–2014) projects in which all authors were involved. SM was the Leader of JAMSTEC Maritime Continent Research Group (2005–2014), of which HJI is a member. JM was the Leader of JAMSTEC Monsoon Meteorology Research Group (2005–2014), of which PMW and SYO are members. JM was also the Chairman of the Steering Committee for the MAHASRI project (2005–2016), which also involved all the authors. FS was the Head of the Climate Dynamics Laboratory (2005–2015) of BPPT, the Principal Investigator of the Indonesian Team for JEPP-HARIMAU and SATREPS-MCCOE projects, and the Indonesian representative for MAHASRI.

Competing interests
The authors declare that they have no competing interest.

Author details
[1]Japan Agency for Marine–Earth Science and Technology (JAMSTEC), Natsushima–cho, Yokosuka 237-0061, Japan. [2]Professor Emeritus, Kobe University, Rokkodai–cho, Nada, Kobe 657-8501, Japan. [3]Research Center for Climatology and Department of Geography, Tokyo Metropolitan University, Minami–Osawa, Hachioji 192-0397, Japan. [4]Center for Regional Resources Development Technology (PTPSW), Agency for the Assessment and Application of Technology (BPPT), Jl. Raya Puspiptek, South Tangerang 15314, Indonesia.

References

Aldrian E, Susanto RD (2003) Identification of three dominant rainfall regions within Indonesia and their relationship to sea surface temperature. Int J Climatol 23:1435–1452. https://doi.org/10.1002/joc.950

Ando K, Syamsudin F, Ishihara Y, Pandoe W, Yamanaka MD, Masumoto Y, Mizuno K (2010) Development of new international research laboratory for maritime-continent seas climate research and contributions to global surface moored buoy array. Hal J, Harrison DE, Stammer D Proceedings of OceanObs'09: Sustained Ocean Observations and Information for Society (Annex), Venice, Italy, 21–25 September 2009, ESA publication WPP-306. https://doi.org/10.5270/OceanObs09

Araki R, Yamanaka MD, Murata F, Hashiguchi H, Oku Y, Sribimawati T, Kudsy M, Renggono F (2006) Seasonal and interannual variations of diurnal cycles of wind and cloud activity observed at Serpong, West Jawa, Indonesia. J Meteor Soc Japan 84A:171–194. https://doi.org/10.2151/jmsj.84A.171

As-syakur AR, Osawa T, Miura F, Nuarsa IW, Ekayanti NW, Dharma IGBS, Adnyana IWS, Arthana IW, Tanaka T (2016) Maritime continent rainfall variability during the TRMM era: the role of monsoon, topography and El Niño Modoki. Dyn Atmos Oceans 75:58–77. https://doi.org/10.1016/j.dynatmoce.2016.05.004

Baumgartner A, Reichel E (1975) The World Water Balance: Mean Annual Global, Continental and Maritime Precipitation, Evaporation and Run-Off. Elsevier, Amsterdam

Belgaman HA, Ichiyanagi K, Tanoue M, Suwarman R, Yoshimura K, Mori S, Kurita N, Yamanaka MD, Syamsudin F (2016) Intraseasonal variability of δ18O of precipitation over the Indonesian maritime continent related to the Madden –Julian oscillation. SOLA 12:192–197. https://doi.org/10.2151/sola.2016-039

Braconnot P, Otto-Bliesner B, Harrison S, Joussaume S, Peterchmitt JY, Abe-Ouchi A, Crucifix M, Driesschaert E, Fichefet T, Hewitt CD, Kageyama M, Kitoh A, Laîné A, Loutre MF, Marti O, Merkel U, Ramstein G, Valdes P, Weber SL, Yu Y, Zhao Y (2007a) Results of PMIP2 coupled simulations of the mid-Holocene and last glacial maximum, part 1: experiments and large-scale features. Clim Past 3:261–277. https://doi.org/10.5194/cp-3-261-2007

Braconnot P, Otto-Bliesner B, Harrison S, Joussaume S, Peterchmitt JY, Abe-Ouchi A, Crucifix M, Driesschaert E, Fichefet T, Hewitt CD, Kageyama M, Kitoh A, Loutre MF, Marti O, Merkel U, Ramstein G, Valdes P, Weber SL, Yu Y, Zhao Y (2007b) Results of PMIP2 coupled simulations of the mid-Holocene and last glacial maximum, part 2: feedbacks with emphasis on the location of the ITCZ and mid- and high latitudes heat budget. Clim Past 3:279–296. https://doi.org/10.5194/cp-3-279-2007

Chan JCL, Li C (2004) The east Asia winter monsoon. In: East Asian Monsoon, Chang CP (eds) World scientific series on Asia–Pacific weather and climate, vol 2. World Scientific Publication Company, pp 54–106. https://doi.org/10.1142/9789812701411_0002

Chang CP, Harr PA, McBride JC, Hsu HH (2004) Maritime continent monsoon: annual cycle and boreal winter variability. In: East Asian Monsoon, Chang CP (eds) World scientific series on Asia–Pacific weather and climate, vol 2. World Scientific Publication Company, pp 107–152. https://doi.org/10.1142/9789812701411_0003

Chang CP, Lu MM, Lim H (2016) Monsoon convection in the maritime continent: interaction of large-scale motion and complex terrain. In: Multiscale convection-coupled Systems in the Tropics: a tribute to Dr. Michio Yanai, Meteorological Monograph, vol 56, pp 6.1–6.29. https://doi.org/10.1175/AMSMONOGRAPHS-D-15-0011.1

Chen TC (2002) A North Pacific short-wave train during the extreme phases of ENSO. J Clim 15:2359–2376. https://doi.org/10.1175/1520-0442(2002)015<2359:ANPSWT>2.0.CO;2

Chen TC, Tsay JD, Matsumoto J, Alpert J (2015) Development and formation mechanism of the Southeast Asian winter heavy rainfall events around the South China Sea. Part I: formation and propagation of cold surge vortex. J Clim 28:1417–1443. https://doi.org/10.1175/JCLI-D-14-00170.1

Chen TC, Tsay JD, Yen MC, Matsumoto J (2013) The winter rainfall of Malaysia. J Clim 26:936–958. https://doi.org/10.1175/JCLI-D-12-00174.1

Collins WD, Bitz CM, Blackmon ML, Bonan GB, Bretherton CS, Carton JA, Chang P, Doney SC, Hack JJ, Henderson TB, Kiehl JT, Large WG, McKenna DS, Santer BD, Smith RD (2006) The community climate system model version 3 (CCSM3). J Clim 19:2122–2143. https://doi.org/10.1175/JCLI3761.1

Emanuel KA (1987) An air-sea interaction model of intraseasonal oscillations in the tropics. J Atmos Sci 44:2324–2430. https://doi.org/10.1175/1520-0469(1987)044<2324:AASIMO>2.0.CO;2

Endo N, Matsumoto J, Lwin T (2009) Trends in precipitation extremes over Southeast Asia. SOLA 5:168–171. https://doi.org/10.2151/sola.2009-043

Frankel HR (2012) The continental drift controversy, Vol 1: Wegener and the early debate. Cambridge University Press, Cambridge. https://books.google.co.jp/books?isbn=1316616045

Fudeyasu H, Ichiyanagi K, Yoshimura K, Mori S, Hamada JI, Sakurai N, Yamanaka MD, Matsumoto J, Syamsudin F (2011) Effects of large-scale moisture transport and mesoscale processes on precipitation isotope ratios observed at Sumatera, Indonesia. J Meteor Soc Japan 89A:49–59. https://doi.org/10.2151/jmsj.2011-A03

Fujita M, Yoneyama K, Mori S, Nasuno T, Satoh M (2011) Diurnal convection peaks over the eastern Indian Ocean off Sumatra during different MJO phases. J Meteor Soc Japan 89A:317–330. https://doi.org/10.2151/jmsj.2011-A22

Fukao S (2006) Coupling processes in the equatorial atmosphere (CPEA): a project overview. J Meteor Soc Japan 84A:1–18. https://doi.org/10.2151/jmsj.84A.1

Gage KS, Nastrom GD (1986) Theoretical interpretation of atmospheric wavenumber spectra of wind and temperature observed by commercial aircraft during GASP. J Atmos Sci 43:729–740. https://doi.org/10.1175/1520-0469(1986)043<0729:TIOAWS>2.0.CO;2

GEO (Group on Earth Observations) (2007) The First 100 Steps to GEOSS, Annex of Early Achievements to the Report on Progress, 2007 Cape Town Ministerial Summit: Earth Observations for Sustainable Growth and Development, 30 November 2007, GEO Secretariat, p. 212. https://www.earthobservations.org/documents/the_first_100_steps_to_geoss.pdf. Accessed 20 Feb 2018

Gill AE (1980) Some simple solutions for heat-induced tropical circulation. Quart J Roy Meteor Soc 106:447–462. https://doi.org/10.1002/qj.49710644905

Gordon AL (2005) Oceanography of the Indonesian seas and their throughflow. Oceanography 18(4):14–27. https://doi.org/10.5670/oceanog.2005.01

Hadi TW, Horinouchi T, Tsuda T, Hashiguchi H, Fukao S (2002) Sea-breeze circulation over Jakarta, Indonesia: a climatology based on boundary layer radar observations. Mon Wea Rev 130:2153–2166. https://doi.org/10.1175/1520-0493(2002)130<2153:SBCOJI>2.0.CO;2

Hadley G (1735) Concerning the cause of the general trade-winds. Phil Trans 39: 58–62. https://doi.org/10.1098/rstl.1735.0014

Halley E (1686) An historical account of the trade winds, and monsoons, observable in the seas between and near the Tropicks, with an attempt to assign the physical cause of the said winds. Phil Trans 16:153–168. https://doi.org/10.1098/rstl.1686.0026

Hamada JI, Mori S, Yamanaka MD, Haryoko U, Lestari S, Sulistyowati R, Syamsudin F (2012) Interannual rainfall variability over northwestern Jawa and its relation to the Indian Ocean dipole and El Niño–southern oscillation events. SOLA 8:69–72. https://doi.org/10.2151/sola.2012-018

Hamada JI, Yamanaka MD, Matsumoto J, Fukao S, Winarso PA, Sribimawati T (2002) Spatial and temporal variations of the rainy season over Indonesia and their link to ENSO. J Meteor Soc Japan 80:285–310. https://doi.org/10.2151/jmsj.80.285

Hamada JI, Yamanaka MD, Mori S, Tauhid YI, Sribimawati T (2008) Differences of rainfall characteristics between coastal and interior areas of central western Sumatera, Indonesia. J Meteor Soc Japan 86:593–611. https://doi.org/10.2151/jmsj.86.593

Hartmann DL (1994) Global physical climatology. Academic Press, San Diego

Hashiguchi H, Fukao S, Tsuda T, Yamanaka MD, Tobing DL, Sribimawati T, Harijono SWB, Wiryosumarto H (1995) Observations of the planetary boundary layer over equatorial Indonesia with an L-band clear-air Doppler radar: initial results. Radio Sci 30:1043–1054. https://doi.org/10.1029/95RS00653

Hashiguchi NO, Yamanaka MD, Ogino SY, Shiotani M, Sribimawati T (2006) Seasonal and interannual variations of temperature in tropical tropopause layer (TTL) over Indonesia based on operational rawinsonde data during 1992–1999. J Geophys Res 111:D15110. https://doi.org/10.1029/2005JD006501

Hattori M, Mori S, Matsumoto J (2011) The cross-equatorial northerly surge over the maritime continent and its relationship to precipitation patterns. J Meteor Soc Japan 89A:27–47. https://doi.org/10.2151/jmsj.2011-A02

Hayashi YY, Sumi A (1986) The 30–40 day oscillations simulated in an "aqua planet" model. J Meteor Soc Japan 64:451–467. https://doi.org/10.2151/jmsj1965.64.4_451

Haylock M, McBride J (2001) Spatial coherence and predictability of Indonesian wet season rainfall. J Clim 4:3882–3887. https://doi.org/10.1175/1520-0442(2001)014<3882:SCAPOI>2.0.CO;2

Hendon HH, Woodberry K (1993) The diurnal cycle of tropical convection. J Geophys Res 98:16623–16637. https://doi.org/10.1029/93JD00525

Hidayat R, Ando K, Masumoto Y, Luo JJ (2016) Interannual variability of rainfall over Indonesia: impacts of ENSO and IOD and their predictability. IOP Conf Series Earth Environ Sci 31:012043. https://doi.org/10.1088/1755-1315/31/1/012043

Hidayat R, Kizu S (2010) Influence of the Madden–Julian oscillation on Indonesian rainfall variability in austral summer. Int J Climatol 30:1816–1825. https://doi.org/10.1002/joc.2005

Hirose M, Takayabu YN, Hamada A, Shige S, Yamamoto MK (2017) Spatial contrast of geographically induced rainfall observed by TRMM PR. J Clim 30: 4165–4184. https://doi.org/10.1175/JCLI-D-16-0442.1

Jaeger L (1983) Monthly and areal patterns of mean global precipitation. In: Street-Perrott A, Beran M, Ratcliffe R (eds) Variations in the global water budget. Springer, pp 129–140. https://doi.org/10.1007/978-94-009-6954-4_9

Kamimera H, Mori S, Yamanaka MD, Syamsudin F (2012) Modulation of diurnal rainfall cycle by the Madden–Julian oscillation based on one-year continuous observations with a meteorological radar in West Sumatera. SOLA 8:111–114. https://doi.org/10.2151/sola.2012-028

Kawashima M, Fujiyoshi Y, Ohi M, Honda T, Kozu T, Shimomai T, Hashiguchi H (2006) Overview of Doppler radar observations of precipitating cloud systems in Sumatera Island during the first CPEA campaign. J Meteor Soc Japan 84A:33–56. https://doi.org/10.2151/jmsj.84A.33

Kawashima M, Fujiyoshi Y, Ohi M, Honda T, Mori S, Sakurai N, Abe Y, Harjupa W, Syamsudin F, Yamanaka MD (2011) Case study of an intense wind event associated with a mesoscale convective system in west Sumatera during the HARIMAU2006 campaign. J Meteor Soc Japan 89A:239–257. https://doi.org/10.2151/jmsj.2011-A15

Kikuchi K, Wang B (2008) Diurnal precipitation regimes in the global tropics. J Clim 21:2680–2696. https://doi.org/10.1175/2007JCLI2051.1

Kimura R (1975) Dynamics of steady convections over heat and cool islands. J Meteor Soc Japan 53:440–457 https://doi.org/10.2151/jmsj1965.53.6_440

Kitoh A, Arakawa O (2005) Reduction in tropical rainfall diurnal variation by global warming simulated by a 20-km mesh climate model. Geophys Res Lett 32:L18709. https://doi.org/10.1029/2005GL023350

Kitoh A, Murakami S (2002) Tropical Pacific climate at the mid-Holocene and the last glacial maximum simulated by a coupled ocean–atmosphere general circulation model. Paleoceanography 17(3):1047. https://doi.org/10.1029/2001PA000724

Kobayashi S, Ota Y, Harada Y, Ebita A, Moriya M, Onoda H, Onogi K, Kamahori H, Kobayashi C, Endo H, Miyaoka K, Takahashi K (2015) The JRA-55 reanalysis: general specifications and basic characteristics. J Meteor Soc Japan 93:5–48. https://doi.org/10.2151/jmsj.2015-001

Koh TY, Teo CK (2009) Toward a mesoscale observation network in Southeast Asia. Bull Amer Meteor Soc 90:481–488. https://doi.org/10.1175/2008BAMS2561.1

Köppen W (1918) Klassifikation der Klimate nach Temperatur, Niederschlag und Jahresablauf (classification of climates according to temperature, precipitation and seasonal cycle). Petermanns Geogr Mitt 64:193–203 243–248. http://koeppen-geiger.vu-wien.ac.at/pdf/Koppen_1918.pdf. Accessed 20 Feb 2018

Köppen W, Geiger R (1954) Klima der Erde (climate of the earth). Wall map 1:16 mill. Klett-Perthes, Gotha http://koeppen-geiger.vu-wien.ac.at/pics/Geiger_1954_Map.jpg. Accessed 20 Feb 2018

Köppen W, Wegener A (1924) Die Klimate der geologischen Vorzeit, Gebrüder Borntraeger, p 256 [Thiede J, Lochte, K., Dummermuth, A (eds); translated by Oelkers B (2015) The Climates of the Geological Past, Schweizerbart Science Publishers, pp. 657 http://www.schweizerbart.de/publications/detail/isbn/9783443010881. Accessed 20 Feb 2018]

Kubota H, Shirooka R, Hamada JI, Syamsudin F (2011) Interannual rainfall variability over the eastern maritime continent. J Meteor Soc Japan 89A: 111–122. https://doi.org/10.2151/jmsj.2011-A07

Kubota H, Yoneyama K, Hamada JI, Wu PM, Sudaryanto A, Wahyono IB (2015) Role of maritime continent convection during the preconditioning stage of the Madden–Julian oscillation observed in CINDY2011/DYNAMO. J Meteor Soc Japan 93A:101–114. https://doi.org/10.2151/jmsj.2015-050

Kurita N, Ichiyanagi K, Matsumoto J, Yamanaka MD, Ohata T (2009) The relationship between the isotopic content of precipitation and the precipitation amount in tropical regions. J Geochem Explor 102:113–122. https://doi.org/10.1016/j.gexplo.2009.03.002

Lean J, Rottman G, Harder J, Kopp G (2005) Source contributions to new understanding of global change and solar variability. Solar Phys 230:27–53. https://doi.org/10.1007/s11207-005-1527-2

Lestari S, Hamada JI, Syamsudin F, Sunaryo, Matsumoto J, Yamanaka MD (2016) ENSO influences on rainfall extremes around Sulawesi and Maluku Islands in the eastern Indonesian maritime continent. SOLA 12:37–41. https://doi.org/10.2151/sola.2016-008

Lindzen RS (1974) Wave CISK in tropics. J Atmos Sci 31:156–179. https://doi.org/10.1175/1520-0469(1974)031<0156:WCITT>2.0.CO;2

Lukas R, Yamagata T, McCreary JP (1996) Pacific low-latitude western boundary currents and the Indonesian throughflow. J Geophys Res 101:12209–12216. https://doi.org/10.1029/96JC01204

Madden RA, Julian PR (1971) Detection of a 40–50 day oscillation in the zonal wind in the tropical Pacific. J Atmos Sci 28:702–708. https://doi.org/10.1175/1520-0469(1971)028<0702:DOADOI>2.0.CO;2

Madden RA, Julian PR (1994) Observations of the 40–50-day tropical oscillation: a review. Mon Wea Rev 122:814–837. https://doi.org/10.1175/1520-0493(1994)122<0814:OOTDTO>2.0.CO;2

Mandelbrot B (1967) How long is the coast of Britain?: Statistical self-similarity and fractional dimension. Science 156:636–638. https://doi.org/10.1126/science.156.3775.636

Marjuki, van der Schrier G, Klein Tank AMG, van den Besselaar EJM, Nurhayati, Swarinoto YS (2016) Observed trends and variability in climate indices relevant for crop yields in Southeast Asia. J Clim 29:2651–2669. https://doi.org/10.1175/JCLI-D-14-00574.1

Marzuki, Hashiguchi H, Yamamoto MK, Mori S, Yamanaka MD (2013) Regional variability of raindrop size distribution over Indonesia. Ann Geophys 31:1941–1948. https://doi.org/10.5194/angeo-31-1941-2013

Marzuki, Hashiguchi H, Yamamoto MK, Yamamoto M, Mori S, Yamanaka MD, Carbone RE, Tuttle JD (2013) Cloud episode propagation over the Indonesian maritime continent from 10 years of infrared brightness temperature observations. Atmos Res 120–121:268–286. https://doi.org/10.1016/j.atmosres.2012.09.004

Marzuki, Kozu T, Shimomai T, Randeu WL, Hashiguchi H, Shibagaki Y (2009) Diurnal variation of rain attenuation obtained from measurement of raindrop size distribution in equatorial Indonesia. IEEE Trans Antennas Propag 57:1191–1196. https://doi.org/10.1109/TAP.2009.2015812

Matsumoto J, Oki T, Yamanaka MD, Hayashi T, Asanuma J (2016) 10 years of MAHASRI: accomplishments and the international science conference wrap-up. GEWEX NEWS 26(4):10–15 http://www.gewex.org/gewex-content/files_mf/1480533350Nov2016_final_opt.pdf. Accessed 20 Feb 2018

Matsumoto J, Wang B, Wu GX, Li JP, Wu PM, Hattori M, Mori S, Yamanaka MD, Hamada JI, Syamsudin F, Koike T, Tamagawa K, Ikoma E, Kinutani H, Kamahori H, Kamiguchi K, Harada Y (2017) An overview of the Asian monsoon years 2007–2012 (AMY) and multi-scale interactions in the extreme rainfall events over the Indonesian maritime continent. Chang CP, Kuo HC, Lau NC, Johnson RH, Wang B, Wheeler MC, The global monsoon system: research and forecast, 3rd edn., world scientific series on Asia–Pacific weather and climate: Vol. 9, World Scientific Publication Company: 365–386. https://doi.org/10.1142/9789813200913_0029

Matsuno T (1966) Quasi-geostrophic motions in the equatorial area. J Meteor Soc Japan 44:25–43. https://doi.org/10.2151/jmsj1965.44.1_25

McBride JL, Haylock MR, Nicholls N (2003) Relationships between the maritime continent heat source and the El Nino–southern oscillation phenomenon. J Clim 16:2905–2914. https://doi.org/10.1175/1520-0442(2003)016<2905:RBTMCH>2.0.CO;2

McPhaden MJ, Meyers G, Ando K, Masumoto Y, Murty VSN, Ravichandran M, Syamsudin F, Vialard J, Yu L, Yu W (2009) RAMA: the research moored array for African–Asian–Australian monsoon analysis and prediction. Bull Amer Meteor Soc 90:459–480. https://doi.org/10.1175/2008BAMS2608.1

Milanković M (1941) Kanon der Erdbestrahlung und seine Anwendung auf das Eiszeitenproblem (canon of insolation and the ice-age problem), Poseńa izdanja Srpska akademija nauka (Royal Serbian Academy special publications, section of mathematical and natural sciences), p 633 [Israel Program for Scientific Translations; (available from U.S. Dept. of Commerce, Clearinghouse for Federal Scientific and Technical Information, Springfield, Va.) (1969), 484pp]

Miura H, Satoh M, Nasuno T, Noda AT, Oouchi K (2007) A Madden–Julian oscillation event realistically simulated by a global cloud-resolving model. Science 318:1763–1765. https://doi.org/10.1126/science.1148443

Mori S, Hamada JI, Sakurai N, Fudeyasu H, Kawashima M, Hashiguchi H, Syamsudin F, Arbain AA, Sulistyowati R, Matsumoto J, Yamanaka MD (2011) Convective systems developed along the coastline of Sumatera Island, Indonesia observed with an X-band Doppler radar during the HARIMAU2006 campaign. J Meteor Soc Japan 89A:61–81. https://doi.org/10.2151/jmsj.2011-A04

Mori S, Hamada JI, Tauhid YI, Yamanaka MD, Okamoto N, Murata F, Sakurai N, Sribimawati T (2004) Diurnal rainfall peak migrations around Sumatera Island, Indonesian maritime continent observed by TRMM satellite and intensive rawinsonde soundings. Mon Wea Rev 132:2021–2039. https://doi.org/10.1175/1520-0493(2004)132<2021:DLRPMO>2.0.CO;2

Murakami T, Matsumoto J (1994) Summer monsoon over the Asian continent and western North Pacific. J Meteor Soc Japan 72:719–745. https://doi.org/10.2151/jmsj1965.72.5_719

Murata F, Yamanaka MD, Fujiwara M, Ogino SY, Hashiguchi H, Fukao S, Kudsy M, Sribimawati T, Harijono SWB, Kelana E (2002) Relationship between wind and precipitation observed with a UHF radar, GPS rawinsonde and surface meteorological instruments at Kototabang, West Sumatera during September–October 1998. J Meteor Soc Japan 80:347–360. https://doi.org/10.2151/jmsj.80.347

Nakazawa T (1988) Tropical super clusters within intraseasonal variations over the western Pacific. J Meteor Soc Japan 66:823–839. https://doi.org/10.2151/jmsj1965.66.6_823

Neale R, Slingo J (2003) The maritime continent and its role in the global climate: a GCM study. J Clim 16:834–848. https://doi.org/10.1175/1520-0442(2003)016<0834:TMCAIR>2.0.CO;2

Neelin JD, Held IM, Cook KH (1987) Evaporation-wind feedback and low-frequency variability in the tropical atmosphere. J Atmos Sci 44:2341–2348. https://doi.org/10.1175/1520-0469(1987)044<2341:EWFALF>2.0.CO;2

Niino H (1987) The linear theory of sea–land breeze circulation. J Meteor Soc Japan 65:901–921. https://doi.org/10.2151/jmsj1965.65.6_901

Niino H, Mori A, Satomura T, Akiba S (2006) Flow regimes of nonlinear heat island circulation. J Atmos Sci 63:1538–1547. https://doi.org/10.1175/JAS3700.1

Nitta TS (1987) Convective activities in the tropical western pacific and their impact on the northern hemisphere summer circulation. J Meteor Soc Japan 65:373–390. https://doi.org/10.2151/jmsj1965.65.3_373

Nitta TS, Mizuno T, Takahashi K (1992) Multi-scale convective systems during the initial phase of the 1986/87 El Niño. J Meteor Soc Japan 70:447–466. https://doi.org/10.2151/jmsj1965.70.1B_447

Nitta TS, Sekine S (1994) Diurnal variation of convective activity over the tropical western Pacific. J Meteor Soc Japan 72:627–641. https://doi.org/10.2151/jmsj1965.72.5_627

Ogino SY, Yamanaka MD, Fukao S (1995) Meridional variation of lower stratospheric gravity wave activity: a quick look at Hakuho-Maru J–COARE cruise rawinsonde data. J Meteor Soc Japan 73:407–413. https://doi.org/10.2151/jmsj1965.73.2B_407

Ogino SY, Yamanaka MD, Mori S, Matsumoto J (2016) How much is the precipitation amount over the tropical coastal region? J Clim 29:1231–1236. https://doi.org/10.1175/JCLI-D-15-0484.1

Ogino SY, Yamanaka MD, Mori S, Matsumoto J (2017) Tropical coastal dehydrator in global atmospheric water circulation. Geophys Res Lett 44:11636–11643. https://doi.org/10.1002/2017GL075760

Ogura Y, Yoshizaki M (1988) Numerical study of orographic–convective precipitation over the eastern Arabian Sea and the Ghat Mountains during the summer monsoon. J Atmos Sci 45:2097–2122. https://doi.org/10.1175/1520-0469(1988)045<2097:NSOOCP>2.0.CO;2

Ohsawa T, Ueda H, Hayashi T, Watanabe A, Matsumoto J (2001) Diurnal variations of convective activity and rainfall in tropical Asia. J Meteor Soc Japan 79:333–352. https://doi.org/10.2151/jmsj.79.333

Onogi K, Tsutsui J, Koide H, Sakamoto M, Kobayashi S, Hatsushika H, Matsumoto T, Yamazaki N, Kamahori H, Takahashi K, Kadokura S, Wada K, Kato K, Oyama R, Ose T, Mannoji N, Taira R (2007) The JRA-25 reanalysis. J Meteor Soc Japan 85:369–432. https://doi.org/10.2151/jmsj.85.369

Ooyama K (1971) A theory on parameterization of cumulus convection. J Meteor Soc Japan 49A:744–756. https://doi.org/10.2151/jmsj1965.49A.0_744

Osprey SM, Butchart N, Knight JR, Scaife AA, Hamilton K, Anstey JA, Schenzinger V, Zhang C (2016) An unexpected disruption of the atmospheric quasi-biennial oscillation. Science 353:1424–1427. https://doi.org/10.1126/science.aah4156

Parrish JT (1993) Climate of the supercontinent Pangea. J Geol 101:215–233. http://www.jstor.org/stable/30081148. Accessed 20 Feb 2018

Peatman SC, Matthews AJ, Stevens DP (2014) Propagation of the Madden–Julian oscillation through the maritime continent and scale interaction with the diurnal cycle of precipitation. Quart J Roy Meteor Soc 140:814–825. https://doi.org/10.1002/qj.2161

Peel MC, Finlayson BL, McMahon TA (2007) Updated world map of the Köppen–Geiger climate classification. Hydrol Earth Syst Sci 11:1633–1644. https://doi.org/10.5194/hess-11-1633-2007

Permana DS, Thompson LG, Setyadi G (2016) Tropical West Pacific moisture dynamics and climate controls on rainfall isotopic ratios in southern Papua, Indonesia. J Geophys Res Atmos 121:2222–2245. https://doi.org/10.1002/2015JD023893

Qian JH (2008) Why precipitation is mostly concentrated over islands in the maritime continent. J Atmos Sci 65:1428–1441. https://doi.org/10.1175/2007JAS2422.1

Qian JH, Robertson AW, Moron V (2013) Diurnal cycle in different weather regimes and rainfall variability over Borneo associated with ENSO. J Clim 26: 1772–1790. https://doi.org/10.1175/JCLI-D-12-00178.1

Ramage CS (1968) Role of a tropical "maritime continent" in the atmospheric circulation. Mon Wea Rev 96:365–369. https://doi.org/10.1175/1520-0493(1968)096<0365:ROATMC>2.0.CO;2

Randall DA, Wood RA, Bony S, Colman R, Fichefet T, Fyfe J, Kattsov V, Pitman A, Shukla J, Srinivasan J, Stouffer RJ, Sumi A, Taylor KE (2007) Climate models and their evaluation. Solomon S, Qin D, Manning M, Chen Z, Marquis M, Averyt KB, Tignor M, Miller HL, Climate Change 2007: The Physical Science Basis. Contribution of Working Group I to the Fourth Assessment Report of the Intergovernmental Panel on Climate Change, Cambridge University Press, Chapter 8. p. 589–662. https://www.ipcc.ch/publications_and_data/ar4/wg1/en/ch8.html. Accessed 20 Feb 2018

Reid JS, Xian P, Holben BN, Hyer EJ, Reid EA, Salinas SV, Zhang J, Campbell JR, Chew BN, Holz RE, Kuciauskas AP, Lagrosas N, Posselt DJ, Sampson CR, Walker AL, Welton EJ, Zhang C (2016) Aerosol meteorology of the maritime continent for the 2012 7SEAS southwest monsoon intensive study, part 1: regional-scale phenomena. Atmos Chem Phys 16:14041–14056. https://doi.org/10.5194/acp-16-14041-2016

Renggono F, Hashiguchi H, Fukao S, Yamanaka MD, Ogino SY, Okamoto N, Murata F, Sitorus BP, Kudsy M, Kartasasmita M, Ibrahim G (2001) Precipitating clouds observed by 1.3-GHz boundary layer radars in equatorial Indonesia. Ann Geophys 19:889–897. https://doi.org/10.5194/angeo-19-889-2001

Rotunno R (1983) On the linear theory of the land and sea breeze. J Atmos Sci 40: 1999–2009. https://doi.org/10.1175/1520-0469(1983)040<1999:OTLTOT>2.0.CO;2

Sakurai N, Kawashima M, Fujiyoshi Y, Hashiguchi H, Shimomai T, Mori S, Hamada JI, Murata F, Yamanaka MD, Tauhid YI, Sribimawati T, Suhardi B (2009) Internal structures of migratory cloud systems with diurnal cycle over Sumatera Island during CPEA–I campaign. J Meteor Soc Japan 87:157–170. https://doi.org/10.2151/jmsj.87.157

Sakurai N, Mori S, Kawashima M, Fujiyoshi Y, Hamada JI, Shimizu S, Fudeyasu H, Tabata Y, Harjupa W, Hashiguchi H, Yamanaka MD, Matsumoto J, Emrizal, Syamsudin F (2011) Migration process and 3D wind field of precipitation systems associated with a diurnal cycle in west Sumatera: dual Doppler radar analysis during the HARIMAU2006 campaign. J Meteor Soc Japan 89:341–361. https://doi.org/10.2151/jmsj.2011-404

Sakurai N, Murata F, Yamanaka MD, Hashiguchi H, Mori S, Hamada JI, Tauhid YI, Sribimawati T, Suhardi B (2005) Diurnal cycle of migration of convective cloud systems over Sumatera Island. J Meteor Soc Japan 83:835–850. https://doi.org/10.2151/jmsj.83.835

Sato T, Miura H, Satoh M, Takayabu YN, Wang Y (2009) Diurnal cycle of precipitation in the tropics simulated in a global cloud-resolving model. J Clim 22:4809–4826. https://doi.org/10.1175/2009JCLI2890.1

Seto TH, Tabata Y, Yamamoto MK, Hashiguchi H, Mega T, Kudsy M, Yamanaka MD, Fukao S (2009) Comparison study of lower-tropospheric horizontal wind over Sumatera, Indonesia using NCEP/NCAR reanalysis, operational radiosonde, and the equatorial atmosphere radar. SOLA 5:21–24. https://doi.org/10.2151/sola.2009-006

Shibagaki Y, Shimomai T, Kozu T, Mori S, Fujiyoshi Y, Hashiguchi H, Yamamoto MK, Fukao S, Yamanaka MD (2006) Multi-scale convective systems associated with an intraseasonal oscillation over the Indonesian maritime continent. Mon Wea Rev 134:1682–1696. https://doi.org/10.1175/MWR3152.1

Simpson RH (1947) Synoptic aspects of the intertropical convergence near Central and South America. Bull Amer Meteor Soc 28:335–346. http://www.jstor.org/stable/26257787. Accessed 20 Feb 2018

Siswanto S, van Oldenborgh GJ, van der Schrier G, Jilderda R, van den Hurk B (2016) Temperature, extreme precipitation, and diurnal rainfall changes in the urbanized Jakarta city during the past 130 years. Int J Climatol 36:3207–3225. https://doi.org/10.1002/joc.4548

Slingo J, Belcher S, Scaife A, McCarthy M, Saulter A, McBeath K, Jenkins A, Huntingford C, Marsh T, Hannaford J, Parry S (2014) The recent storms and floods in the UK. UK Met Office, p 27. http://nora.nerc.ac.uk/505192/. Accessed 20 Feb 2018

Sorooshian S, Gao X, Hsu K, Maddox RA, Hong Y, Gupta HV, Imam B (2002) Diurnal variability of tropical rainfall retrieved from combined GOES and TRMM satellite information. J Clim 15:983–1001. https://doi.org/10.1175/1520-0442(2002)015<0983:DVOTRR>2.0.CO;2

Stull RB (1988) An introduction to boundary layer meteorology. Kluwer Academic, Dordrecht

Sui CH, Lau KM, Takayabu YN, Short DA (1997) Diurnal variations in tropical oceanic cumulus convection during TOGA COARE. J Atmos Sci 54:639–655. https://doi.org/10.1175/1520-0469(1997)054<0639:DVITOC>2.0.CO;2

Sulistyowati R, Hapsari RI, Syamsudin F, Mori S, Oishi ST, Yamanaka MD (2014) Rainfall-driven diurnal variations of water level in the Ciliwung River, West Jawa, Indonesia. SOLA 10:141–144. https://doi.org/10.2151/sola.2014-029

Supari, Tangang F, Juneng L, Aldrian E (2016) Observed changes in extreme temperature and precipitation over Indonesia. Int J Climatol 37:1979–97. https://doi.org/10.1002/joc.4829

Suwarman R, Ichiyanagi K, Tanoue M, Yoshimura K, Mori S, Yamanaka MD, Kurita N, Belgaman HA, Syamsudin F (2017) El Niño southern oscillation signature in atmospheric water stable isotopes over maritime continent during wet season. J Meteor Soc Japan 95:49–66. https://doi.org/10.2151/jmsj.2017-003

Suwarman R, Ichiyanagi K, Tanoue M, Yoshimura K, Mori S, Yamanaka MD, Kurita N, Syamsudin F (2013) The variability of stable isotopes and water origin of precipitation over the maritime continent. SOLA 9:74–78. https://doi.org/10.2151/sola.2013-017

Tabata Y, Hashiguchi H, Yamamoto MK, Yamamoto M, Yamanaka MD, Mori S, Syamsudin F, Manik T (2011a) Lower tropospheric horizontal wind over Indonesia: a comparison of wind-profiler network observations with global reanalyses. J Atmos Solar Terr Phys 73:986–995. https://doi.org/10.1016/j.jastp.2010.09.016

Tabata Y, Hashiguchi H, Yamamoto MK, Yamamoto M, Yamanaka MD, Mori S, Syamsudin F, Manik T (2011b) Observational study on diurnal precipitation cycle in equatorial Indonesia using 1.3-GHz wind profiling radar network and TRMM precipitation radar. J Atmos Solar Terr Phys 73:1031–1042. https://doi.org/10.1016/j.jastp.2010.10.003

Takasuka D, Miyakawa T, Satoh M, Miura H (2015) Topographical effects on internally produced MJO-like disturbances in an aqua-planet version of NICAM. SOLA 11:170–176. https://doi.org/10.2151/sola.2015-038

Trenberth KE, Fasullo JT, Kiehl J (2009) Earth's global energy budget. Bull Amer Meteor Soc 90:311–323. https://doi.org/10.1175/2008BAMS2634.1

Tsuda T, Fukao S, Yamamoto M, Nakamura T, Yamanaka MD, Adachi T, Hashiguchi H, Fujioka N, Tsutsumi M, Kato S, Harijono SWB, Sribimawati T, Sitorus BP, Yahya RB, Karmini M, Renggono F, Parapat BL, Djojonegoro W, Mardio P, Adikusumah N, Endi HT, Wiryosumarto H (1995) A preliminary report on observations of equatorial atmosphere dynamics in Indonesia with radars and radiosondes. J Meteor Soc Japan 73:393–406. https://doi.org/10.2151/jmsj1965.73.2B_393

Tsuda T, Nishida M, Rocken C, Ware RH (2000) A global morphology of gravity wave activity in the stratosphere revealed by the GPS occultation data (GPS/MET). J Geophys Res 105:7257–7273. https://doi.org/10.1029/1999JD901005

van Bemmelen W (1922) Land- und seebrise in Batavia. Beitr Phys Frei Atmos 10:169–177

Villafuerte MQ, Matsumoto J (2015) Significant influences of global mean temperature and ENSO on extreme rainfall in Southeast Asia. J Clim 28:1905–1919. https://doi.org/10.1175/JCLI-D-14-00531.1

Voris HK (2000) Maps of Pleistocene sea levels in Southeast Asia: shorelines, river systems and time durations. J Biogeogr 27:1153–1167. https://doi.org/10.1046/j.1365-2699.2000.00489.x

Wallace AR (1869) The Malay archipelago: the land of the orang-Utan, and the bird of paradise, A narrative of travel, with studies of man and nature, vol 1. Macmillan and Co. http://darwin-online.org.uk/converted/Ancillary/1869_MalayArchipelago_A1013.1/1869_MalayArchipelago_A1013.1.1.html. Accessed 20 Feb 2018

Wallace JM, Hobbs PV (2006) Atmospheric science: an introductory survey, 2nd edn. Academic Press, San Diego. https://www.sciencedirect.com/science/book/9780127329512. Accessed 20 Feb 2018

Wang B (2012) Theories. Lau WKM, Waliser DE, Intraseasonal variability in the Atmosphere–Ocean climate system, 2nd edn, Springer-Verlag: 335–398. doi: https://doi.org/10.1007/978-3-642-13914-7_10

Wang B, Kim HJ, Kikuchi K, Kitoh A (2012) Diagnostic metrics for evaluation of annual and diurnal cycles. Clim Dyn 37:941–955. https://doi.org/10.1007/s00382-010-0877-0

Wegener A (1929) Die Entstehung der Kontinente und Ozeane, Verte umgearbeitete Auflage. https://archive.org/details/Entstehung1929. Accessed 20 Feb 2018. [Translated by Biram J (1966) The Origin of Continents and Oceans, 4th Ed.]

Wheeler MC, Hendon HH (2004) An all-season real-time multivariate MJO index: development of an index for monitoring and prediction. Mon Wea Rev 132:1917–1932. https://doi.org/10.1175/1520-0493(2004)132<1917:AARMMI>2.0.CO;2

Widiyatmi I, Hashiguchi H, Fukao S, Yamanaka MD, Ogino SY, Gage KS, Harijono SWB, Diharto S, Djojodihardjo H (2001) Examination of 3–6 day disturbances over equatorial Indonesia based on boundary layer radar observations during 1996–1999 at Serpong, Biak and Bukittinggi. J Meteor Soc Japan 79:317–331. https://doi.org/10.2151/jmsj.79.317

Wilson JT (1963) Continental drift. Sci Amer 208(4):86–103. http://www.jstor.org/stable/24936535. Accessed 20 Feb 2018

Wu PM, Arbain AA, Mori S, Hamada JI, Hattori M, Syamsudin F, Yamanaka MD (2013) The effects of an active phase of the Madden–Julian oscillation on the extreme precipitation event over western Jawa Island in January 2013. SOLA 9:79–83. https://doi.org/10.2151/sola.2013-018

Wu PM, Ardiansyah D, Yokoi S, Mori S, Syamsudin F, Yoneyama K (2017) Why torrential rain occurs on the western coast of Sumatra Island at the leading edge of the MJO westerly wind bursts. SOLA 13:36–40. https://doi.org/10.2151/sola.2017-007

Wu PM, Hamada JI, Mori S, Tauhid YI, Yamanaka MD, Kimura F (2003) Diurnal variation of precipitable water over a mountainous area in Sumatera Island. J Appl Meteorol 42:1107–1115. https://doi.org/10.1175/1520-0450(2003)042<1107:DVOPWO>2.0.CO;2

Wu PM, Hamada JI, Yamanaka MD, Matsumoto J, Hara M (2009) The impact of orographically-induced gravity wave on the diurnal cycle of rainfall over southeast Kalimantan Island. Atmos Ocean Sci Lett 2:35–39. https://doi.org/10.1080/16742834.2009.11446773

Wu PM, Hara M, Fudeyasu H, Yamanaka MD, Matsumoto J, Syamsudin F, Sulistyowati R, Djajadihardja YS (2007) The impact of trans-equatorial monsoon flow on the formation of repeated torrential rains over Java Island. SOLA 3:93–96. https://doi.org/10.2151/sola.2007-024

Wu PM, Hara M, Hamada JI, Yamanaka MD, Kimura F (2009) Why a large amount of rain falls over the sea in the vicinity of western Sumatera Island during nighttime. J Appl Meteor Climatol 48:1345–1361. https://doi.org/10.1175/2009JAMC2052.1

Wu PM, Mori S, Hamada JI, Yamanaka MD, Matsumoto J, Kimura F (2008) Diurnal variation of rainfall and precipitable water over Siberut Island off the western coast of Sumatera Island. SOLA 4:125–128. https://doi.org/10.2151/sola.2008-032

Wu PM, Mori S, Syamsudin F (2018) Land–sea surface air temperature contrast on the western coast of Sumatera Island during an active phase of the Madden–Julian oscillation. Prog Earth Planet Sci 5:4. https://doi.org/10.1186/s40645-017-0160-7

Wu PM, Yamanaka MD, Matsumoto J (2008) The formation of nocturnal rainfall offshore from convection over western Kalimantan (Borneo) Island. J Meteor Soc Japan 86A:187–203. https://doi.org/10.2151/jmsj.86A.187

Xie P, Arkin PA (1997) Global precipitation: a 17-year monthly analysis based on gauge observations, satellite estimates, and numerical model outputs. Bull Amer Meteor Soc 78:2539–2558. https://doi.org/10.1175/1520-0477(1997)078<2539:GPAYMA>2.0.CO;2

Xie SP, Xu H, Saji NH, Wang Y, Liu WT (2006) Role of narrow mountains in large-scale organization of Asian monsoon convection. J Clim 19:3420–3429. https://doi.org/10.1175/JCLI3777.1

Yamanaka MD (2016) Physical climatology of Indonesian maritime continent: an outline to comprehend observational studies. Atmos Res 178–179:231–259. https://doi.org/10.1016/j.atmosres.2016.03.017

Yamanaka MD, Mori S, Syamsudin F (2016) MAHASRI activities related to the Indonesian maritime continent. GEWEX NEWS 26(4):18–22. http://www.gewex.org/gewex-content/files_mf/1480533350Nov2016_final_opt.pdf. Accessed 20 Feb 2018

Yamanaka MD, Mori S, Wu PM, Hamada JI, Sakurai N, Hashiguchi H, Yamamoto MK, Shibagaki Y, Kawashima M, Fujiyoshi Y, Shimomai T, Manik T, Erlansyah, Setiawan W, Tejasukmana B, Syamsudin F, Djajadihardia YS, Anggadiredja JT (2008) HARIMAU radar–profiler network over Indonesian maritime continent: a GEOSS early achievement for hydrological cycle and disaster prevention. J Disaster Res 3:78–88. https://doi.org/10.20965/jdr.2008.p0078

Yang GY, Slingo J (2001) The diurnal cycle in the tropics. Mon Wea Rev 129:784–801. https://doi.org/10.1175/1520-0493(2001)129<0784:TDCITT>2.0.CO;2

Yang L, Wang DX, Huang J, Wang X, Zeng L, Shi R, He Y, Xie Q, Wang SG, Chen R, Jinnan Yuan J, Wang Q, Chen J, Zu T, Li J, Sui D, Peng S (2015) Toward a mesoscale hydrological and marine meteorological observation network in the South China Sea. Bull Amer Meteor Soc 96:1117–1135. https://doi.org/10.1175/BAMS-D-14-00159.1

Yanto, Rajagopalan B, Zagona E (2016) Space–time variability of Indonesian rainfall at inter-annual and multi-decadal time scales. Clim Dyn 47:2975–2989. https://doi.org/10.1007/s00382-016-3008-8

Yatagai A, Kamiguchi K, Arakawa O, Hamada A, Yasutomi A, Kitoh A (2012) APHRODITE: constructing a long-term daily gridded precipitation dataset for Asia based on a dense network of rain gauges. Bull Amer Meteor Soc 93: 1401–1415. https://doi.org/10.1175/BAMS-D-11-00122.1

YMC (Years of the Maritime Continent) (2017) websites: http://www.bmkg.go.id/ymc/ http://www.jamstec.go.jp/ymc/. Accessed 20 Feb 2018

Yokoi S, Matsumoto J (2008) Collaborative effects of cold surge and tropical depression–type disturbance on heavy rainfall in central Vietnam. Mon Wea Rev 136:3275–3287. https://doi.org/10.1175/2008MWR2456.1

Yokoi S, Mori S, Katsumata M, Geng B, Yasunaga K, Syamsudin F, Nurhayati, Yoneyama K (2017) Diurnal cycle of precipitation observed in the western coastal area of Sumatra Island: offshore preconditioning by gravity waves. Mon Wea Rev 145:3745–3761. https://doi.org/10.1175/MWR-D-16-0468.1

Yoneyama K, Masumoto Y, Kuroda Y, Katsumata M, Mizuno K, Takayabu YN, Yoshizaki M, Shareef A, Fujiyoshi Y, McPhaden MJ, Murty VSN, Shirooka R, Yasunaga K, Yamada H, Sato N, Ushiyama T, Moteki Q, Seiki A, Fujita M, Ando K, Hase H, Ueki I, Horii T, Yokoyama C, Miyakawa T (2008) MISMO field experiment in the equatorial Indian Ocean. Bull Amer Meteor Soc 89:1889–1903. https://doi.org/10.1175/2008BAMS2519.1

Yoneyama K, Zhang C, Long CN (2013) Tracking pulses of the Madden–Julian oscillation. Bull Amer Meteor Soc 94:1871–1891. https://doi.org/10.1175/BAMS-D-12-00157.1

Yoshida Y, Maruyama K, Tsutsui J, Nakashiki N, Bryan FO, Blackmon M, Boville BA, Smith RD (2005) Multi-century ensemble global warming projections using the Community Climate System Model (CCSM3). J Earth Simulator 3:2–10. http://www.jamstec.go.jp/esc/publication/journal/jes_vol.3/pdf/JES3-21yoshida.pdf. Accessed 20 Feb 2018

Zhang C (2005) Madden–Julian oscillation. Rev Geophys 43:RG2003. https://doi.org/10.1029/2004RG000158

Zhang C (2013) Madden–Julian oscillation bridging weather and climate. Bull Amer Meteor Soc 94:1849–1870. https://doi.org/10.1175/BAMS-D-12-00026.1

Zhang C, Ling J (2017) Barrier effect of the Indo–Pacific maritime continent on the MJO: perspectives from tracking MJO precipitation. J Clim 30:3439–3459. https://doi.org/10.1175/JCLI-D-16-0614.1

Zhang R, Sumi A, Kimoto M (1996) Impact of El Nino on the East Asian monsoon. J Meteor Soc Japan 74:49–62. https://doi.org/10.2151/jmsj1965.74.1_49.

9

Seasonal variation in isotopic composition and the origin of precipitation over Bangladesh

Masahiro Tanoue[1*], Kimpei Ichiyanagi[2], Kei Yoshimura[3], Masashi Kiguchi[4], Toru Terao[5] and Taiichi Hayashi[6]

Abstract

Water isotopic composition ($\delta^{18}O$ and δD) in terrestrial proxies of past precipitation enable us to better understand and interpret variation in the Indian Summer Monsoon (ISM). Previous studies have suggested that the origin of precipitation is an important factor controlling the isotopic composition of precipitation around the Indian subcontinent; however, it is difficult to quantify using the Lagrangian approach because the approach does not satisfy the assumption of an adiabatic process over a convective area. We investigated the isotopic composition of precipitation at three sites over Bangladesh in 2010 and estimated the origins of precipitation by the Eulerian approach using an isotope-incorporating Atmospheric General Circulation Model. Our observations showed similar seasonal and intraseasonal variations in the $\delta^{18}O$ values of precipitation among the sites, whereas the temporal characteristics of the precipitation amount differed among the sites. The isotopic composition was linked to the migration of organized convective activity around the region. The model showed that the pre-monsoon season (from mid-March to early June) was characterized by high $\delta^{18}O$ values of precipitation originating from the Bay of Bengal and the Arabian Sea. In the ISM season (from mid-June to early October), the contribution of these sources to precipitation gradually decreased, while the contribution from the Indian Ocean increased, resulting in decreasing $\delta^{18}O$ values of precipitation due to the enhanced rainout process during the transportation. These moisture contributed less to precipitation over Bangladesh in the post-monsoon season (from mid-October to November), whereas moisture originating from the Pacific Ocean and land surface (i.e., recycling of water) contributed to precipitation in the season. Because the recycling of water originated from past precipitation with low $\delta^{18}O$ values in the ISM season, its contribution to precipitation reduced the $\delta^{18}O$ values of precipitation in the ISM and post-monsoon seasons. These results suggest that the origins of precipitation and the migration of organized convective activity are the dominant factors controlling the isotopic composition of precipitation in the region. These characteristics can be used to identify monsoon onset and withdrawal based on water isotopic composition.

Keywords: Isotopic composition of precipitation, Bangladesh, Origins of precipitation, Isotope-incorporating atmospheric general circulation model, Indian summer monsoon

Introduction

The Indian Summer Monsoon (ISM), which is characterized by strong seasonality of rainfall and floods, affects regional society and agriculture around the Indian subcontinent. An understanding of the ISM is needed to forecast its onset, withdrawal, duration, intensity, and variation. To further understand and forecast the ISM, many studies have attempted to reconstruct paleo-ISM variation from oxygen isotope ($\delta^{18}O$) variations in terrestrial proxies of past precipitation (e.g., stalagmites, tree rings, and sediment) in the region (e.g., Fleitmann et al. 2003; Wang et al. 2005; Yancheva et al. 2007; Wang et al. 2008; Zhang et al. 2008).

The interpretation of paleo-ISM variation from $\delta^{18}O$ values of terrestrial proxies is mainly based on the "amount effect," i.e., the negative relation between the amount and isotopic composition of precipitation, which has been observed in tropical regions at monthly or daily

* Correspondence: tanoue.masahiro.s6@sic.shibaura-it.ac.jp
[1]Department of Civil Engineering, Shibaura Institute of Technology, 3-7-5 Toyosu, Koto-ku, Tokyo 135-8548, Japan
Full list of author information is available at the end of the article

time scales (e.g., Dansgaard 1964; Araguás-Araguás et al. 1998; Risi et al. 2008; Kurita et al. 2009). The amount effect is due to the increase in the removal of heavier water isotopes ($H_2{}^{18}O$ and $HD^{16}O$) from cloud vapor, as heavier water isotopes are preferentially removed from cloud vapor (e.g., Dansgaard 1964). However, it has not been observed at stations around the Indian subcontinent (e.g., Datta et al. 1991; Araguás-Araguás et al. 1998; Bhattacharya et al. 2003; Breitenbach et al. 2010; Midhun and Ramesh 2016), except in the southeastern part of peninsular India and Sri Lanka (Lekshmy et al. 2015).

Because the amount effect cannot be applied to reconstruct $\delta^{18}O$ variation in terrestrial proxies around the Indian subcontinent, many studies have explored the factors controlling the isotopic composition of precipitation in the region (e.g., Datta et al. 1991; Bhattacharya et al. 2003; Sengupta and Sarkar 2006; Breitenbach et al. 2010; Lekshmy et al. 2014, 2015; Chakraborty et al. 2016; Midhun and Ramesh 2016). For example, Bhattacharya et al. (2003) investigated the isotopic composition of precipitation in Bombay and New Delhi and concluded that it reflected oceanic moisture, with low $\delta^{18}O$ values, during the ISM season and continental moisture, with high $\delta^{18}O$ values, during the winter monsoon season. In addition, Breitenbach et al. (2010) assessed the isotopic composition of precipitation for every rainfall event in the northeastern part of India and estimated the travel distance of air parcels during their transport from the vapor source regions to the observation site using the Lagrangian approach (i.e., backward trajectory analysis). They found that low $\delta^{18}O$ values of precipitation were associated with long-distance transport. Similarly, Lekshmy et al. (2014) and Chakraborty et al. (2016) found low $\delta^{18}O$ values of precipitation when air parcels passed over a convective area over the Indian Ocean. This finding was supported by a study based on multiple simulations of isotope-incorporating Atmospheric General Circulation Models (AGCMs) (Midhun and Ramesh 2016), which showed that the inter-annual variability in $\delta^{18}O$ values of precipitation around the Indian subcontinent was controlled by rainout in the upstream area of the moisture flux. This suggests that the origins of moisture and rainout in the upstream area are the dominant controlling factors of the isotopic composition of precipitation in the region.

Although some studies have indicated that the recycling of water is an important factor for precipitation variability around the Indian subcontinent based on land–atmosphere coupling simulations (e.g., Guimberteau et al. 2012; Yamashima et al. 2014) and satellite observations (e.g., Ono and Takahashi 2016), the recycling of water has not been discussed as a possible factor controlling the isotopic composition of precipitation around the region, as it is difficult in the Lagrangian approach to quantify the contribution of the recycling of water to precipitation (Breitenbach et al. 2010). In the Lagrangian approach, an individual air parcel trajectory can be observed; however, the uncertainties of the method can be large when an air parcel passes over a convective area due to an unsatisfied assumption of the adiabatic process. On the other hand, the Eulerian approach focuses on specific locations through which the fluid flows as time passes. In addition, the Eulerian approach is easily combined with an isotope-incorporating AGCM. This approach can be used to estimate the origins of moisture and precipitation in many regions (e.g., Yoshimura et al. 2004; Ichiyanagi et al. 2005; Shinoda et al. 2005; Sato et al. 2007; Hiraoka et al. 2011; Suwarman et al. 2013; Kudo et al. 2014; Belgaman et al. 2016) and to evaluate the atmospheric hydrological cycle in the model by comparing the simulated and observed isotopic composition of precipitation. In this study, we estimated the origins of precipitation using an isotope-incorporating AGCM over Bangladesh. The dominant factors controlling seasonal variation in the isotopic composition of precipitation in the region were further investigated. Origins of moisture estimated by the Eulerian and Lagrangian approaches were compared to understand the uncertainty of both approaches in the region. The results of the study will enhance our understanding of the atmospheric hydrological cycle associated with the ISM over Bangladesh and provide a way to identify the timing of monsoon onset and withdrawal using water isotopic composition.

Methods/Experimental

The isotopic composition of precipitation was observed at three sites over Bangladesh. We used an isotope-incorporating Global Spectral Model (IsoGSM, Yoshimura et al. 2008) to simulate seasonal variation in the isotopic composition of precipitation and then estimated the origins of precipitation using the Eulerian approach based. The estimated origins of moisture were compared to the estimation based on the Lagrangian approach.

Study area

We observed the amount and isotopic composition of precipitation at three sites over Bangladesh: Sylhet (91.88° E, 24.91°N), Dhaka (90.38°E, 23.78°N), and Chittagong (91.81°E, 22.35°N) (Fig. 1). In Bangladesh, the precipitation is high during summer, and it is low during winter. In this study, we defined the dry winter (pentad 1–13, January to early March), pre-monsoon (pentad 14–31, mid-March to early June), ISM (pentad 32–56, mid-June to early October), and post-monsoon (pentad 57–67, mid-October to November) seasons based on the analysis by Hoque et al. (2011) and the observed seasonal variation in the isotopic

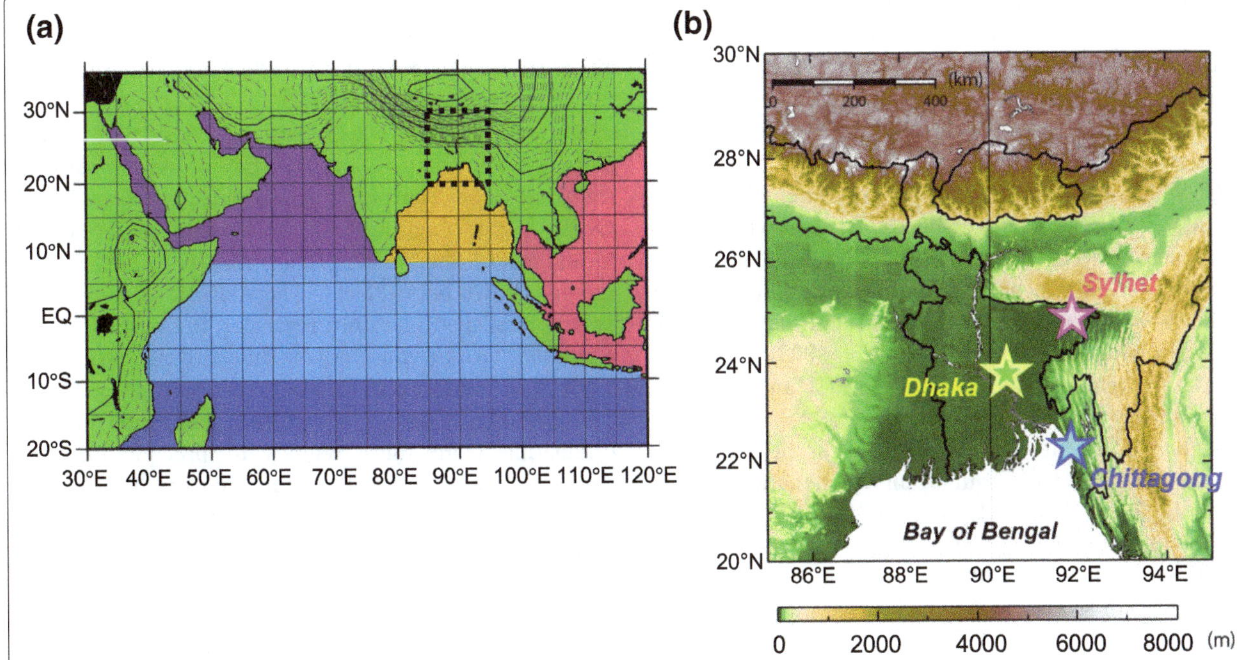

Fig. 1 **a** Surface geo-potential height (contour) in this study. The solid and dashed contour intervals are 1000 and 200 gpm, respectively. Vapor source regions were divided into seven regions: the Pacific Ocean (red), the Bay of Bengal (Yellow), the Arabian Sea (purple), northern part of the Indian Ocean (light blue), southern part of the Indian Ocean (dark blue), land surface (green), and others (black). The dotted line indicates the area of (**b**). **b** Locations of the observation sites over Bangladesh. The topographic data (GTOPO30) were downloaded from the United States Geological Survey (USGS)

composition of precipitation. The pre-monsoon season was characterized by dominant westerly winds and unorganized deep convection (Matsumoto 1997; Islam and Uyeda 2008). In the ISM season, a southwesterly moisture flux is dominant over Bangladesh, and precipitation is caused by deep convection and cyclone activity over the Indian Ocean and the Bay of Bengal (Islam and Uyeda 2008; Hatsuzuka et al. 2014). In this season, a huge amount of water vapor is transported from the Bay of Bengal to Bangladesh (Hoque et al. 2011), and it precipitates on the southern slope of the Meghalaya Plateau due to orographic upward currents (e.g., Fujinami et al. 2017). Precipitation in the northeastern part of Bangladesh (e.g., Sylhet) is relatively high in the country. In the post-monsoon season, dry northwesterly and southwesterly winds are dominant in the region (Hoque et al. 2011), and precipitation is frequently caused by atmospheric disturbances over the Bay of Bengal (Breitenbach et al. 2010).

Observation of precipitation samples and analysis methods

We observed the amount and isotopic composition of precipitation at Sylhet, Dhaka, and Chittagong at 3 h intervals (00UTC, 03UTC, 06UTC, 09UTC, 12UTC, 15UTC, 18UTC, and 21UTC) from January to December 2010. The Bangladesh Meteorological Department operates a pluviometer with precipitation water and reads the amount of precipitation every 3 h. We arranged for part of the collected precipitation water to be transferred to a 15 ml bottle. We collected a total of 1416 precipitation samples. These samples were stored in dark and cool locations prior to the analysis of isotopic composition.

The precipitation samples were analyzed by a mass spectrometer (DELTA-V, Thermo Fisher Scientific, Waltham, MA, USA) in the Graduate School of Science and Technology, Kumamoto University, Japan. The equilibrium of CO_2 and H_2 gases with precipitation samples was prepared by H_2–H_2O (Horita et al. 1989) and H_2O–CO_2 (Epstein and Mayeda 1953) equilibrium techniques, respectively, with an automatic equilibration setup. The automatic setup was connected to a dual-inlet system on the DELTA-V instrument. We measured the isotopic ratios of the CO_2 and H_2 gases and calculated the relative difference between sample water and Standard Mean Ocean water using two laboratory standards (AMW, $\delta D = 0.59‰$ and $\delta^{18}O = -0.262‰$; CIW, $\delta D = -89.88‰$ and $\delta^{18}O = -12.730‰$). To validate the measurements of the isotopic ratio of precipitation samples, we included one laboratory standard (KTW, $\delta D = -47.32‰$ and $\delta^{18}O = -7.300‰$) and checked the δD and $\delta^{18}O$ values every analysis. The analytical errors for δD and $\delta^{18}O$ were less than $\pm 0.5‰$ and $\pm 0.05‰$, respectively. The measured

3 h isotopic composition of precipitation was averaged to a daily value (from 00UTC to 21UTC) and weighted by the precipitation amount to compare the simulation results.

Description of IsoGSM

IsoGSM is based on the Scripps Experimental Climate Prediction Center's global spectral model with heavy water isotopes ($H_2{}^{18}O$ and HDO) as tracers. The major physical processes included in the model are the relaxed Arakawa–Schubert scheme (Moorthi and Suarez 1992), the Noah land surface model (Ek et al. 2003), the Chou radiation scheme (Chou and Suarez 1994), and a planetary boundary scheme (Hong and Pan 1996). The physical schemes of water isotopes included in the model are the parameterization of equilibrium fractionation among vapor, liquid, and ice (Majoube 1971a, 1971b), kinetic fractionation for surface evaporation from open water (Merlivat and Jouzel 1979), condensation from vapor to ice under super-saturated conditions at temperatures lower than − 20 °C (Jouzel and Merlivat 1984), and evaporation and isotopic exchange between falling raindrops and vapor in surrounding air (Stewart 1975). Midhun and Ramesh (2016) validated the amount and $\delta^{18}O$ value of precipitation derived from multi-GCM simulations, and IsoGSM showed good temporal and spatial variations in the simulations.

Simulation design

We selected the T62 horizontal resolution (approximately 180 km) and 28 vertical layers in this study. Although the IsoGSM simulation was performed for the period of 1979–2010, we used only the model output for the year 2010. The simulation output was generated at daily intervals and interpolated into 17 vertical pressure levels (1000, 925, 850, 700, 600, 500, 400, 300, 250, 200, 150, 100, 70, 50, 30, 20, and 10 hPa).

Daily sea surface temperatures and ice distributions were obtained from the National Oceanic and Atmospheric Administration (NOAA) Optimum Interpolation Sea Surface Temperature V2 (Reynolds et al. 2002) at 1.0° × 1.0° horizontal resolution. The National Centers for Environmental Prediction/Department of Energy Reanalysis 2 (NCEP/DOE R2), with a horizontal resolution of 250 km (Kanamitsu et al. 2002), was used for the initial condition of the IsoGSM simulation.

IsoGSM applied a global downscaling technique, which nudged temperature and wind speed at scales larger than 1000 km toward the NCEP/DOE R2 for every 6 h for all sigma levels (Yoshimura and Kanamitsu 2008). This method did not nudge the fields of isotopic composition and humidity. Several validation studies using this method for a limited number of observations have shown that the nudging technique downscales global climate fields in the model reasonably well at various time scales (e.g., Yoshimura et al. 2008; Uemura et al. 2008; Frankenberg et al. 2009; Berkelhammer et al. 2011; Yoshimura et al. 2011).

The seven vapor source regions considered in this study are shown in Fig. 1a: the Bay of Bengal (brown), the Arabian Sea (purple), the northern part of the Indian Ocean (light blue), the southern parts of the Indian Ocean (blue), the Pacific Ocean (magenta), land surface (green), and others (Black). These divisions were based on the published literature regarding the origins of precipitation around the Indian subcontinent (Numaguti 1999; Yoshimura et al. 2004; Breitenbach et al. 2010). We confirmed that the relative contribution of moisture originating from these regions to precipitation over Bangladesh was above 80%, indicating that the selected vapor source regions were sufficient to represent the dominant sources of precipitation in the region.

Backward trajectory analysis

We conducted a 10 day backward trajectory analysis to estimate the probable origins of moisture over Bangladesh. We assumed that air parcels transported at a constant pressure level (850 hPa) and had the same horizontal wind speed. The location of air parcel at the next time step ($\lambda+\Delta\lambda$, $\phi+\Delta\phi$) could be predicted by the following equation:

$$\Delta\lambda = \frac{u(\lambda,\phi)\times 60\times 60\times 30}{2\pi(H+R)\cos\phi}$$
$$\Delta\phi = \frac{v(\lambda,\phi)\times 60\times 60\times 30}{2\pi(H+R)},$$

where λ and ϕ indicate the longitude and latitude of air parcels, respectively, u and v indicate the horizontal and meridional wind speed (m/s), respectively, R is the radius of the earth (6.3773 × 10^6 m), and H is the geopotential height at the constant pressure level. Horizontal and meridional wind speed were taken from 6 h nudged IsoGSM simulations and a Japanese 55 year reanalysis (JRA-55) (Kobayashi et al. 2015; Harada et al. 2016) and interpolated into 0.5° × 0.5° resolution and 30 min intervals using a bilinear method. A backward trajectory analysis was used to estimate the general origin of moisture at the synoptic scale, although it could not reflect local moisture sources (i.e., the recycling of water) (e.g., Breitenbach et al. 2010). In addition, this method assumes an adiabatic process, which is not maintained when an air parcel passes over a convective area. Trajectories were calculated for the sampling days.

Data

We used the NOAA interpolated outgoing longwave radiation (OLR) with a spatial horizontal resolution of

2.5° × 2.5° at a daily time interval in the year 2010 to compare organized convective activity with variations in the isotopic composition of precipitation. A Hovmöller diagram of the averaged OLR was calculated over 85–97.5° E. We identified convective activity using the criterion OLR < 220 W/m^2.

Results

Seasonal variation in the amount and isotopic composition of precipitation

The temporal characteristics of the daily precipitation were different among the three sites (Fig. 2). Heavy precipitation events (> 50 mm/day) occurred frequently at Sylhet from April to October, while they were less frequent at Dhaka and Chittagong. This difference reflects the precipitation system.

The δ^{18}O value of precipitation displayed a clear and similar seasonal variation among the sites. The δ^{18}O values gradually decreased from mid-May and increased from early October, corresponded to the ISM season. In addition, intraseasonal variation (about 2~ 4 weeks) was found, with depletion occurring in mid-May, mid-June, early and late August, and early October. Fujinami et al. (2011) and Murata et al. (2017) showed that the intraseasonal variation of precipitation was dominant around Cherapunji in the northeastern part of India. These results indicate that the δ^{18}O value of precipitation over Bangladesh can be controlled by seasonal and intraseasonal

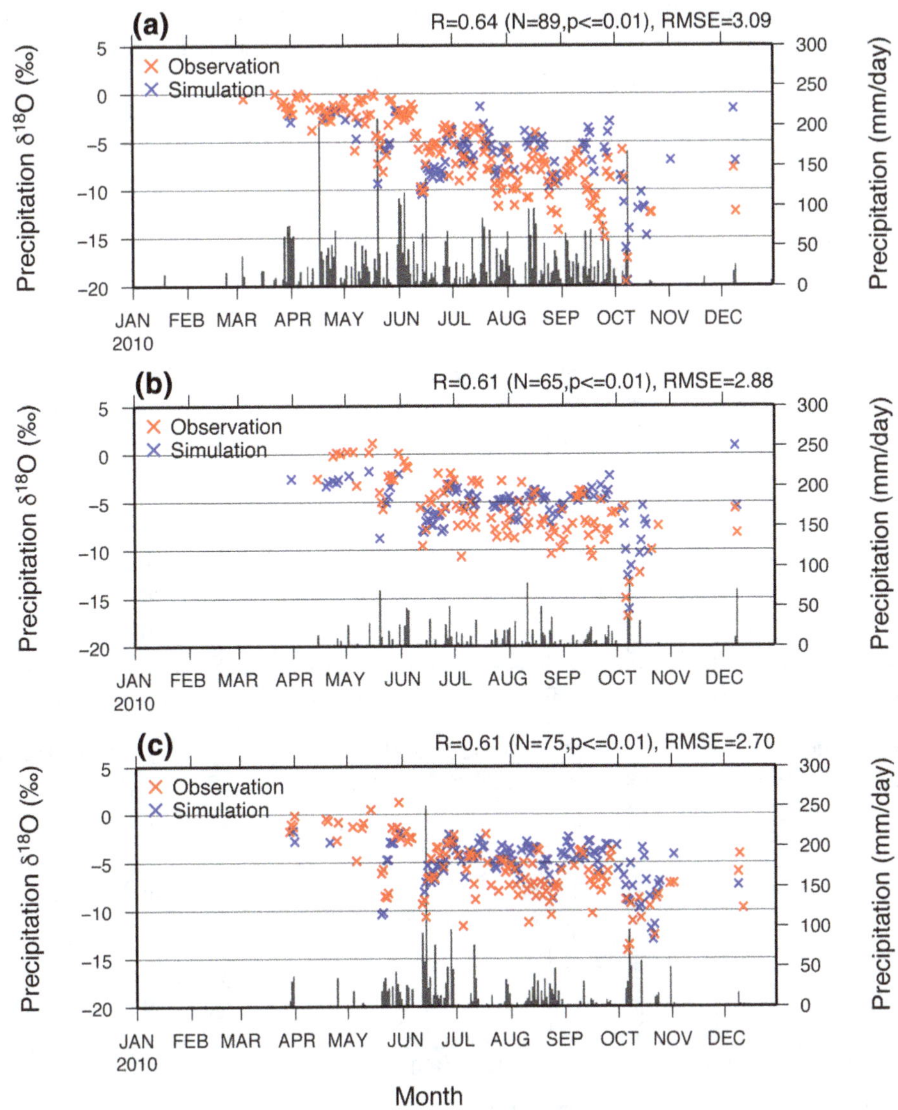

Fig. 2 Daily variations in the amount (bar) and δ^{18}O values (cross) of precipitation in 2010 at Sylhet (**a**), Dhaka (**b**), and Chittagong (**c**). Red and blue crosses indicate observations and IsoGSM simulations, respectively. The correlation coefficient (*R*), number of samples (*N*), probability (*p*), and root mean square error (RMSE) between the observed and simulated δ^{18}O values of precipitation are shown

variations (e.g., the ISM and monsoon low-pressure system).

Validation of IsoGSM simulation

The isotopic composition of precipitation reflects the components of the atmospheric hydrological cycle, including the rainout history of water vapor (e.g., Yoshimura et al. 2003; Ichiyanagi et al. 2005; Fudeyasu et al. 2011) and phase changes in water (e.g., Pfahl et al. 2012). A comparison between the simulated and observed isotopic composition of precipitation is used to validate the atmospheric hydrological cycle in the model, as well as to indirectly assess the reliability of the Eulerian approach (Yoshimura et al. 2003, 2004). Overall, the model successfully replicated the observed seasonal and intraseasonal variations in the isotopic composition over Bangladesh (Fig. 2). The correlation coefficients between the observed and simulated $\delta^{18}O$ values of precipitation at Sylhet, Dhaka, and Chittagong were 0.64, 0.61, and 0.61, respectively, with a significance level exceeding 95%. The root mean square errors (RMSEs) between the observed and simulated $\delta^{18}O$ values of precipitation at Sylhet, Dhaka, and Chittagong were 3.09‰, 2.88‰, and 2.70‰, respectively. This large RMSE was due to the overestimation of the simulated $\delta^{18}O$ values in the depletion events, especially in mid-June, early and late August, and early October.

Figure 3 shows the simulated and observed amount effect (i.e., the relation between the amount and $\delta^{18}O$ value of precipitation) at three sites. We did not find amount effect, except in Chittagong, but the simulation shows the amount effect at the three sites. The correlation coefficients between the simulated amount and $\delta^{18}O$ values of precipitation at Sylhet, Dhaka, and Chittagong were − 0.47, − 0.42, and − 0.26, respectively, with a significance level exceeding 95%. This discrepancy is due to the scale difference between the amount and isotopic composition of precipitation (Kurita et al. 2009). Variation in the isotopic composition of precipitation was mainly associated with large-scale disturbances, whereas that in the local precipitation amount was associated not only with large-scale disturbances but also regional circulation (e.g., nocturnal jet). The model cannot fully resolve the regional circulation due to the rough horizontal resolution, resulting in the stronger simulated amount effect.

A Hovmöller diagram of the NOAA interpolated and simulated OLR are shown in Fig. 4. An organized convective area (OLR < 220 W/m^2) was found at a latitude

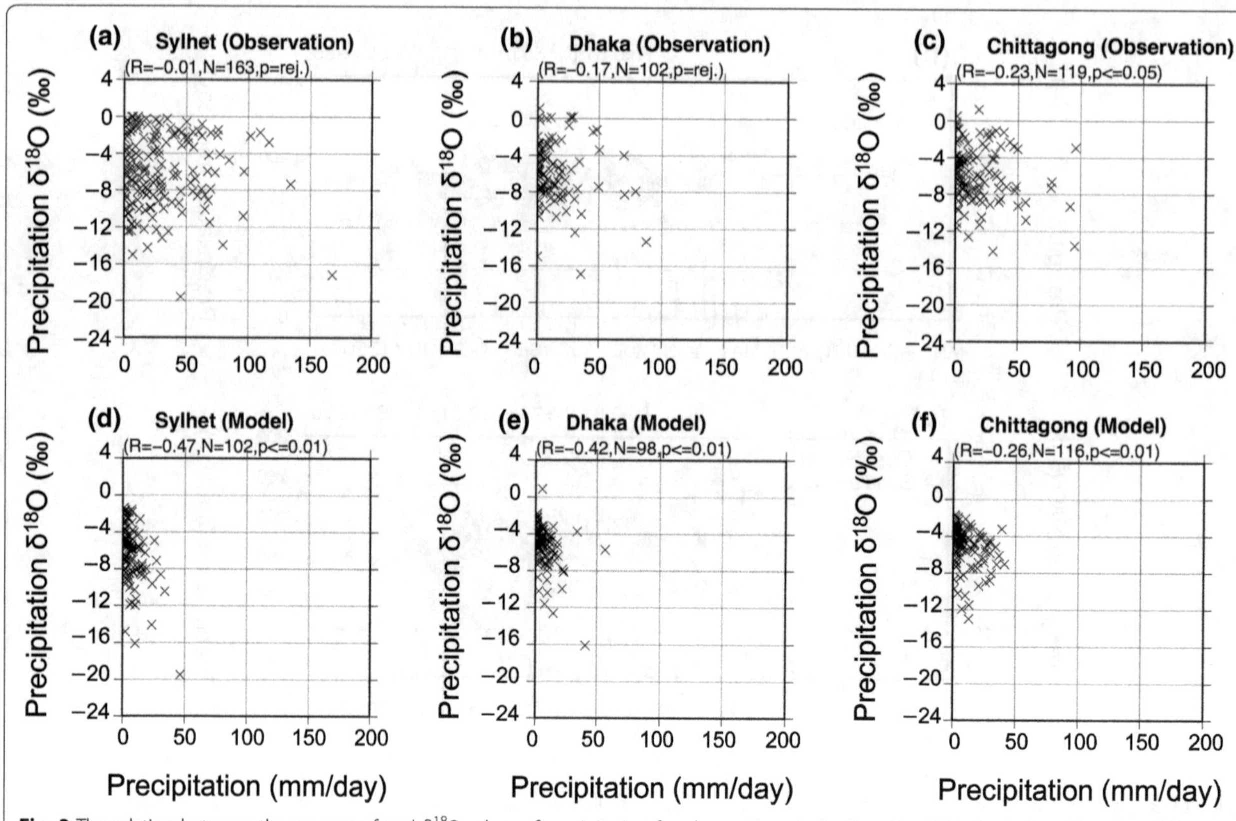

Fig. 3 The relation between the amount of and $\delta^{18}O$ values of precipitation for observations (**a**, **b**, **c**) and IsoGSM simulations (**d**, **e**, **f**) at Sylhet (**a**, **d**), Dhaka (**b**, **e**), and Chittagong (**c**, **f**). Values with a simulated precipitation amount of < 1.0 mm/day were excluded

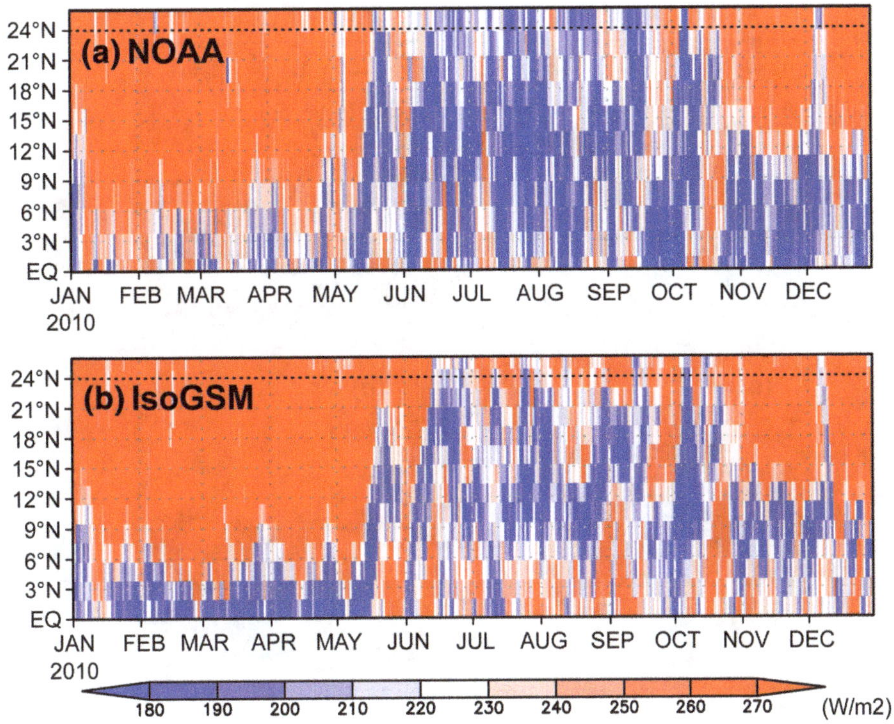

Fig. 4 Hovmöller diagram of the NOAA interpolated (**a**) and simulated (**b**) outgoing longwave radiation (OLR, W/m^2) averaged over 85°E–97.5°E. Black dotted lines represent the latitude of Dhaka

below 10°N until mid-May, after which it moved northward (Fig. 4a). The convective activity frequently appeared around Bangladesh and the Bay of Bengal in late May, mid-June, mid-July, late August, mid-September, and late October. The timing corresponded to depletion events in the $\delta^{18}O$ values of precipitation (Fig. 2). The IsoGSM displayed a similar migration of the organized convective area, which appeared around Bangladesh and the Bay of Bengal in mid-June, late August, early September, and late October (Fig. 4b). However, some discrepancies were found; for example, the convective area failed to appear around Bangladesh and the Bay of Bengal in late May and from late June to early September. Moreover, the values derived from the IsoGSM simulation were still underestimated at latitudes below 24°N in the ISM season. The discrepancy and underestimation cause a small convective precipitation amount around Bangladesh, resulting in an overestimation of the simulated $\delta^{18}O$ values of precipitation because of a clear positive correlation between the $\delta^{18}O$ values of precipitation and the proportional contribution of the convective precipitation amount (Aggarwal et al. 2016).

Origins of moisture and precipitation

Figure 5 shows the spatial patterns of the vertically integrated moisture flux and its divergence averaged for the dry winter, pre-monsoon, ISM, and post-monsoon seasons, as derived from the IsoGSM simulation. The westerly moisture flux was dominant over Bangladesh and the northern part of the Bay of Bengal in the dry winter and pre-monsoon seasons (Fig. 5a, b). A moisture divergence zone appeared in the Bay of Bengal and the Arabian Sea in these seasons, the divergence in the pre-monsoon season was weaker than that in the dry winter monsoon season. The convergence zone was located in the region over the Bay of Bengal, the Indian subcontinent, and the Indochina Peninsula in the ISM season (Fig. 5c). A large moisture flux from the southern part of the Indian Ocean to the Indian subcontinent was dominant. In the post-monsoon season, the convergence zone returned to the southern part of the Bay of Bengal, and a westerly moisture flux gradually became dominant in the northern part of India (Fig. 5d). In the Indonesia peninsula, an easterly moisture flux was dominant in this season.

Figure 6 shows the daily variation in the simulated total precipitable water (TPW) and its origins at Sylhet, Dhaka, and Chittagong. The temporal characteristics of the TPW and its origins were similar among the sites. The TPW increased in mid-March and decreased in mid-October. The oceans close to Bangladesh (the Bay of Bengal and the Arabian Sea) were the dominant source of moisture among the sites in the pre-monsoon

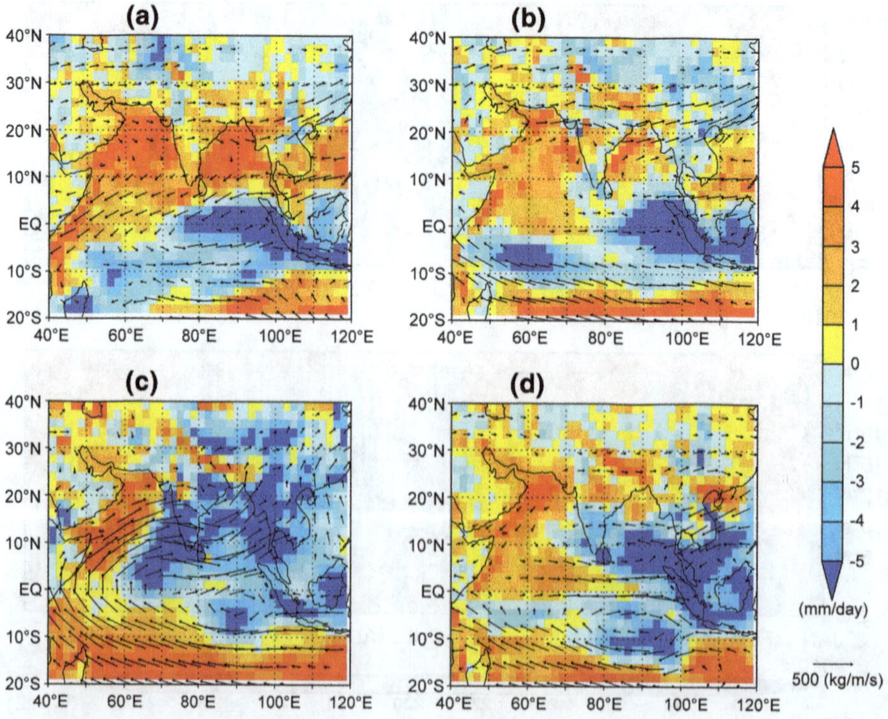

Fig. 5 Spatial distributions of the vertical integrated moisture flux (arrows, kg/m/s) and its divergence (colored, mm/d) in the dry winter (**a**), pre-monsoon (**b**), Indian Summer Monsoon (ISM) (**c**), and post-monsoon (**d**) seasons

season, contributing > 54% of the moisture in the TPW (Table 1). These sources and the Indian Ocean contributed > 61% of the moisture in the TPW in the ISM season. In the post-monsoon season, the contribution of moisture originating from the Bay of Bengal, the Arabian Sea, and the Indian Ocean to the TPW was abruptly reduced (< 12%), whereas moisture originating from the Pacific Ocean and land surface (i.e., the recycling of water) were the main sources of moisture over Bangladesh (> 42%).

The daily variations in the simulated amount and origin of precipitation are shown in Fig. 7. The simulated amount of precipitation was overall underestimated compared with the observed amount (Fig. 2); for example, in the pre-monsoon season at Sylhet, the model could capture heavy precipitation events in late June and early October. Table 2 shows the relative contribution of vapor source regions to precipitation at the sites. The main sources of precipitation in each season were almost identical among the sites. The main sources of precipitation were moisture originating from the oceans close to Bangladesh in the pre-monsoon season (49~60%). The contribution of these moisture sources decreased in the ISM season (36~39%), while the contribution of sources originating from the Indian Ocean increased (26~30%). In the post-monsoon season, the Pacific Ocean and recycling of water were the main sources of precipitation over Bangladesh, contributing approximately half of the precipitation. The recycling of water was the main source of precipitation in the inland region (Sylhet). The contribution of the recycling of water to precipitation seemed to increase from coastal to inland regions due to the high evaporation rate over the Indian subcontinent in the ISM season (4 mm/d; not shown). The contribution of the recycling of water could not be neglected in the inland region, even in the ISM season.

Discussion

We estimated the origins of precipitation using the Eulerian approach combined with IsoGSM. Because the model could capture the observed seasonal and intraseasonal variations in the isotopic composition of precipitation (Fig. 2), the simulated atmospheric hydrological cycle was considered reliable. We summarized the $\delta^{18}O$ values and origins of precipitation in each season. The observed weighted mean of $\delta^{18}O$ value and the total amount of precipitation are summarized in Tables 3. The pre-monsoon season was characterized by high $\delta^{18}O$ values of precipitation (− 3.45 to − 1.82‰) (Fig. 2), with the Arabian Sea and the Bay of Bengal being the main sources of precipitation (Fig. 7). The organized convective activity moved northward and appeared around Bangladesh in the ISM season (Fig. 4). In this season, the $\delta^{18}O$ values of precipitation decreased (− 8.49 to − 7.28‰) (Fig. 2), and the Indian Ocean

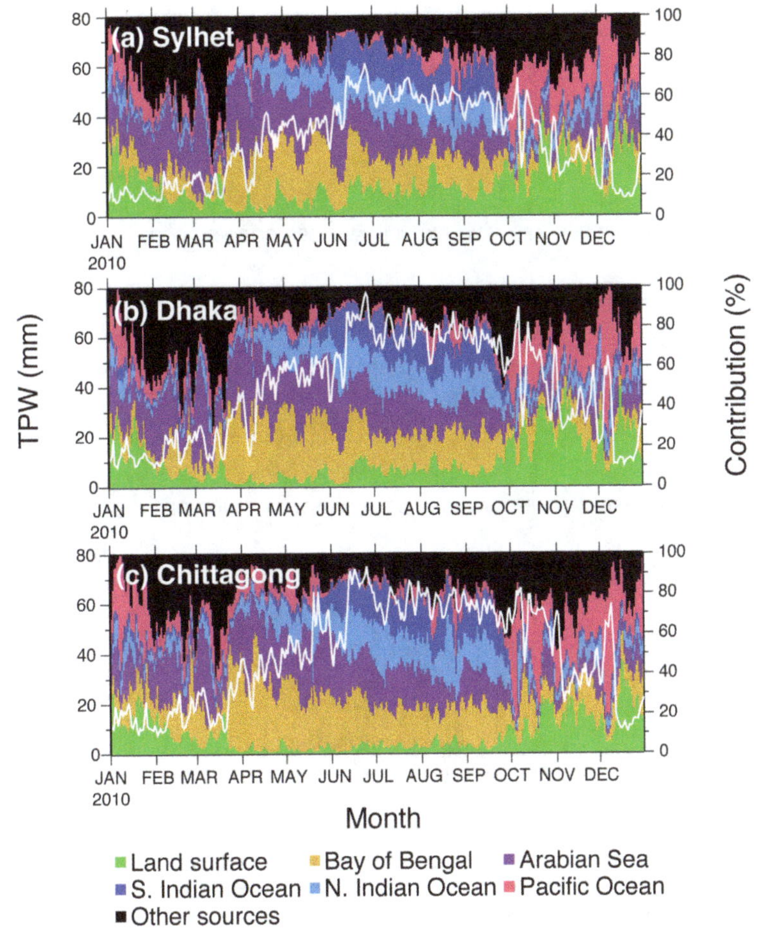

Fig. 6 Daily variation in the total precipitable water (TPW) (white solid line) and its origins at Sylhet (**a**), Dhaka (**b**), and Chittagong (**c**) simulated by IsoGSM. Differences in shading color denote variation in the different vapor source regions shown in Fig. 1a

contributed to precipitation over Bangladesh (Fig. 7). In the post-monsoon season, the convective activity shifted southward (Fig. 4), and the $\delta^{18}O$ values of precipitation were the lowest (– 15.94 to – 9.29‰) (Fig. 2). The Pacific Ocean, the Bay of Bengal, and land surface were the main sources of precipitation in this season (Fig. 7).

Our results suggest that the origins of precipitation and migration of organized convective activity are the main factors controlling the isotopic composition of precipitation over Bangladesh. The observed amount effect was not significant (Fig. 3), indicating that the contribution of the local precipitation amount to the variation in

Table 1 Relative contribution of vapor source regions to moisture in total precipitable water (TPW) for each season

Season	Site	LND	BOB	ARS	NID	SID	PAC
Pre-monsoon	Sylhet		26%	28%	10%		
	Dhaka		29%	28%	11%		
	Chittagong		31%	27%	11%		
Indian Summer Monsoon	Sylhet	14%	16%	16%	14%	15%	
	Dhaka	10%	18%	17%	15%	16%	
	Chittagong		20%	17%	17%	17%	
Post-monsoon	Sylhet	24%					23%
	Dhaka	26%	10%				21%
	Chittagong	17%	12%				25%

Values less than 10% were omitted. *LND* land surface (the recycling of water), *BOB* the Bay of Bengal, *ARS* the Arabian Sea, *NID* the northern part of the Indian Ocean, *SID* the southern part of the Indian Ocean, *PAC* the Pacific Ocean

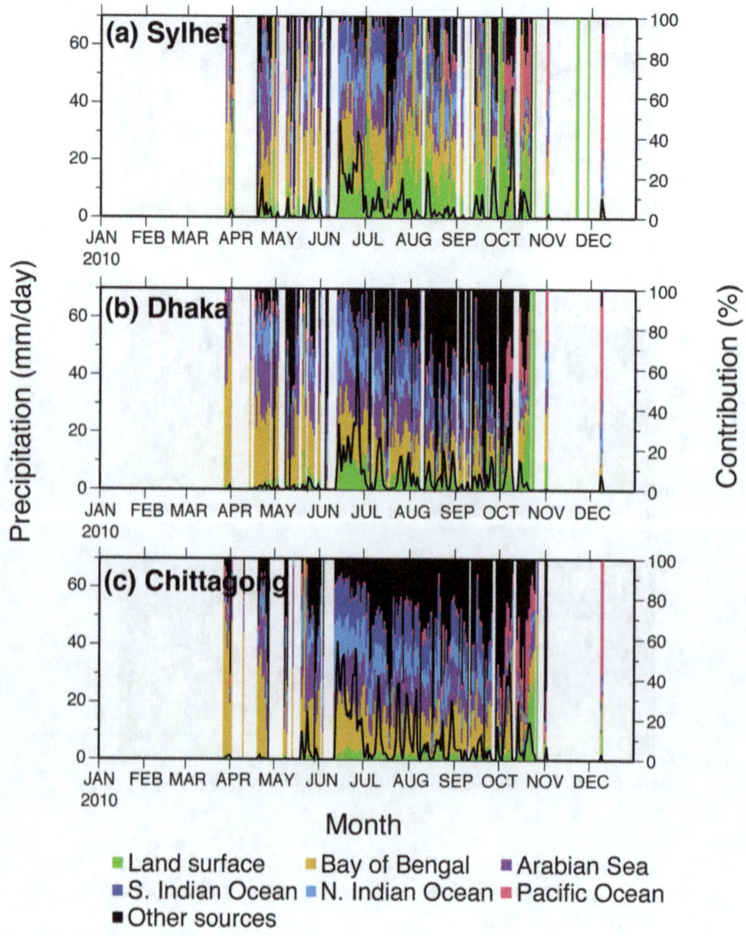

Fig. 7 Daily variation in the simulated amount of precipitation (black line) and its origins at Sylhet (**a**), Dhaka (**b**), and Chittagong (**c**) simulated by IsoGSM. Differences in shading color denote variation in the different vapor source regions shown in Fig. 1a. Values with a simulated precipitation amount of < 1.0 mm/day were excluded

the isotopic composition of precipitation was small. The origin of precipitation was the main factor controlling the isotopic composition of precipitation at the seasonal scale. This indirectly represents the travel distance during transport from the vapor source regions to the sites and the contribution of the recycling of water to precipitation. Long travel distances increase rainout, resulting in decreasing $\delta^{18}O$ values of water vapor and precipitation along the trajectory (e.g., Breitenbach et al. 2010; Midhun et al. 2013). High $\delta^{18}O$ values of precipitation

Table 2 Relative contributions of vapor source regions to precipitation for each season

		LND	BOB	ARS	NID	SID	PAC
Pre-monsoon	Sylhet	10%	34%	26%	11%		
	Dhaka		36%	24%	13%		
	Chittagong		28%	21%	15%		
Indian Summer Monsoon	Sylhet	19%	23%	16%	15%	14%	
	Dhaka	10%	20%	16%	13%	13%	
	Chittagong		22%	16%	15%	15%	
Post-monsoon	Sylhet	24%	12%				37%
	Dhaka	19%	18%				26%
	Chittagong	20%	11%				29%

Values less than 10% were omitted. *LND* land surface (the recycling of water), *BOB* the Bay of Bengal, *ARS* the Arabian Sea, *NID* the northern part of the Indian Ocean, *SID* the southern part of the Indian Ocean, *PAC* the Pacific Ocean

Table 3 Summary of the weighted mean of $\delta^{18}O$ value and the total amount of precipitation at three sites in each monsoon season

Season	Site	$\delta^{18}O$ ‰	Precipitation (mm)
Dry winter	Chittagong	–	0
	Dhaka	–	0
	Sylhet	− 0.08	84
Pre-monsoon	Chittagong	− 3.45	454
	Dhaka	− 1.82	311
	Sylhet	− 2.56	2175
Indian Summer Monsoon	Chittagong	− 7.62	1712
	Dhaka	− 7.28	935
	Sylhet	− 8.49	2909
Post-monsoon	Chittagong	− 9.29	205
	Dhaka	− 13.11	122
	Sylhet	− 15.94	188

For calculation of the weighted means, weighted by the amount of precipitation has been adopted

were observed in the pre-monsoon season, with the main source being identified as the ocean close to Bangladesh (i.e., the Arabian Sea and the Bay of Bengal). In contrast, remote vapor source regions (i.e., the Indian Ocean) contributed to precipitation over Bangladesh in the ISM season. The effect of the travel distance contributed to a reduction in $\delta^{18}O$ values of precipitation over Bangladesh. Additionally, the recycling of water reduced the $\delta^{18}O$ values of precipitation in the ISM and post-monsoon seasons because the values in the recycling of water were lower than those in precipitation. This effect resulted in a seasonally declining trend in the $\delta^{18}O$ values of precipitation (Fig. 2). The migration of organized convective activity caused an intraseasonal variation in the $\delta^{18}O$ values of precipitation around the Indian subcontinent (e.g., Lekshmy et al. 2014, 2015). The observed $\delta^{18}O$ values of precipitation dropped drastically in mid-May, late June, late July, late August, late September, and late October, with these events being linked to the existence of convective activity over Bangladesh (Fig. 4). Lekshmy et al. (2014) found similar depletion events in the $\delta^{18}O$ values of precipitation across nine sites in southern India. The migration of the convective activity affected the $\delta^{18}O$ values of precipitation around the Indian subcontinent. The origins of precipitation and migration of organized convective activity were important factors for better understanding and interpreting the terrestrial proxies of past precipitation around the Indian subcontinent.

Our simulation showed drastic changes in the vapor source regions in late March (pentad 17), early June (pentad 32), and late September (pentad 54). To investigate the factors that induce drastic changes around Bangladesh, Fig. 8 shows the spatial distribution of the difference between the amount of moisture originating from the ocean (the Bay of Bengal, the Arabian Sea, and the Indian Ocean) and the land surface divided by the TPW. Reddish shading indicated that ocean moisture was dominant. In pentads before the first drastic change event (pentad 16), the oceanic moisture was less than the land moisture, but this increased due to the transport of the oceanic moisture by a southerly moisture flux in pentad 17. The second drastic change event occurred in early June (pentad 32). The event corresponded to the time when the strong southwesterly moisture flux arrived in Bangladesh. The third drastic change event occurred in late September (pentad 54), when the land moisture gradually increased from pentad 53 to 55. The southwesterly moisture flux gradually weakened, while the easterly moisture flux was enhanced in pentad 55. These drastic change events were related to changes in atmospheric circulation patterns.

Previous studies have determined the timing of the monsoon onset and withdrawal using variations in the local precipitation amount (e.g., Matsumoto 1997). However, determining this timing is difficult in Bangladesh, especially in Sylhet (Fig. 2a). Some previous studies identified the onset and withdrawal of the monsoon using the origins of moisture in the Asian monsoon region (e.g., Yoshimura et al. 2004; Ichiyanagi et al. 2005; Hiraoka et al. 2011). For example, Ichiyanagi et al. (2005) defined the monsoon withdrawal in Thailand to be when the amount of moisture originating from the Java Sea and the Pacific Ocean exceeded that originating from the Indian Ocean. In Bangladesh, we found clear seasonal variation in the origins of moisture and precipitation (Figs. 4 and 5), and the timing of monsoon onset and withdrawal could be identified by comparing the amounts of the ocean and land moisture (e.g., Fig. 8). In addition, seasonal variations in the isotopic composition of precipitation also have the potential to identify the timing of monsoon onset and withdrawal because the ISM season is characterized by a seasonally declining trend in the $\delta^{18}O$ values of precipitation. Isotopic composition and the origins of precipitation are useful for identifying the timing of monsoon onset and withdrawal and for estimating its duration and long-term trend.

To compare the estimations of the origins of moisture based on the Lagrangian and Eulerian approaches, we conducted a backward trajectory analysis and investigated the *E–P* distribution for each season (Fig. 9). The trajectories of the IsoGSM simulation and JRA-55 were generally similar in the pre-monsoon and ISM seasons. In the pre-monsoon season, most trajectories originated from the region around the Red Sea and passed over the northern part of India or the south coast of peninsular India (Fig. 9a). A positive *E–P* value was found at the

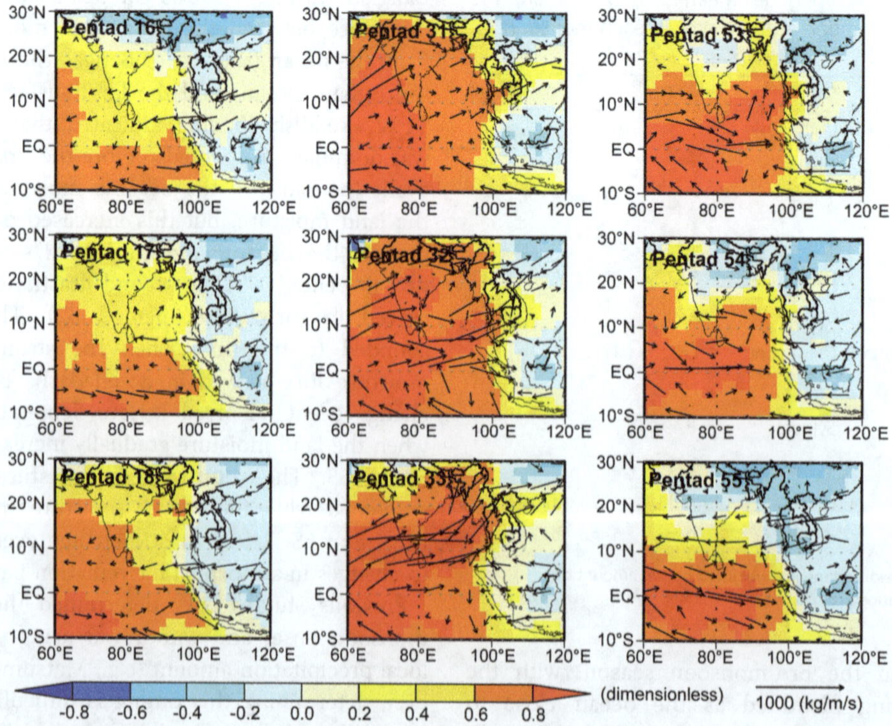

Fig. 8 The spatial distributions of the vertically integrated moisture flux and differences between the amount of moisture originating from the ocean (the Arabian Sea, the Bay of Bengal, the Indian Ocean) and land surface divided by the TPW in late March (pentad 16–18), early June (pentad 33–35), and late September (pentad 53–55). Red (blue) shaded indicated that moisture originating from the ocean (the land surface) was dominant

Bay of Bengal, the Arabian Sea, and the southern part of the Indian Ocean, while a negative $E–P$ value was found in the northern part of the Indian Ocean (Fig. 9b). The main sources of moisture were identified as the Bay of Bengal and the Arabian Sea in this season. In the ISM season, air parcels transported from the western part of the Indian Ocean passed over the Arabian Sea and the Bay of Bengal (Fig. 9c). A positive $E–P$ value was recognized in the western part of the Arabian Sea and the southern part of the Indian Ocean, and these oceans were determined to be the main sources of moisture in the ISM season (Fig. 9d). Because the value in the Bay of Bengal was less than − 2 mm/day, this region was clearly not the main source of moisture in the ISM season. In the post-monsoon season, most trajectories originated from East Asia and the Pacific Ocean in the JRA-55 simulation, whereas some trajectories originated from the northern part of the Indian Ocean and the Bay of Bengal in the IsoGSM simulation (Fig. 9e). The Arabian Sea and the western part of the Indian Ocean can be the main source of moisture over Bangladesh because a high positive $E–P$ value was found in these regions, while the south coast of Vietnam, the Indian Ocean, and the Pacific Ocean are not the main vapor source regions over Bangladesh because a negative $E–P$ value was found in these regions (Fig. 9f).

The origins of moisture estimated by the Eulerian and Lagrangian approaches were almost identical, except for the post-monsoon season. In the post-monsoon season, the Eulerian approach identified the Pacific Ocean and recycling of water as the main sources of moisture, while the Lagrangian approach identified the Pacific Ocean and the Indian Ocean as the main sources of moisture. Although we could not directly identify the main sources of moisture based on any observations, this discrepancy likely arose from the inability of the Lagrangian approach to explain the local moisture source. In addition, because the approach adopted the assumption of an adiabatic process, it may be not adequate for analyzing the isotopic composition of precipitation. The Eulerian approach has the advantage of enabling investigation of the effect of recycling of water on the isotopic composition of precipitation.

Although the model could capture the seasonal and intraseasonal variations in the isotopic composition of precipitation reasonably well, there was room for improvement in the calculation of simulated isotopic composition. For example, the simulated precipitation amount was underestimated due to the insufficient horizontal resolution of IsoGSM for capturing the local precipitation system over Bangladesh. Recently, a regional isotope circulation model was developed (Yoshimura et al. 2010), which can resolve line-type precipitation systems over the western part of the

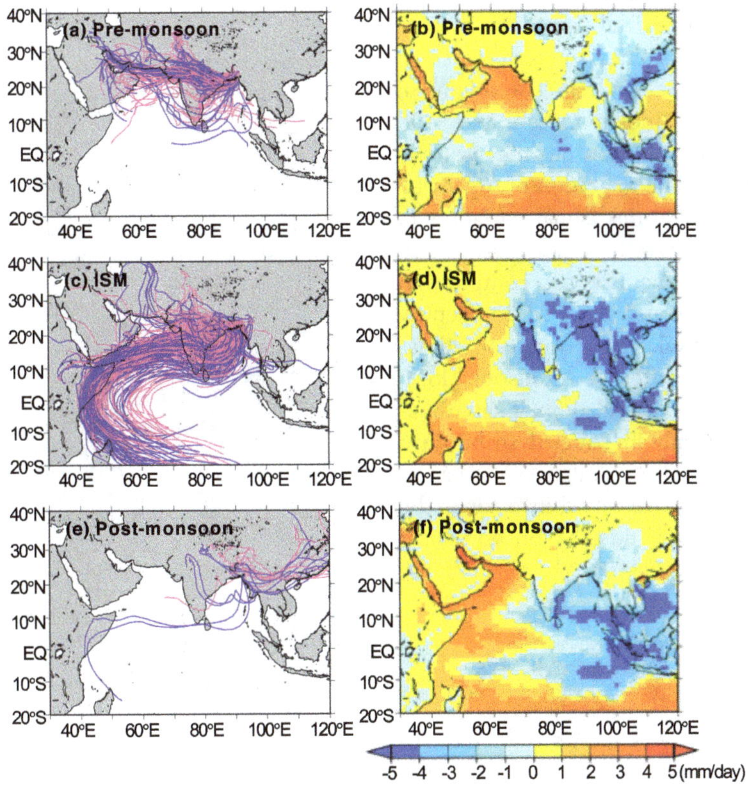

Fig. 9 Backward trajectory analysis for every precipitation sampling day in the pre-monsoon (**a**), Indian Summer Monsoon (ISM) (**c**), and post-monsoon seasons (**e**). Red and blue solid lines indicate the results of an IsoGSM simulation and JRA-55, respectively. The spatial pattern of the averaged *E–P* (mm/day) in the pre-monsoon (**b**), ISM (**d**), and post-monsoon seasons (**f**) derived from an IsoGSM simulation

USA (Yoshimura et al. 2010) and orographic winter precipitation patterns over Japan (Tanoue et al. 2016, 2017). The Eulerian approach combined with a regional isotope circulation model can obtain the origins of moisture and precipitation at a fine horizontal resolution (10 to 180 km). The RMSE between the observed and simulated isotopic composition of precipitation in the coastal region (i.e., Chittagong) was smaller than that in the inland region (i.e., Sylhet), with the large RMSE in the inland region likely caused by the recycling of water. The IsoGSM assumes no fractionation when water vapor is evaporated from the land surface (Yoshimura et al. 2008). Assuming isotopic fractionation, the isotopic values could be low because the isotopic enrichment of water during evaporation is lower than that during transpiration. This suggests that understanding the isotopic variation associated with the recycling of water could improve the overestimation of the isotopic enrichment of precipitation simulated by IsoGSM in the ISM and post-monsoon seasons over Bangladesh.

Summary and conclusions

We observed the isotopic composition of precipitation at three sites (Sylhet, Dhaka, and Chittagong) over Bangladesh in 2010 and estimated the origin of precipitation using the Eulerian approach combined with IsoGSM.

The temporal characteristics of the observed precipitation amount differed among sites, while the observed $\delta^{18}O$ values of precipitation displayed clear and similar seasonal variation. Across the region, the intraseasonal variation in the $\delta^{18}O$ values of precipitation was found, which was linked to the migration of organized convective activity. The model could capture seasonal and intraseasonal variations in the observed $\delta^{18}O$ values of precipitation reasonably well; however, the overestimation of the simulated $\delta^{18}O$ values was found in the ISM and post-monsoon seasons, especially in the inland region (i.e., Sylhet).

The main sources of precipitation were the Bay of Bengal and the Arabian Sea in the pre-monsoon season. In the ISM season, these sources gradually decreased, while the Indian Ocean contributed more to precipitation over Bangladesh. These oceanic moisture sources contributed less to precipitation in the post-monsoon season, as the Pacific Ocean, the Bay of Bengal, and land surface were the main sources of precipitation at that time.

Our results indicated that the main sources of precipitation and the migration of organized convective activity were the dominant factors controlling the isotopic composition of precipitation over Bangladesh. The origins of precipitation, particularly the Indian Ocean and land surface, controlled seasonal variation in the isotopic composition of precipitation. The migration of organized convective activity controlled intraseasonal variation in the isotopic composition.

Our results indicated that the isotopic composition and origins of precipitation could be used to identify monsoon onset and withdrawal. These variations drastically changed in the timing after pre-monsoon, ISM, and post-monsoon seasons over Bangladesh. These drastic changes were found not only in Bangladesh but also in the Asian monsoon region (e.g., Araguás-Araguás et al. 1998; Ichiyanagi et al. 2005); we would investigate the relation between these variations and the monsoon onset and withdrawal for future study.

Abbreviations

AGCM: Atmospheric General Circulation Model; ISM: Indian summer monsoon; IsoGSM: Isotope-enabled Global Spectral Model; NCEP/DOE R2: National Centers for Environmental Prediction/Department of Energy Reanalysis 2; NOAA: The National Oceanic and Atmospheric Administration; OLR: Outgoing Longwave Radiation; TPW: Total precipitable water; USGS: The United States Geological Survey

Acknowledgements

We thank Dr. Yusuke Yamane, Dr. Fumie Murata, Dr. Azusa Fukushima, Mr. Sayeed Ahmed Choudhury, Mr. Quamrul Alam, and staffs of the Bangladesh Meteorological Department for their assistance with our observations.

Funding

This study was supported by a Japan Society for the Promotion of Science (JSPS) KAKENHI Grant Numbers 21403005 and 23231057, and the Environment Research and Technology Development Fund (S-5) of the Ministry of the Environment, Japan.

This work was conducted by the joint research program of the Institute for Space-Earth Environmental Research (ISEE), Nagoya University, Japan.

Authors' contributions

MT carried out the analysis and wrote the manuscript. KI planned the observations. MK, TT, and TH helped conduct the observations. KY supported the model experiment. All authors read and approved the final manuscript.

Competing interests

The authors declare that they have no competing interests.

Author details

[1]Department of Civil Engineering, Shibaura Institute of Technology, 3-7-5 Toyosu, Koto-ku, Tokyo 135-8548, Japan. [2]Faculty of Advanced Science and Technology, Kumamoto University, 2-39-1, Kurokami, Chuo-ku, Kumamoto 860-8555, Japan. [3]Institute of Industrial Science Chiba Experimental Station, The University of Tokyo, 5-1-5 Kashiwanoha, Kashiwa-city, Chiba 277-8574, Japan. [4]Institute of Industrial Science, The University of Tokyo, 4-6-1 Komaba, Meguro-ku, Tokyo 153-8505, Japan. [5]Faculty of Education, Kagawa University, 1-1, Saiwaicho, Takamatsu, Kagawa 760-8522, Japan. [6]Center for Southeast Asian Studies, Kyoto University, 46, Shimoadachi-cho, Yoshida, Sakyo-ku, Kyoto 606-8501, Japan.

References

Aggarwal PK, Romatschke U, Araguas-Araguas L, Belachew D, Longstaffe FJ, Berg P, Schumacher C, Funk A (2016) Proportions of convective and stratiform precipitation revealed in water isotope ratios. Nat Geosci 9:624–629. https://doi.org/10.1038/ngeo2739

Araguás-Araguás L, Froehlich K, Rozanski K (1998) Stable isotope composition of precipitation over Southeast Asia. J Geophys Res 103:28721–28742. https://doi.org/10.1029/98JD02582

Belgaman HA, Ichiyanagi K, Tanoue M, Suwarman R, Yoshimura K, Mori S, Kurita N, Yamanaka MD, Syamsudin F (2016) Intraseasonal variability of $\delta^{18}O$ of precipitation over the Indonesian maritime continent related to the Madden–Julian oscillation. SOLA 12:192–197. https://doi.org/10.2151/sola.2016-039

Berkelhammer M, Stott L, Yoshimura K, Johnson K, Sinha A (2011) Synoptic and mesoscale controls on the isotopic composition of precipitation in the western United States. Clim Dyn 38:433–454. https://doi.org/10.1007/s00382-011-1262-3

Bhattacharya SK, Froehlich K, Aggarwal PK, Kulkarni KM (2003) Isotopic variation in Indian monsoon precipitation: records from Bombay and New Delhi. Geophys Res Lett 30:2285. https://doi.org/10.1029/2003GL018453

Breitenbach SFM, Adkins JF, Meyer H, Marwan N, Kumar KK, Haug GH (2010) Strong influence of water vapor source dynamics on stable isotopes in precipitation observed in southern Meghalaya, NE India. Earth Planet Sci Lett 292:212–220. https://doi.org/10.1016/j.epsl.2010.01.038

Chakraborty S, Sinha N, Chattopadhyay R, Sengupta S, Mohan PM, Datye A (2016) Atmospheric controls on the precipitation isotopes over the Andaman Islands, Bay of Bengal. Sci Rep 6:19555. https://doi.org/10.1038/srep19555

Chou MD, Suarez MJ (1994) An Efficient Thermal Infrared Radiation Parameterization for Use in General Circulation Models. NASA Technical Memorandum No. 104606, Vol. 3, p. 85. http://citeseerx.ist.psu.edu/viewdoc/download?doi=10.1.1.26.4850&rep=rep1&type=pdf

Dansgaard W (1964) Stable isotopes in precipitation. Tellus 16:436–468

Datta PS, Tyagi SK, Chandrasekharan H (1991) Factors controlling stable isotope composition of rainfall in New Delhi, India. J Hydrol 128:223–236. https://doi.org/10.1016/0022-1694(91)90139-9

Ek MB, Mitchell KE, Lin Y, Rogers E, Grunmann P (2003) Implementation of Noah land surface model advances in the National Centers for Environmental Prediction operational mesoscale Eta model. J Geophys Res 108:8851. https://doi.org/10.1029/2002JD003296

Epstein S, Mayeda T (1953) Variations of the $^{18}O/^{16}O$ ratios in natural waters. Geochim Cosmochim Acta 4:213–214

Fleitmann D, Burns SJ, Mudelsee M, Neff U, Kramers J, Mangini A, Matter A (2003) Holocene forcing of the Indian monsoon recorded in a stalagmite from southern Oman. Science 300:1737–1739. https://doi.org/10.1126/science.1083130

Frankenberg C, Yoshimura K, Warneke T, Aben I, Butz A, Deutscher N, Griffith D, Hase F, Notholt J, Schneider M, Schrijver H, Röckmann T (2009) Dynamic processes governing lower-tropospheric HDO/H_2O ratios as observed from space and ground. Science 325:1374–1377. https://doi.org/10.1126/science.1173791

Fudeyasu H, Ichiyanagi K, Yoshimura K, Mori S, Hamada JI, Sakurai N, Yamanaka MD, Matsumoto J, Syamsudin F (2011) Effects of large-scale moisture transport and mesoscale processes on precipitation isotope ratios observed at Sumatera, Indonesia. J Meteor Soc Japan 89A:49–59 https://doi.org/10.2151/jmsj.2011-A03

Fujinami H, Hatsuzuka D, Yasunari T, Hayashi T, Terao T, Murata F, Kiguchi M, Yamane Y, Matsumoto J, Islam MN, Habib A (2011) Characteristic intraseasonal oscillation of rainfall and its effect on interannual variability over Bangladesh during boreal summer. Int J Climatol 31:1192–1204 https://doi.org/10.1002/joc.2146

Fujinami H, Sato T, Kanamori H, Murata F (2017) Contrasting features of monsoon precipitation around the Meghalaya plateau under westerly and easterly regimes. J Geophys Res Atmos 122:9591–9610. https://doi.org/10.1002/2016JD026116

Guimberteau M, Laval K, Perrier A, Polcher J (2012) Global effect of irrigation and its impact on the onset of the Indian summer monsoon. Clim Dyn 39:1329–1348. https://doi.org/10.1007/s00382-011-1252-5

Harada Y, Kamahori H, Kobayashi C, Endo H, Kobayashi S, Ota Y, Onoda H, Onogi K, Miyaoka K, Takahashi K (2016) The JRA-55 reanalysis: representation of atmospheric circulation and climate variability. J Meteor Soc Japan 94:269–302. https://doi.org/10.2151/jmsj.2016-015

Hatsuzuka D, Yasunari T, Fujinami H (2014) Characteristics of low pressure systems associated with intraseasonal oscillation of rainfall over Bangladesh during boreal summer. Mon Weather Rev 142:4758–4774. https://doi.org/10.1175/MWR-D-13-00307.1

Hiraoka A, Kawamura R, Ichiyanagi K, Tanoue M, Yoshimura K (2011) Water origins over central and southern Japan during the early summer rainy season as simulated with an isotope circulation model. SOLA 7:141–144. https://doi.org/10.2151/sola.2011-036

Hong SY, Pan HL (1996) Nonlocal boundary layer vertical diffusion in a medium-range forecast model. Mon Weather Rev 124:2322–2339. https://doi.org/10.1175/1520-0493(1996)124<2322:NBLVDI>2.0.CO;2

Hoque R, Matsumoto J, Hirano J (2011) Climatological characteristics of monsoon seasonal transitions over Bangladesh. Geogr Reports Tokyo Metrop Univ 46:31–41

Horita J, Ueda A, Mizukami K, Takatori I (1989) Automatic δD and $\delta^{18}O$ analysis of multi-water samples using H_2- and CO_2- water equilibration methods with common equilibration set-up. Appl Radiat Isot 40:801–805

Ichiyanagi K, Yoshimura K, Yamanaka MD (2005) Validation of changing water origins over Indochina during the withdrawal of the Asian monsoon using stable isotopes. SOLA 1:113–116. https://doi.org/10.2151/sola.2005-030

Islam N, Uyeda H (2008) Vertical variations of rain intensity in different rainy periods in and around Bangladesh derived from TRMM observations. Int J Climatol 28:273–279. https://doi.org/10.1002/joc.1585

Jouzel J, Merlivat L (1984) Deuterium and oxygen 18 in precipitation: modeling of the isotopic effects during snow formation. J Geophys Res 89:11749–11757. https://doi.org/10.1029/JD089iD07p11749

Kanamitsu M, Ebisuzaki W, Woollen J, Yang SK, Hnilo JJ, Fiorino M, Potter GL (2002) NCEP-DEO AMIP-II reanalysis (R-2). Bull Am Meteorol Soc 83:1631–1643

Kobayashi S, Ota Y, Harada Y, Ebita A, Moriya M, Onoda H, Onogi K, Kamahori H, Kobayashi C, Endo H, Miyaoka K, Takahashi K (2015) The JRA-55 reanalysis: general specifications and basic characteristics. J Meteor Soc Japan 93:5–48. https://doi.org/10.2151/jmsj.2015-001

Kudo T, Kawamura R, Hirata H, Ichiyanagi K, Tanoue M, Yoshimura K (2014) Large-scale vapor transport of remotely evaporated seawater by a Rossby wave response to typhoon forcing during the Baiu/Meiyu season as revealed by the JRA-55 reanalysis. J Geophys Res Atmos 119:8825–8838. https://doi.org/10.1002/2014JD021999

Kurita N, Ichiyanagi K, Matsumoto J, Yamanaka MD, Ohata T (2009) The relationship between the isotopic content of precipitation and the precipitation amount in tropical regions. J Geochemical Explor 102:113–122. https://doi.org/10.1016/j.gexplo.2009.03.002

Lekshmy PR, Midhun M, Ramesh R (2015) Spatial variation of amount effect over peninsular India and Sri Lanka: role of seasonality. Geophys Res Lett 42:5500–5507. https://doi.org/10.1002/2015GL064517

Lekshmy PR, Midhun M, Ramesh R, Jani RA (2014) ^{18}O depletion in monsoon rain relates to large scale organized convection rather than the amount of rainfall. Sci Rep 4:5661 https://doi.org/10.1038/srep05661

Majoube M (1971a) Oxygen-18 and deuterium fractionation between water and steam (in French). J Chim Phys Phys Chim Biol 68:1423–1436

Majoube M (1971b) Fractionation in O-18 between ice and water vapor (in French). J Chim Phys Phys Chim Biol 68:625–636

Matsumoto J (1997) Seasonal transition of summer rainy season over land adjacent monsoon region. Adv Atmos Sci 14:231–245

Merlivat L, Jouzel J (1979) Global climatic interpretation of the deuterium – oxygen 18 relationship for precipitation. J Geophys Res 84: 5029–5033. https://doi.org/10.1029/JC084iC08p05029

Midhun M, Lekshmy PR, Ramesh R (2013) Hydrogen and oxygen isotopic compositions of water vapor over the Bay of Bengal during monsoon. Geophys Res Lett 40:6324–6328

Midhun M, Ramesh R (2016) Validation of $\delta^{18}O$ as a proxy for past monsoon rain by multi-GCM simulations. Clim Dyn 46:1371–1385

Moorthi S, Suarez MJ (1992) Relaxed Arakawa–Schubert: a parameterization of moist convection for general circulation models. Mon Weather Rev 120:978–1002. https://doi.org/10.1175/1520-0493(1992)120<0978:RASAPO>2.0.CO;2

Murata F, Terao T, Fujinami H, Hayashi T, Asada H, Matsumoto J, Syiemlieh HJ (2017) Dominant synoptic disturbance in the extreme rainfall at Cherrapunji, Northeast India, based on 104 years of rainfall data (1902-2005). J Clim 30: 8237–8251. https://doi.org/10.1175/JCLI-D-16-0435.1

Numaguti A (1999) Origin and recycling processes of precipitating water over the Eurasian continent: experiments using an atmospheric general circulation model. J Geophys Res 104:1957–1972

Ono M, Takahashi HG (2016) Seasonal transition of precipitation characteristics associated with land surface conditions in and around Bangladesh. J Geophys Res Atmos 121:11190–11200 https://doi.org/10.1002/2015JD024524

Pfahl S, Wernli H, Yoshimura K (2012) The isotopic composition of precipitation from a winter storm-a case study with the limited-area model COSMOiso. Atmos Chem Phys 12:1629–1648. https://doi.org/10.5194/acp-12-1629-2012

Reynolds RW, Rayner NA, Smith TM, Stokes DC, Wang W (2002) An improved in situ and satellite SST analysis for climate. J Climate 15:1609–1625. https://doi.org/10.1175/1520-0442(2002)015<1609:AIISAS>2.0.CO;2

Risi C, Bony S, Vimeux F (2008) Influence of convective processes on the isotopic composition ($\delta^{18}O$ and δD) of precipitation and water vapor in the tropics: 2. Physical interpretation of the amount effect. J Geophys Res 113:D19306. https://doi.org/10.1029/2008JD009943

Sato T, Tsujimura M, Yamanaka T, Iwasaki H, Sugimoto A, Sugita M, Oyunbaatar D (2007) Water sources in semiarid Northeast Asia as revealed by field observations and isotope transport model. J Geophys Res 112:D17112. https://doi.org/10.1029/2006JD008321

Sengupta S, Sarkar A (2006) Stable isotope evidence of dual (Arabian Sea and Bay of Bengal) vapour sources in monsoonal precipitation over North India. Earth Planet Sci Lett 250:511–521. https://doi.org/10.1016/j.epsl.2006.08.011

Shinoda T, Uyeda H, Yoshimura K (2005) Structure of moist layer and sources of water over the southern region far from the Meiyu/Baiu front. J Meteor Soc Japan 83:137–152

Stewart MK (1975) Stable isotope fractionation due to evaporation and isotopic exchange of falling water drops. J Geophys Res 80:1133–1146. https://doi.org/10.1029/JC080i009p01133

Suwarman R, Ichiyanagi K, Tanoue M, Yoshimura K, Mori S, Yamanaka MD, Syamsudin F (2013) The variability of stable isotopes and water origin of precipitation over the Maritime Continent. SOLA 9:74–78. https://doi.org/10.2151/sola.2013-017

Tanoue M, Ichiyanagi K, Yoshimura K (2016) Verification of the isotopic composition of precipitation simulated by a regional isotope circulation model over Japan. Isotopes Environ Health Stud:52. https://doi.org/10.1080/10256016.2016.1148695

Tanoue M, Ichiyanagi K, Yoshimura K, Shimada J, Hirabayashi Y (2017) Estimation of the isotopic composition and origins of winter precipitation over Japan using a regional isotope circulation model. J Geophys Res Atmos. https://doi.org/10.1002/2017JD026751

Uemura R, Matsui Y, Yoshimura K, Motoyama H, Yoshida N (2008) Evidence of deuterium excess in water vapor as an indicator of ocean surface conditions. J Geophys Res 113:D19114. https://doi.org/10.1029/2008JD010209

Wang Y, Cheng H, Edwards RL, Kong X, Shao X, Chen S, Wu J, Jiang X, Wang X, An Z (2008) Millennial- and orbital-scale changes in the East Asian monsoon over the past 224,000 years. Nature 451:1090–1093. https://doi.org/10.1038/nature06692

Wang YJ, Cheng H, Edwards RL, He YQ, Kong XG, An ZS, Wu JY, Kelly MJ, Dykoski CA, Li XD (2005) The Holocene Asian monsoon: links to solar changes and North Atlantic climate. Science 308:854–857. https://doi.org/10.1126/science.1106296

Yamashima R, Matsumoto J, Takata K, Takahashi HG (2014) Impact of historical land-use changes on the Indian summer monsoon onset. Int J Climatol 35: 2419–2430. https://doi.org/10.1002/joc.4132

Yancheva G, Nowaczyk NR, Mingram J, Dulski P, Schettler G, Negendank JFW, Liu J, Sigman DM, Peterson LC, Haug GH (2007) Influence of the intertropical convergence zone on the East Asian monsoon. Nature 445:74–77. https://doi.org/10.1038/nature05431

Yoshimura K, Frankenberg C, Lee J, Kanamitsu M, Worden J, Röckmann T (2011) Comparison of an isotopic atmospheric general circulation model with new quasi-global satellite measurements of water vapor isotopologues. J Geophys Res 116:D19118. https://doi.org/10.1029/2011JD016035

Yoshimura K, Kanamitsu M (2008) Dynamical global downscaling of global reanalysis. Mon Weather Rev 136:2983–2998. https://doi.org/10.1175/2008MWR2281.1

Yoshimura K, Kanamitsu M, Dettinger M (2010) Regional downscaling for stable water isotopes: a case study of an atmospheric river event. J Geophys Res 115:D18114. https://doi.org/10.1029/2010JD014032

Yoshimura K, Kanamitsu M, Noone D, Oki T (2008) Historical isotope simulation using reanalysis atmospheric data. J Geophys Res 113:D19108 doi: 1029.10/2008JD010074

Yoshimura K, Oki T, Ohte N, Kanae S (2003) A quantitative analysis of short-term $\delta^{18}O$ variability with a Rayleigh-type isotope circulation model. J Geophys Res 108:4647. https://doi.org/10.1029/2003JD003477

Yoshimura K, Oki T, Ohte N, Kanae S (2004) Colored moisture analysis estimates of variations in 1998 Asian monsoon water sources. J Meteor Soc Japan 82: 1315–1329

10

Potential impact of sea surface temperature on rainfall over the western Philippines

Julie Mae B. Dado[1*] and Hiroshi G. Takahashi[1,2]

Abstract

The study used a 5 km-resolution regional climate model, the Advanced Research Weather Research and Forecasting Model, to quantify the potential impact of sea surface temperature (SST) west of the Philippines on summer monsoon rainfall on the northwestern coast of the country. A set of control simulations (CTL) driven by ERA-Interim reanalysis data and the monthly National Oceanic and Atmospheric Administration Optimum Interpolation SST dataset was performed for the months of June to August of 1982–2012. A second set of simulations driven by climatological SST values was performed for the same period. The difference between these two simulation sets is analyzed to determine the sensitivity of rainfall to interannual variations in local SST, not remote SST, via a regional climate model. The CTL simulations represented spatial and temporal variations in rainfall well, yielding realistic climatological rainfall values with high spatial correlations with observations. The interannual correlation of monthly rainfall over the northwestern region of the Philippines was also high when compared to observations. The results showed that positive SST anomalies west of the Philippines induced positive rainfall anomalies in the northwestern Philippines via an increase in latent heat flux from the sea surface, implying that summer monsoon rainfall in the northwestern Philippines is modulated by interannual variations in SST west of the Philippines. The impact of SST on latent heat flux and rainfall were 20–40%, greatly exceeding the 7% approximation from the Clausius–Clapeyron equation, which can be explained by the enhancement of low-level winds and a weak warming of surface air temperature over the ocean.

Keywords: Summer monsoon, Rainfall, Sea surface temperature, Regional climate model

Introduction

The Philippines is located in an area primarily affected by the Asian summer monsoon. Specifically, the northwestern side of the country is affected by this system, which is known locally as the southwest monsoon (SWM). The SWM is part of the western North Pacific Monsoon within the Asian-Pacific summer monsoon system (Wang and LinHo 2002). The subsystems within the Asian summer monsoon are known to interact with each other. The monsoon onset of regions to the west of the Philippines, specifically Indochina Peninsula and South China Sea (SCS), is associated with the development in the circulation and convective features in the tropical East Indian Ocean (Ding and Chan 2005). The Asian monsoon is a complex system, but the past decade have seen significant advances in understanding climate over the region through the Monsoon Asian Hydro-Atmosphere Scientific Research and Prediction Initiative (MAHASRI) project (Matsumoto et al. 2016).

Rainfall is an important characteristic of the SWM season because these winds carry warm, moist air, which has a higher potential for convective activity. The northwestern regions of the Philippines experience a rainy season during June to September. A significant percentage (43%) of the mean annual rainfall in the Philippines is associated with the SWM season, which is essential for supplying water for agriculture, energy, and domestic use (Asuncion and Jose 1980; Cayanan et al. 2011). Although there has been a general decline in the SWM mean rainfall over northwestern Philippines (Cruz et al. 2013), intense rainfall episodes still occur during the SWM season, which can result in heavy flooding.

The oceans surrounding the Philippines have a major impact on its regional climate. Large-scale oceanic impacts, including El Niño–Southern Oscillation (ENSO, Lyon and Camargo 2009; Villafuerte and

* Correspondence: dado-juliemaeborejon@ed.tmu.ac.jp; jbdado@gmail.com
[1]Department of Geography, Tokyo Metropolitan University, Minami-Osawa 1-1, Hachioji, Tokyo 192-0397, Japan
Full list of author information is available at the end of the article

Matsumoto 2015), have been examined in many studies. However, the impact of local sea surface temperature (SST) on the country is not well understood. Thus, there is a need to examine the local SST impact on regional rainfall over the Philippines.

Studies have noted the relationship between SST and atmosphere through the exchange of surface fluxes (Deser et al. 2010). The amount of near-surface moisture for convective rainfall is sensitive to SST through the Clausius–Clapeyron relationship and is strongly associated on interannual timescales over oceanic regions (Wentz and Schabel 2000). In general, higher SSTs correspond to increases in precipitation (Vecchi and Harrison 2002; Trenberth and Shea 2005). In addition, a recent study showed that precipitation tends to increase linearly with increasing SST over tropical monsoon basins, suggesting a noticeable impact of SST on tropical precipitation (Roxy 2014).

In the western North Pacific (WNP) region, the correlation between local rainfall and SST in the boreal summer is negative indicating that atmospheric dynamics are more dominant than ocean conditions in determining the rainfall variability of this region (Trenberth and Shea 2005; Wang et al. 2005). This relationship may be explained by the monsoon activity during this season. The summer monsoon drives rainfall but it also leads to SST cooling, hence the negative correlation between local rainfall and SST. Moreover, Trenberth and Shea (2005) also showed the covariability between surface temperature and precipitation, and although relationship is negative in WNP, it would be interesting to isolate the SST effect alone. Note that impact of local SST on rainfall variation in this region has not been evaluated yet, although these previous studies have indicated that SST cannot be the main driver of rainfall in this region.

However, it is difficult to quantify the impact of SST on rainfall from observational datasets because of strong air–sea interactions (Peralta and Narisma 2014). This is similar to a mid-latitude winter monsoon case over the Sea of Japan (Takahashi and Idenaga 2013; Takahashi et al. 2013), although the range in surface air temperature is quite different. Higher SSTs bring higher latent heat fluxes under the same atmospheric conditions due to higher water vapor pressures on the sea surface, which is explained by the Clausius–Clapeyron equation. At the same time, the evaporation over the sea surface decreases the SST. In SWM condition, the strong monsoon westerly also increases latent heat flux which brings abundant rainfall to the northwestern Philippines. And since rainfall on the northwestern Philippines is predominantly controlled by the SWM airflow, the SST effect is seemingly hidden in all these dynamics. These dynamics are highly coupled, and while it is difficult to isolate the one-way impacts from observational data alone, numerical experiments are useful for understanding and quantifying such one-way impacts.

The SWM months also coincide with high tropical cyclone (TC) activity in the WNP basin including the Philippines (Cinco et al. 2016; Lyon and Camargo 2009; Takahashi and Yasunari 2006). This high frequency of TC makes a substantial contribution on rainfall over the monsoon trough (Takahashi et al. 2015a), especially in the northern island of Luzon (Cinco et al. 2016). It has also been shown that SWM rainfall is enhanced in the presence of TCs (Cayanan et al. 2011; Kubota and Wang 2009), and as such, this study also attempts to isolate the TC effect in the analysis of SWM rainfall.

To determine the contribution of SST to rainfall in a statistically robust manner, long-term or multiple-case experiments are required, as the atmospheric response to SST anomalies can vary under different atmospheric conditions (Peng et al. 1997). Moreover, long-term climate simulations can be used to investigate the extent to which interannual variations in SST affect regional climate.

In this study, we examined the response of rainfall associated with the SWM from SST forcing located west of the Philippines. We attempted to isolate the impact of SST alone, neglecting feedback from the atmosphere to the ocean, by applying SST as the boundary condition in a regional climate model. We focused on the effects of SST located west of the Philippines, as SWM winds pass through this area before arriving on the northwestern coast of the country. The SST–rainfall relationship is important in further understanding the hydroclimate within this region which translates in its significance in economic sectors such as agriculture. The methodology is discussed in detail in the "Methods/Experimental" section, and an evaluation of the performance of the model and quantification of the SST impact is described in the "Results" section. Changes in latent heating and the lack of a tropical cyclone (TC) effect are discussed in the "Discussion" section, and a summary is presented in the "Conclusions" section.

Methods/Experimental

Model and experimental setup

We used a non-hydrostatic regional climate model, the Advanced Research Weather Research and Forecasting Model (WRF) ver. 3.6.1 with the Advanced Research WRF dynamical core (Skamarock et al. 2008) to examine the impact of SST on precipitation. The initial and lateral boundary conditions used were from the ERA-Interim reanalysis data (Berrisford et al. 2009) of the European Centre for Medium-Range Weather Forecasts, and SST data were from the Optimum Interpolation Sea Surface Temperature (OISST) ver. 2 (Reynolds et al. 2002). A set of control simulations (CTL) was performed for the months of June to August for the years 1982–

2012. Temporally interpolated monthly values of SST were prescribed every 6 h. Multi-year simulations were performed to investigate the robustness of the atmospheric response to SST forcing under various atmospheric conditions.

The simulation period was from May 27 to September 1 of each year. The first 5 days served as the simulation warm-up time and were not included in the analysis. The horizontal grid resolution of the first domain was 25 km, while that of the two-way nested domain was 5 km. Both domains had 33 terrain-following levels. Figure 1 shows the model domains. The cumulus convective parameterization (CCP) scheme was not applied to the two domains in this study because a series of test simulations with CCP yielded unrealistic amounts of rainfall. Previous studies have shown that WRF is able to capture rainfall in Asian monsoon regions without convective parameterization (Sugimoto and Takahashi 2017; Sugimoto and Takahashi 2016; Takahashi et al. 2010). Three schemes were used: the WRF single-moment six-class microphysics scheme (Hong and Lim 2006), the Mellor–Yamada–Janjić planetary boundary layer scheme (Janjić 1994), and the unified Noah land surface model (Tewari et al. 2004). High-resolution model can represent topography and land–sea contrast in detail; thus, analysis of the SST impact was carried out using the 5 km-model results.

Sensitivity experiments

To analyze the effect of interannual variations in SST, we performed a set of sensitivity experiments. In this experiment, the 31-year mean monthly climatological SST was computed using OISST, and then applied to the same 31-year time period as the CTL simulations (hereinafter referred to as the CLM runs). The numerical designs of the CTL and CLM runs were the same, except for application of the SST. The

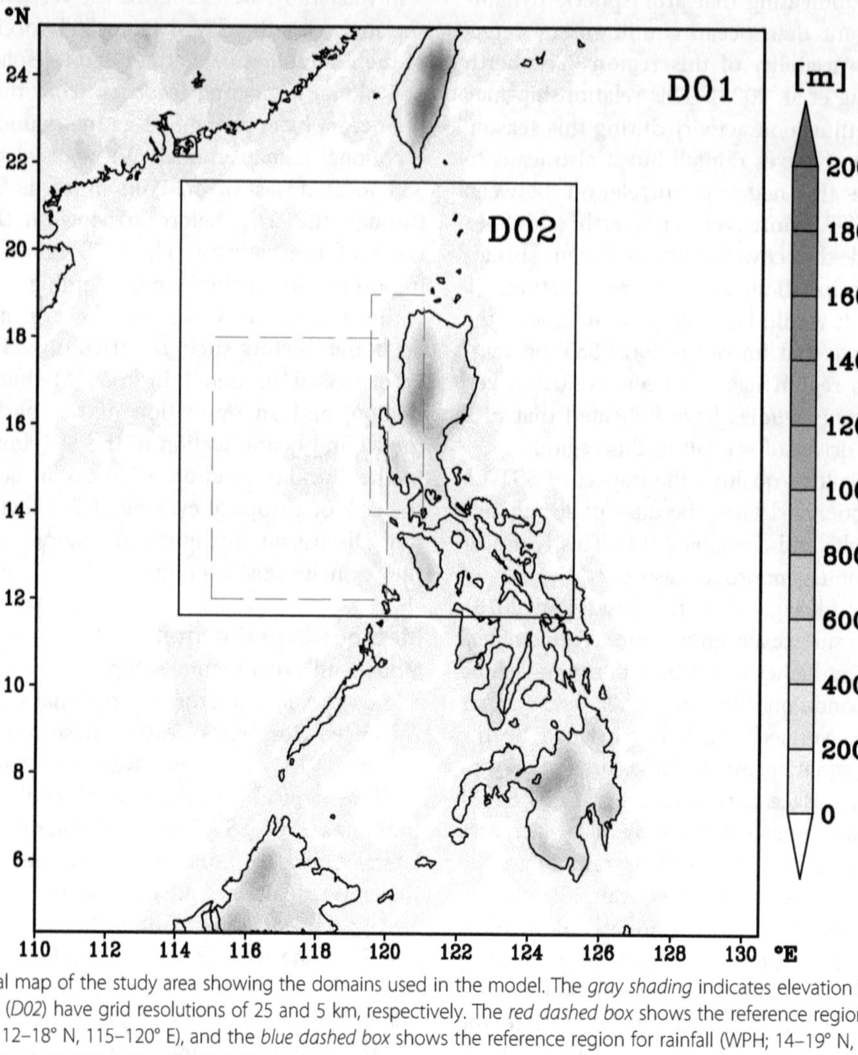

Fig. 1 Topographical map of the study area showing the domains used in the model. The *gray shading* indicates elevation in meters. Domain 1 (*D01*) and domain 2 (*D02*) have grid resolutions of 25 and 5 km, respectively. The *red dashed box* shows the reference region for sea surface temperature (WSST; 12–18° N, 115–120° E), and the *blue dashed box* shows the reference region for rainfall (WPH; 14–19° N, 119.5–121° E)

difference between the CTL and CLM runs represented the atmospheric response to the interannual SST variability. This experimental design was based on a previous study to detect local SST impact (Takahashi et al. 2015b). The local SST impact can be isolated using this design with the remote SST effect already included in the large-scale atmospheric conditions used as lateral boundary forcing for both the CTL and CLM experiments.

Observational datasets used for the evaluation

The simulated rainfall from the CTL runs was evaluated using data from the Asian Precipitation Highly-Resolved Observational Data Integration Towards Evaluation (APHRODITE) of Water Resources (Yatagai et al. 2012) and the Global Precipitation Climatology Project (GPCP) ver. 2.2 (Adler et al. 2003), and the low-level circulations were compared with data from the Japanese 55-year Reanalysis (JRA55; Kobayashi et al. 2015), which differed from the driving atmospheric datasets used in the simulations. Figure 1 shows the regions selected for a targeted analysis of the effect of SST on the climate of the northwestern Philippines. The red dashed box shows the reference region for SST (hereinafter referred as WSST; 12–18° N, 115–120° E), and the blue dashed box shows the reference region for rainfall (hereinafter referred as the northwestern region of the Philippines (WPH); 14–19° N, 119.5–121° E).

Tropical cyclone (TC) track data

The number of unique TCs crossing the first domain of the simulation (D01) in Fig. 1 is counted for the months of June–August of 1982–2012 based on the best track data from the Joint Typhoon Warning Center (JTWC). For each month, years with TC counts below the mean were considered as low-TC years, which we assumed to be more representative of the SWM climate. The SST impact under this low-TC condition is explained in detail in the "Discussion" section.

Results

Evaluation of CTL simulations

First, we evaluated the performance of the simulated regional climate over the northwestern Philippines. The simulated 850 hPa horizontal winds in domain 2 were compared with JRA55, which differed from the ERA-Interim boundary forcing data. The low-level monsoon southwesterlies were

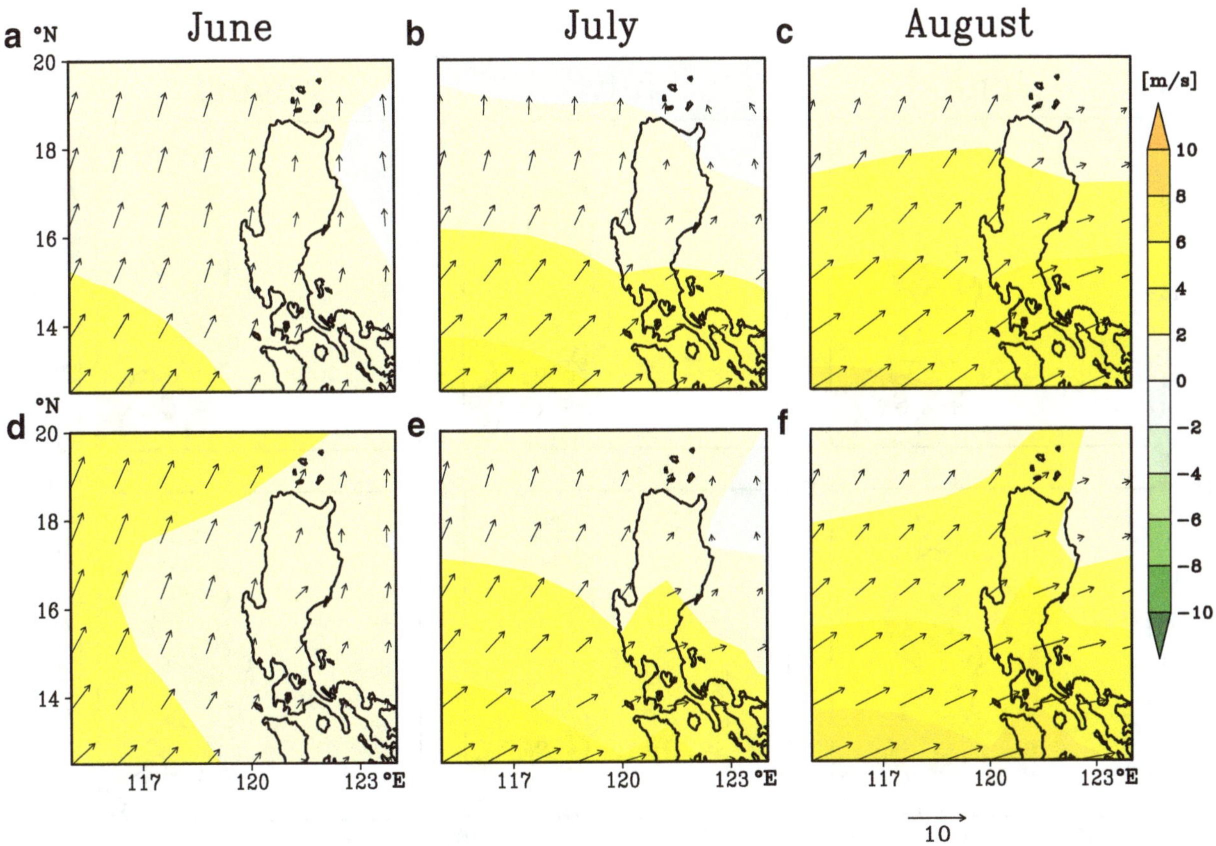

Fig. 2 Climatological monthly 850 hPa level winds (m s^{-1}) from 1982 to 2012 over domain 2, calculated by the CTL simulation (**d**–**f**) and compared with the JRA55 (**a**–**c**). Monsoon strength, denoted by the zonal wind component, is shown as *shaded contours* in the figure

well simulated for the months of June, July, and August (Fig. 2). Furthermore, the model captured the monsoon strength, denoted by the zonal wind component.

The spatial pattern of the simulated rainfall in the CTL experiment was evaluated by comparing the climatological monthly total for each month with the APHRODITE dataset (Fig. 3). APHRODITE has data for 1950–2007; however, we only considered the years 1982–2007 to remain consistent in reproducing climatological values for both simulations and observations. We verified that the results were nearly the same when using climatological rainfall values for 1982–2012. To facilitate the comparison between the observations and simulations, the simulated rainfall over the ocean was masked because the APHRODITE dataset was limited to the terrestrial region.

The spatial distribution of climatological rainfall was represented well by the CTL. It was able to show the west–east gradient in rainfall in the assessed months. In addition, the spatial distribution of higher rainfall values in the northwestern portion was noted in both the observational and simulated data. However, peaks in rainfall differed slightly, and the model produced higher rainfall values, predominantly in areas with high elevations. This may represent a positive bias of the simulated rainfall over high-elevation regions. In addition, the APHRODITE data were mostly obtained from coastal observation stations which can result in its underestimation of rainfall in high-elevated regions. In addition, we validated the climatological simulated rainfall against the GPCP data, which were produced from different observations, and noted a spatial distribution similar to that for APHRODITE (not shown).

The CTL monthly rainfall values in the northwestern Philippines (WPH in Fig. 1) were compared with APHRODITE monthly rainfall values for 1982–2007. Similarly, the GPCP monthly mean rainfall was obtained for the same WPH region for the years 1982–2012. Ocean grid points in the WPH region were included when comparing the CTL rainfall with the GPCP rainfall. The seasonal variation in the climatological monthly rainfall was simulated well by the model when compared with APHRODITE and GPCP data, and the WRF rainfall showed a similar seasonal peak in August (not shown).

Furthermore, we analyzed the interannual rainfall variations in the region west of the Philippines (WPH in Fig. 1) for both the observational and simulated data as scatter plots of the observed and simulated monthly precipitation for the WPH region (Fig. 4). The comparison revealed a high correlation between the monthly total

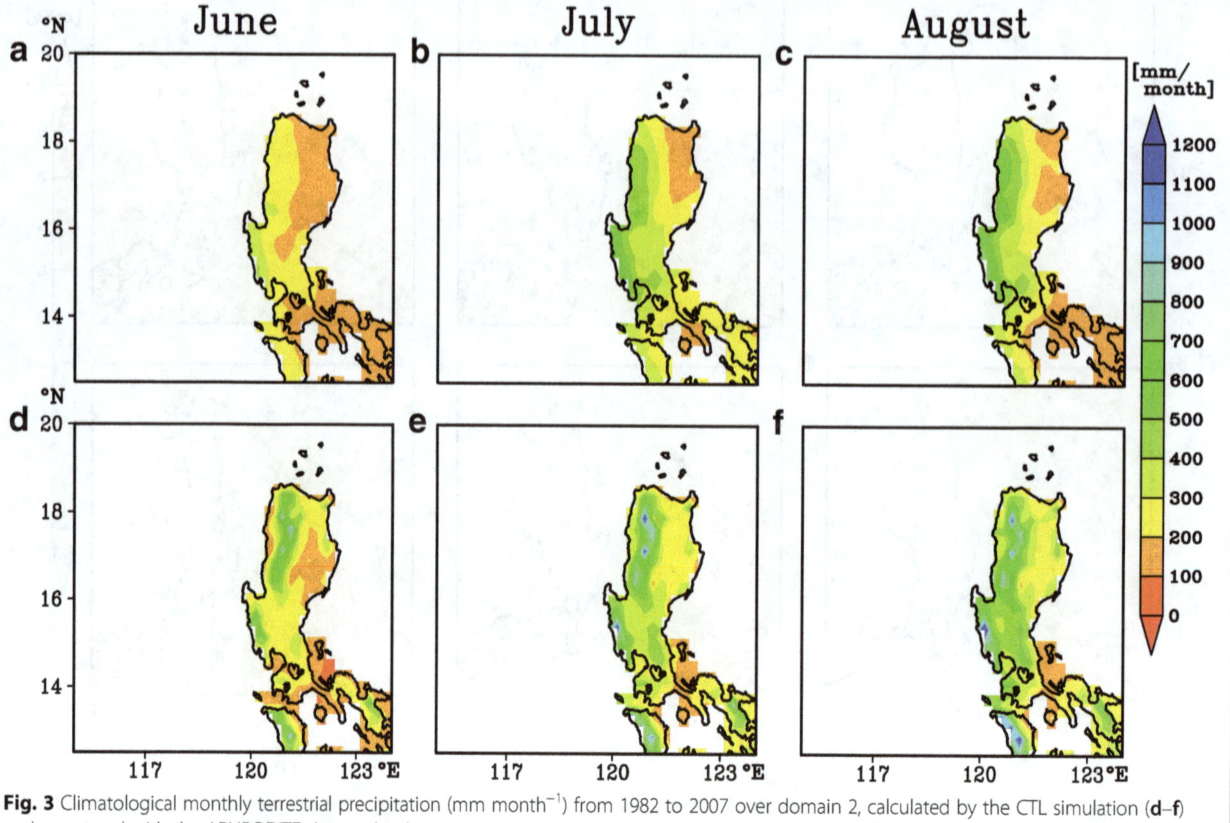

Fig. 3 Climatological monthly terrestrial precipitation (mm month^{-1}) from 1982 to 2007 over domain 2, calculated by the CTL simulation (**d**–**f**) and compared with the APHRODITE dataset (**a**–**c**)

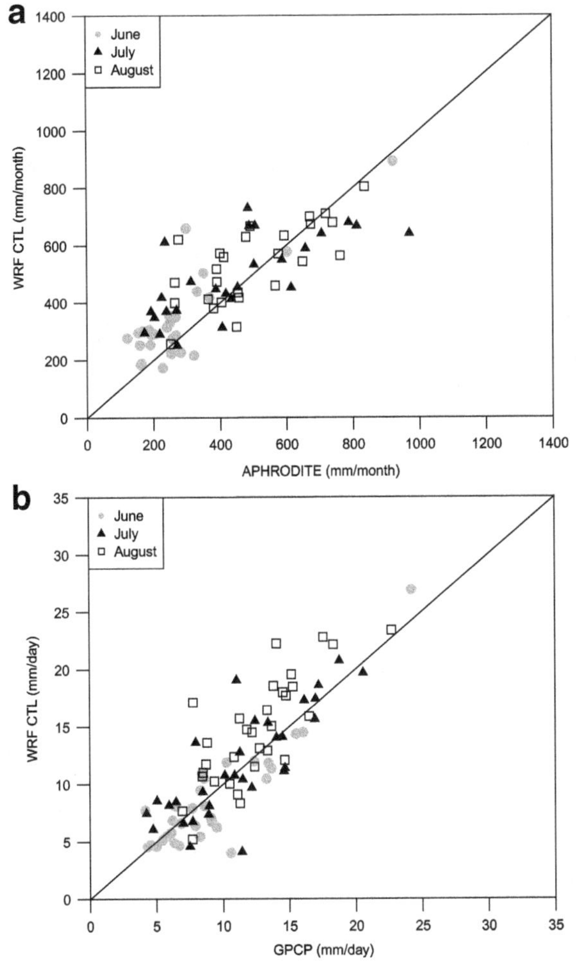

Fig. 4 **a** Scatter plot of the monthly total rainfall (mm month^{-1}) from the CTL simulations compared with APHRODITE data for 1982–2007. **b** Scatter plot of monthly mean of daily rainfall (mm day^{-1}) from the CTL simulations compared with GPCP data for 1982–2012

rainfall from APHRODITE and the CTL simulation for the years 1982–2007 for the WPH region (Fig. 4a). The interannual seasonal mean (June to August) correlation coefficient was 0.69. In addition, the comparison revealed a high correlation between the monthly means of daily rainfall from the GPCP data and the CTL simulation (Fig. 4b) for the WPH region, with an interannual correlation of the seasonal mean of 0.80. These interannual correlations were statistically significant at a 99.9% confidence limit, determined using Student's t test.

These results indicate that the CTL runs captured the summer rainfall in the WPH region well. Despite differences in the amount of monthly rainfall, the WRF CTL runs were able to capture the interannual and seasonal variations in precipitation in the WPH region; for the purpose of this study, the model demonstrated acceptable performance.

Sensitivity of rainfall to SST variability

To estimate the local impact of SST on rainfall spatially, we generated regression maps using Theil–Sen method to show the interannual variation in the sensitivity of monthly rainfall with area-averaged SST anomalies in the WSST region for the months of June, July, and August (Fig. 5). The sensitivity of rainfall to SST was defined as the difference between the CTL and CLM experiments. The regression coefficient was calculated using the sensitivity of the meteorological variables and the area-averaged SST anomalies. Because we used the same atmospheric boundary conditions for the CTL and CLM experiments, we could not extract the impact of remote SST. Therefore, we only evaluated the impact of SST within the model domain. The simulated results were transformed into coarser values at each 1-degree grid point from the original 5 km output to minimize noise. Similarly, the SST anomaly was defined as the difference between the CTL and CLM values averaged over the WSST region (Fig. 1).

Positive rainfall sensitivity to 1 K SST warming was noted over the whole calculation domain (Fig. 5), which can be explained by the increase in latent heat flux due to the increase in SST. A detailed description of the process underlying rainfall sensitivity to changes in SST is given in the "Discussion" section. Higher and statistically significant sensitivity was observed for oceanic rainfall near the WSST region. Statistically significant values were noted in the WPH region, and August had the highest regression coefficient (Fig. 5).

These maps show that monthly rainfall in the WPH increased by ~100 mm K^{-1} SST warming in the WSST. Because the interannual standard deviation of SST over the WSST was between 0.4 and 0.6 K, 40–60 mm of interannual variation in monthly rainfall could be attributed to the interannual variation of SST over the WSST (Fig. 5). Because the average climatological rainfall was ~500 mm per month in this region, the contribution of SST to interannual rainfall variation was not negligible.

To understand the interannual relationship between SST and rainfall over and around the Philippines, we examined the time series of the SST anomaly over the WSST region and rainfall sensitivity (CTL - CLM) over the WPH region for June, July, and August (Fig. 6). A positive or negative SST anomaly generally corresponded to positive or negative rainfall sensitivity, respectively. This relationship is evident in the scatter plots (Fig. 6d–f), which clearly show that the sensitivity of rainfall in the WPH region increased with SST in the WSST region. Because the relationship is almost linear, the impact of SST on rainfall can be quantified on an interannual timescale.

Using the Theil–Sen slope estimator method, the calculated regression coefficient of rainfall in the WPH region on the area-averaged SST over the WSST region increased

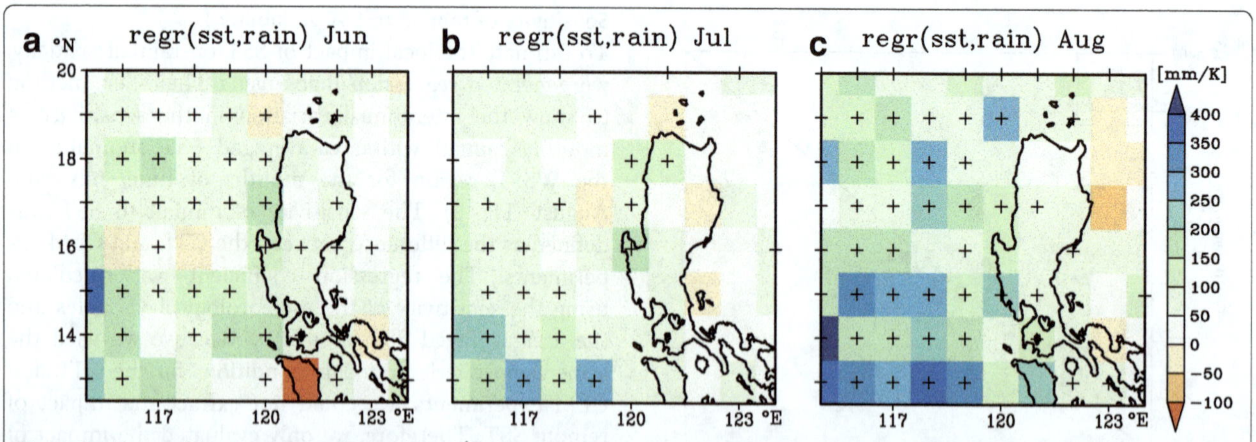

Fig. 5 Spatial map of the regression coefficients (mm K^{-1}) of the sensitivity of monthly rainfall to the monthly SST anomaly of the WSST reference region for 1982–2012 for the months of **a** June, **b** July, and **c** August. Significant values at the 90% significance level are marked with *plus signs*, as determined by the correlation coefficients based on 29 degrees of freedom. The rainfall anomalies were remapped to a 1° grid resolution

by 80.65, 78.86, and 99.69 mm K^{-1} SST warming in the WSST region for the months of June, July, and August, respectively. This accounted for 23–37% of the climatological monthly rainfall in the WPH region, depending on the month considered. These results indicate that it is important to account for the impact of SST on rainfall in the WPH region. The mechanism behind this relationship is discussed in the following section.

Discussion

Quantification of the sensitivity of latent heat flux to SST

The major sources of water vapor for rainfall in the WPH region are the evaporation from the western seas of the Philippines, which is windward of the Philippines in the summer, and water vapor transport due to the monsoon southwesterlies, especially during the SWM season. In our experiment, water vapor transport from the lateral boundary (see the western lateral boundary of Fig. 1) should have been the same, because the same lateral boundary conditions were used. Therefore, any changes in water vapor should be explained by changes in the evaporation, which is directly related to latent heat flux, within the calculation domains.

Similarly, we regressed the interannual latent heat flux sensitivity to area-averaged SST anomaly in the WSST

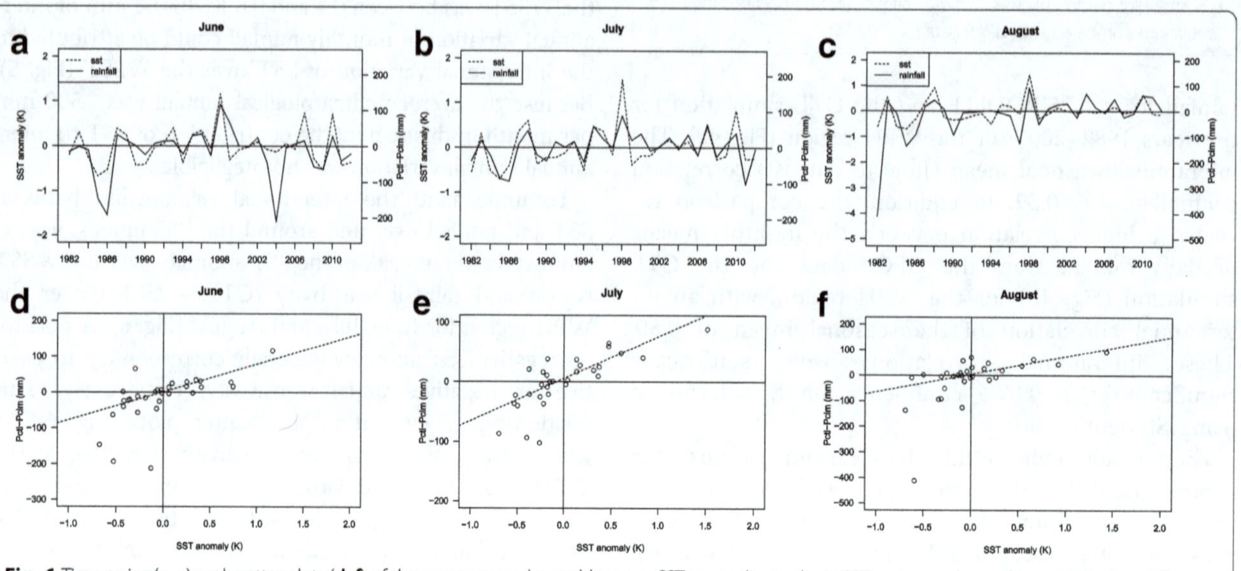

Fig. 6 Time series (**a**–**c**) and scatter plots (**d**–**f**) of the area-averaged monthly mean SST anomalies in the WSST region and monthly precipitation difference ($P_{CTL} - P_{CLM}$) in the WPH region for June, July, and August, respectively. The *dashed line* in the scatter plot indicates the linear relationship between SST anomaly and rainfall sensitivity estimated using the Theil–Sen method

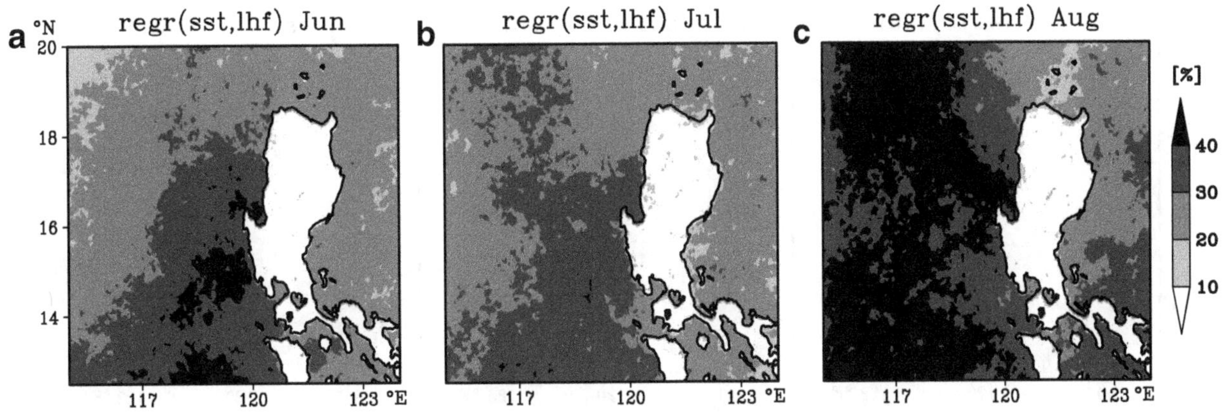

Fig. 7 Spatial map of the regression coefficients of the sensitivity of latent heat flux (lhf) to the monthly SST anomaly of the WSST reference region for 1982–2012 for the months of **a** June, **b** July, and **c** August. The values are shown in percentage relative to the 31 yr.-mean lhf from CTL

region for the months of June, July, and August. The values are expressed in percentage relative to the 31-year mean latent heat from CTL experiment.

The latent heat flux increased by 20–40% per 1 K SST warming in the WPH region, with larger increases in the seas west of the Philippines (Fig. 7). This increase in the latent heat flux was higher than the simulated change in rainfall in the WPH region. Because the relative humidity change at 2 m height was very small (within −2–2%) in most of the domain (Fig. 8), we can infer that the increase in water vapor due to the increase in latent heat was mostly converted into rainfall.

To analyze this large increase in latent heat in the domain, the bulk formula for the latent heat flux from the sea into the atmosphere can be given as.

$lE = l\rho\beta C_H U(q_{sat}(T_s) - q)$,

where l is the latent heat of vaporization, E is the amount of evaporation, ρ is the density of air, β is the efficiency of evaporation which is 1 over water, C_H is a mixing constant, U is the wind velocity, and T_s is the SST over the western seas of the Philippines. q_{sat} is the saturated specific humidity at T_s, and q is the specific humidity of the lower atmosphere. Therefore, latent heating can be explained by wind velocity and the difference between the saturated specific humidity at SST and the surface specific humidity.

Changes in wind speed and specific humidity are likely critical in explaining the much larger sensitivities in latent heat flux over the WSST regions than expected from the Clausius–Clapeyron equation. The changes in the wind speed per 1 K SST warming were mostly positive (within 10%), which increased the latent heat flux over the WSST region by 5–10% percent (Fig. 9). This increase in wind speed was associated with a downward transport of momentum due to the development of the planetary boundary layer height (e.g., Stull 2006) and enhancement of regional circulation that is accompanied by latent heating and convective activity. The 2 m air temperature increase was <1 K across most of the

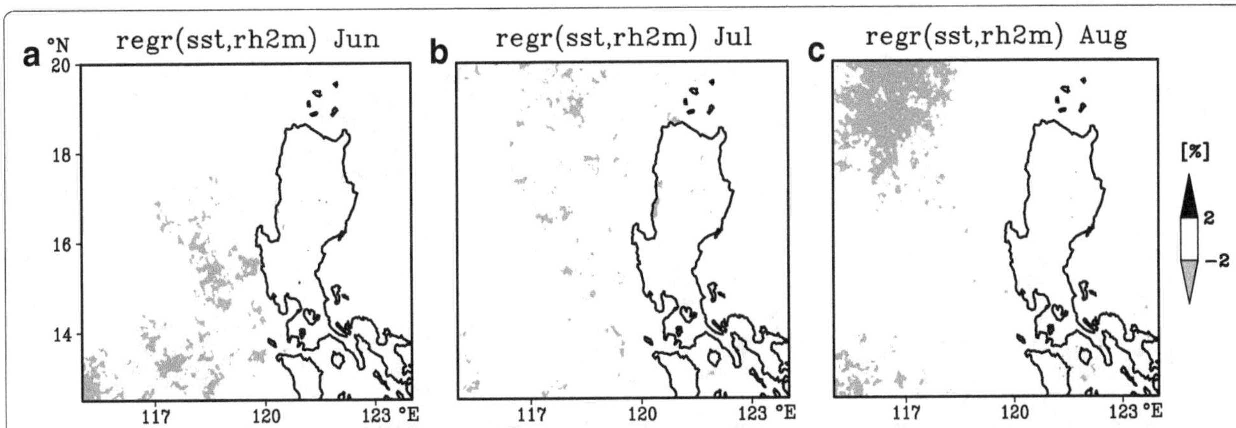

Fig. 8 Spatial map of the regression coefficients of the sensitivity of 2 m relative humidity (rh2m) to the monthly SST anomaly of the WSST reference region for 1982–2012 for the months of **a** June, **b** July, and **c** August. The values are shown in percentage relative to the 31 yr.-mean rh2m from CTL

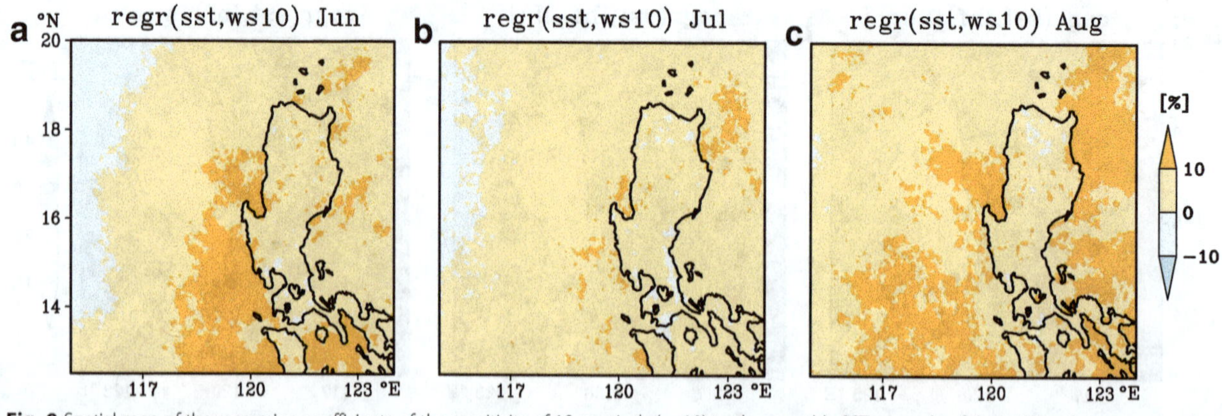

Fig. 9 Spatial map of the regression coefficients of the sensitivity of 10 m winds (ws10) to the monthly SST anomaly of the WSST reference region for 1982–2012 for the months of **a** June, **b** July, and **c** August. The values are shown in percentage relative to the 31 yr.-mean ws10 from CTL

domain (Fig. 10). Due to the small change in relative humidity, the rate of regressed sensitivity in q between the CTL and CLM simulations onto the area-averaged SST warming was expected to be <7%, which is the theoretical value in an idealized situation under the Clausius–Clapeyron equation. Based on this, the change in the term $(q_{\mathrm{sat}}(T_s) - q)$ has a larger value than 7%. The small sensitivity in the 2 m air temperature can contribute to the positive sensitivity in latent heat flux by a few percent. Moreover, the changes in relative humidity were small, but mostly exhibited negative sensitivity to SST warming over the WSST region, which can also contribute 1–2% to the large increase in latent heat flux.

There are additional possible effects that can increase latent heat flux or rainfall. Takahashi et al. (2013) suggested that an increase in the height of the planetary boundary layer could contribute to the increase in latent heat flux. In addition, an increase in precipitable water of <7% in the Clausius–Clapeyron equation suggested that the increased water vapor due to the increase of latent heat flux was excessively used for rainfall, decreasing the precipitable water over the domain.

Table 1 summarizes the changes in the atmospheric variables due to 1 K SST warming in the WSST region. These results suggest that the large changes in the latent heating can be explained mainly by the increase in surface wind speeds coupled with weak air temperature warming near the surface.

Isolating the effect of tropical cyclones

The western North Pacific basin is highly active for TCs and many TCs pass near or through the Philippines, especially during the months of July and August in the SWM season (Cinco et al. 2016); therefore, this subsection discusses the effects of TCs on the sensitivity to local SST changes. We examined whether there was a systematic difference in sensitivity between years of high and low TC activity, so as to only include years representative of a SWM effect.

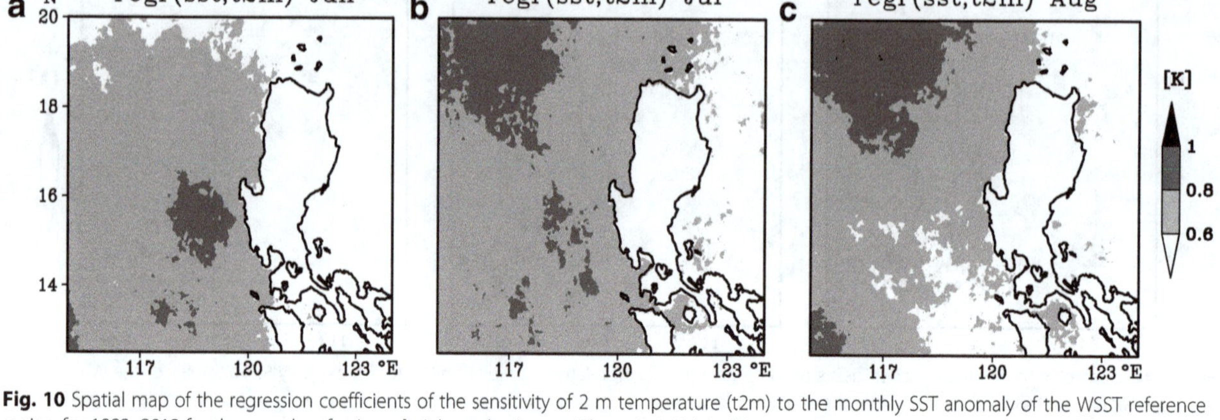

Fig. 10 Spatial map of the regression coefficients of the sensitivity of 2 m temperature (t2m) to the monthly SST anomaly of the WSST reference region for 1982–2012 for the months of **a** June, **b** July, and **c** August. The unit is Kelvin (K)

Table 1 Change in atmospheric variables due to a 1 K SST warming in the WSST region from 1982 to 2012. The actual values are shown with the percentage change relative to the mean CTL value (in parentheses). All variables were obtained from an average of the WSST region, except for rainfall, which was from the WPH region

	June	July	August
Rainfall (mm)	80.65 (37.95%)	78.86 (24.3%)	99.69 (23.19%)
Latent heat flux (W/m^2)	35.77 (32.88%)	33 (30.9%)	45.5 (41.58%)
2 m relative humidity	−1.54 (−1.74%)	−1.44 (−1.62%)	−1.42 (−1.58%)
10 m wind speed (m/s)	0.28 (5.73%)	0.24 (4.56%)	0.53 (8.68%)
2 m temperature (K)	0.73 (0.24%)	0.74 (0.25%)	0.67 (0.22%)
Precipitable water vapor (mm)	2.81 (5.25%)	2.32 (4.23%)	2.65 (4.75%)

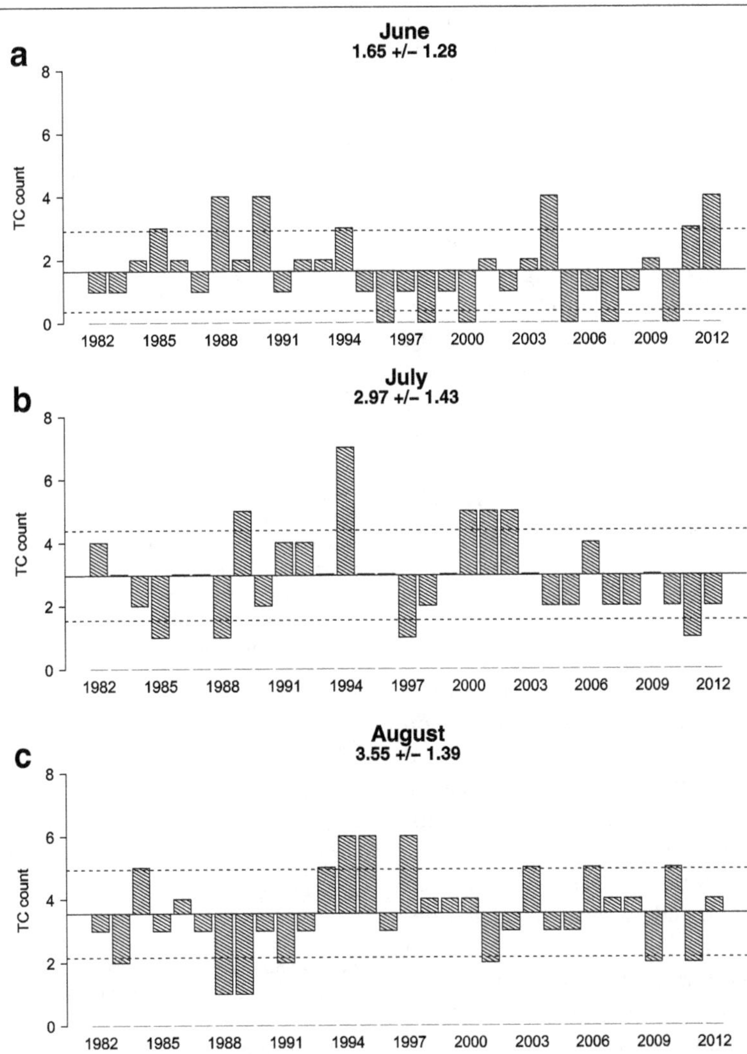

Fig. 11 Number of tropical cyclones passing through domain 1 (D01 in Fig. 1) for each month (**a**–**c**) from 1982 to 2012 based on JTWC best track data. The counts are plotted around the mean, while the *dashed lines* indicate the standard deviation. The means and standard deviations are written in the graph for each month

Figure 11 shows the number of TCs entering the region bounded by D01 in Fig. 1. TC counts are based on the JTWC best track data, not on the actual TC simulated by the model. To confirm the presence of TCs in the model, we compared the 850 hPa zonal winds from the boundary data, ERA-Interim, with WRF for years with high TC counts. The easterly wind footprints signify the presence of a TC in our model domain, and these footprints are consistent with the TC count from the JTWC data (not shown).

The mean count from 1982 to 2012 was calculated for each month, with years below the mean considered as low-TC years (Fig. 11). The relationships between the variables shown in Table 1 were recalculated for these low-TC years, which were assumed to more representative of the SWM climate (Additional file 1: Table S2).

The relationships held true in both low- and high-TC years (Additional file 1: Table S2), and the results were similar to those in Table 1, wherein the large changes in latent heating were mainly attributed to enhancement of low-level winds and weak warming at the surface. Notably, the differences in the sensitivity between low- and high-TC years were not systematic, implying that the impact of TCs around the Philippines is complicated; therefore, we concluded that we cannot easily determine whether TCs intensify or weaken the SST sensitivity. Although this may be largely due to the intensity of the TCs and their corresponding locations, we were unable to determine this in our experiments because of the non-linearity of TC behavior. As previous studies have noted, the presence of a TC in the vicinity of the Philippines may enhance southwesterly winds (Cayanan et al. 2011); however, future studies should statistically analyze the interaction between SWM and TCs around the Philippines.

Conclusions

In this study, we explored the potential impact of SST west of the Philippines on summer monsoon rainfall on the northwestern coast of the country. We performed two sets of simulations using the same initial and lateral boundary conditions, but with different SST conditions, from June to August for 1982–2012.

The spatial pattern of monthly rainfall was simulated well by the CTL simulation, with a clear horizontal delineation between rainfall in the western and eastern regions of the Philippines during the summer monsoon months. The monthly rainfall values were in close agreement with the two observational datasets used.

Positive SST anomalies in the WSST region corresponded to positive rainfall anomalies in the WPH region. Based on the regression analysis, rainfall in the northwestern region of the Philippines is modulated by interannual variations in SST west of the Philippines by 80~100 mm K^{-1} SST warming. This translates to 20~40% of the rainfall change in the WPH region relative to climatological values in the region.

The change in rainfall exceeded the theoretical approximation of 7% due to the Clausius–Clapeyron contribution, which may be attributed to the increase in the latent heat flux in the WSST region. This latent heat flux increase is associated with an increase in surface winds coupled with a weak near-surface warming. We isolated the impact of TCs and noted that this mechanism holds true for years with low or no TC effects, which are more representative of the SWM climate.

Abbreviations

APHRODITE: Asian Precipitation Highly-Resolved Observational Data Integration Towards Evaluation of Water Resources; CCP: Cumulus convective parameterization; CLM: Climatological SST-driven simulation; CTL: Monthly SST-driven simulation; ENSO: El Niño–Southern Oscillation; GPCP: Global Precipitation Climatology Project; JRA55: Japanese 55-year Reanalysis; JTWC: Joint Typhoon Warning Center; lhf: Latent heat flux; OISST: Optimum Interpolation Sea Surface Temperature; rh2m: 2 m relative humidity; SCS: South China Sea; SST: Sea surface temperature; SWM: Southwest monsoon; t2m: 2 m temperature; TC: Tropical cyclone; WNP: Western North Pacific; WPH: Reference region for rainfall northwest of the Philippines; WRF: Advanced Research Weather Research and Forecasting Model ver. 3.6.1; ws10: 10 m wind speed; WSST: Reference region for sea surface temperature west of the Philippines

Acknowledgments

We thank Dr. Shiori Sugimoto for preparing the ERA-Interim boundary datasets used in the WRF simulations. This study is partly supported by the 8th Japan Aerospace Exploration Agency (JAXA) Precipitation Measuring Mission (PMM) Project No. 309. JM Dado is a recipient of the "Asian Human Resources Fund" scholarship from the Tokyo Metropolitan Government. We are also grateful for the constructive comments from the editor and reviewers in further improving this manuscript.

Funding

JM Dado is a recipient of the "Asian Human Resources Fund" scholarship from the Tokyo Metropolitan Government. This study is also partly supported by the 8th Japan Aerospace Exploration Agency (JAXA) Precipitation Measuring Mission (PMM) Project No. 309.

Authors' contributions

JMD proposed the topic and carried out the experiments in the study. JMD and HGT designed the study and analyzed the data and their interpretation. Both authors collaborated in the construction of the manuscript and both read and approved the final manuscript.

Authors' information

JMD is a former researcher at the Manila Observatory in the Philippines and is currently doing her doctoral studies at Tokyo Metropolitan University. HGT is an assistant professor at Tokyo Metropolitan University and a scientist at Japan Agency for Marine-Earth Science and Technology (JAMSTEC).

Competing interests

The authors declare that they have no competing interest.

Author details

[1]Department of Geography, Tokyo Metropolitan University, Minami-Osawa 1-1, Hachioji, Tokyo 192-0397, Japan. [2]Japan Agency for Marine-Earth Science and Technology, 3173-25, Showa-machi, Kanazawa-ku, Yokohama, Kanagawa 236-0001, Japan.

References

Adler RF, Huffman GJ, Chang A, Ferraro R, Xie P, Janowiak J, Rudolf B, Schneider U, Curtis S, Bolvin D, Gruber A, Susskind J, Arkin P, Nelkin E (2003) The version 2 Global Precipitation Climatology Project (GPCP) monthly precipitation analysis (1979-present). J Hydrometeorol 4:1147–1167

Asuncion JF, Jose AM (1980) A study of the characteristics of the northeast and southwest monsoons in the Philippines. National Research Council of the Philippines (NRCP) assisted project, p 49 (available from Philippine Atmospheric Geophysical and Astronomical Services Administration (PAGASA))

Berrisford P, Dee D, Fielding K, Fuentes M, Kallberg P, Kobayashi S, Uppala S (2009) The ERA-Interim archive. ERA report series 1, European Centre for Medium-Range Weather Forecasts

Cayanan EO, Chen T-C, Argete JC, Yen M-C, Nilo PD (2011) The effect of tropical cyclones on southwest monsoon rainfall in the Philippines. J Meteorol Soc Japan 89A:123–139. doi:10.2151/jmsj.2011-A08

Cinco TA, de Guzman RG, Ortiz AM, Delfino RJ, Lasco RD, Ares E (2016) Observed trends and impacts of tropical cyclones in the Philippines. Int J Climatol 36: 4638–4650

Cruz FT, Narisma GT, Villafuerte MQ II, Cheng Chua KU, Olaguera LM (2013) A climatological analysis of the southwest monsoon rainfall in the Philippines. Atmos Res 122:609–616. doi:10.1016/j.atmosres.2012.06.010

Deser C, Alexander MA, Xie SP, Phillips AS (2010) Sea surface temperature variability: patterns and mechanisms. Annu Rev Mar Sci 2:115–143. doi:10.1146/annurev-marine-120408-151453

Ding Y, Chan JCL (2005) The east Asian summer monsoon: an overview. Meteorog Atmos Phys 89:117–142. doi:10.1007/s00703-005-0125-z

Hong SY, Lim JOJ (2006) The WRF single–moment 6–class microphysics scheme (WSM6). J Korean Meteor Soc 42:129–151

Janjić ZI (1994) The step–mountain eta coordinate model: further developments of the convection, viscous sublayer, and turbulence closure schemes. Mon Wea Rev 122:927–945

Kobayashi S, Ota Y, Harada Y, Ebita A, Moriya M, Onoda H, Onogi K, Kamahori H, Kobayashi C, Endo H, Miyaoka K, Takahashi K (2015) The JRA-55 reanalysis: general specifications and basic characteristics. J Meteorol Soc Jpn 93(1):5–48. doi:10.2151/jmsj.2015-001

Kubota H, Wang B (2009) How much do tropical cyclones affect seasonal and interannual rainfall variability over the western North Pacific? J. Clim. 22(20): 5495–5510, doi:10.1175/2009JCLI2646.1.

Lyon B, Camargo SJ (2009) The seasonally-varying influence of ENSO on rainfall and tropical cyclone activity in the Philippines. Clim Dyn 32:125–141. doi:10. 1007/s00382-008-0380-z

Matsumoto J, Oki T, Yamanaka MD, Hayashi T, Asanuma J (2016) 10 years of MAHASRI: accomplishments and the international science conference wrap-up. GEWEX Newsletter 26(4):10–23

Peng S, Robinson WA, Hoerling MP (1997) The modeled atmospheric response to mid-latitude SST anomalies and its dependence on background circulation states. J Clim 10:971–987. doi:10.1175/1520-0442(1997)010<0971:TMARTM>2.0.CO;2

Peralta JA I, Narisma G (2014) The influence of West Philippine Sea surface temperatures on southwest monsoon rainfall in the Philippines. Undergraduate thesis. Philippines: Ateneo de Manila University.

Reynolds RW, Rayner NA, Smith TM, Stokes DC, Wang W (2002) An improved in situ and satellite SST analysis for climate. J Clim 15:1609–1625. doi:10.1175/ 1520-0442(2002)015<1609:AIISAS>2.0.CO;2

Roxy M (2014) Sensitivity of precipitation to sea surface temperature over the tropical summer monsoon region—and its quantification. Clim Dyn 43:1159–1169. doi:10.1007/s00382-013-1881-y

Skamarock WC, Klemp JB, Dudhia J, Gill DO, Barker DM, Duda MG, Huang X-Y, Wang W, Powers JG (2008) A description of the advanced research WRF version 3. NCAR Tech. Note NCAR/TN-475+STR. 113. doi:10.5065/D68S4MVH

Stull R (2006) The atmospheric boundary layer. In: Wallace JM, Hobbs PV (ed) Atmospheric science: an introductory survey, 2nd edn. USA: Elsevier Inc.

Sugimoto S, Takahashi HG (2016) Effect of spatial resolution and cumulus parameterization on simulated precipitation over South Asia. SOLA 12A:7–12. doi:10.2151/sola.12A-002

Sugimoto S, Takahashi HG (2017) Seasonal differences in precipitation sensitivity to soil moisture in Bangladesh and surrounding regions. J Clim 30:921–938. doi:10.1175/jcli-d-15-0800.1

Takahashi HG, Idenaga T (2013) Impact of SST on precipitation and snowfall on the sea of Japan side in the winter monsoon season: timescale dependency. J Meteorol Soc Jpn 91:639–653. doi:10.2151/jmsj.2013-506

Takahashi HG, Yasunari T (2006) A climatological monsoon break in rainfall over Indochina - a singularity in the seasonal march of the Asian summer monsoon. J Clim 19:1545–1556. doi:10.1175/JCLI3724.1

Takahashi HG, Yoshikane T, Hara M, Takata K, Yasunari T (2010) High-resolution modelling of the potential impact of land surface conditions on regional climate over Indochina associated with the diurnal precipitation cycle. Int J Climatol 30:2004–2020. doi:10.1002/joc.2119

Takahashi HG, Ishizaki NN, Kawase H, Hara M, Yoshikane T, Ma X, Kimura F (2013) Potential impact of sea surface temperature on winter precipitation over the Japan Sea side of Japan: a regional climate modeling study. J Meteorol Soc Jpn 91:471–488. doi:10.2151/jmsj.2013-404

Takahashi HG, Adachi SA, Sato T, Hara M, Ma X, Kimura F (2015b) An oceanic impact of the Kuroshio on surface air temperature on the Pacific coast of Japan in summer: regional H_2O greenhouse gas effect. J Clim 28:7128–7144. doi:10.1175/JCLI-D-14-00763.1

Takahashi HG, Fujinami H, Yasunari T, Matsumoto J, Baimoung S (2015a) Role of tropical cyclones along the monsoon trough in the 2011 Thai flood and interannual variability. J Clim 28:1465–1476. doi:10.1175/jcli-d-14-00147.1

Tewari M, Chen F, Wang W, Dudhia J, LeMone MA, Mitchell K, Ek M, Gayno G, Wegiel J, Cuenca R (2004) Implementation and verification of the unified NOAH land surface model in the WRF model. 20th conference on weather analysis and forecasting/16th conference on numerical weather prediction, pp. 11–15

Trenberth KE, Shea DJ (2005) Relationships between precipitation and surface temperature. Geophys Res Lett 32(L14703). doi:10.1029/2005GL022760

Vecchi GA, Harrison DE (2002) Monsoon breaks and subseasonal sea surface temperature variability in the bay of Bengal. J Clim 15:1485–1493. doi:10. 1175/1520-0442(2002)015<1485:MBASSS>2.0.CO;2

Villafuerte MQ II, Matsumoto J (2015) Significant influences of global mean temperature and ENSO on extreme rainfall in Southeast Asia. J Clim 28:1905–1919. doi:10.1175/JCLI-D-14-00531.1

Wang B, LinHo L (2002) Rainy season of the Asian-Pacific summer monsoon. J Clim 15:386–398. doi:10.1175/1520-0442(2002)015<0386:RSOTAP>2.0.CO;2

Wang B, Ding Q, Fu X, Kang I-S, Jin K, Shukla J, Doblas-Reyes F (2005) Fundamental challenge in simulation and prediction of summer monsoon rainfall. Geophys Res Lett 32:L15711. doi:10.1029/2005GL022734

Wentz F, Schabel M (2000) Precise climate monitoring using complementary satellite data sets. Nature 403:414–416. doi:10.1038/35000184

Yatagai A, Kamiguchi K, Arakawa O, Hamada A, Yasutomi N, Kitoh A (2012) APHRODITE: constructing a long-term daily gridded precipitation dataset for Asia based on a dense network of rain gauges. Bull Amer Meteor Soc 93: 1401–1415. doi:10.1175/BAMS-D-11-00122.1

11

A review of progress towards understanding the transient global mean surface temperature response to radiative perturbation

Masakazu Yoshimori[1,2*], Masahiro Watanabe[3], Hideo Shiogama[4], Akira Oka[3], Ayako Abe-Ouchi[3,5], Rumi Ohgaito[5] and Youichi Kamae[6]

Abstract

The correct understanding of the transient response to external radiative perturbation is important for the interpretation of observed climate change, the prediction of near-future climate change, and committed warming under climate stabilization scenarios, as well as the estimation of equilibrium climate sensitivity based on observation data. It has been known for some time that the radiative damping rate per unit of global mean surface temperature increase varies with time, and this inconstancy affects the transient response. Knowledge of the equilibrium response alone is insufficient, but understanding the transient response of the global mean surface temperature has made rapid progress. The recent progress accompanies the relatively new concept of the efficacies of ocean heat uptake and forcing. The ocean heat uptake efficacy associates the temperature response induced by ocean heat uptake with equilibrium temperature response, and the efficacy of forcing compares the temperature response caused by non-CO_2 forcing with that by CO_2 forcing.

In this review article, recent studies on these efficacies are discussed, starting from the classical global feedback framework and basis of the transient response. An attempt is made to structure different studies that emphasize different aspects of the transient response and to stress the relevance of those individual studies. The implications on the definition and computation of forcing and on the estimation of the equilibrium response in climate models are also discussed. Along with these discussions, examples are provided with MIROC climate model multi-millennial simulations.

Keywords: Transient climate response, Equilibrium climate sensitivity, Climate feedback, Ocean heat uptake efficacy, Efficacy of forcing

Introduction

A climate system is said to be in equilibrium when it exhibits a statistically stable state for an extended period of time under a given set of boundary conditions. When external radiative perturbation is added to the system (radiative forcing), it begins to reveal different statistical behavior, which includes the mean state. The statistical behavior evolves as time progresses, and the system eventually reaches a new equilibrium that is constrained by the newly imposed boundary conditions. Historically, the equilibrium response has received much attention; the equilibrium climate sensitivity (ECS) is the most notable example (Knutti and Hegerl 2008; Maslin and Austin 2012). The ECS is defined as the globally and annually averaged value of the equilibrium surface temperature response to a doubling of atmospheric CO_2 concentration. The ECS is an idealized yet useful concept, because it represents the final boundary of the system, and gives a measure of how much warming would occur in the future even if the current CO_2 concentration is sustained and not raised. We note that changes in

* Correspondence: myoshimo@ees.hokudai.ac.jp
[1]Faculty of Environmental Earth Science, Hokkaido University, Kita 10, Nishi 5, Kita-ku, Sapporo 060-0810, Japan
[2]Arctic Research Center, Hokkaido University, Sapporo, Japan
Full list of author information is available at the end of the article

vegetation and ice-sheet distributions and in atmospheric gas concentrations by a biogeochemical response are excluded from the formal definition of ECS, although the effect of vegetation change is sometimes included in recent model-based estimates. While ice sheets exhibit multiple equilibria (e.g., Abe-Ouchi et al. 2013), the ECS is an inherent property of the climate system, which is independent of the history of radiative forcing.

The transient response, which generally includes the effect of time-varying boundary conditions, is more complicated to characterize and understand than the equilibrium response. In addition, the system does not approach equilibrium at a constant rate even under constant radiative forcing. Nevertheless, the transient response has received more and more attention recently, because it is more relevant to near-future climate change and thus has more practical implications (Allen and Frame 2007). From the viewpoint of climate stabilization scenarios, understanding the transient response is also important for evaluating the magnitude of "committed" warming and the rate of the change accurately (Armour and Roe 2011; Meehl et al. 2005; Plattner et al. 2008; Solomon et al. 2009; Zickfeld et al. 2013). Here, the committed warming is referred to an additional surface warming that is expected to emerge when constant radiative forcing is continuously applied after a stabilization point.

In this review article, we focus on the transient response of the global mean surface temperature to external radiative perturbation, because the magnitudes of many climatic impacts can be scaled proportionally to this global mean variable relatively well (e.g., Ishizaki et al. 2012; Ishizaki et al. 2013; Yoshimori and Abe-Ouchi 2012). One of the motivations for this review arises from the recent, rapid conceptual development on climate feedbacks. Here, climate feedbacks are loosely defined as climate change processes that cause the additional radiative adjustment to the climate system in response to the initial radiative perturbation. For example, previously underappreciated factors such as spatiotemporally varying climate feedbacks related to the ocean heat uptake and the dependence of climate sensitivity on forcing agents have been identified as important. These factors are essential for a quantitative understanding of the transient response. Nevertheless, these factors are rather confusing and not necessarily understood widely by the climate science community. In addition, the recent progress requires further clarification of the definitions of forcing and feedback and discussions on how they are to be computed. It is becoming clear that the omission of such factors would introduce a bias into the estimate of ECS based on observation data. The understanding of the background long-term response to anthropogenic radiative perturbation must also be well established for clarifying observed contemporary climate variations, such as the warming pause or "hiatus" during the early twenty-first century (e.g., Kosaka and Xie 2013; Meehl et al. 2011; Watanabe et al. 2013; Watanabe et al. 2014).

Transient climate response in a global feedback framework

In this section, the transient response of the global mean surface temperature to radiative perturbation is reviewed in a classical global feedback framework. In this framework, the loss of radiative energy from the climate system, i.e., radiative damping, is formulated in proportion to the global mean surface temperature change. In the following, all variables represent global mean values unless noted explicitly. We start with an equilibrium state in which the global mean net radiation at the top of the atmosphere (TOA) is zero.

A classical definition of radiative forcing is instantaneous or stratosphere adjusted forcing in which the radiative perturbation is evaluated at the tropopause instantaneously (on the time scale of radiative transfer) after the perturbation is applied or after the stratospheric temperature responds radiatively to the perturbation. The tropopause is chosen because the troposphere and surface are tightly coupled thermally through radiation and convection. The adjustments of stratospheric temperature are included in the forcing because its response time is on the order of months, and the subsequent forcing thus represents a more effective measure of radiative effect on the troposphere-surface system (Hansen et al. 1997; Stuber et al. 2001). We note that the magnitude of radiative forcing at the TOA is the same as that at the tropopause after the stratospheric adjustments, because any vertical convergence of radiative fluxes in the stratosphere would cause local temperature to adjust (Hansen et al. 1997).

When radiative forcing F is applied to the TOA, the energy budget equation with the net TOA radiation, called N, may be written in the simplest form as

$$N = F - \lambda \Delta T \qquad (1)$$

where λ is termed the *climate feedback parameter* and represents how much energy is lost to space in accordance with the unit increase of the global mean surface temperature T (e.g., Gregory et al. 2015; Gregory et al. 2004; Knutti and Hegerl 2008; Winton et al. 2010). Symbol Δ denotes the deviation (or anomaly) from the unperturbed climate. Parameter λ must be positive in order for the system to eventually reach a stable state. We note that a "signed" feedback parameter, e.g., $\Lambda \equiv -\lambda$, which has a positive/negative value for the amplification/suppression of initial perturbation (i.e., positive/negative feedback), is also widely used in other literature (e.g., Armour et al. 2013; Boer and Yu 2003; Winton et al. 2013a). N reflects mostly ocean heat uptake with other forms of energy consumption being latent heat for snow/ice melting, for example (Rhein et al. 2013), and eventually diminishes to zero under constant radiative forcing.

When the feedback is distinguished from the forcing as the surface temperature-mediated response of the climate system as in Eq. (1), it is more consistent to include the so-called rapid tropospheric adjustments in the definition of forcing. The troposphere may respond to forcing directly and rapidly (typically, «1 year) without invoking a surface temperature change. Examples of the tropospheric adjustments include cloud response to lower tropospheric warming due to the increase in atmospheric CO_2 concentration (cloud adjustments, Gregory and Webb 2008), cloud albedo and lifetime changes due to the increase in aerosols (indirect effect of aerosols), and cloud amount changes due to the increase in light-absorbing aerosols (semi-direct effect of aerosols). The resulting forcing is referred to the stratosphere-troposphere-adjusted radiative forcing or *effective radiative forcing* (ERF). For details on the mechanism of tropospheric adjustments, readers are referred to Andrews et al. (2012a), Kamae et al. (2015), and Sherwood et al. (2015).

It is often assumed that λ is constant, i.e., $\lambda = \lambda_{eq}$ at equilibrium, which leads to the conclusion that the transient response (ΔT) to constant radiative forcing is always smaller than the equilibrium response (ΔT_{eq}). Therefore, a rewrite of Eq. (1) indeed gives

$$\Delta T = \frac{F-N}{\lambda} < \Delta T_{eq} = \frac{F}{\lambda_{eq}} \tag{2}$$

The time variation of λ is extensively discussed later. The global mean surface temperature change at the time of the doubled atmospheric CO_2 concentration (average of 61 to 80 years in practice) under a 1 % compound annual increase (1pctCO2 experiment of the Coupled Model Intercomparison Project 5 (CMIP5)) is termed *transient climate response* (TCR, Cubasch et al. 2001). The TCR satisfies

$$\text{TCR} = \frac{F_{2\times}-N}{\lambda} < \text{ECS} = \frac{F_{2\times}}{\lambda_{eq}} \tag{3}$$

where $F_{2\times}$ denotes the radiative forcing due to a doubling of the atmospheric CO_2 concentration. Figure 1 shows the TCR against ECS for the CMIP5 atmosphere-ocean general circulation models (AOGCMs) (Taylor et al. 2012), in which the TCR is indeed smaller than the ECS and a high correlation exists between the two (Flato et al. 2013).

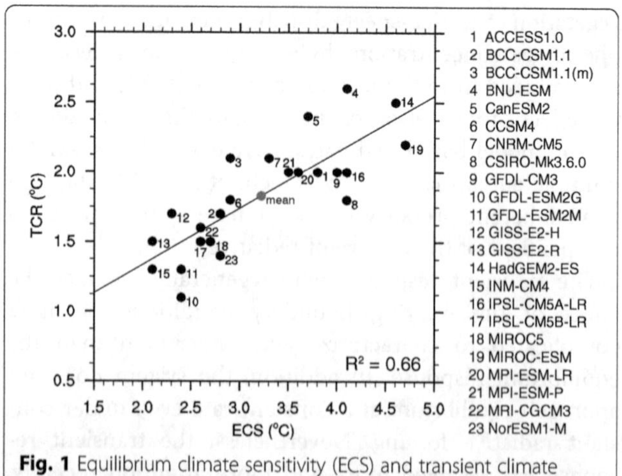

Fig. 1 Equilibrium climate sensitivity (ECS) and transient climate response (TCR) of 23 CMIP5 models. Redrawn from the data in Table 9.5 and Figure 9.42b in Flato et al. (2013). The linear regression line is also plotted. The *number at the lower right corner of the figure* indicates the fractional variance shared by ECS and TCR. The *right column* lists model names

Time scale of the response

In this section, the response time of the climate system is discussed, while still assuming that λ is constant. As stated above, N essentially reflects ocean heat uptake and may thus be expressed as

$$N = \frac{d(C\Delta T)}{dt} \tag{4}$$

where $C = \rho c_p h$ (ρ: density of seawater, c_p: specific heat of seawater at constant pressure, h: effective depth) is the effective heat capacity. Here, the term "effective" is used to represent the fact that the extent of penetration of the heat anomaly with respect to the unperturbed climate varies with time and space. A substitution of Eq. (4) with constant heat capacity into Eq. (1) yields

$$C\frac{d(\Delta T)}{dt} = F-\lambda\Delta T \tag{5}$$

At equilibrium, we obtain

$$\Delta T_{eq} \equiv \Delta T(t\to\infty) = \frac{F}{\lambda}, \tag{6}$$

and at the transient state, we obtain

$$\Delta T = \frac{F}{\lambda}\left(1-e^{-t/\tau}\right) \text{or} \frac{\Delta T}{\Delta T_{eq}} = 1-e^{-t/\tau} \tag{7}$$

where $\tau \equiv C/\lambda$ (Hansen et al. 1981; Hansen et al. 1984; Hansen et al. 1985; Wigley and Schlesinger 1985). Equation (7) means that ΔT reaches 63 % of ΔT_{eq} at $t = \tau$, and the rate depends on the effective heat capacity and the climate feedback parameter. Substitutions of $C = 2.1 \times 10^8$ or $C = 1.5 \times 10^{10}$ J K^{-1} m^{-2}, corresponding to a typical ocean mixed-layer depth (50 m) or the average depth of world oceans (3700 m), respectively, and a nominal value of $\lambda = 0.75$ W m^{-2} K^{-1} (corresponding to the ECS of about 3 °C), yield $\tau \cong 5$ or $\tau \cong 370$ years for the two ocean depths, respectively. The value of $\tau \cong 5$ does not include the effect of the heat exchange between the mixed layer and the deep ocean, and the $\tau \cong 370$ value assumes that the heat anomaly is communicated instantly throughout the water column. Both cases are unrealistic but give some insight into the response time. In reality, the heat anomaly spreads gradually from the surface to depth, and thus C varies with time. Schwartz (2007, 2008) applied a single-layer model with a similar formulation to estimate the ECS from the observation data, but their use of constant effective heat capacity for the entire climate system on all time scales of response was criticized (Foster et al. 2008; Knutti et al. 2008; Scafetta 2009). Donohoe et al. (2014) reported that the effective heat capacity is equivalent to about 50-m ocean depth during the first decade (consistent with our estimate of 5 years for the mixed-layer ocean) and equivalent to several hundred meters ocean depth after a century following an instantaneous quadrupling of atmospheric CO_2 concentration (abrupt4xCO2 experiment) for the CMIP5 AOGCMs.

As an intermediate model that ties the two limiting cases of box-ocean models with a mixed layer or with full depth, Gregory (2000) introduced the following two-layer ocean model to investigate the transient temperature response:

$$C\frac{d(\Delta T)}{dt} = F - \lambda\Delta T - H \text{ (upper ocean)} \tag{8}$$

$$C_D\frac{d(\Delta T_D)}{dt} = H \text{ (deep ocean)} \tag{9}$$

where the heat exchange between the two layers is assumed to be proportional to their difference in anomalous temperature with respect to the unperturbed climate:

$$H = \gamma(\Delta T - \Delta T_D) \tag{10}$$

where γ is the proportionality coefficient. The heat capacity of the atmosphere, land, and other surface components of the climate system is small and implicitly included in the heat capacity of the upper ocean (C), and C is much smaller than the heat capacity of the deep ocean (C_D), i.e., $C << C_D$. We note that the addition of Eqs. (8) and (9) yields Eq. (1). That is, the surface temperature change and the ocean heat uptake hold a linear relation under constant radiative forcing in the two-layer model. Geoffroy et al. (2013a) applied this model to analyze the CMIP5 AOGCM simulations and showed that the two-layer models tuned to emulate the global mean surface temperature response of AOGCMs in the abrupt4xCO2 experiment can also emulate the result of the slowly increasing CO_2 (1pctCO2) experiment reasonably well. This supports the assumption that the two-layer model is effective in capturing the minimal physics of the global mean surface temperature transient response.

The upper bound of the ocean response time can be estimated by the following scaling argument based on the fact that the important slow process of heat transport is vertical eddy diffusion:

$$\rho c_p\frac{d(\Delta T)}{dt} = \kappa_T\frac{\partial^2 T}{\partial z^2} \text{ or } \frac{d(\Delta T)}{dt} = \tilde{\kappa}_T\frac{\partial^2 T}{\partial z^2} \tag{11}$$

where κ_T is the *thermal conductivity* and $\tilde{\kappa}_T = \kappa_T/(\rho c_p)$ is the *thermal diffusivity* by eddies. A substitution of the effective value of $\tilde{\kappa}_T \sim 10^{-4}$ m^2 s^{-1} and $h = 3700$ m yields a time scale of about 4000 years ($\tau \sim h^2/\tilde{\kappa}_T$). Figure 2 shows the long-term integration of the MIROC3.2 AOGCM (K-1 model developers 2004) for the first 140 years of the 1pctCO2 experiment and the continuous integration under the four times constant CO_2 level (const4xCO2) experiment thereafter. For reference, the two times constant CO_2 level (const2xCO2) experiment, starting from 70 years of the 1pctCO2 experiment, is also plotted. We note that the equilibrium response of these experiments was studied in Yoshimori et al. (2014) and the entire simulations were used to study marine biogeochemical response in Yamamoto et al. (2014). While the surface air temperature responds to the forcing rather quickly, the ocean temperature changes much more slowly. The order of the time scale for the

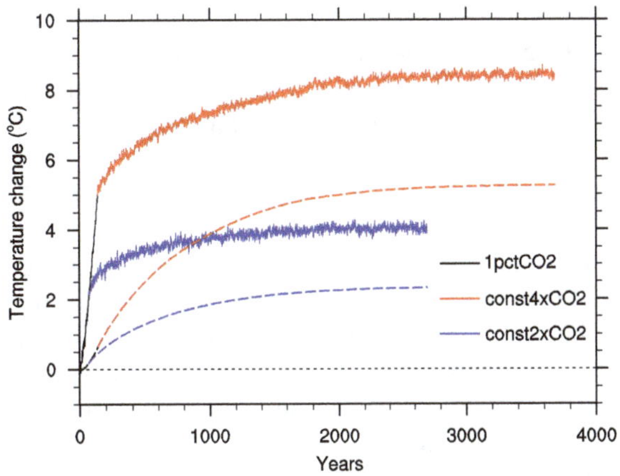

Fig. 2 Global and annual mean temperature changes from the preindustrial control simulation in the MIROC3.2 (medres) AOGCM experiments. *Solid lines*: surface air temperature; *dashed lines*: potential temperature averaged over the full ocean depth; *black lines*: 1 % compound annual increase of CO_2 (1pctCO2, 1–140 years); *blue lines*: constant 2xCO_2 forcing starting from year 70 of 1pctCO2 (const2xCO2, 71–2699 years); *red lines*: constant 4xCO_2 forcing starting from year 140 of 1pctCO2 (const4xCO2, 141–3685 years)

ocean to reach a new steady state in Fig. 2 is loosely consistent with the above estimate based on the scaling argument of 4000 years (i.e., 10^3 years). We note that multi-millennial simulations for the increased CO_2 concentration are not available for the CMIPs.

Traditionally, one-dimensional box-diffusion ocean models (Hansen et al. 1985; Long and Collins 2013; Siegenthaler and Oeschger 1984; Wigley and Schlesinger 1985) or upwelling-diffusion (UD) ocean models (Andronova and Schlesinger 2001; Baker and Roe 2009; Johansson et al. 2015; Wigley and Raper 2001) have been used. AOGCM studies, however, point out the following limitations of such models: (a) the deep convection at high latitudes identified as an important process in AOGCMs is not properly represented (Gregory 2000); (b) the vertical structure of the heat anomaly stored in the UD ocean models is biased to a shallower depth as compared to an AOGCM (Li et al. 2013); and (c) the total heat anomaly stored in the UD ocean models, which corresponds to the thermosteric sea level rise, is underestimated as compared to an AOGCM (Li et al. 2013).

Kostov et al. (2014) showed that the time scale of the penetration depth of the heat anomaly with respect to the unperturbed climate is positively correlated with the depth and strength of the Atlantic meridional overturning circulation (AMOC) on a century time scale in the eight CMIP5 AOGCMs. This result suggests that the AMOC plays an important role in setting the rate of global ocean heat uptake. In addition to the AMOC, Marshall and Zanna (2014) used a conceptual model to point out the importance of ocean eddy mixing in the Southern Ocean, with larger mixing leading to a faster (yet smaller) global heat uptake response.

Ocean heat uptake efficiency

The ocean heat uptake under the monotonically increasing radiative forcing is often approximated by

$$N = \kappa \Delta T \qquad (12)$$

where κ is called the *ocean heat uptake efficiency* (Gregory and Mitchell 1997; Raper et al. 2002). It is assumed that κ is either constant or slowly varies with ΔT. Physically, Eq. (12) represents that heat exchange between the upper and deep oceans occurs in proportion to the surface temperature change and that the deep ocean is implicitly assumed to have an infinite heat capacity with a constant temperature. This assumption imposes a major limitation on the use of ocean heat uptake efficiency; however, it should be noted that this limitation does not affect the two-layer ocean model introduced earlier. Indeed, Eq. (12) can be derived from Eqs. (8)–(10) by assuming that the total ocean heat uptake is dominated by the deep ocean ($N \approx H$) and by replacing γ with κ. Equation (12) is invalid for a transient response immediately after a rapid change in radiative forcing ($N \approx F, \Delta T \approx 0$; e.g., abrupt4xCO2 experiment) as it requires an infinite κ. Equation (12) is also invalid on long time scales and in particular near equilibrium because κ approaches zero when the transient response becomes close to the equilibrium ($N \approx 0, \Delta T \neq 0$). Figure 3 shows a schematic diagram of the relation between ΔT and N after a step increase in radiative forcing, where κ represents the slope of the line (thin black line) pointing from the origin to a state in the simulation (a point in the thick black curve). An example of a monotonically increasing CO_2 (1pctCO2) experiment is presented by the black crosses in Fig. 4. Approximate linearity is observed except for the initial warming period, thereby supporting the general validity of Eq. (12) in this case.

The CMIP5 AOGCM simulations of recent climate change showed that κ decreases with time (Watanabe et al. 2013). This result could reflect less heat uptake in more stratified ocean, i.e., faster warming in the upper ocean that suppresses vertical mixing, or warming of the deep ocean that weakens the effect of mixing. Figure 5a displays the relation between the surface air temperature change and the temperature changes for four oceanic layers of different depths. The figure shows that the upper 300 m of ocean warms at half the speed of the surface air and the ocean below 1000 m warms slightly.

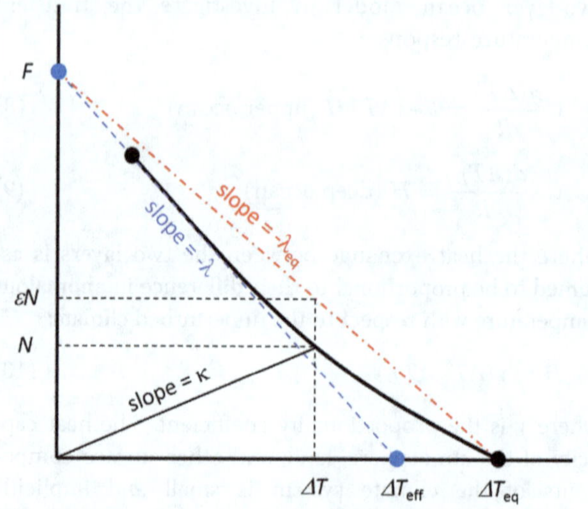

Fig. 3 Schematic diagram of the summary of concepts developed for understanding of the transient response of the global mean surface temperature. The *horizontal* and *vertical axes* are surface temperature and TOA net radiation changes after a step increase in radiative forcing, respectively. Figure A1 of Winton et al. (2013a) is referenced. *N*: TOA net radiation change; *F*: effective radiative forcing; ΔT: transient temperature change; λ: (effective) climate feedback parameter; λ_{eq}: equilibrium climate feedback parameter; κ: ocean heat uptake efficiency; ε: ocean heat uptake efficacy; ΔT_{eff}: effective climate sensitivity (in the case of $2xCO_2$ forcing); ΔT_{eq}: equilibrium temperature change. See the text for the details

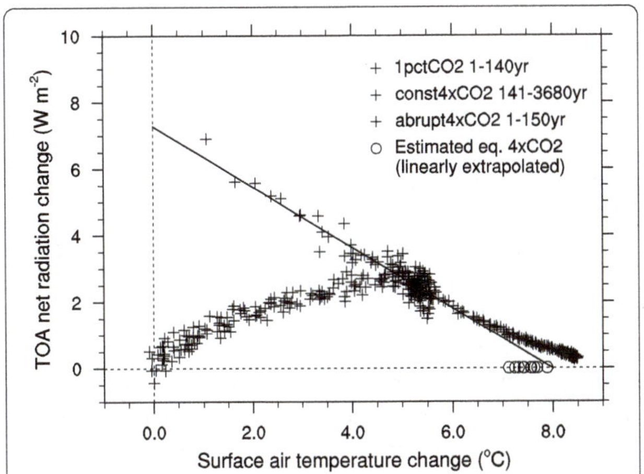

Fig. 4 Relation between global mean surface air temperature and TOA net radiation changes from the preindustrial control simulation in the MIROC3.2 (medres) AOGCM experiments. *Black crosses*: 1pctCO2 experiment (every year for 1–140 years); *blue crosses*: abrutp4xCO2 experiment (every year for 1–150 years); *red crosses*: const4xCO2 experiment (every 20-year average for 141–3680 years); *blue line*: linear regression for the 150-year abrupt4xCO2 experiment; *blue circles*: equilibrium temperature response to 4xCO_2 estimated by Gregory's method for 10 abrupt4xCO2 20-year-long ensemble experiments (that start from different initial conditions in the preindustrial control simulation)

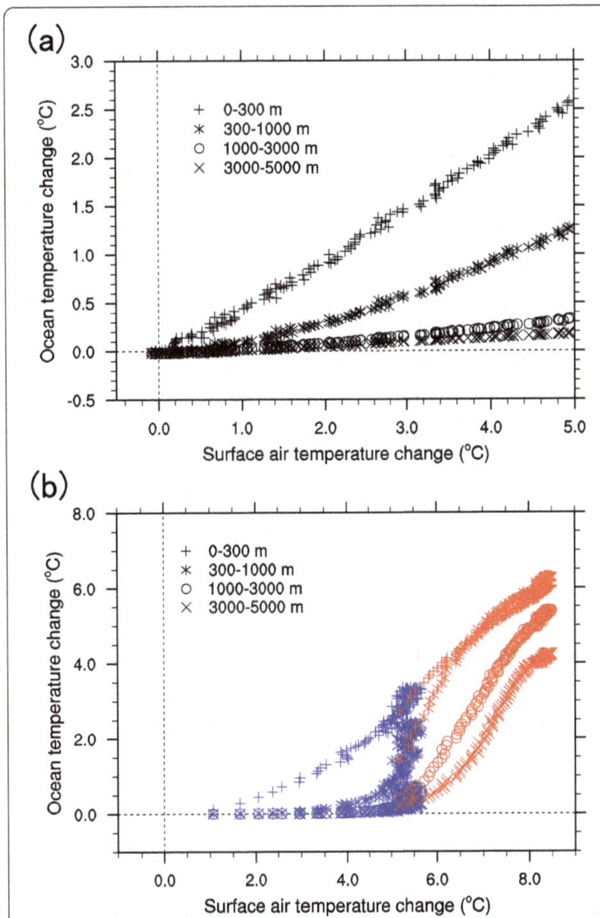

Fig. 5 Global mean ocean temperature changes at four different depth layers from the preindustrial control simulation in the MIROC3.2 (medres) AOGCM experiments. The changes are plotted as functions of the surface air temperature change. **a** 1pctCO2 experiment (every year for 1–140 years). **b** abrupt4xCO2 experiment (*blue*, every year for 1–150 years) and const4xCO2 experiment (*red*: every 20-year average for 141–3680 years)

Thus, Fig. 5a indicates that the vertical stratification is reduced on a centennial scale. Similarly, Fig. 5b displays the millennial scale of the ocean temperature change with respect to the surface air temperature change. The figure shows that the ocean temperature below 1000 m reaches the upper ocean warming after about a century, although temperature changes never reach the same magnitude. This result suggests that the effect of vertical mixing might become smaller. It also shows that the upper ocean does not warm as much as the surface air at the equilibrium.

The ocean heat uptake efficiency is contrasted with the climate feedback parameter. The substitution of Eq. (12) into Eq. (1) yields

$$F = \rho \Delta T \tag{13}$$

where $\rho \equiv \kappa + \lambda$ is termed the *climate resistance* (Gregory and Forster 2008; Gregory et al. 2009, 2010). Here, κ reflects primarily oceanic processes and a larger value indicates that more of the heat anomaly is absorbed in the ocean and hence surface warming is suppressed. On the other hand, λ reflects primarily the atmospheric processes (and sea ice-albedo feedback) and the larger value indicates that excessive energy is more effectively lost to space. The resistance ρ depends on κ (oceanic process) and λ (atmospheric process), and as a result of both processes, the surface temperature change remains small when resistance ρ is large.

By assuming $\lambda = \lambda_{eq}$, arranging Eq. (2) and Eq. (12) yields

$$\frac{\Delta T}{\Delta T_{eq}} = \frac{1}{1 + \kappa/\lambda} \tag{14}$$

As indicated by Eq. (14), the ratio of transient warming to equilibrium warming is larger when κ is smaller or λ is larger. We note that a case of large λ is consistent with Eq. (7). Raper et al. (2002) found a negative correlation between κ and λ in nine CMIP2 AOGCMs. They noted that larger surface warming caused by the smaller radiative damping rate of λ accompanies stronger ocean stratification and less heat release from the interior of the ocean through convection at high latitudes. This relation corresponds to a larger net heat uptake of the global ocean, i.e., a larger κ In the unperturbed climate, the AMOC transports heat to the northern high latitudes, where heat is released to the atmosphere. The AMOC acting as a heat sink of the ocean was reported by Knutti and Stocker (2000) and Marshall and Zanna (2014). The negative correlation found by Raper et al. (2002) was not verified in later studies with a larger number of models, however (Geoffroy et al. 2013a; Gregory and Forster 2008; Plattner et al. 2008). We note that Kuhlbrodt and Gregory (2012) emphasized the dominant role of the Southern Ocean and found the importance of eddy-induced transport in the model discrepancies of global ocean heat uptake.

Dufresne and Bony (2008) used Eq. (12) to quantify the relative contribution of radiative forcing, ocean heat uptake, and climate feedbacks to the spread of the TCR as well as the ECS in the CMIP3 AOGCMs. They concluded that cloud feedback is the predominant factor for the spread of both quantities. Boe et al. (2010), on the other hand, found a positive correlation in the CMIP3 AOGCMs between present-day ocean mixed-layer depths and future deep ocean warming and concluded that ocean heat uptake plays a major role in the uncertainty of transient warming, given that deep ocean warming is negatively correlated with surface temperature. Their analysis was based on the correlation of a single factor and so does not compare quantitatively with other factors, such as radiative forcing and climate feedbacks.

Effective climate sensitivity and ocean heat uptake efficacy

To discriminate the temperature change estimated by using the feedback parameter during the transient period from the ECS, the term *effective climate sensitivity* was introduced (Murphy 1995; Senior and Mitchell 2000) and is defined as follows:

$$\Delta T_{\text{eff}} \equiv \frac{F_{2\times}}{\lambda} = \frac{F_{2\times}\Delta T}{F - N} \tag{15}$$

While Watterson (2000) showed that ΔT_{eff} is stable over time, numerous studies reported that ΔT_{eff} varies with time or is different from the ECS (e.g., Armour et al. 2013; Bitz et al. 2012; Boer and Yu 2003; Gregory et al. 2004; Kiehl et al. 2006; Li et al. 2013; Long and Collins 2013; Murphy 1995; Senior and Mitchell 2000; Williams et al. 2008; Winton et al. 2010).

In applying Eq. (15), stratosphere adjusted radiative forcing or ERF was used for F. One way to estimate the ERF is the regression method by Gregory et al. (2004), in which the ERF is given by the intercept at $\Delta T = 0$ of the ΔT-N linear regression after a step increase of forcing, if assuming Eq. (1). The blue colors in Fig. 4 represent the result of the abrupt4xCO2 experiment, and the blue crosses represent the global and annual mean anomalies for the initial 150 years. The regression line by Gregory's method is indicated by a blue line, which gives an ERF of 7.3 W m^{-2} at the y-intercept. In Fig. 3, the regression line by Gregory's method is indicated by a blue dashed line, and the ERF and effective climate sensitivity are denoted by F and ΔT_{eff}, respectively.

Williams et al. (2008) proposed *effective forcing*, in which the feedback parameter remains constant in time. Their effective forcing is different from the ERF by Gregory's method in that Williams applied linear regression for the stabilization period under constant forcing after 70 years of a monotonic CO_2 increase (1pctCO2) experiment and thus included an initial adjustment of the climate system on the decade level. Winton et al. (2010) argued that the decadal scale adjustment clearly contains the oceanic changes, and hence, it is more appropriate to regard the adjustment as a part of the feedback, rather than as a part of the forcing. Winton introduced a new parameter termed *ocean heat uptake efficacy* (ε), which is described in detail below.

In general, the TOA radiative anomaly induced by a unit surface temperature change during the transient period (λ) is different from that at equilibrium (λ_{eq}). This led Winton et al. (2010) to decompose the temperature response into equilibrium (ΔT_{eq}) and disequilibrium (or remaining) components, such that $\Delta T' \equiv \Delta T_{\text{eq}} - \Delta T$

$$\Delta T_{\text{eq}} = F/\lambda_{\text{eq}} \tag{16}$$

$$\Delta T' = \varepsilon N/\lambda_{\text{eq}} \tag{17}$$

without using the time-variant λ. Combining Eqs. (16) and (17) yields

$$\varepsilon N = F - \lambda_{\text{eq}}\Delta T \tag{18}$$

Therefore, ΔT for a given ocean heat uptake N under a known forcing is determined by a single time-varying parameter ε. We note that Eqs. (16)–(18) correspond to the red dash-dot line in Fig. 3. The feedback parameter is introduced for the ocean heat uptake, λ_o, such that

$$N \equiv \lambda_o \Delta T', \tag{19}$$

and substitution into Eq. (17) (Rose et al. 2014) yields

$$\varepsilon = \lambda_{\text{eq}}/\lambda_o \tag{20}$$

Here, Eq. (20) means that the ocean heat uptake efficacy parameter represents the ratio between feedbacks operating under radiative forcing and ocean heat uptake alone.

The ocean heat uptake efficacy is analogous to the *efficacy of forcing*. The efficacy of forcing is defined as the ratio of ΔT caused by a specific forcing agent to that induced by the CO_2 of the equivalent radiative forcing

(Hansen et al. 1997; Hansen et al. 2005; Yoshimori and Broccoli 2008):

$$\varepsilon_f \equiv \frac{\Delta T_{\text{non-CO}_2}/F_{\text{non-CO}_2}}{\Delta T_{\text{CO}_2}/F_{\text{CO}_2}} = \frac{\lambda_{\text{CO}_2}}{\lambda_{\text{non-CO}_2}} \quad (21)$$

In other words, the efficacy parameter of forcing represents the ratio between feedbacks operating under CO_2 and non-CO_2 forcing alone. To emphasize the similarity to Eqs. (16) and (17), Eq. (21) is written as

$$\Delta T_{\text{CO}_2} = F_{\text{CO}_2}/\lambda_{\text{CO}_2}, \quad (22)$$

$$\Delta T_{\text{non-CO}_2} = \varepsilon_f F_{\text{non-CO}_2}/\lambda_{\text{CO}_2} \quad (23)$$

Combining Eqs. (16) and (17), we obtain

$$\frac{\Delta T}{\Delta T_{\text{eq}}} = 1 - \frac{\varepsilon N}{F} \quad (24)$$

As long as Eq. (12) is valid, Eq. (24) is also expressed as

$$\frac{\Delta T}{\Delta T_{\text{eq}}} = \left(1 + \frac{\varepsilon \kappa}{\lambda_{\text{eq}}}\right)^{-1} \quad (25)$$

Therefore, Eq. (25) indicates that larger ocean heat uptake efficacy or efficiency leads to a smaller ratio of transient warming to equilibrium warming (Winton et al. 2010).

Geoffroy et al. (2013a) extended the study of Geoffroy et al. (2013b) by applying the following modified version of the two-layer model proposed by Held et al. (2010):

$$C\frac{d(\Delta T)}{dt} = F - \lambda_{\text{eq}}\Delta T - \varepsilon_0 H \text{ (upper ocean)} \quad (26)$$

$$C_D\frac{d(\Delta T_D)}{dt} = H \text{ (deep ocean)} \quad (27)$$

where ε_0 is a constant parameter of ocean heat uptake efficacy. By rewriting, Eq. (26) is more easily compared with Eq. (8)

$$C\frac{d(\Delta T)}{dt} = F - \lambda_{\text{eq}}\Delta T - H + (1 - \varepsilon_0)H \quad (28)$$

The last term on the right side represents the time-varying effect of λ, as explained below.

By combining Eqs. (26) and (27), the ocean heat uptake is written as

$$\begin{aligned} N &= C\frac{d(\Delta T)}{dt} + C_D\frac{d(\Delta T_D)}{dt} \\ &= F - \lambda_{\text{eq}}\Delta T - (\varepsilon_0 - 1)H \end{aligned} \quad (29)$$

In Fig. 3, Eq. (29) ties the two limiting cases for the slope α of the ΔT-N relation: Eq. (29) represents $\alpha = -\lambda_{eq} - \varepsilon_0\gamma$ while the surface temperature is initially adjusting to the imposed forcing ($\Delta T_D << \Delta T$) and the upper ocean dominates the total ocean heat uptake. Then, $\alpha = -\lambda_{eq}/\varepsilon_0$ while the deep ocean temperature is catching up with the surface warming ($\Delta T_D \propto \Delta T$) and the deep ocean dominates the total ocean heat uptake. Again, the efficacy parameter (ε_0) is assumed to be constant. We note the similarity of Eq. (29) to Eq. (18) in Eq. (30):

$$N = F - \lambda_{\text{eq}}\Delta T - (\varepsilon - 1)N \quad (30)$$

However, Eqs. (29) and (30) are equivalent ($N = H$) only when the heat uptake by the upper ocean Eq. (26) is much smaller than that by the deep ocean Eq. (27). We also note that no piecewise behavior is assumed in Eq. (18); the time-varying efficacy parameter (ε) captures the curvature of the slope in Fig. 3. According to Geoffroy et al. (2013a), the model with ocean heat uptake efficacy mimics the CMIP5 AOGCM simulations more closely because it can represent nonlinearity between the surface temperature change and the ocean heat uptake under constant radiative forcing. An example of the nonlinear curvature of the ΔT-N relation in an AOGCM is denoted by the blue and red crosses in Fig. 4. The equilibrium temperature response estimated by Gregory's method using the initial 20-year (blue circles) or 150-year integration (blue line) in the abrupt4xCO2 experiment tends to be smaller than the "true" equilibrium temperature response obtained from the long millennial integration.

Geoffroy et al. (2012) demonstrated that the spread of the TCR in the CMIP5 AOGCMs is caused by climate feedbacks (λ_{eq}), radiative forcing (F), and ocean heat uptake efficacy (ε_0), in the order of large to small contributions. Their result is consistent with that of Dufresne and Bony (2008) and leads to the conclusion that the uncertainty in cloud feedback is a dominant factor to the uncertainty in both ECS and TCR.

Time dependence of global feedbacks and physical meaning of ocean heat uptake efficacy

Armour et al. (2013) emphasized the advantage of a local/regional feedback framework, in which feedbacks are formulated as the TOA radiative anomaly per unit local surface temperature change, rather than per unit global mean surface temperature change (i.e., global feedback framework). In their local feedback framework, radiative damping is approximated by

$$\lambda(t)\Delta T(t) = \overline{\lambda(r,t)\Delta T(r,t)} \approx \overline{\lambda(r)\Delta T(r,t)}, \quad (31)$$

where r denotes location and the overbar represents the global average of local variables. The time variation of global mean radiative damping is therefore determined by the time-invariant local feedback parameter $\lambda(r)$ and the time-varying spatial pattern of surface temperature change $\Delta T(r,t)$. If there is no change in the horizontal energy transport of the atmosphere to compensate, the local ocean heat uptake and the corresponding net TOA radiation must be balanced by the local radiative damping to reach equilibrium. Thus, the shift of ocean heat uptake to the region of a less positive local feedback parameter means greater surface warming remains to occur before the equilibrium. In this local feedback framework, an increase of ocean heat uptake efficacy is interpreted as the region of ocean heat uptake shifting to the region of the less positive local feedback parameter, such as that for subpolar oceans. Figure 6 shows the anomalous sea surface temperature (SST) patterns at different times of the increasing CO_2 experiment in (a) and (b) and their difference in (c). The surface warming is more dominant in high latitudes at later times presumably because of the large effective heat capacity in the Southern Ocean and the progress in the Arctic sea ice melting. The difference in the local feedback parameters between low and high latitudes is the time-dependent global mean radiative feedback parameter. As an extension, a larger efficacy of forcing is interpreted as the radiative forcing applied to the region having the less positive local feedback parameter. In an idealized model configuration, Rose et al. (2014), however, showed that local feedback parameters can be sensitive to the latitudinal difference in ocean heat uptake. Because the pattern of ocean heat uptake varies with time, their result challenges the validity of the assumption that local feedback parameters are fixed in time. By conducting atmospheric general circulation model (AGCM) experiments with a fixed SST pattern (scaled by the global mean surface temperature change), Andrews et al. (2015) verified that the time variation of the global feedback parameter develops from the evolving pattern of surface temperature change. They also questioned whether that the local feedback framework explains everything. The dynamical effect of the atmosphere on local feedback parameters, therefore, requires further study (Feldl and Roe 2013; Rose et al. 2014).

Winton et al. (2013b) proposed an indirect radiative influence of the AMOC. They argued that weakening of the AMOC under global warming reduces the heat transport to the northern high latitudes and that corresponds to a shift of ocean heat uptake regions from low to high latitudes. This shift results in the increase of the ocean heat uptake efficacy and slows down the global surface temperature rise. As mentioned already, Kostov et al. (2014), on the other hand, proposed that the stronger AMOC enhances the penetration of the heat anomaly into deeper ocean in the perturbed climate, and the resulting increase in the effective heat capacity of the climate system slows down the global surface temperature rise. Therefore, the quantitative contribution of the AMOC still needs to be established.

Meraner et al. (2013) argued another aspect of inconstancy of the global feedback parameter, in which the

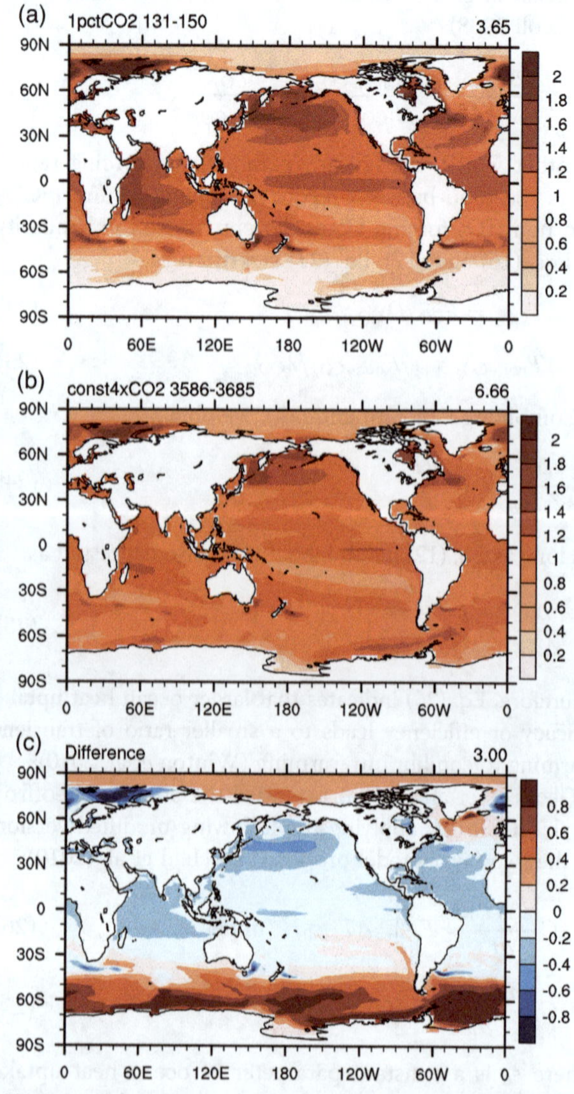

Fig. 6 SST changes normalized by the global mean SST changes from the control: **a** at the time of quadrupled CO_2 concentration in the 1pctCO2 experiment (average of 131–150 years); **b** at equilibrium in the const4xCO2 experiment (average of 3586–3685 years); and **c** the difference between **b** and **a**. The *number at the upper right corner of each figure* indicates the global mean SST changes from the control in degree Celsius

parameter becomes less positive and hence the ECS becomes larger in a warmer climate through enhanced water vapor feedback. The state dependency of climate feedbacks has been discussed in many previous studies (e.g., Colman and McAvaney 2009; Colman et al. 1997; Feldl and Roe 2013; Hansen et al. 2005; Jonko et al. 2013; Yoshimori et al. 2011), but the result appears to depend on the model and so the effect of the dependency on the transient response still needs to be established. In a similar context, the possibility of inconstancy of the climate resistance parameter (as well as the global feedback parameter) in the CMIP5 AOGCMs was reported by

Gregory et al. (2015), although they also listed errors in the estimated radiative forcing as an alternate or additional possibility.

In summary, time-varying global feedbacks or radiative damping may arise from (a) the state dependency of climate feedbacks, that is, the global feedback parameter depends on the background temperature T; (b) changes in the spatial pattern of the surface temperature anomaly $\Delta T(r,t)$ associated with ocean heat uptake and dynamics (ocean heat uptake efficacy) or forcing agent (efficacy of forcing); and (c) the non-local effect of atmospheric dynamics resulting from (b) on the local feedback parameter $\lambda(r,t)$.

Definitions and estimates of forcing

The time variation of the global climate feedback parameter has implications for how forcing and feedback are to be defined and computed. As discussed already, the nonlinearity of the relation between surface temperature and net radiation changes may introduce an error or arbitrariness into the estimate by Gregory's method. By using Eq. (1), a more general time-varying radiative forcing was estimated by Forster et al. (2013), but the global climate feedback parameter was necessarily assumed to be independent of the forcing agent and also time invariant. In addition to Gregory's method, the ERF was computed by using the fixed SST method by Hansen et al. (1997) and Hansen et al. (2005). In Hansen's method, the AGCM was run to equilibrium under constant forcing without allowing the SST to change. Because the typical time scale of adjustments was shorter than a year, Hansen's method might be applicable to time-varying forcing as well. The notable difference between Gregory's and Hansen's methods is that only the global surface temperature change is constrained to zero in Gregory's method. On the other hand, a local SST change is not allowed in Hansen's method, although it allows the land surface temperature to respond (Shine et al. (2003) suppressed the land surface response as well as SST change). A high correlation exists between the two methods, but Gregory's method tends to estimate a smaller ERF than does Hansen's method (Fig. 7). In both cases, the precise time scale of the adjustments is somewhat ambiguous. While the regression method is useful to investigate the relative spread among models (Andrews et al. 2015; Andrews et al. 2012b), it is expected that magnitude of forcing depends on the choice of period during the transient response, because the ΔT-N relation is nonlinear in general (Andrews et al. 2012b; Chen et al. 2014; Gregory et al. 2004; Winton et al. 2010). We note that the dominant factor of the model spread of the CO_2 ERF is still instantaneous forcing, not tropospheric adjustments (Chung and Soden 2015). As mentioned already, the effective forcing introduced by Williams et al. (2008) precludes the definition of feedback as a surface temperature-mediated response, because the decadal adjustments include the SST change.

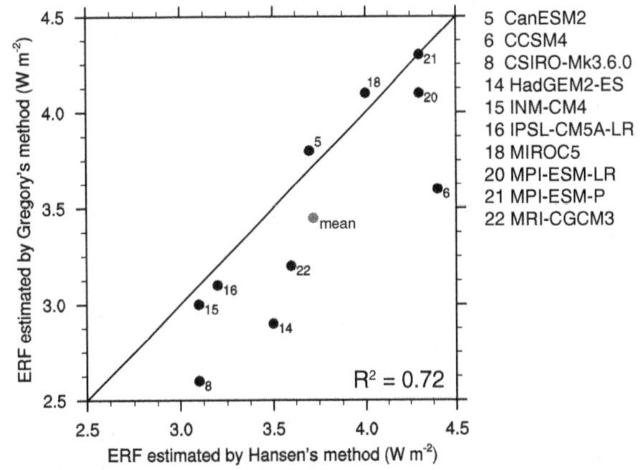

Fig. 7 Effective radiative forcing for the doubling of atmospheric CO_2 concentration estimated with Hansen's method and Gregory's method for 10 CMIP5 models. See the text for the details. The data taken from Table 9.5 of Flato et al. (2013). The one-to-one line is also plotted. The *number at the lower right corner of the figure* indicates the fractional variance shared by the two variables. The *right column* lists the model names with their numbers, which are identical to those in Fig. 1

Implication for the estimate of ECS

As stated in an earlier section, it takes a couple of millennia for a comprehensive climate model or AOGCM to reach a new equilibrium. Therefore, AOGCMs are not routinely run to equilibrium, and the ECS is not obtained precisely by following the definition. Traditionally, atmospheric GCMs coupled to "slab" mixed-layer ocean models (ASGCMs) are used to obtain the ECS. The mixed layer usually has a small yet sufficiently large heat capacity to simulate the annual cycle, and the steady state without considering the change in ocean dynamics is obtained after a few decades of integration. While some ASGCMs exhibit reasonably close ECSs to those obtained by the long integration of AOGCMs, other studies point out the difference (Danabasoglu and Gent 2009; Li et al. 2013; Shell 2013; Stouffer and Manabe 1999; Yokohata et al. 2008). Gregory's method was adapted by Andrews et al. (2012b) for the CMIP5 AOGCM analysis. It is expected that the linear part of the ΔT-N relation at the later stage of integration (fitting a line to the simulated trajectory closer to the equilibrium point in Figs. 3 and 4) yields an accurate ECS (Armour et al. 2013). Geoffroy et al. (2013a) applied the two-layer energy balance model with the ocean heat uptake efficacy parameter to the CMIP5 AOGCMs and estimated the optimal ECS and other parameters simultaneously. As a consequence, the model captures the nonlinearity of the ΔT-N relations. Although the mean of 16 estimated ECSs differs by only about 8 % compared to that estimated by the ocean heat uptake efficacy parameter of one (Geoffroy et al. 2013b), the difference is more than 20 % for the two AOGCM cases.

The nonlinearity occurring from the time variation of the global climate feedback parameter or the ocean heat uptake efficacy has a very important implication when the ECS is estimated from observation data. A general tendency of the global climate feedback parameter that is decreasing or the ocean heat uptake efficacy that is larger than one indicates that the ECS based on the energy budget may be underestimated when the transient data are used (Geoffroy et al. 2013a; Winton et al. 2010). In other words, the deviation of the effective climate sensitivity from the equilibrium climate sensitivity illustrated in Fig. 3 is not negligible, as pointed out in many studies (e.g., Armour et al. 2013; Rose et al. 2014; Winton et al. 2010). Because the observed data represent the transient stage of the response under increasing greenhouse gas forcing, this deviation is a concern for the observation-based estimate of the ECS (e.g., Forster and Gregory 2006; Gregory et al. 2002; Lewis and Curry 2015; Otto et al. 2013). The inconstancy of the climate resistance parameter yields similar concerns if the ECS estimate is made based on the assumption of its constancy (Gregory et al. 2015; Gregory and Forster 2008). We note that the efficacy of forcing is another important factor that may distort the estimate of the ECS from observations, as pointed out by Kummer and Dessler (2014), Shindell (2014), and Shindell et al. (2015). Marvel et al. (2015) argued that the best estimate of the TCR moves upward from 1.3 to 1.8 °C and the ECS from 2.0 to 2.9–3.0 °C if the efficacy of forcing based on a single model is taken into account. Therefore, even though the time variation of the climate feedback parameter is taken into account, the lack of the forcing efficacy factor (e.g., Masters 2014) may introduce bias in the ECS estimate. So far, the time variation of the forcing efficacy has not been well established. While climate models of reduced complexity enable us to constrain the ECS and TCR with statistical inference by conducting a large ensemble of parameter perturbations, we must not overlook the fact that the effect of the ocean heat uptake efficacy and the forcing efficacy are usually not taken into account (e.g., Aldrin et al. 2012; Skeie et al. 2014). This is not only limited to the ECS estimate based on historical observations but also is closely related to the ECS estimate based on a volcanic event (Merlis et al. 2014) or a past climate (Yoshimori et al. 2009).

Conclusions

Correct understanding of the transient response to radiative perturbation is important for the interpretation of observed climate change, the prediction of near-future climate change, and committed warming under climate stabilization scenarios, as well as the estimation of ECS based on observation data. It has been known for some time that the radiative damping rate per unit of global mean surface temperature increase (i.e., global climate feedback parameter) varies with time. This inconstancy affects the transient global mean surface temperature change, and different studies have emphasized the various different aspects of this inconstancy.

Recent progress has focused not only on the speed of the ocean heat uptake but also the spatial pattern of the ocean heat uptake. The spatial pattern influences the global radiative damping and thus the global climate feedback parameter. These influences arise from the evolving spatial pattern of the surface temperature anomaly and/or changes in the local feedback parameter. The inconstancy of the global climate feedback may be formulated by a single parameter, called ocean heat uptake efficacy. In addition, the dependency of the global climate feedback parameter on the background climate has also been pointed out. The physical mechanisms behind the behavior of the global climate feedback parameter still need to be established. Detailed analysis of the atmosphere-ocean interaction as well as perturbed physics ensemble experiments including the ocean component might be useful for identifying the physical mechanisms (Collins et al. 2006). Ocean heat uptake processes associated with the AMOC and those operating in the Southern Ocean are of great importance in understanding both speed and spatial pattern of ocean heat uptake and require further study.

The efficacy of forcing is another important factor determining the transient response to radiative perturbation. The current understanding of the forcing efficacy is very limited, and its time variation is particularly not well understood. Nevertheless, its inconstancy is presumed to be in the latitudinal distribution of forcing that varies with time.

All of these studies require the estimate of radiative forcing in a consistent manner. As proposed by Radiative Forcing Model Intercomparison Project,[1] ERF estimated by using AGCMs with a fixed SST, which does not require extrapolation, would greatly advance the study in a more quantitative way. Because the extrapolation using the regression technique in estimating the ECS would introduce ambiguity, we emphasize the usefulness of the integration of AOGCMs to equilibrium under doubled and quadrupled CO_2 forcing as presented in this review article and other studies with different ocean parameterizations (Yamamoto et al. 2015; Yamamoto et al. 2014).

In the fifth intergovernmental panel on climate change (IPCC) assessment report (IPCC, 2013; IPCC-AR5), the lower bound of the ECS estimate was revised from the fourth IPCC assessment report from 2 to 1.5 °C. In addition, the dispersion in the ECS estimate based on the energy budget and the emergent constraint (the empirical relation found in models between the ECS and observable variable) (Fasullo et al. 2015; Klein and Hall 2015) by using observation data precludes the IPCC-AR5 from providing the best estimate. These issues might be resolved at least partially by taking both ocean heat uptake and forcing efficacies into account.

Abbreviation

AGCM, atmospheric GCM; AMOC, Atlantic meridional overturning circulation; AOGCM, atmosphere- ocean GCM; ASGCM, atmosphere-slab ocean GCM; CMIP, coupled model intercomparison project; ECS, equilibrium climate sensitivity; ERF, effective radiative forcing; GCM, general circulation model; IPCC, intergovernmental panel on climate change; IPCC-AR5, the fifth assessment report of the IPCC; UD, upwelling-diffusion; SST, sea surface temperature; TCR, transient climate response; TOA, top of the atmosphere

Acknowledgements

We thank Dr. Kyle Armour and one anonymous reviewer whose suggestions helped to greatly improve the manuscript. Dr. Armour pointed out the piecewise interpretation of Eq. (29) and drew the authors' attention to the importance of AMOC. Figure 4 was inspired by Dr. Isaac Held's blog. The authors are also thankful to the MIROC model developing team. The developers of NCL are also appreciated. Simulation data were produced by using the JAMSTEC Earth Simulator and the NIES super computer system. This work was supported by JSPS KAKENHI Grant Number 15K05280.

Authors' contributions

MY drafted the manuscript. AAO, RO, and HS provided the MIROC simulation data, and MY conducted the analysis. All of the authors made intellectual contributions to the contents and structuring of the manuscript and approved the final manuscript.

Competing interests

The authors declare that they have no competing interests.

Author details

[1]Faculty of Environmental Earth Science, Hokkaido University, Kita 10, Nishi 5, Kita-ku, Sapporo 060-0810, Japan. [2]Arctic Research Center, Hokkaido University, Sapporo, Japan. [3]Atmosphere and Ocean Research Institute, The University of Tokyo, Kashiwa, Japan. [4]Center for Global Environmental Research, National Institute for Environmental Studies, Tsukuba, Japan. [5]Japan Agency for Marine-Earth Science and Technology, Yokohama, Japan. [6]Faculty of Life and Environmental Sciences, University of Tsukuba, Tsukuba, Japan.

References

Abe Ouchi A, Saito F, Kawamura K, Raymo ME, Okuno J, Takahashi K, Blatter H (2013) Insolation-driven 100,000-year glacial cycles and hysteresis of ice-sheet volume. Nature 500:190. doi:10.1038/nature12374

Aldrin M, Holden M, Guttorp P, Skeie RB, Myhre G, Berntsen TK (2012) Bayesian estimation of climate sensitivity based on a simple climate model fitted to observations of hemispheric temperatures and global ocean heat content. Environmetrics 23:253–271. doi:10.1002/env.2140

Allen MR, Frame DJ (2007) Call off the quest. Science 318:582–583

Andrews T, Gregory JM, Forster PM, Webb MJ (2012a) Cloud adjustment and its role in CO_2 radiative forcing and climate sensitivity: a review. Surv Geophys 33:619–635

Andrews T, Gregory JM, Webb MJ (2015) The dependence of radiative forcing and feedback on evolving patterns of surface temperature change in climate models. J Clim 28:1630–1648

Andrews T, Gregory JM, Webb MJ and Taylor KE (2012b) Forcing, feedbacks and climate sensitivity in CMIP5 coupled atmosphere-ocean climate models. Geophysical Research Letters, 39:doi:Artn L09712 10.1029/2012gl051607

Andronova NG, Schlesinger ME (2001) Objective estimation of the probability density function for climate sensitivity. J Geophys Res-Atmos 106:22605–22611

Armour KC, Bitz CM, Roe GH (2013) Time-varying climate sensitivity from regional feedbacks. J Clim 26:4518–4534

Armour KC and Roe GH (2011) Climate commitment in an uncertain world. Geophysical Research Letters, 38:doi:Artn L01707 10.1029/2010gl045850

Baker MB, Roe GH (2009) The shape of things to come: why is climate change so predictable? J Clim 22:4574–4589

Bitz CM, Shell KM, Gent PR, Bailey DA, Danabasoglue G, Armour KC, Holland MM, Kiehl JT (2012) Climate sensitivity of the community climate system model, version 4. J Clim 25:3053–3070

Boe J, Hall A and Qu X (2010) Deep ocean heat uptake as a major source of spread in transient climate change simulations (vol 37, L17701, 2010). Geophysical Research Letters, 37:doi:Artn L17701 10.1029/2010gl044726

Boer GJ, Yu B (2003) Climate sensitivity and response. Clim Dyn 20:415–429

Chen XL, Zhou TJ, Guo Z (2014) Climate sensitivities of two versions of FGOALS model to idealized radiative forcing. Science China-Earth Sciences 57:1363–1373

Chung E-S, Soden BJ (2015) An assessment of methods for computing radiative forcing in climate models. Environ Res Lett 10:074004. doi:10.1088/1748-9326/10/7/074004

Collins M, Booth BBB, Harris GR, Murphy JM, Sexton DMH, Webb MJ (2006) Towards quantifying uncertainty in transient climate change. Clim Dyn 27:127–147

Colman R and McAvaney B (2009) Climate feedbacks under a very broad range of forcing. Geophysical Research Letters, 36:doi:Artn L01702 10.1029/2008gl036268

Colman RA, Power SB, McAvaney BJ (1997) Non-linear climate feedback analysis in an atmospheric general circulation model. Clim Dyn 13:717–731

Cubasch U, Meehl GA, Boer GJ, Stouffer RJ, Dix M, Noda A, Senior CA, Raper S, Yap KS (2001) Projections of future climate change. In: Houghton JT, Ding Y, Griggs DJ, Noguer M, Van Der Linden PJ, Dai X, Maskell K, Johnson CA (eds) Climate Change 2001: The Physical Science Basis. Contribution of Working Group I to the Third Assessment Report of the Intergovernmental Panel on Climate Change. Cambridge University Press, Cambridge, United Kingdom and New York, NY, USA, pp 525–582

Danabasoglu G, Gent PR (2009) Equilibrium climate sensitivity: is it accurate to use a slab ocean model? J Clim 22:2494–2499

Donohoe A, Armour KC, Pendergrass AG, Battisti DS (2014) Shortwave and longwave radiative contributions to global warming under increasing CO_2. Proc Natl Acad Sci U S A 111:16700–16705

Dufresne JL, Bony S (2008) An assessment of the primary sources of spread of global warming estimates from coupled atmosphere-ocean models. J Clim 21:5135–5144

Fasullo JT, Sanderson BM, Trenberth KE (2015) Recent progress in constraining climate sensitivity with model ensembles. Curr Climate Change Reports. doi:10.1007/s40641-015-0021-7

Feldl N, Roe GH (2013) The nonlinear and nonlocal nature of climate feedbacks. J Clim 26:8289–8304

Flato G, Marotzke J, Abiodun B, Braconnot P, Chou SC, Collins W, Cox P, Driouech F, Emori S, Eyring V, Forest C, Gleckler P, Guilyardi E, Jakob C, Kattsov V, Reason C, Rummukainen M (2013) Evaluation of climate models. In: Stocker TF, Qin D, Plattner G-K, Tignor M, Allen SK, Boschung J, Nauels A, Xia Y, Bex V, Midgley PM (eds) Climate Change 2013: The Physical Science Basis. Contribution of Working Group I to the Fifth Assessment Report of the Intergovernmental Panel on Climate Change. Cambridge University Press, Cambridge, United Kingdom and New York, NY, USA, pp 741–866

Forster PM, Andrews T, Good P, Gregory JM, Jackson LS, Zelinka M (2013) Evaluating adjusted forcing and model spread for historical and future scenarios in the CMIP5 generation of climate models. J Geophys Res-Atmos 118:1139–1150

Forster PMD, Gregory JM (2006) The climate sensitivity and its components diagnosed from Earth radiation budget data. J Clim 19:39–52

Foster G, Annan JD, Schmidt GA and Mann ME (2008) Comment on "Heat capacity, time constant, and sensitivity of Earth's climate system" by S. E. Schwartz. Journal of Geophysical Research, 113:doi:10.1029/2007jd009373

Geoffroy O, Saint-Martin D, Bellon G, Voldoire A, Olivie DJL, Tyteca S (2013a) Transient climate response in a two-layer energy-balance model. Part II: Representation of the efficacy of deep-ocean heat uptake and validation for CMIP5 AOGCMs. J Clim 26:1859–1876

Geoffroy O, Saint-Martin D, Olivie DJL, Voldoire A, Bellon G, Tyteca S (2013b) Transient climate response in a two-layer energy-balance model. Part I: analytical solution and parameter calibration using CMIP5 AOGCM experiments. J Clim 26:1841–1857

Geoffroy O, Saint-Martin D and Ribes A (2012) Quantifying the sources of spread in climate change experiments. Geophysical Research Letters, 39:doi:Artn L24703 10.1029/2012gl054172

Gregory J, Webb M (2008) Tropospheric adjustment induces a cloud component in CO_2 forcing. J Clim 21:58–71

Gregory JM (2000) Vertical heat transports in the ocean and their effect an time-dependent climate change. Clim Dyn 16:501–515

Gregory JM, Andrews T, Good P (2015) The inconsistency of the transient climate response parameter under increasing CO_2. Phil Trans R Soc A 373:20140417

Gregory JM and Forster PM (2008) Transient climate response estimated from radiative forcing and observed temperature change. Journal of Geophysical Research-Atmospheres, 113:doi:Artn D23105 10.1029/2008jd010405

Gregory JM, Ingram WJ, Palmer MA, Jones GS, Stott PA, Thorpe RB, Lowe JA, Johns TC and Williams KD (2004) A new method for diagnosing radiative forcing and climate sensitivity. Geophysical Research Letters, 31:doi:Artn L03205 10.1029/2003gl018747

Gregory JM, Jones CD, Cadule P, Friedlingstein P (2009) Quantifying carbon cycle feedbacks. J Clim 22:5232–5250

Gregory JM, Jones CD, Cadule P, Friedlingstein P (2010) Quantifying carbon cycle feedbacks (vol 22, pg 5232, 2009). J Clim 23:6706–6707

Gregory JM, Mitchell JFB (1997) The climate response to CO_2 of the Hadley Centre coupled AOGCM with and without flux adjustment. Geophys Res Lett 24:1943–1946

Gregory JM, Stouffer RJ, Raper SCB, Stott PA, Rayner NA (2002) An observationally based estimate of the climate sensitivity. J Clim 15:3117–3121

Hansen J, Johnson D, Lacis A, Lebedeff S, Lee P, Rind D, Russell G (1981) Climate impact of increasing atmospheric carbon dioxide. Science 213:957–966

Hansen J, Lacis A, Rind D, Russell G, Stone P, Fung I, Ruedy R and Lerner J (1984) Climate sensitivity: analysis of feedback mechanisms. Hansen, J.E. and T. Takahashi (eds.): Climate Processes and Climate Sensitivity, Geophysical Monograph Series, 29, AGU, Washington, D. C., USA, 130-163.

Hansen J, Russell G, Lacis A, Fung I, Rind D, Stone P (1985) Climate response times: dependence on climate sensitivity and ocean mixing. Science 229:857–859

Hansen J, Sato M, Ruedy R (1997) Radiative forcing and climate response. J Geophys Res-Atmos 102:6831–6864

Hansen J, Sato M, Ruedy R, Nazarenko L, Lacis A, Schmidt GA, Russell G, Aleinov I, Bauer M, Bauer S, Bell N, Cairns B, Canuto V, Chandler M, Cheng Y, Del Genio A, Faluvegi G, Fleming E, Friend A, Hall T, Jackman C, Kelley M, Kiang N, Koch D, Lean J, Lerner J, Lo K, Menon S, Miller R, Minnis P, Novakov T, Oinas V, Perlwitz J, Perlwitz J, Rind D, Romanou A, Shindell D, Stone P, Sun S, Tausnev N, Thresher D, Wielicki B, Wong T, Yao M and Zhang S (2005) Efficacy of climate forcings. Journal of Geophysical Research-Atmospheres, 110:doi:Artn D18104 10.1029/2005jd005776

Held IM, Winton M, Takahashi K, Delworth T, Zeng FR, Vallis GK (2010) Probing the fast and slow components of global warming by returning abruptly to preindustrial forcing. J Clim 23:2418–2427

IPCC (2013) Climate change 2013: the physical science basis. In: Stocker TF, Qin D, Plattner G-K, Tignor M, Allen SK, Boschung J, Nauels A, Xia Y, Bex V, Midgley PM (eds) Contribution of Working Group I to the Fifth Assessment Report of the Intergovernmental Panel on Climate Change. Cambridge University Press, Cambridge, United Kingdom and New York, NY, USA, p 1535

Ishizaki Y, Shiogama H, Emori S, Yokohata T, Nozawa T, Ogura T, Abe M, Yoshimori M, Takahashi K (2012) Temperature scaling pattern dependence on representative concentration pathway emission scenarios. Clim Chang 112:535–546

Ishizaki Y, Shiogama H, Emori S, Yokohata T, Nozawa T, Takahashi K, Ogura T, Yoshimori M, Nagashima T (2013) Dependence of precipitation scaling patterns on emission scenarios for representative concentration pathways. J Clim 26:8868–8879

Johansson DJA, O'Neill BC, Tebaldi C, Häggström O (2015) Equilibrium climate sensitivity in light of observations over the warming hiatus. Nat Clim Chang 5:449–453

Jonko AK, Shell KM, Sanderson BM, Danabasoglu G (2013) Climate feedbacks in CCSM3 under changing CO_2 forcing. Part II: variation of climate feedbacks and sensitivity with forcing. J Climate 26:2784–2795

K-1 model developers (2004) K-1 coupled GCM (MIROC) description. Tech. rep., Center for Climate System Research, The University of Tokyo, Tokyo, Japan.

Kamae Y, Watanabe M, Ogura T, Yoshimori M, Shiogama H (2015) Rapid adjustments of cloud and hydrological cycle to increasing CO2: a review. Curr Clim Change Rep 1:103–113

Kiehl JT, Shields CA, Hack JJ, Collins WD (2006) The climate sensitivity of the community climate system model version 3 (CCSM3). J Clim 19:2584–2596

Klein SA, Hall A (2015) Emergent constraints for cloud feedbacks. Curr Clim Change Rep. doi:10.1007/s40641-015-0027-1

Knutti R, Hegerl GC (2008) The equilibrium sensitivity of the Earth's temperature to radiation changes. Nat Geosci 1:735–743

Knutti R, Krähenmann S, Frame DJ and Allen MR (2008) Comment on "Heat capacity, time constant, and sensitivity of Earth's climate system" by S. E. Schwartz. Journal of Geophysical Research, 113:doi:10.1029/2007jd009473

Knutti R, Stocker TF (2000) Influence of the thermohaline circulation on projected sea level rise. J Clim 13:1997–2001

Kosaka Y, Xie SP (2013) Recent global-warming hiatus tied to equatorial Pacific surface cooling. Nature 501:403–407

Kostov Y, Armour KC, Marshall J (2014) Impact of the Atlantic meridional overturning circulation on ocean heat storage and transient climate change. Geophys Res Lett 41:2108–2116

Kuhlbrodt T and Gregory JM (2012) Ocean heat uptake and its consequences for the magnitude of sea level rise and climate change. Geophysical Research Letters, 39:doi:Artn L18608, 10.1029/2012gl052952

Kummer JR, Dessler AE (2014) The impact of forcing efficacy on the equilibrium climate sensitivity. Geophys Res Lett 41:3565–3568. doi:10.1002/2014gl060046

Lewis N, Curry JA (2015) The implications for climate sensitivity of AR5 forcing and heat uptake estimates. Clim Dyn 45:1009–1023

Li C, von Storch JS, Marotzke J (2013) Deep-ocean heat uptake and equilibrium climate response. Clim Dyn 40:1071–1086

Long DJ, Collins M (2013) Quantifying global climate feedbacks, responses and forcing under abrupt and gradual CO_2 forcing. Clim Dyn 41:2471–2479

Marshall DP, Zanna L (2014) A conceptual model of ocean heat uptake under climate change. J Clim 27:8444–8465

Marvel K, Schmidt GA, Miller RK, Nazarenko LS (2015) Implications for climate sensitivity from the response to individual forcings. Nat Clim Chang. doi:10. 1038/NCLIMATE2888

Maslin M, Austin P (2012) Climate models at their limit? Nature 486:183–184

Masters T (2014) Observational estimate of climate sensitivity from changes in the rate of ocean heat uptake and comparison to CMIP5 models. Clim Dyn 42:2173–2181

Meehl GA, Arblaster JM, Fasullo JT, Hu AX, Trenberth KE (2011) Model-based evidence of deep-ocean heat uptake during surface-temperature hiatus periods. Nat Clim Chang 1:360–364

Meehl GA, Washington WM, Collins WD, Arblaster JM, Hu A, Buja LE, Strand WG, Teng H (2005) How much more global warming and sea level rise? Science 307:1769–1772

Meraner K, Mauritsen T, Voigt A (2013) Robust increase in equilibrium climate sensitivity under global warming. Geophys Res Lett 40:5944–5948

Merlis TM, Held IM, Stenchikov GL, Zeng FR, Horowitz LW (2014) Constraining transient climate sensitivity using coupled climate model simulations of volcanic eruptions. J Clim 27:7781–7795

Murphy JM (1995) Transient-response of the Hadley Center coupled ocean-atmosphere model to increasing carbon-dioxide.3. Analysis of global-mean response using simple-models. J Clim 8:496–514

Otto A, Otto FEL, Boucher O, Church J, Hegerl G, Forster PM, Gillett NP, Gregory J, Johnson GC, Knutti R, Lewis N, Lohmann U, Marotzke J, Myhre G, Shindell D, Stevens B, Allen MR (2013) Energy budget constraints on climate response. Nat Geosci 6:415–416

Plattner GK, Knutti R, Joos F, Stocker TF, von Bloh W, Brovkin V, Cameron D, Driesschaert E, Dutkiewicz S, Eby M, Edwards NR, Fichefet T, Hargreaves JC, Jones CD, Loutre MF, Matthews HD, Mouchet A, Muller SA, Nawrath S, Price A, Sokolov A, Strassmann KM, Weaver AJ (2008) Long-term climate commitments projected with climate-carbon cycle models. J Clim 21:2721–2751

Raper SCB, Gregory JM, Stouffer RJ (2002) The role of climate sensitivity and ocean heat uptake on AOGCM transient temperature response. J Clim 15:124–130

Rhein M, Rintoul SR, Aoki S, Campos E, Chambers D, Feely RA, Gulev S, Johnson GC, Josey SA, Kostianoy A, Mauritzen C, Roemmich D, Talley LD, Wang F (2013) Observations: ocean. In: Stocker TF, Qin D, Plattner G-K, Tignor M, Allen SK, Boschung J, Nauels A, Xia Y, Bex V, Midgley PM (eds) Climate Change 2013: The Physical Science Basis. Contribution of Working Group I to the Fifth Assessment Report of the Intergovernmental Panel on Climate Change. Cambridge University Press, Cambridge, United Kingdom and New York, NY, USA, pp 255–315

Rose BEJ, Armour KC, Battisti DS, Feldl N, Koll DDB (2014) The dependence of transient climate sensitivity and radiative feedbacks on the spatial pattern of ocean heat uptake. Geophys Res Lett 41:1071–1078

Scafetta N (2009) Empirical analysis of the solar contribution to global mean air surface temperature change. J Atmospheric Solar-Terrestrial Physics 71:1916–1923

Schwartz SE (2007) Heat capacity, time constant, and sensitivity of Earth's climate system. Journal of Geophysical Research-Atmospheres, 112:doi:Artn D24s05 10.1029/2007jd008746

Schwartz SE (2008) Reply to comments by G. Foster et al., R. Knutti et al., and N. Scafetta on "Heat capacity, time constant, and sensitivity of Earth's climate system". Journal of Geophysical Research, 113:doi:10.1029/2008jd009872

Senior CA, Mitchell JFB (2000) The time-dependence of climate sensitivity. Geophys Res Lett 27:2685–2688

Shell KM (2013) Consistent differences in climate feedbacks between atmosphere-ocean GCMs and atmospheric GCMs with slab-ocean models. J Clim 26:4264–4281

Sherwood SC, Bony S, Boucher O, Bretherton C, Forster PM, Gregory JM, Stevens B (2015) Adjustments in the forcing-feedback framework for understanding climate change. Bull Am Meteorol Soc 96:217–228

Shindell DT (2014) Inhomogeneous forcing and transient climate sensitivity. Nat Clim Chang 4:274–277

Shindell DT, Faluvegi G, Rotstayn L, Milly G (2015) Spatial patterns of radiative forcing and surface temperature response. J Geophys Res-Atmos 120:5385–5403

Shine KP, Cook J, Highwood EJ and Joshi MM (2003) An alternative to radiative forcing for estimating the relative importance of climate change mechanisms. Geophysical Research Letters, 30:doi:Artn 2047 10.1029/2003gl018141

Siegenthaler U, Oeschger H (1984) Transient temperature-changes due to increasing CO_2 using simple-models. Ann Glaciol 5:153–159

Skeie RB, Berntsen T, Aldrin M, Holden M, Myhre G (2014) A lower and more constrained estimate of climate sensitivity using updated observations and detailed radiative forcing time series. Earth System Dynamics 5:139–175

Solomon S, Plattner GK, Knutti R, Friedlingstein P (2009) Irreversible climate change due to carbon dioxide emissions. Proc Natl Acad Sci U S A 106:1704–1709

Stouffer RJ, Manabe S (1999) Response of a coupled ocean-atmosphere model to increasing atmospheric carbon dioxide: sensitivity to the rate of increase. J Clim 12:2224–2237

Stuber N, Sausen R, Ponater M (2001) Stratosphere adjusted radiative forcing calculations in a comprehensive climate model. Theor Appl Climatol 68:125–135

Taylor KE, Stouffer RJ, Meehl GA (2012) An overview of CMIP5 and the experiment design. Bull Am Meteorol Soc 93:485–498

Watanabe M, Kamae Y, Yoshimori M, Oka A, Sato M, Ishii M, Mochizuki T, Kimoto M (2013) Strengthening of ocean heat uptake efficiency associated with the recent climate hiatus. Geophys Res Lett 40:3175–3179

Watanabe M, Shiogama H, Tatebe H, Hayashi M, Ishii M, Kimoto M (2014) Contribution of natural decadal variability to global warming acceleration and hiatus. Nat Clim Chang 4:893–897

Watterson IG (2000) Interpretation of simulated global warming using a simple model. J Clim 13:202–215

Wigley TM, Raper SC (2001) Interpretation of high projections for global-mean warming. Science 293:451–454

Wigley TML, Schlesinger ME (1985) Analytical solution for the effect of increasing CO_2 on global mean temperature. Nature 315:649–652

Williams KD, Ingram WJ, Gregory JM (2008) Time variation of effective climate sensitivity in GCMs. J Clim 21:5076–5090

Winton M, Adcroft A, Griffies SM, Hallberg RW, Horowitz LW, Stouffer RJ (2013a) Influence of ocean and atmosphere components on simulated climate sensitivities. J Clim 26:231–245

Winton M, Griffies SM, Samuels BL, Sarmiento JL, Frolicher TL (2013b) Connecting changing ocean circulation with changing climate. J Clim 26:2268–2278

Winton M, Takahashi K, Held IM (2010) Importance of ocean heat uptake efficacy to transient climate change. J Clim 23:2333–2344

Yamamoto A, Abe-Ouchi A, Shigemitsu M, Oka A, Takahashi K, Ohgaito R, Yamanaka Y (2015) Global deep ocean oxygenation by enhanced ventilation in the Southern Ocean under long-term global warming. Glob Biogeochem Cycles 29:1801–1815

Yamamoto A, Yamanaka Y, Oka A, Abe-Ouchi A (2014) Ocean oxygen depletion due to decomposition of submarine methane hydrate. Geophys Res Lett 41: 5075–5083

Yokohata T, Emori S, Nozawa T, Ogura T, Kawamiya M, Tsushima Y, Suzuki T, Yukimoto S, Abe-Ouchi A, Hasumi H, Sumi A, Kimoto M (2008) Comparison of equilibrium and transient responses to CO_2 increase in eight state-of-the-art climate models. Tellus Series a-Dynamic Meteorol Oceanography 60:946–961

Yoshimori M, Abe-Ouchi A (2012) Sources of spread in multimodel projections of the Greenland Ice Sheet surface mass balance. J Clim 25:1157–1175

Yoshimori M, Broccoli AJ (2008) Equilibrium response of an atmosphere-mixed layer ocean model to different radiative forcing agents: global and zonal mean response. J Clim 21:4399–4423

Yoshimori M, Hargreaves JC, Annan JD, Yokohata T, Abe-Ouchi A (2011) Dependency of feedbacks on forcing and climate state in physics parameter ensembles. J Clim 24:6440–6455

Yoshimori M, Yokohata T, Abe-Ouchi A (2009) A comparison of climate feedback strength between CO_2 doubling and LGM experiments. J Clim 22:3374–3395

Yoshimori M, Watanabe M, Abe-Ouchi A, Shiogama H, Ogura T (2014) Relative contribution of feedback processes to Arctic amplification of temperature change in MIROC GCM. Clim Dyn 42:1613–1630

Zickfeld K, Eby M, Weaver AJ, Alexander K, Crespin E, Edwards NR, Eliseev AV, Feulner G, Fichefet T, Forest CE, Friedlingstein P, Goosse H, Holden PB, Joos F, Kawamiya M, Kicklighter D, Kienert H, Matsumoto K, Mokhov II, Monier E, Olsen SM, Pedersen JOP, Perrette M, Philippon-Berthier G, Ridgwell A, Schlosser A, Von Deimling TS, Shaffer G, Sokolov A, Spahni R, Steinacher M, Tachiiri K, Tokos KS, Yoshimori M, Zeng N, Zhao F (2013) Long-term climate change commitment and reversibility: an EMIC intercomparison. J Clim 26:5782–5809

12

$\delta^{18}O$ and SST signal decomposition and dynamic of the Pliocene-Pleistocene climate system: new insights on orbital nonlinear behavior vs. long-term trend

Paolo Viaggi

Abstract

The global LR04 $\delta^{18}O$, the tropical ODP Site 846 sea surface temperature (SST), and the global ΔSST stack records were investigated using the advanced method for time-series decomposition singular spectrum analysis to outline the quantitative role of orbital forcings and to investigate the nonlinear dynamics of the Pliocene and Pleistocene climate system. For the first time, a detailed quantitative evaluation is provided of the $\delta^{18}O$ and SST variance paced by long-period orbital modulation, short eccentricity, obliquity, precession, and half-precession cycles. New insights into the nonlinear dynamic of the orbital components suggest considering astronomical signals as composite feedback lagged responses paced by orbitals and damped (Early Pliocene) or amplified (Mid-Late Pleistocene) in a range of − 100 to + 400% the forcing. The Early Pliocene asymptotic decay of the $\delta^{18}O$ and SST response sensitivity up to − 100% observed for the first time in all orbital responses is interpreted as damping effect of a wide global forest cover along with a possible high ocean primary productivity, through the CO_2-related negative feedbacks during time of global greenhouse. An anomalous post-Mid-Pleistocene Transition (MPT) sharply declines to near-zero in obliquity response sensitivity observed in both global $\delta^{18}O$ and tropical SST, suggesting an attenuation mechanism of the obliquity driving force and a reduction of the related feedback amplification processes. It is hypothesized the post-MPT obliquity damping has contributed to the strengthening of the short eccentricity response by mitigating the obliquity "ice killing", favoring a long-life ice sheet sensitive to a synergistic ~ 100-kyr amplification of positive feedback processes during the time of a global icy state. The global $\delta^{18}O$, the tropical SST, and the global ΔSST trend components, all explaining ~ 76% of the Plio-Pleistocene variance and significantly modifying the mean climate state, appear to be related to the long-term pCO_2 proxies, supposedly controlled by plate tectonics through the global carbon cycle (CO_2 outgassing, explosive volcanism, orography and erosion, paleogeography, oceanic paleocirculation, and ocean fertilization). Finally, singular spectrum analysis provides a valuable tool in cyclostratigraphy with the remarkable advantage of separating full-resolution time series by variance strength.

Keywords: Singular spectrum analysis, Nonlinear, Chaotic, Damping, Paleoclimate, Milankovitch, Pliocene-Pleistocene, Greenhouse

Correspondence: paolov6363@yahoo.it
Museo Geologico Giovanni Capellini, SMA, Bologna University, Via Zamboni 63, I-40126 Bologna, Italy

Introduction

The Milankovitch theory elegantly describes the collective effects of changes in the Earth's movements on its climate. Such orbital changes (eccentricity, obliquity, and precession) alter the amount and the distribution of solar radiation reaching the Earth. However, despite the studies of Kominz and Pisias (1979) and Wunsch (2004), in some textbooks or papers, there is still the misleading notion that the insolation changes are a major element controlling climate. A lot of studies have analyzed a variety of records finding the Milankovitch frequencies, but few have quantitatively investigated the hypothesis that most of the energy conveyed in paleoclimate records is not related to orbital forcing (Kominz and Pisias 1979; Raymo 1994; Wunsch 2004; Mudelsee and Raymo 2005; Hansen et al. 2013). This study applies a novel nested analytical approach based on the advanced method for signal decomposition singular spectrum analysis (SSA) with the aim of separating full-resolution time series by signal variance, the latter a useful "proxy" of the energy conveyed in paleoclimate records. SSA is a relatively new and powerful technique to processing time-series data, applicable to broad areas such as climatology, marine science, geophysics, engineering, image processing, and medicine (Vautard and Ghil 1989; Elsner and Tsonis 1996; Ghil et al. 2002; Hassani 2007). However, the SSA approach is little used in the study of paleoclimate records compared to conventional Fourier analysis, despite its capability to estimate frequencies with higher resolution than Fourier approach and to detect signal structural changes, as highlighted in the present study. Plio-Pleistocene long-term climatic cooling and the increasing glacial cycle amplitude towards the Late Pleistocene are evidence of significant changes in the climate system dynamic described in the literature, but the cause of these features remains poorly understood. The aim of this study is to investigate these topics by decomposing two selected $\delta^{18}O$ and sea surface temperature (SST) records to outline the quantitative role of astronomical forcings and the nonlinear behavior in the orbital responses and to assess the Earth's long-term controlling factors as a "stand-alone" climate system and its interaction with the astronomical one. Through the SSA, it provides for the first time a detailed quantitative evaluation of the $\delta^{18}O$ and SST signal variance in a wide spectrum of orbital components from long-period orbital modulation cycles to the half-precession, both unidentified in the previous studies of the original records. The nonlinear dynamic of the astronomical climate responses is discussed regarding new insights obtained by SSA time-series processing. A link among the atmospheric pCO_2, the sea surface temperature, and the $\delta^{18}O$ long-term trend components is also provided to support the hypothesis of a dominant non-astronomical control on the Earth's climate.

Methods/Experimental

This work contributes to the research issues by applying SSA to the global LR04 $\delta^{18}O$ stack (Lisiecki and Raymo 2005b) and the tropical Ocean Drilling Program (ODP) Site 846 SST record (Herbert et al. 2010b) (referred to as HT846 in this study), both high-resolution time series covering ~ 5 million years to capture long-term climate dynamics. The global LR04 benthic stack was obtained from 57 oceanic series globally distributed over a wide latitudinal range, including the Atlantic, Pacific, and Indian Oceans (Lisiecki and Raymo 2005a). Benthic $\delta^{18}O$ measures changes in global ice volume and deep water temperatures, which are controlled by high-latitude surface temperatures (Lisiecki and Raymo 2007). The ice volume component lags the deep water temperature signal, and the net phase of the $\delta^{18}O$ record most likely is weighted towards the slow ice volume component (Lisiecki and Raymo 2005a; Herbert et al. 2010a; Berger et al. 2016). Because the LR04 stack incorporates information from so many sites, it has a high signal-to-noise ratio and more accurately reflects changes in global climate (Lisiecki and Raymo 2005a). The original resolution of the LR04 record ranges from 1 to 5 kyr within four time intervals with constant data spacing (Lisiecki and Raymo 2005a). The average resolution is 2.5 ± 1.4 kyr. The HT846 is an alkenone-based tropical SST record located in the Eastern Equatorial Pacific (EEP). The ODP Site 846 is considered not "typical" tropical SST signal because it cools more than the ODP 662 and 722 sites, probably as the result of upwelling changes in the EEP (Herbert et al. 2010a; Lisiecki 2010b); nevertheless, its high-resolution SST record extending ~ 5 Myr makes it suitable to capture both long-term dynamics and high-frequency cycles. The original data spacing of the HT846 record is uneven throughout the 5070 kyr of the time series (Herbert et al. 2010b), with an average value of 2.3 ± 1.2 kyr. SSA requires evenly spaced records. Therefore, both records were resampled at constant time intervals of 1.0 kyr. Further details on the methodology used in this study and a discussion of the quality and reliability of the time series can be found in Additional file 1. The valuable global ΔSST stack of Martinez-Boti et al. (2015) extends 3.4 Myr with a resolution lower than that of HT846 (which is one of the ten sites included in the stack). Thus, the ΔSST stack was used in this study as a comparison record to assess some common features between the tropical Site 846 and the global SST records. The LR04 stack is orbitally tuned to an ice model driven by June 21 insolation at 65° N from La93(1,1) orbital solutions (Laskar et al. 1993), with correction of the sedimentation rate (Lisiecki and Raymo 2005a). The HT846 record is synchronized using benthic foraminiferal $\delta^{18}O$ measured in the same sediments at the study site and aligned to the LR04 stack (Herbert et al. 2010a). Unfortunately, high-resolution extended records of CO_2 proxies to study the long-term dynamics are not currently available. The valuable simulated time series of Van de Wal et al. (2011) and Stap et al. (2016) are not suitable for the purposes of the present study because the data are not $\delta^{18}O$ independent. Nevertheless, the long-term links among climate proxies can also be studied with low-resolution records, provided that the data are temporally extended and in sufficient number and quality to ensure reliable equation fitting. In this study, the composite atmospheric boron-based pCO_2 proxy of the ODP Sites 668B-999A (Hönisch et al. 2009; Bartoli et al. 2011) and the atmospheric alkenone-based pCO_2 and SST proxies of the ODP Site 999A (Seki et al. 2010) were selected. Seki et al. (2010) proxies are rather poor in resolution but cover a long period of time. The high-resolution atmospheric CO_2 reconstructions of Martinez-Boti et al. (2015), based on multi-site boron-isotope records, were considered despite the short time series from 3.3 to 2.3 million years ago. The composite European Project for Ice Coring in Antarctica (EPICA) pCO_2 record (Luthi et al. 2008; Berger et al. 2016) and the $\delta^{13}C$-based pCO_2 proxy (Lisiecki 2010b) are also considered. In order to evaluate the trend relationships among $\delta^{18}O$, SST, and pCO_2 proxies, the high-resolution LR04 and HT846 SSA-component time series were downsampled at the approximate age of the low-resolution pCO_2 proxies from the ODP Sites 668B and 999A (Hönisch et al. 2009; Seki et al. 2010; Bartoli et al. 2011). By applying the advanced method for signal decomposition singular spectrum analysis (Vautard and Ghil 1989; Elsner and Tsonis 1996; Ghil et al. 2002; Hassani 2007), the 1.0-kyr resampled LR04 and HT846 records are partitioned in a new time series of its data components. The SSA is achieved by the computation of forward-backward covariance matrices with an order of 50 (embedding dimension) using the singular value decomposition (SVD) procedure. This method for time-series analysis can be used to isolate full-resolution independent signal components based on signal strength (variance). In contrast with Fourier analysis with fixed basis of sine and cosine functions, SSA uses an adaptive basis generated by the time series itself. As a result, the underlying model in SSA is more general and can extract amplitude-modulated wave components with higher resolution than Fourier analysis. Moreover, SSA can be effectively used as a non-parametric method for detection of structural changes in the time series, reflecting the main physical phenomena underlying the data (Vautard and Ghil 1989; Hassani 2007). In the present study, the SST and $\delta^{18}O$ SSA-components capture important aspects of the nonlinear dynamic response of the climate system in a variance-oriented way. However, some high-variance $\delta^{18}O$ and SST signal components can remain characterized by a still complex frequency spectrum with some bands dispersed in different components. A novel nested-SSA approach has been applied to these 50-order components to better separate the frequency bands and correctly estimate the variance fraction, by increasing the embedding dimension up to 300 and partitioning the component into subcomponents. More details on the SSA methodology and the complete SSA-components dataset are provided in Additional files 1 and 2. The SSA-components are then investigated by a Fourier frequency spectrum (FFS) with a data window to analyze the frequency power and to reduce spectral leakage in the Fourier spectra. The data tapering used in this study is the cs2-Hann window and the best exact N fast Fourier transform (FFT) algorithm. The peak detection uses peak-based critical limit significance levels. In this type of confidence limit, one seeks to disprove the null hypothesis where one postulates either a white noise signal or a red noise signal. The frequency peaks shown in this study are the most significant peaks above a 99.9% critical limit, unless otherwise indicated. For each peak, the time-integral squared amplitude (TISA) power is computed in relative percent. The summed power reported in the frequency analysis is merely the sum of the component powers, not the power of the composite signal that would result from the addition of the components. Because clear non-stationarity in some components complicates the Fourier representation, these signals are investigated by FFS for time intervals to capture dynamics over time. Finally, by using the La93 orbital solutions (Laskar et al. 1993) (see Additional file 1), the phase relationships to the related orbitally driven LR04-SST components were examined by cross-spectral analysis. The data were standardized on the entire Plio-Pleistocene record (mean = 0 and standard deviation = 1), and some time series were inverted in order to have the same paleoclimatic polarity (positive forcing or warming for positive standardized values). The software used in this research is AutoSignal™.

The chronostratigraphic reference charts used in this study are the ICS International Chronostratigraphic Chart (2017-02) for the Pliocene and Pleistocene and the ICS global chronostratigraphical correlation table for the last 2.7 million years (v. 2016a) for the Quaternary.

Results

Analysis of $\delta^{18}O$ SSA-components

The SSA allowed the resampled LR04 signal to be decomposed into five $\delta^{18}O$ time series, which cumulatively explain 99.5% of the original variance (LR04-filtered). The spectral framework of these $\delta^{18}O$ components (abbreviated as comp.) is diversified, with fluctuations ranging from 1330-kyr to 11-kyr periodicity (Fig. 1, Table 1). The features of the $\delta^{18}O$ components are described as follows.

$\delta^{18}O$ component-1

This component explains 81.5% of the total variance (eigenvalue 10.89) and constitutes the most significant fraction of the LR04 signal associated with a strong trend of isotopic enrichment (long-term LR04 component, LT-LR04). The FFS shows a broad spectrum of nonsignificant peaks

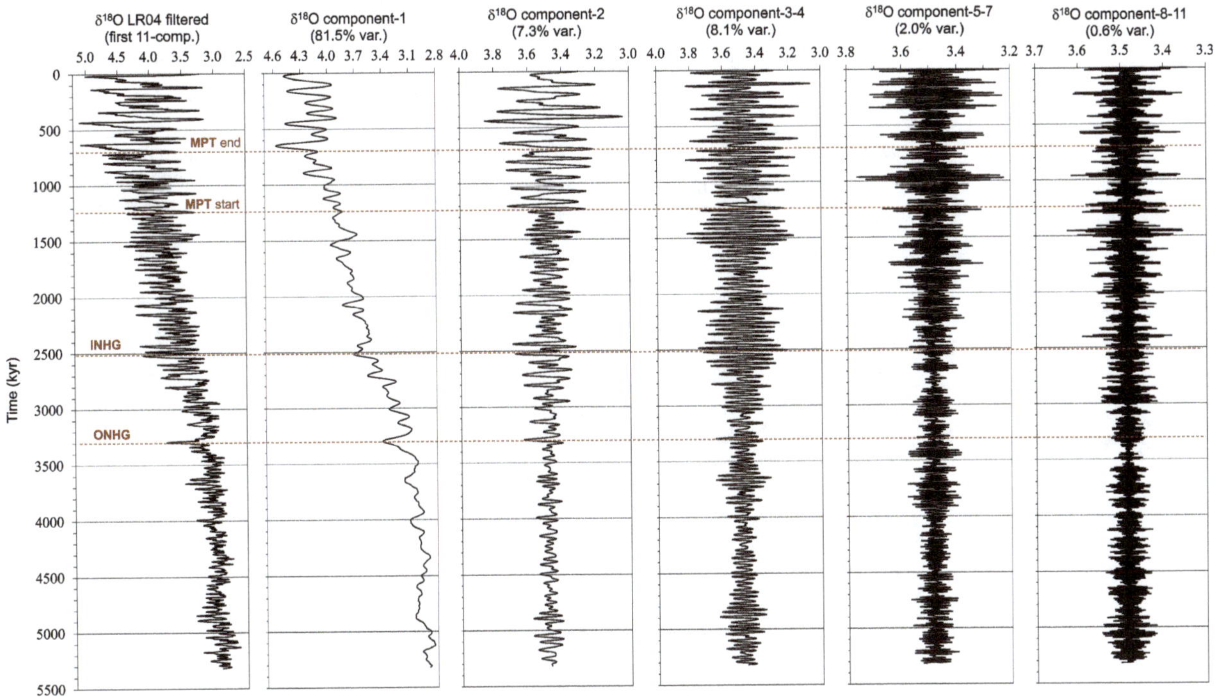

Fig. 1 Time-series panel of the five $\delta^{18}O$ components isolated from the LR04 signal by singular spectrum analysis, together with the LR04-filtered record (left) obtained by combining these data components (99.5% total variance). The *x*-axis scale (‰) is variable in relation to the variance fraction of the data components. Original LR04 $\delta^{18}O$ data from Lisiecki and Raymo (2005b) resampled at 1 kyr. Time scale in kyr

(< 50% of critical significance level) with periods equal to ~ 1.33 Myr (TISA power 77.2%), 409 kyr (10.8%), 294 kyr (6.9%), and 195 kyr (5.1%), overlapping the general trend (Fig. 1, Table 1). To separate the oscillatory terms from component-1 and estimate the variance fractions, an SSA was performed by increasing the embedding dimension to 300. Three main subcomponents (abbreviated as subcomp.) were identified equal to a reconstruction R^2 of 0.999 (Table 2).

The component-1 subcomp-1 incorporates 96.2% of the component-1 variance (81.5%), equivalent to an absolute fraction of 78.4%. It exhibits the non-Milankovitch feature of a strong isotopic trend with mild oscillations and, together with component-1, allows the subdivision of the main drift into four subtrends (Fig. 2).

These subtrends are characterized by a mild curvilinear shape caused by a terminal acceleration of the isotopic enrichment, separated by a late warming step that temporarily interrupts the drift. The first subtrend in chronological order (subtrend-I) is in the range of 5.33–3.30 Myr (Zanclean-Early Piacenzian) and culminates with a major event of isotopic enrichment at approximately 3.30 Myr, related to the first Pliocene transition of climatic deterioration known as the onset of the Northern Hemisphere Glaciation (ONHG) (Shackleton and Opdyke 1977; Blanc et al. 1983; Lourens and Hilgen 1994; Lisiecki and Raymo 2007). This subtrend is interrupted by a rapid warming at approximately 3.2 Myr, when a new isotopic drift (subtrend-II) begins and culminates at approximately 2.5 Myr (Middle Piacenzian to Gelasian) in a second major climatic event known as the intensification of the Northern Hemisphere Glaciation (INHG) (Zagwijn 1974; Lisiecki and Raymo 2007; Seki et al. 2010). Because of its global significance, this episode of climatic deterioration at 2.5 Myr was recently used to ratify the new Plio-Pleistocene limit to the base of the Gelasian (GSSP 2.58 Myr) which is formally assigned to the Pleistocene (Gibbard and Head 2009). Subtrend-III begins with warming at approximately 2.35 Myr and reaches its peak at 1.54 Myr at the end of the Santernian (Middle Gelasian-Santernian). The most recent segment of isotopic enrichment (subtrend-IV) starts with the warming at 1.45 Myr (Early Emilian) and culminates with a noticeable acceleration at 0.65 Myr, close to the beginning of the Middle Pleistocene, correlating to Marine Isotope Stage (MIS) 16. The Mid-Pleistocene Transition (MPT)—recently dated to start at approximately 1.2 Myr (Head and Gibbard 2005; Clark et al. 2006; Head et al. 2008)—which marks the beginning of the classic “Pleistocene Ice Age”, falls within subtrend-IV but is not directly visible in this component. The end of the subtrend-IV marks the final stage of the MPT at about 0.7 Myr (Clark et al. 2006). After MIS-16, the cooling trend is broken by a wide swing of $\delta^{18}O$ depletion that peaks at approximately 0.40–0.30 Myr (MIS-11 and MIS-9e), referred in this study

Table 1 Main descriptive parameters and Fourier frequency results of the five δ^{18}O components isolated from the LR04 signal by singular spectrum analysis. The spectral peaks of the trend component-1 are not significant (< 50% of critical significance level) and are shown only for the purpose of discussion. Original LR04 data from Lisiecki and Raymo (2005b) are resampled at 1 kyr

$\delta^{18}O$ component rank	$\delta^{18}O$ comp. variance (%)	Frequency (kyr^{-1})	TISA power (%)	Period (kyr)	Weighted mean period (kyr)	Forcing
1	81.5	0.0007519	77.2	1330	Not significant peaks (< 50% of critical significance level)	Long-term trend + long-period orbital modulation and eccentricity cycles
		0.0024474	10.8	409		
		0.0034051	6.9	294		
		0.0051187	5.1	195		
2	7.3	0.0105636	20.9	95		Transition obliquity to short eccentricity
		0.0140889	16.8	71		
		0.0186578	22.3	54		
		0.0244330	40.1	41		
3–4	8.1	0.0244339	83.1	41	41	Obliquity
		0.0253170	12.2	39		
		0.0186593	4.7	54		
5–7	2.0	0.0422565	25.7	24	22	Precession
		0.0447325	57.6	22		
		0.0487779	16.7	21		
8–11	0.6	0.0700410	27.0	14	13	"Half-precession" (equatorial)
		0.0764743	46.5	13		
		0.0893457	26.5	11		
> 11	0.5					Noise

as the Mid-Brunhes Oscillation (MBO), containing the Mid-Brunhes Event (MBE) at approximately 0.43 Myr (Berger and Yin 2012). This subtrend pattern shows four steps of gradual buildup of the NHG, followed by a relative stasis of temperature cooling and ice growth during the last 600 kyr (MBO).

The frequency spectrum of comp-1 subcomp-1 still shows a nonsignificant residual peak at 1.33 Myr, very close to the ~ 1.2 Myr obliquity amplitude modulation cycle (Laskar et al. 2004, 2011; Boulila et al. 2011), tightly enveloped to the long-term trend, and to which the subtrend's mild terminal oscillations may be related

Table 2 Descriptive parameters and Fourier frequency results of the δ^{18}O subcomponents isolated from the LR04 component-1 by singular spectrum analysis. The processing of the subcomponents was conducted to better separate and estimate the variance fraction of specific frequency bands from still complex data components. The italicized percentage values are the absolute variance of the subcomponents (referred to the total LR04 variance)

$\delta^{18}O$ comp. rank	$\delta^{18}O$ comp. variance (%)	$\delta^{18}O$ subcomp. rank	$\delta^{18}O$ subcomp. var. (rel.) (%)	$\delta^{18}O$ subcomp. var. (abs) (%)	Frequency (kyr^{-1})	TISA power (%)	Period (kyr)		Forcing
1	81.5	1	96.2	*78.4*	0.0007519	96.6	1330	Not significant peak (< 50% of critical significance level)	Long-term trend + obliquity modulation cycle
		2	1.0	*0.8*	0.0006351	94.9	1575		Obliquity modulation cycle
					0.0024431	5.1	409		Eccentricity (long)
		3–10	2.8	*2.3*	0.0034061	16.0	294		Earth's orbit secular frequency
					0.0051192	24.5	195		Obliquity modulation cycle
					0.0061178	16.6	163		
					0.0075215	15.6	133		Eccentricity (short)
					0.0093870	12.8	107		
					0.0105656	14.5	95		

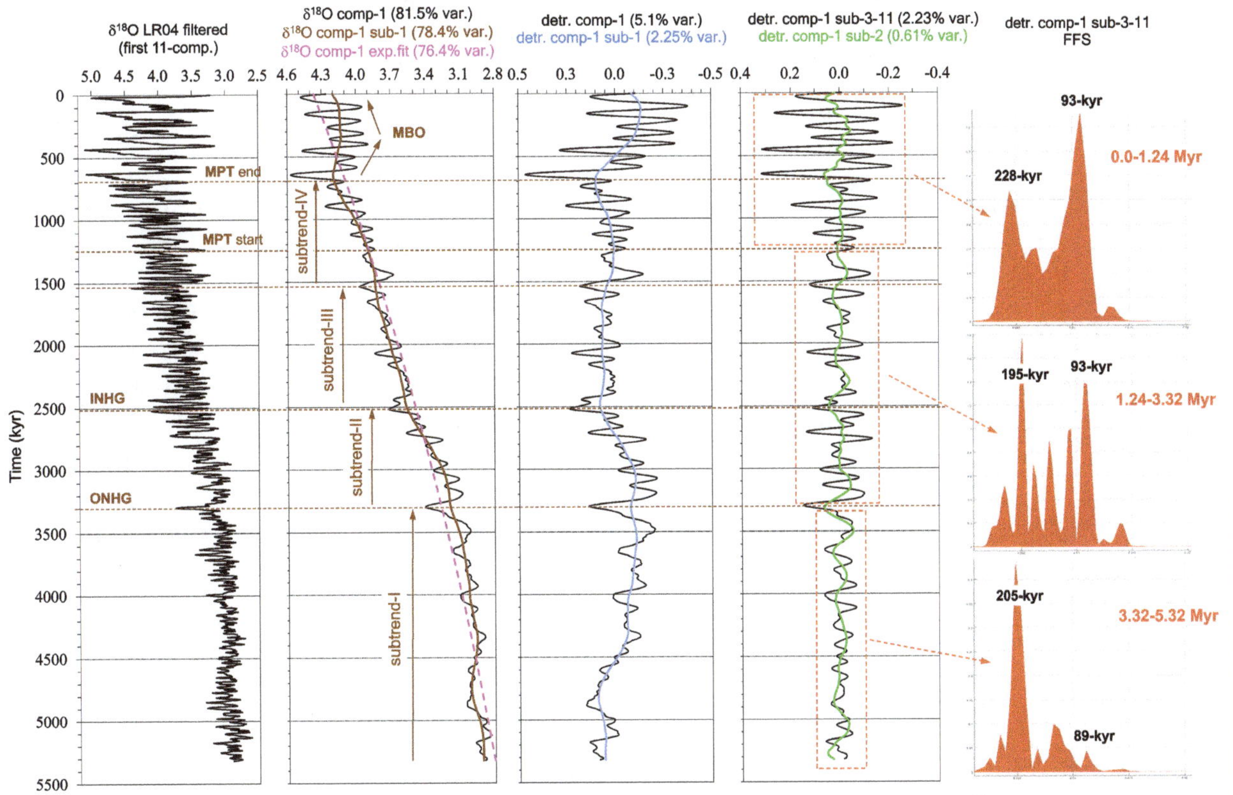

Fig. 2 Time-series panel of the $\delta^{18}O$ component-1 and its subcomponents isolated by singular spectrum analysis. $\delta^{18}O$ comp-1 exponential trend (‰) = $1.44194 + 2.92378\ e^{(-t/6963.268)}$. The vertical brown arrows show four subtrends of component-1 and their correlation to the NHG transitions known in the literature (MBO this study). To the right, the Fourier frequency spectrum of $\delta^{18}O$ detrended comp-1 subcomp-3-11 (2.23% variance) for three time intervals are highlighted by red boxes. Note the increase in amplitude after the ONHG and MPT and the power increase of the ~ 93-kyr eccentricity cycle. *x*-axis scale variable (‰). The *y*-axis scale of the frequency spectrum increases upwards. Time scale in kyr

(Fig. 2, Table 2). The comp-1 subcomp-2 contains 0.8% only of the $\delta^{18}O$ variance dominated by a 1.57-Myr periodicity (power 94.9%), which could be related to the 1.33 obliquity modulation cycle, and an ancillary term of 409 kyr. A wide spectrum of long-term cycles characterized the comp-1 subcomp-3-10, which explains 2.3% of the $\delta^{18}O$ variance (294, 195, 163, 133, 107, and 95 kyr).

To estimate the variance fraction of the "pure" long-term trend, component-1 was detrended by subtracting an exponential fitting with least squares minimization ($R^2 = 0.94$) and the oscillatory signal was processed by a 300-order SSA (Fig. 2, Table 3). The LT-LR04 exponential fit contains 93.8% of the comp-1 variance, equivalent to an absolute contribution of 76.4%, whereas the detrended oscillatory terms explain an absolute 5.1% variance. The 2% variance difference between the comp-1 subcomp-1 (78.4%) and the comp-1 exponential fit (76.4%) may be attributed to the 1.33-Myr cycle still superimposed to the comp-1 subcomp-1. Considering the variance fraction of the comp-1 subcomp-2 (0.8%, Table 2), a ~ 2.8% contribution of the 1.57–1.33-Myr cycle may be estimated. Three subcomponents were isolated from the detrended comp-1 (Table 3). The detrended comp-1 subcomp-1 explains 2.25% of the $\delta^{18}O$ variance and exhibits a spectrum dominated by a 3.75-Myr periodicity (power 95.8%) with an ancillary term of 855 kyr. Oddly, the cycle at about 1.57–1.33 Myr disappears in the detrended component-1, while a new 3.75-Myr cycle is ascribable to the 3.7-Myr g9-s9 terms of the Earth's orbit secular frequency (Laskar et al. 2011). Prevailing 409-kyr (80.6%) and 294-kyr (19.4%) cycles characterized the detrended comp-1 subcomp-2, explaining only 0.61% of the $\delta^{18}O$ variance. These periodicities may be related to the astronomically stable 405-kyr-long eccentricity cycle and to the 304-kyr Earth orbit secular frequency, respectively (Laskar et al. 2004, 2011; Boulila et al. 2011).

The detrended comp-1 subcomp-3-11 explains 2.23% of the $\delta^{18}O$ variance and shows a wide spectrum of frequency bands. A 294-kyr cycle (power 15.2%) may be related to the 304-kyr Earth orbit secular frequency. Oscillations of 163 kyr (17.2%) and 195 kyr (24.9%) may be linked to the long-period obliquity modulation cycles (~ 160–200 kyr) (Boulila et al. 2011). Finally, the short eccentricity cycles of 133 kyr (15.9%), 107 kyr (12.7%), and 95 kyr (14.1%) are also recognizable. The FFS of this subcomponent by time intervals (Fig. 2) shows two time horizons of amplitude shifting at about 3.32 Myr

Table 3 Main descriptive parameters and Fourier frequency results of the $\delta^{18}O$ subcomponents isolated from the detrended component-1 and component-2 by singular spectrum analysis. The processing of the subcomponents was conducted to better separate and estimate the variance fraction of specific frequency bands from still complex data components. The italicized percentage values are the absolute variance of the subcomponents (referred to the total LR04 variance)

$\delta^{18}O$ comp. rank	$\delta^{18}O$ comp. variance (%)	$\delta^{18}O$ subcomp. rank and var. (abs.)	$\delta^{18}O$ subcomp. var. (rel.) (%)	$\delta^{18}O$ subcomp. var. (abs) (%)	Frequency (kyr^{-1})	TISA power (%)	Period (kyr)	Forcing
1	81.5	exp.fit	93.8	*76.4*	–	–	–	Long-term trend
		detrend 6.2% (*5.1%*) 1	44.2	*2.25*	0.0002668	95.8	3748	Earth's orbit secular frequency
					0.0011701	4.2	855	Eccentricity (long)
		2	11.9	*0.61*	0.0024443	80.6	409	
					0.0034017	19.4	294	Earth's orbit secular frequency
		3–11	43.8	*2.23*	0.0034062	15.2	294	
					0.0051199	24.9	195	Obliquity modulation cycle
					0.0061180	17.2	163	
					0.0075215	15.9	133	Eccentricity (short)
					0.0093867	12.7	107	
					0.0105655	14.1	95	
2	7.3	1–4, 9–15	75.3	*5.5*	0.0105637	56.6	95	Eccentricity (short)
					0.0140872	43.4	71	
		5–8	24.4	*1.8*	0.0244321	67.5	41	Obliquity
					0.0186582	32.5	54	

(ONHG) and 1.24–0.70 Myr (MPT) that delineate three intervals, in which occur a progressive increase in signal amplitude and a relative strengthening of the short eccentricity cycle (93 kyr) at the expense of the obliquity modulation cycles (195–205 kyr), revealing a transient nonlinear behavior of this small fraction of isotopic response.

$\delta^{18}O$ component-2

Component-2 explains 7.3% of the total variance (eigenvalue 0.97) and exhibits four significant oscillatory components in the 41-kyr (power 40.1%) and 54-kyr (22.3%) obliquity bands and in the 95-kyr (20.9%) and 71-kyr (16.8%) short eccentricity bands (Fig. 3, Table 1).

Two shifting horizons of isotopic amplitude (3.32 and 1.24 Myr) are distinctly observed, correlating to the ONHG and MPT start, respectively, and defining three time intervals of increasing amplitude and a significant change in orbital periodicity. The oldest segment between 5.32 and 3.32 Myr (Zanclean-Early Piacenzian) exhibits the lowest amplitude and power and a relative prevalence of the obliquity cycle (41-kyr and 54-kyr). The eccentricity bands at 89 kyr and 102 kyr are particularly weak (Fig. 3a). From 3.32 to 1.24 Myr (Early Piacenzian-Emilian), a relative increase of the short eccentricity power (71-kyr and 91-kyr) over the obliquity cycle occurs (Fig. 3b). The most recent segment from 1.24 Myr exhibits a further increase in the average amplitude and an overwhelming power prevalence of the eccentricity 93-kyr cycle over the 41-kyr response, which maintains about the same power (Fig. 3c). To further isolate the two orbital terms from component-2, the SSA was performed by increasing the embedding dimension to 300, producing two subcomponents equal to a reconstruction R^2 of 0.998 (Fig. 3; Table 3). Subcomponent-1-4, 9-15 explains 75.3% of the component-2 variance, equivalent to an absolute variance fraction of 5.5%. It exhibits frequencies in the short eccentricity band (95-kyr, power 56.6%; 71-kyr, 43.4%). Subcomponent-5-8 explains 24.4% of the component-2 variance (1.8% absolute variance) in the obliquity frequency band (41-kyr, 67.5%, 54-kyr, 32.5%). Interestingly, the amplitude of this obliquity's subcomponent is decreasing after the MPT (Fig. 3). The change in the nonlinear dynamic of the component-2 response (transition from obliquity to short eccentricity) is configured as a sort of "switching" of the orbital power and amplitude between the 41-kyr obliquity and the 93-kyr eccentricity cycles. From this perspective, due to the change in periodicity depending on the amplitude, component-2 may be the expression of a nonlinear phase-locking mechanism between obliquity and eccentricity astronomical forcing (Tziperman et al. 2006; Ruggieri et al. 2009). Notably, by modeling the Northern Hemisphere ice volume over the last 3 Myr, Berger et al. (1999) simulated similar power spectrum changes using a combination of insolation and linearly decreasing CO_2 concentration. Imbrie et al. (2011) simulated the same spectral behavior using the Pleistocene LR04 record with a

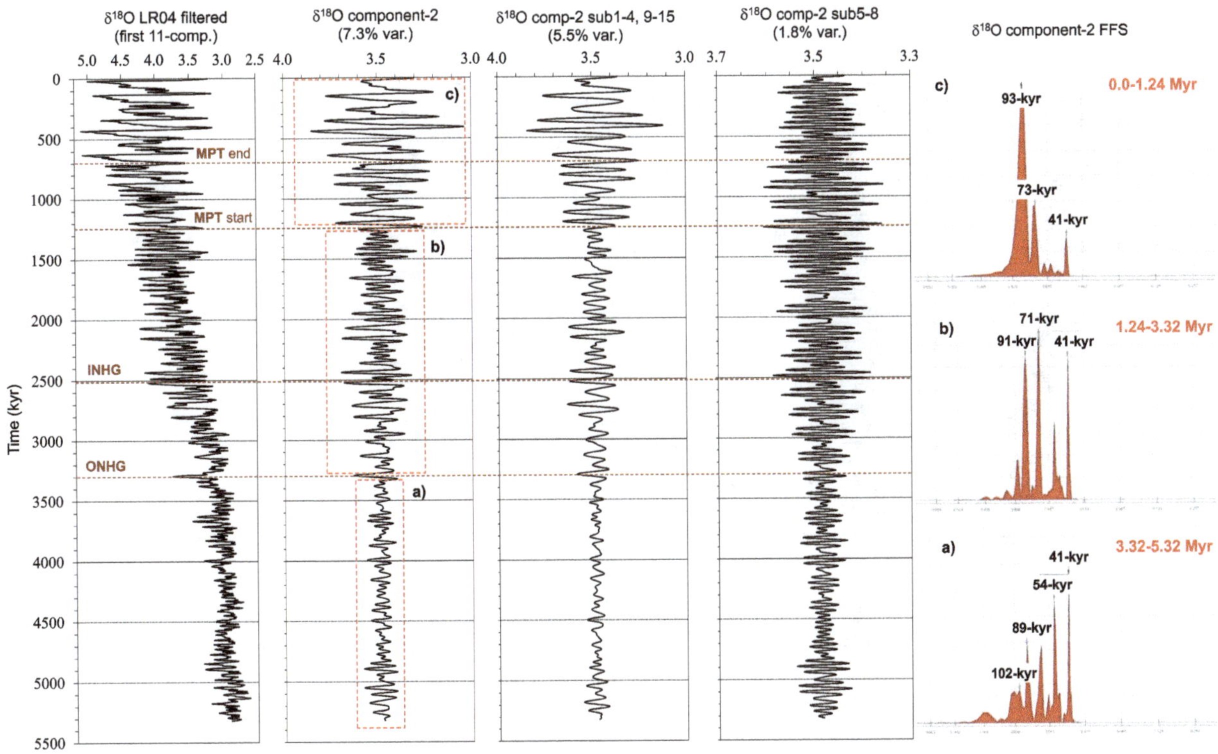

Fig. 3 Time-series panel of δ^{18}O component-2 and its subcomponents and Fourier frequency spectrum of comp-2 for three time intervals **a**, **b** and **c**, highlighted by red-labeled boxes. Note the evident increase in comp-2 amplitude after the ONHG and MPT, and the transition in periodicity from **a** obliquity to **c** eccentricity with a gradual upwards strengthening of the short eccentricity power. Interestingly, the amplitudes of the eccentricity in comp-2 subcomp-1-4, 9-15 are increasing, whereas the amplitude of the obliquity in comp-2 subcomp-5-8 is decreasing after the MPT. *x*-axis scale variable (‰). The *y*-axis scale of the frequency spectrum increases upwards. Time scale in kyr

new phase-space model of orbital forcings but with a state threshold function of the ice volume, showing that ice volume plays an important role in pacing the orbital response of the climate system. These simulated data, together with the observed features of the SSA comp-2 and detrended comp-1 subcomp-3-11 (Fig. 2), strengthen the evidence of a paleoclimatological transition where the ONHG marks the beginning of a weak and gradual sensibility of the climate system to a 93-kyr eccentricity forcing, supporting the hypothesis of an ice volume CO_2-induced control on eccentricity sensibility (Raymo 1998; Berger et al. 1999; Hansen et al. 2007; Lisiecki and Raymo 2007; Imbrie et al. 2011; Berger and Yin 2012).

δ^{18}O component-3-4

This signal explains 8.1% of the δ^{18}O variance (eigenvalue 1.08) with a 95.3% power at an obliquity-driven 41–39-kyr period (Table 1). Component-3-4 is characterized by a clear increase in the average 41-kyr amplitude and power towards the beginning of the MPT (Fig. 4a, b), followed by an amplitude stabilization and power reduction during the MPT and post-MPT (Fig. 4c). Nonstationary behavior is observed at the beginning of the MPT, where a change in the shape of the 41-kyr cycle that appears enveloped in a very weak 93-kyr signal can be noted (Fig. 4c), suggesting an apparent influence between the eccentricity and the obliquity cycles after the beginning of the MPT. The ONHG transition at 3.32 Myr does not seem to be associated with any particular features of component-3-4, except for a strong power increase in the 41-kyr cycle after the transition.

δ^{18}O component-5-7

This component explains 2.0% of the δ^{18}O variance (eigenvalue 0.27) driven by precession-related bands at ~ 22 kyr (Table 1). This precession δ^{18}O component also shows the nonlinear shape of the upwardly increasing isotopic amplitude (Fig. 1).

δ^{18}O component-8-11

Component-8-11 exhibits 0.6% of the δ^{18}O variance (eigenvalue 0.07) and contains a weight mean period at ~ 13 kyr (Table 1). Furthermore, the feature of increasing mean amplitude in this weak component is still observed (Fig. 1). From an interpretative viewpoint, component-8-11 is related to the "half-precessional" cycle of equatorial origin modulated at ~ 11 kyr (Berger et al. 2006). Hagelberg et al. (1994) found precession-related

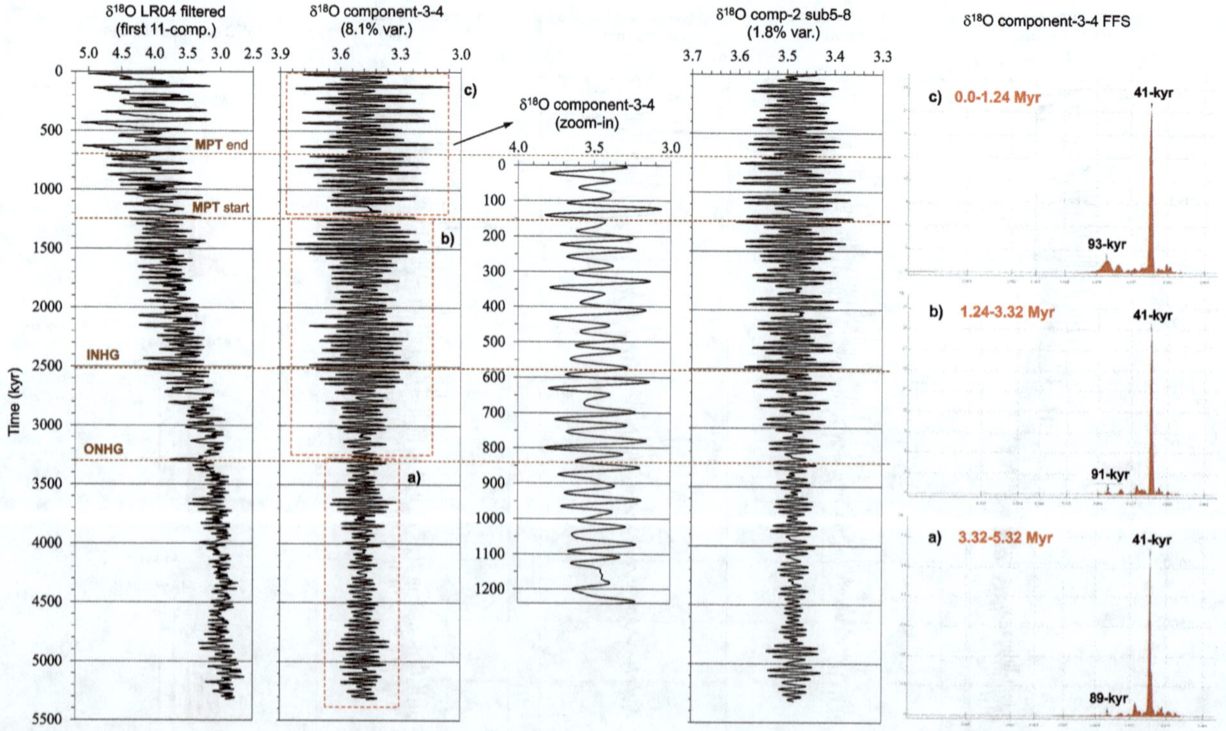

Fig. 4 Time-series panel of the δ^{18}O component-3-4 (8.1% var.) and component-2 subcomp-5-8 (1.8% var.) related to obliquity. The Fourier frequency spectrum of component-3-4 in three time intervals is highlighted by red-labeled boxes. The spectrum is always dominated by obliquity 41-kyr power. Note a clear increase in the average 41-kyr amplitude and power towards the beginning of the MPT (box 4a and 4b), followed by an amplitude stabilization and power reduction (box 4c) and an amplitude reduction (comp-2 subcomp-5-8) during the MPT and post-MPT. The envelope of two to three obliquity cycles in a weak eccentricity 93-kyr framework after the beginning of the MPT is noteworthy (box 4c and magnified time series to the right). FFS scale power increases upwards from **a** to **b** and decreases from **b** to **c**. *x*-axis scale variable (‰). Time scale in kyr

climate variability at periods from 10 to 12 kyr in three sites for the Late Pleistocene (ODP 846, ODP 663, and DSDP 609). They concluded that this variability may be derived from the high sensitivity of the tropics to summertime insolation in both hemispheres, triggering an amplified response of tropical precipitation and temperature that may be transmitted to high latitudes via Atlantic advective transport. The results of magnetic susceptibility and particle size analysis from the Northwestern Chinese Loess Plateau reveal well-defined half-precessional cycles during the last interglaciation, interpreted as a direct response to low-latitude forcing through its modulation of the East Asian summer monsoon (Sun and Huang 2006). However, despite this interesting evidence of the semi-precessional cycle of equatorial origin in the δ^{18}O deep-temperature and ice-volume signal, its contribution to the total variance of the global δ^{18}O record is less than 1%.

Analysis of the SST SSA-components

The resampled HT846 record was partitioned by SSA into five SST time series, which cumulatively explain 99.6% of the original variance (HT846-filtered) (Fig. 5, Table 4). The spectral frameworks of these SST components, ranging from 1266-kyr to 11-kyr periodicity, are very similar to that of the LR04 record. The features of the SST components are described and compared to the δ^{18}O ones as follows.

SST component-1

This component explains 87.8% of the original variance (eigenvalue 128.1) and contains the most significant fraction of the SST signal associated with a strong cooling trend (long-term HT846 component, LT-HT846). The FFS shows a broad spectrum of significant peaks with periods of ~ 1.27 Myr (TISA power 83.3%), 406 kyr (8.6%), 298 kyr (5.0%), and 193 kyr (3.1%), overlapping the general trend (Fig. 5, Table 4). A high-order SSA was performed to separate the oscillatory terms from the component-1 trend, from which three subcomponents were identified equal to a reconstruction R^2 of 0.999 (Table 5).

The component-1 subcomp-1 explains an absolute variance of 81.7% and exhibits a strong cooling trend with significant mild oscillations. The four LR04s' subtrends are generally identifiable in the tropical SST long-term components, with some differences especially in subtrend-III due to a strong cooling event at ~ 2.1 Myr (Herbert et al. 2010a), which is less evident in the δ^{18}O signal (Fig. 6). In addition, the MBO feature is

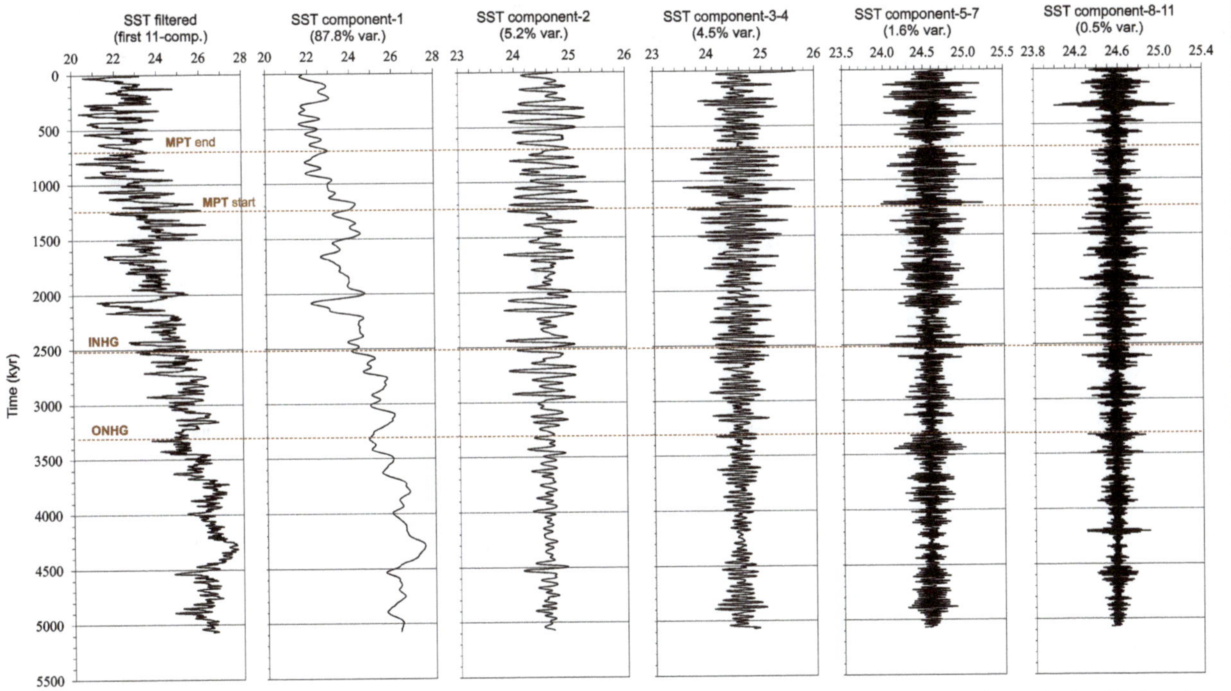

Fig. 5 Time-series panel of the five SST components isolated from the ODP Site 846 record by singular spectrum analysis with the SST-filtered record (left) obtained by combining these data components (99.6% total variance). The *x*-axis scale (°C) is variable in relation to the component's variance fraction. Original SST data from Herbert et al. (2010b) resampled at 1 kyr. Time scale in kyr

less marked in the SST signal. These differences between deep-temperature/ice-volume and sea surface temperature signals could be due to temporal effects of upwelling in the EEP "cool" region (Lisiecki 2010b).

The frequency spectrum of the SST comp-1 subcomp-1 still shows a significant peak of 1.27 Myr period (98.7% power), related to the ~ 1.2-Myr obliquity modulation cycle (Laskar et al. 2004, 2011; Boulila et al. 2011), to which the mild subtrend oscillations may be related (Table 5). The comp-1 subcomp-2 explains 3.0% of the HT846 variance dominated by a 1.50-Myr periodicity (94.1%), probably related to the same obliquity modulation cycle. A wide spectrum of long-term cycles characterized the comp-1 subcomp-3-10, explaining 3.1% of the original SST variance (399, 292, 213–193, 126, and 95 kyr).

To estimate the variance of the SST long-term trend, the component-1 was detrended by subtracting an exponential fitting with least squares minimization ($R^2 = 0.87$), and the oscillatory signal was processed by a high-order SSA (Fig. 6, Table 6). The LT-HT846 exponential fit explains 87.2% of the comp-1 variance, equivalent to an absolute contribution of 76.6%. The detrended oscillatory signal explains an absolute 11.2% variance. The 5.1% variance difference between the comp-1 subcomp-1 (81.7%) and the comp-1 exponential fit (76.6%) may be attributed to the 1.27-Myr cycle still superimposed on the comp-1 subcomp-1. Three subcomponents were isolated from the SST detrended comp-1 (reconstruction $R^2 = 0.999$). The detrended comp-1 subcomp-1 explains 5.56% of the SST original variance and exhibits a spectrum dominated by 1.37-Myr (power 51.9%) and 3.6-Myr (40.6%) cycles, with an ancillary term of 827 kyr. Again, the 3.7- to 3.6-Myr cycle appears in the LR04 and HT846 detrended comp-1 subcomp-1 only; however, in the HT846 record, the 1.37-Myr cycle does not disappear. Therefore, the 5.56% tropical SST variance (detrended comp-1 subcomp-1) related to the long-term cycles (1.37-Myr obliquity modulation cycle and 3.6-Myr Earth orbit secular frequency) is consistent with the 5.1% variance estimate by difference from comp-1 subcomp-1 (which includes the 1.27-Myr cycle only).

Prevailing 405-kyr (70.3%) and 300-kyr (29.7%) cycles characterize the detrended comp-1 subcomp-2 explaining 2.81% of the SST variance. These frequencies may be related to the astronomically stable 405-kyr-long eccentricity cycle and to the 304-kyr Earth orbit secular frequency (Laskar et al. 2004, 2011; Boulila et al. 2011). The detrended comp-1 subcomp-3-11 explains 2.84% of the SST variance and shows a wide spectrum of frequency bands. A 295-kyr cycle (power 15.4%) can be attributed to the 304-kyr Earth orbit secular frequency. Oscillations of 193 kyr (45%) and 148 kyr (6.7%) may be linked to the long-period obliquity modulation cycles (~ 160–200 kyr) (Boulila et al. 2011). Finally, weak short eccentricity cycles of 136 kyr (9.2%), 126 kyr (14%), and 95 kyr (9.7%) are also recognizable. The FFS of this subcomponent by time intervals (Fig. 6) shows a relative strengthening of the short

Table 4 Main descriptive parameters and Fourier frequency results of the five ODP Site 846 SST components isolated by singular spectrum analysis. Original SST data from Herbert et al. (2010b) resampled at 1 kyr

SST component rank	SST comp. variance (%)	Frequency (kyr^{-1})	TISA power (%)	Period (kyr)	Weighted mean period (kyr)	Forcing
1	87.8	0.0007899	83.3	1266		Long-term trend + long-period orbital modulation and eccentricity cycles
		0.0024651	8.6	406		
		0.0033527	5.0	298		
		0.0051684	3.1	193		
2	5.2	0.0105090	32.6	95		Transition obliquity to short eccentricity
		0.0110927	26.3	90		
		0.0144249	41.1	69		
		0.0244798	–	41	Very low power peak	
3–4	4.5	0.0244852	69.3	41	40	Obliquity
		0.0252735	18.4	40		
		0.0268223	12.3	37		
5–7	1.6	0.0439726	31.6	23	22	Precession
		0.0463271	43.1	22		
		0.0487111	25.3	21		
8–11	0.5	0.0732060	36.7	14	12	"Half-precession" (equatorial)
		0.0767562	30.5	13		
		0.0946970	32.8	11		
> 11	0.4					Noise

eccentricity cycle (93-kyr) at the expense of the obliquity modulation cycles (~ 195- to 158-kyr), whereas the increase in signal amplitude by time is not observable post-MPT.

SST component-2

This component explains 5.2% of the total variance (eigenvalue 7.56) and exhibits a basically 95- to 90-kyr short eccentricity periodicity (58.9%) with a 69-kyr term (41.1%) and a very weak but significant 41-kyr obliquity oscillation (Fig. 7; Table 4). Compared to the $\delta^{18}O$ comp-2, the two shifting horizons of increasing amplitude (ONHG and MPT start) and the change in orbital periodicity from 41–54 to ~ 93 kyr by three time segments are less evident (Fig. 7a–c). The transition from obliquity to short eccentricity is less evident but still present as shown also by the subcomponent analysis (Table 6). Three subcomponents were isolated by

Table 5 Descriptive parameters and Fourier frequency results of the ODP Site 846 SST subcomponents isolated from the SST component-1 by singular spectrum analysis. The processing of the subcomponents was done to better separate and estimate the variance fraction of specific frequency bands from still complex data components. The italicized percentage values are the absolute variance of the subcomponents (referred to the total SST variance)

SST comp. rank	SST comp. variance (%)	SST subcomp. rank	SST subcomp. var. (rel.) (%)	SST subcomp. var. (abs) (%)	Frequency (kyr^{-1})	TISA power (%)	Period (kyr)	Forcing
1	87.8	1	93.0	*81.7*	0.0007899	98.7	1266	Long-term trend + obliquity modulation cycle
		2	3.4	*3.0*	0.0006678	94.1	1497	Obliquity modulation cycle
					0.0024657	5.9	406	Eccentricity (long)
		3–10	3.5	*3.1*	0.0025065	8.2	399	
					0.0034190	18.5	292	Earth's orbit secular frequency
					0.0046896	15.3	213	Obliquity modulation cycle
					0.0051832	36.9	193	
					0.0079507	12.7	126	Eccentricity (short)
					0.0105043	8.4	95	

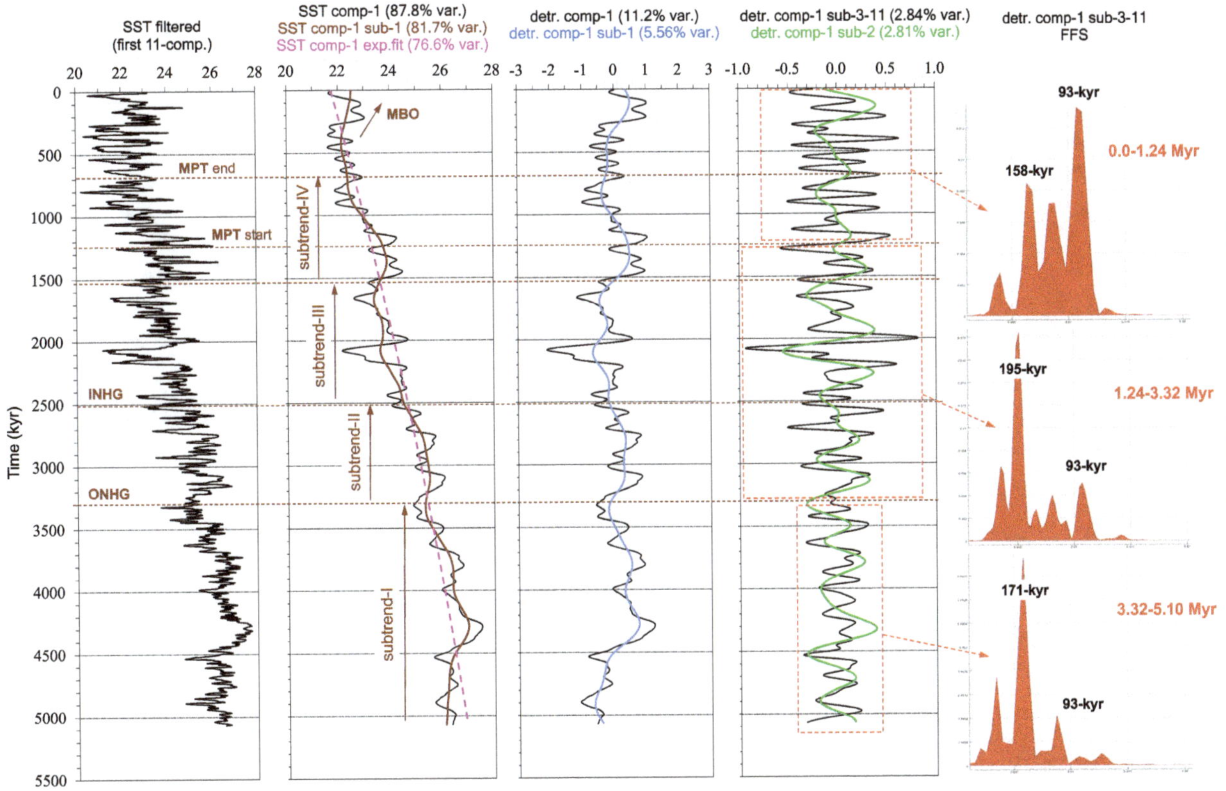

Fig. 6 Time-series panel of the ODP Site 846 SST component-1 and its subcomponents isolated by singular spectrum analysis. SST comp-1 exponential trend (°C) = 33.58041–11.84075 $e^{(-t/8830.503)}$. The vertical brown arrows show four subtrends of component-1 and their correlation to the NHG transitions known in the literature (MBO this study). To the right, Fourier frequency spectrum of SST detrended comp-1 subcomp-3-11 (2.84% variance) for three time intervals are highlighted by red boxes. Note the increase in amplitude after the ONHG (but not after the MPT) and the power increase of the ~ 93-kyr eccentricity cycle. *x*-axis scale variable (°C). The *y*-axis scale of the frequency spectrum increases upwards. Time scale in kyr

a high-order SSA ($R^2 = 0.999$). Subcomponent-1-4 explains 3.65% of the original SST variance and shows a frequency spectrum in the short eccentricity band (95-kyr, power 43.2% and 69-kyr, 56.8%). Subcomponent-5-8, 11-15 explains 1.27% variance in a still complex transition spectrum from the obliquity (53-kyr, 60.2%) to eccentricity band (126-kyr, 39.8%). Finally, subcomp-9-10 explains only 0.27% of the SST variance in the obliquity frequency band (45-kyr, 62.2%; 41-kyr, 37.8%).

SST component-3-4

This component explains 4.5% of the SST variance (eigenvalue 6.56) at the 87.7% obliquity-driven 41–40-kyr period (Table 4). The SST component-3-4 is still characterized by a less evident increase in the average amplitude by time approaching the beginning of the MPT (Fig. 8a, b), followed by a clear decrease in amplitude after the MPT start (Fig. 8c). A slight change in the shape of the 41-kyr cycle that appears enveloped in a very weak ~ 93-kyr signal can be observed after 1.2 Myr (Fig. 8c).

SST component-5-7

This component explains 1.6% of the SST variance (eigenvalue 2.24) driven by precession frequency bands at ~ 22 kyr (Table 4), similar to the precession $\delta^{18}O$ component.

SST component-8-11

Component-8-11 explains a total 0.5% of the SST variance (eigenvalue 0.74) and shows a weight mean period at ~ 12 kyr (Table 4). An increasing mean amplitude in this weak component is still observed (Fig. 5). This component is similar to that of the $\delta^{18}O$ identified in this study and can be related to the "half-precessional" cycle of equatorial origin modulated at ~ 11 kyr (Berger et al. 2006).

Summary of $\delta^{18}O$ and SST forcing variance contribution

The SSA results allow the LR04 $\delta^{18}O$ and ODP Site 846 SST forcing variance contribution to be estimated quantitatively in detail. A complex envelope of long-term orbital cycles accounts for a maximum of 5.1% of the $\delta^{18}O$ variance (due to a small fraction of short eccentricity in detrend

Table 6 Main descriptive parameters and Fourier frequency results of the ODP Site 846 SST subcomponents isolated from the detrended comp-1 and comp-2 by singular spectrum analysis. The processing of the subcomponents was made in order to better separate and estimate the variance fraction of specific frequency bands from still complex data components. The italicized percentage numbers are the absolute variance of the subcomponents (referred to the total SST variance)

SST comp. rank	SST comp. variance (%)	SST subcomp. rank and var. (abs.)	SST subcomp var. (rel.) (%)	SST subcomp var. (abs) (%)	Frequency (kyr^{-1})	TISA power (%)	Period (kyr)	Forcing
1	87.8	exp.fit	87.2	*76.6*	–	–	–	Long-term trend
		detrend 12.8% (*11.2%*) 1	49.6	*5.56*	0.0002758	40.6	3626	Earth's orbit secular frequency
					0.0007313	51.9	1367	Obliquity modulation cycle
					0.0012092	7.5	827	Eccentricity (long)
		2	25.1	*2.81*	0.0024667	70.3	405	
					0.0033343	29.7	300	Earth's orbit secular frequency
		3–11	25.4	*2.84*	0.0033896	15.4	295	
					0.0051716	45.0	193	Obliquity modulation cycle
					0.0067394	6.7	148	
					0.0073694	9.2	136	Eccentricity (short)
					0.0079678	14.0	126	
					0.0104995	9.7	95	
2	5.2	1–4	70.2	*3.65*	0.0105084	43.2	95	Eccentricity (short)
					0.0144299	56.8	69	
		5–8, 11–15	24.5	*1.27*	0.0079524	39.8	126	Transition obliquity to short eccentricity
					0.0187962	60.2	53	
		9–10	5.2	*0.27*	0.0220179	62.2	45	Obliquity
					0.0244820	37.8	41	

comp-1 subcomp-3-11; Table 3). Within the 2.23% variance of the $\delta^{18}O$ detrend comp-1 subcomp-3-11, a ~ 0.95% short eccentricity contribution may be roughly estimated by using the TISA power of the short eccentricity terms (42.7% of 2.23% variance). This approach is justified because of the low variance of the detrend comp-1 subcomp-3-11, causing a negligible error. Therefore, $\delta^{18}O$ long-period orbital modulation cycles (1.57–1.33 Myr and ~ 160–200 kyr obliquity amplitude modulation; 3.75 Myr and ~ 300 kyr Earth orbit secular frequencies) and the ~ 405-kyr-long eccentricity cycles tightly overlapped to the $\delta^{18}O$ trend, influencing the Earth's climate with an estimated variance of ~ 4.2%. Regarding the tropical SST, a similar complex envelope of long-term orbital cycles account for a maximum 11.2% of the HT846 variance (Table 6), and by subtracting the small variance of the short eccentricity term (32.9% of 2.84%, variance = 0.93%), a contribution of ~ 10.3% can be estimated.

Long-term cycles (1.5–1.0 Myr) were previously observed in the planktonic $\delta^{18}O$ record of ODP Site 625B from the Gulf of Mexico and interpreted as a low-frequency component of the climate system, likely unrelated to orbital forcing (Joyce et al. 1990). Instead, according to Lourens and Hilgen (1994), the long-term 1.3-Myr period observed from some Mediterranean and equatorial Atlantic and Pacific sites resulted from a long-term obliquity modulation cycle. Through a correlation between Cenozoic icehouse sequences and astronomical cycles, Boulila et al. (2011) demonstrate a close correlation between ~ 1.2-Myr obliquity modulation cycles and third-order eustatic sequences, suggesting glacioeustasy as the major driving process of third-order sequences in an icehouse world. A complex envelope of Plio-Pleistocene long-term astronomical cycles and especially the 1.57–1.33-Myr obliquity amplitude modulation to which the mild NHG's subtrend terminal oscillations may be related by transient interaction on the long-term trend are further geological evidences of the chaotic behavior of the Solar System (Laskar et al. 2004, 2011; Boulila et al. 2011; Westerhold et al. 2017), and the SSA results show that their influence on Earth's climate, estimated for the first time in this study, have important differences in variance (global $\delta^{18}O$ ~ 4.2%; tropical SST ~ 10.3%). Interestingly, the tight overlap of these long cycles could be the reflection of a close link between these weak orbital forcings and the much more effective forcing at the origin of the long-term trends through a temperature-ice volume-induced control on the long-cycle sensibility. The lack of the ~ 1.2-Myr eustatic cycles in the Mesozoic greenhouse world (ephemeral or no ice sheets) (Boulila et al. 2011) may support this hypothesis.

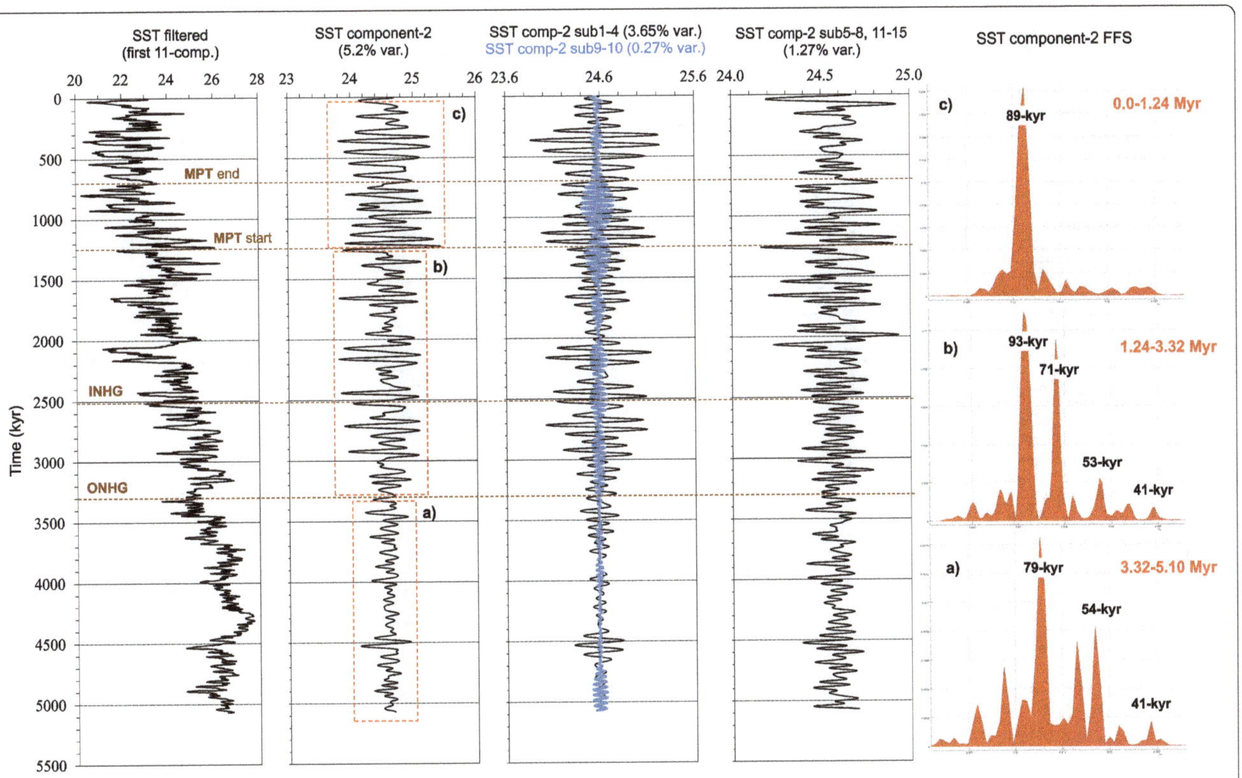

Fig. 7 Time-series panel of ODP Site 846 SST component-2 and its subcomponents and Fourier frequency spectrum of comp-2 for three time intervals **a**, **b** and **c**, highlighted by red-labeled boxes. The transition nature from **a** obliquity to **c** short eccentricity is less evident in this tropical SST site, but still present. The increasing amplitude of the short eccentricity is more clear in subcomp-1-4. The obliquity decrease after the MPT is still shown by subcomp-9-10. *x*-axis scale variable (°C). The *y*-axis scale of the frequency spectrum increases upwards. Time scale in kyr

The short eccentricity does not manifest itself as a "pure" $\delta^{18}O$ component; rather, it appears to move gradually to the Pleistocene climate signal as an interaction subcomponent only, with a long-term response (detrend comp-1 subcomp-3-11, ~ 0.95% variance) and the obliquity-eccentricity transitional component-2 (component-2 subcomp-1-4, 9-15, 5.5% variance). Therefore, the short eccentricity cycle contributes ~ 6.5% to the global Plio-Pleistocene $\delta^{18}O$ signal. Similarly, the SST detrend comp-1 subcomp-3-11 explains ~ 0.93% of short eccentricity variance, and a 3.65% fraction is recorded in the SST comp-2 subcomp-1-4. A small fraction of short eccentricity terms are still recorded in the SST comp-2 subcomp-5-8, 11-15 for which a 0.5% variance can be estimated (39.8% of 1.27% variance), resulting in a SST short eccentricity total variance of ~ 5.1%. These values agree with the consideration that the amount of insolation perturbation at the eccentricity-driven ~ 100-kyr period is too small to directly cause climate change of an ice age amplitude (Imbrie et al. 1993; Berger et al. 1999, 2006; Lisiecki 2010a).

The 41-kyr obliquity cycle appears to transfer to the climate system both as a "pure" $\delta^{18}O$ component, with signal strength of 8.1% (component-3-4), and as an interaction subcomponent like the obliquity-eccentricity transitional component-2 (comp-2 subcomp-5-8, 1.8% var.). Therefore, the total $\delta^{18}O$ signal variance in the 41-kyr obliquity band is estimated to be 9.9%. The SST shows a similar behavior characterized by a "pure" component-3-4 (4.5% variance) plus transitional obliquity-eccentricity subcomponents (comp-2 subcomp-9-10, 0.27% var.; comp-2 subcomp-5-8, 11–15, ~ 0.76%), from which a total obliquity contribution of 5.5% may be estimated.

The precession and semi-precession $\delta^{18}O$ variance can be evaluated at about 2.0% and 0.6%, close to the SST contribution of 1.6% and 0.5%, respectively. In summary, the variance of the $\delta^{18}O$ orbital components related to long-term, short eccentricity, obliquity, precession, and semi-precession cycles can be evaluated at approximately 4.2%, 6.5%, 9.9%, 2.0%, and 0.6%, respectively (total 23.2% variance), whereas the orbital SST components account for 10.3%, 5.1%, 5.5%, 1.6%, and 0.5%, respectively (total 23.0% variance). Within orbital forcings, the impact of the obliquity cycle appears to be the most consistent in the LR04 stack. In the tropical HT846 record, the most consistent orbital signal is related to the long-term cycles. These results agree with those of Kominz and Pisias (1979), which state that the fraction of the record variance related to orbital changes is less

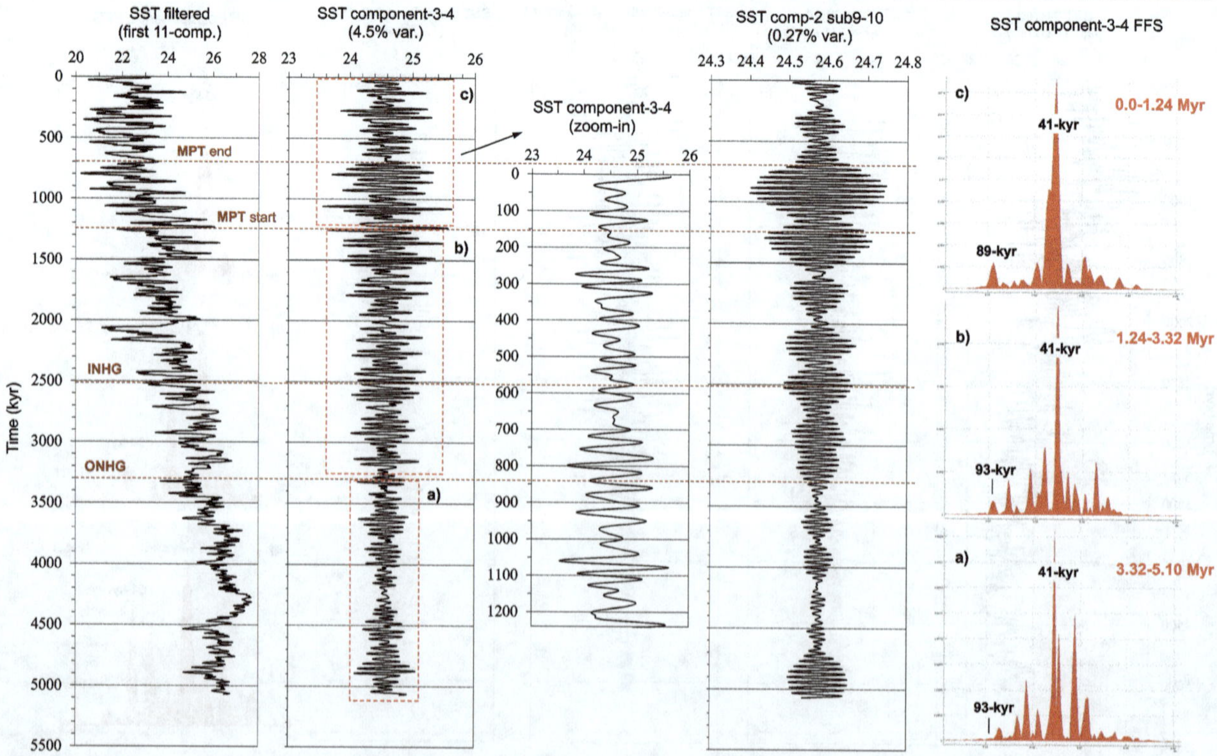

Fig. 8 Time-series panel of the ODP Site 846 SST obliquity's component-3-4 (4.5% var.) and component-2 subcomp-9-10 (0.27% var.). Fourier frequency spectrum of comp-3-4 by three time intervals highlighted by red labeled boxes. The spectrum is always dominated by obliquity 41-kyr power. Note a clear increase in the average 41-kyr amplitude and power towards the beginning of the MPT (box 8a and 8b), followed by a clear decrease in amplitude after the beginning of the MPT (box 8c and comp-2 subcomp-9-10). The envelope of two to three obliquity cycles in a weak eccentricity ~ 93-kyr framework after the beginning of the MPT is less evident (box 8c and magnified time series to the right). FFS scale power increasing from **a** to **b** and was constant from **b** to **c**. *x*-axis scale variable (°C). Timescale in kyr

than 25%, and of Wunsch (2004), which states that this fraction never exceeds 20%. Interestingly, the cumulative influence on the Earth's climate of orbital forcings estimated from the global $\delta^{18}O$ stack is equal to that of the tropical SST single record, while the individual contribution of each astronomical term is quite different. This result could be attributed to the latitudinal gradients of orbital forcings (Berger 1978; Liu and Herbert 2004), while the constancy of the long-term variance might reflect a global latitude-independent nature of the paleoclimatic trend's control factors. The fact that the $\delta^{18}O$ and SST orbital components explain ~ 23% of the total variance clearly shows that astronomical forcing does not play a major role in guiding the response of the climate system in the Plio-Pleistocene, in agreement with Kominz and Pisias (1979), Raymo (1994), Wunsch (2004), and Mudelsee and Raymo (2005). However, ~ 76% of the Plio-Pleistocene climate variance is related to non-Milankovitch long-term controlling factors, drives approximately three fourths of the energy conveyed in paleoclimate records, and thus constitutes the main Earth's climate system. This robust and stable trend signal needs to be explained.

Statistical analysis of the mean climate state through time

The SSA time series allow a quantitative analysis of the mean climate state through time. The Plio-Pleistocene $\delta^{18}O$ and SST components and the solution of orbital forcings (Laskar et al. 1993) were standardized (mean = 0, standard deviation = 1) and binned at three arbitrary constant time intervals (532 kyr, 760 kyr, and 1064 kyr) to avoid data density distortion in descriptive statistics and significance tests of mean differences among groups (Table 7). Some data were inverted in order to have the same paleoclimatic polarity (Fig. 9).

As observed in the comparison of the 95% confidence interval (CI) for the means (Fig. 9a, c) and in the significance tests of the mean differences in the analysis of variance (ANOVA) table (Table 7), the $\delta^{18}O$ and SST trend components (comp-1, comp-1 subcomp-1, and comp-1 exponential fit) can very significantly change the signal mean for different time intervals. Instead, the $\delta^{18}O$ and SST orbital components (comp-2-11) and the astronomical forcings modified the mean with little significance or no significance. These results are similar across all three binned time intervals showing that this feature is not a time frame artifact (Table 7). In more detail, the mean reaches a deviation from the standardized mean up to − 1.51–1.18 standard units through the time segments in the $\delta^{18}O$ component-1, up to − 1.50–1.21 in the $\delta^{18}O$ comp-1 subcomp-1, and in the range − 1.72–1.40 in the $\delta^{18}O$ comp-1 exponential fit. The SST trend components exhibit similar results with deviations up to − 1.45–1.27, − 1.49–1.33, and − 1.67–1.50 for the comp-1, comp-1 subcomp-1, and comp-1 exponential fit, respectively. The $\delta^{18}O$ and SST orbital signals reach maximum values of − 0.08–0.09 and − 0.09–0.04 standard deviations, respectively. Eccentricity reaches a maximum deviation from the standardized mean of − 0.30–0.28 units. Obliquity ranges from − 0.12 to 0.12 and precession from − 0.02 to 0.01. The insolation mean (65° N at June 21) oscillates in the range of − 0.08–0.06. The ANOVA analysis was also done to all single orbital components, obtaining similar results. The histograms of the Plio-Pleistocene $\delta^{18}O$ and SST astronomical components-2-11, compared to that of the trend component-1 subcomp-1 (Fig. 9b, d), show a roughly normal distribution centered on the orbital response mean, while the trend component exhibits a multimodal distribution away from the standardized mean. This analysis suggests Plio-Pleistocene orbital forcing could be considered a random walk at zero-sum (the standardized mean): on long time scales, astronomical forcing oscillates around a basically time-invariant normally distributed mean of the Plio-Pleistocene climate system response. These results, consistent with those of the quantitative estimation of the variance, suggest that changes in the mean climate state during the Plio-Pleistocene were primarily caused by non-Milankovitch first-order controlling factors recorded in the global LR04 and tropical HT846 trend components (~ 76% variance). The orbital forcing acts at a hierarchically lower strength level (~ 23%), with cycles overlapping a primary trend forcing system.

Nonlinear response of $\delta^{18}O$ and SST orbital vs. trend components

Figure 10 exhibits the exponential relationships between the orbital $\delta^{18}O$ and SST components variance vs. the average $\delta^{18}O$-SST component-1 exponential fit trends by time segments binned at 532 kyr. This plot shows that the exponential increase in signal amplitude towards the Mid-Late Pleistocene is observed in all $\delta^{18}O$ and SST orbital components, from semi-precession to the eccentricity band. The exponential growth of the orbital response variance is dependent on the long-term mean climate state (LT-MCS) at a given time (mean ice volume-bottom water temperature, average sea surface-atmospheric temperature), both at a global scale (LR04) and at a local tropical scale (ODP Site 846). The exponential R^2 ranges from 0.7 to 0.9 for the $\delta^{18}O$ components and is slightly lower in the SST components (0.6 to 0.9). The minimum exponential fitting does not fall below an $R^2 = 0.64$ for SST component-2 subcomp-1-4 and $R^2 = 0.72$ for $\delta^{18}O$ component-8-11. The exponential coefficient e^{kx} for the $\delta^{18}O$ semi-precessional component is 1.05 (comp-8-11) and increases with the orbital period; it is 1.78 for the precessional component (comp-5-7), 1.88 for the obliquity component (comp-3-4), and 2.93 for the short eccentricity subcomponent (comp-2 subcomp-1-4; 9-15). The exponential coefficient is high (2.37) for the short eccentricity-obliquity modulation subcomponent (detrended comp-1 subcomp-3-11) (Fig. 10a). These data indicate an increase in $\delta^{18}O$ exponential growth from semi-precession to ~ 100-kyr eccentricity response towards the Mid-Late Pleistocene. The amplitude flattening of the obliquity-driven $\delta^{18}O$ component-3-4 during the MPT and post-MPT time shown in Fig. 4 is clearly evident in Fig. 10a also, a feature related to that observed by Lisiecki and Raymo (2007). Regarding the tropical SST, the absolute exponential coefficient e^{kx} for the semi-precessional component is 0.31 (comp-8-11). It is 0.37 for the precessional component (comp-5-7), 0.48 for the obliquity component (comp-3-4), and 0.67 for the short eccentricity subcomponent (comp-2 subcomp-1-4) (Fig. 10b). These results in tropical SST are similar to those of the global $\delta^{18}O$, although of lesser magnitude. The amplitude contraction of the obliquity-driven SST component-3-4 after the MPT start, shown in Fig. 8, is very evident even in Fig. 10b.

Figure 11 shows a quantitative analysis of the $\delta^{18}O$ and SST response modulation relative to orbital forcing as a function of the LT-MCS. The standardized response sensitivity R_s is given by the following function

$$R_s = \left(\sigma^2{}_{resp} - \sigma^2{}_{forc}\right) / \sigma^2{}_{forc} \times 100$$

where $\sigma^2{}_{resp}$ and $\sigma^2{}_{forc}$ are the variance of the standardized (0—mean; 1—standard deviation) orbital response component and forcing calculated by time segments binned at 532 kyr, respectively. For the purpose of this exercise, the eccentricity and obliquity response are approximated because small variances in these terms cannot be merged in a composite signal, or a small fraction of different frequency terms are included. Since the variance is small and the aim of this analysis is an order of magnitude estimate, the errors should be negligible. The $\delta^{18}O$ short eccentricity response (~ 6.5% variance) was considered both as a single main subcomponent (comp-2 subcomp-1-4, 9-15; 5.5% variance) containing ~ 85% of the short eccentricity response and as the weighted mean by variance of the comp-2 subcomp-1-4, 9-15 (5.5% variance) and the detrend comp-1 subcomp-3-11 (2.23% variance), the latter containing also a small fraction of long-term cycles. The $\delta^{18}O$ obliquity response (9.9% variance) was calculated as the weighted mean of the comp-3-4 (8.1%) and comp-2 subcomp-5-8 (1.8%). Similarly, the SST short eccentricity response (~ 5.1% variance) was

Table 7 Significance tests of the mean differences among $\delta^{18}O$ and SST components and the solution of orbital forcings (Laskar et al. 1993) binned at three arbitrary constant time intervals (532 kyr, 760 kyr, and 1064 kyr). Data standardized on the entire Plio-Pleistocene record (mean = 0, standard deviation = 1) and, where necessary, inverted to have the same paleoclimatic polarity (positive values for positive forcing and warming)

Component/forcing	Time bin (kyr)	Mean[a]	Sum of squares	df	*F*	Sig.
LR04 comp-1	532	− 1.47 ÷ 1.18	5059.52	9	11,504.34	0.00
	760	− 1.51 ÷ 1.13	4996.27	6	13,708.49	0.00
	1064	− 1.43 ÷ 1.14	4917.43	4	16,271.23	0.00
LR04 comp-1 subcomp-1	532	− 1.47 ÷ 1.21	5257.57	9	50,499.31	0.00
	760	− 1.50 ÷ 1.17	5182.42	6	33,599.07	0.00
	1064	− 1.44 ÷ 1.17	5116.42	4	33,559.24	0.00
LR04 comp-1 exp.fit. trend	532	− 1.72 ÷ 1.40	5262.71	9	55,162.47	0.00
	760	− 1.62 ÷ 1.35	5204.25	6	40,161.50	0.00
	1064	− 1.49 ÷ 1.28	5094.35	4	30,132.42	0.00
LR04 comp-2-11 (orbitals)	532	− 0.08 ÷ 0.09	8.80	9	0.98	0.46
	760	− 0.01 ÷ 0.02	0.74	6	0.12	0.99
	1064	− 0.01 ÷ 0.01	0.33	4	0.08	0.99
SST comp-1	532	− 1.45 ÷ 1.26	4644.74	9	6235.99	0.00
	760	− 1.42 ÷ 1.27	4468.87	6	6339.54	0.00
	1064	− 1.38 ÷ 1.16	4389.59	4	8244.21	0.00
SST comp-1 subcomp-1	532	− 1.49 ÷ 1.26	4950.98	9	24,819.72	0.00
	760	− 1.49 ÷ 1.33	4838.36	6	18,152.80	0.00
	1064	− 1.42 ÷ 1.18	4722.96	4	17,566.50	0.00
SST comp-1 exp.fit. trend	532	− 1.67 ÷ 1.50	5006.76	9	49,988.56	0.00
	760	− 1.57 ÷ 1.44	4949.41	6	36,725.39	0.00
	1064	− 1.44 ÷ 1.36	4842.28	4	27,747.29	0.00
SST comp-2-11 (orbitals)	532	− 0.09 ÷ 0.04	6.25	9	0.69	0.71
	760	− 0.01 ÷ 0.02	0.89	6	0.15	0.99
	1064	− 0.02 ÷ 0.01	0.60	4	0.15	0.96
Eccentricity	532	− 0.30 ÷ 0.25	169.60	9	19.43	0.00
	760	− 0.28 ÷ 0.28	143.98	6	24.64	0.00
	1064	− 0.22 ÷ 0.14	77.44	4	19.63	0.00
Obliquity	532	− 0.12 ÷ 0.11	35.42	9	3.96	0.00
	760	− 0.12 ÷ 0.12	37.02	6	6.21	0.00
	1064	− 0.11 ÷ 0.10	31.96	4	8.03	0.00
Precession	532	− 0.02 ÷ 0.01	0.30	9	0.03	1.00
	760	0.00	0.02	6	0.00	1.00
	1064	− 0.01 ÷ 0.01	0.12	4	0.03	1.00
65°N Insolation at June 21	532	− 0.08 ÷ 0.06	8.69	9	0.97	0.47
	760	− 0.05 ÷ 0.06	8.89	6	1.48	0.18
	1064	− 0.05 ÷ 0.04	6.37	4	1.59	0.17

df degrees of freedom, *F* Fisher test, *Sig* statistical significance
[a]Min-max of the mean among time segments

approximated as a single comp-2 subcomp-1-4 (3.65% variance) explaining ~72% of the short eccentricity response and as weighted mean of the comp-2 subcomp-1-4 (3.65% variance) and the detrended comp-1 subcomp-3-11 (2.84% variance), the latter containing a small fraction of long-term cycles. The weighted mean of the SST comp-3-4 (4.5% variance) and comp-2 subcomp-9-10 (0.27% variance) was calculated as the SST obliquity response, covering ~87% of the

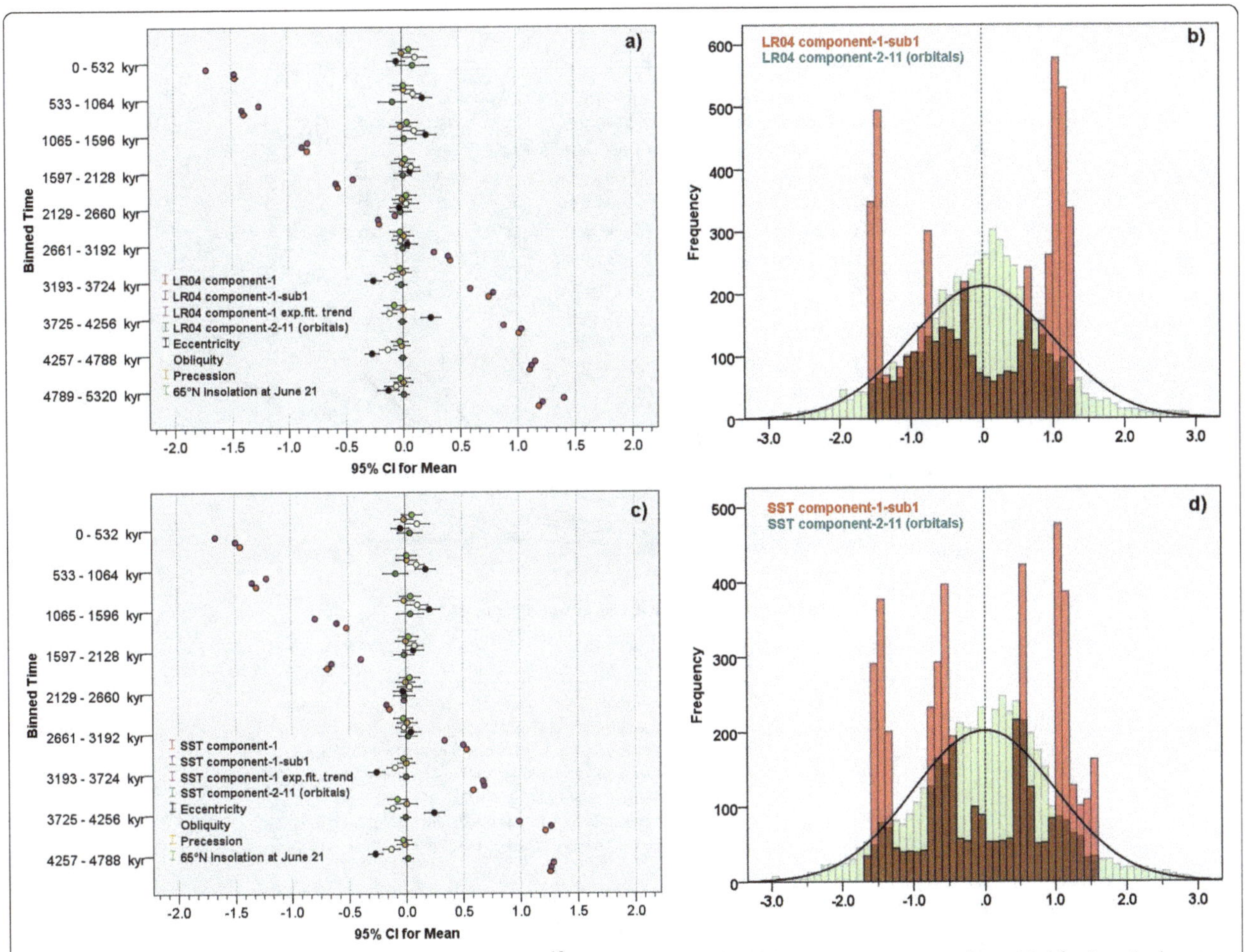

Fig. 9 Plots of the 95% confidence interval for the mean of **a** $\delta^{18}O$ components and **c** SST components compared to orbital forcings (Laskar et al. 1993), binned at 532-kyr constant time intervals. Overlapped histograms of **b** Plio-Pleistocene $\delta^{18}O$ and **d** SST orbital component-2-11 vs. trend component-1 subcomp-1. The averages of the astronomical components oscillate around the standardized mean with little statistically significance differences and a roughly normal distribution. Instead, the averages of the trend components differ very significantly, deviating markedly from the mean with a general multimodal distribution. Data standardized on the entire Plio-Pleistocene record (mean = 0, standard deviation = 1) and, where necessary, inverted to have the same paleoclimatic polarity (positive values for positive forcing and warming)

total signal (4.77% of 5.5%). A self-sustained climate system, which is only paced by orbital cycles, would produce a response sensitivity near zero because the variance of the response would match the variance in forcing, whereas values much larger than zero would indicate a nonlinear reinforcement of the signal response. Negative R_s values show a response variance lower than orbital forcing, suggesting a damping mechanism for the climate system. Figure 11a exhibits a $\delta^{18}O$ nonlinear eccentricity R_s up to 400% towards the Mid-Late Pleistocene similar to that of a precession response in shape but with a magnitude up to 200%. The two eccentricity curves are very similar, indicating a negligible error in the approximations. The $\delta^{18}O$ obliquity R_s shows a nonlinear amplification up to 180% towards the end of the MPT, followed by strong variance depletion near to zero (balance line) post-MPT. This pattern of low $\delta^{18}O$ obliquity variance is similar to that shown in Fig. 5b from Lisiecki and Raymo (2007). Remarkably, the asymptotic decay of the $\delta^{18}O$ R_s towards the Early Pliocene becomes all negative up to − 100%, suggesting a damped orbital response during the time of minimum average ice volume and higher temperature. This R_s pattern is similar considering the SST (Fig. 11b), although more scattered and with a smaller positive magnitude which does not reach 200%. The SST obliquity R_s also shows strong variance depletion up to − 20% post-MPT as well as the new feature of asymptotic decay up to − 100% of all the SST R_s towards the Early Pliocene.

Link between the pCO_2 and the SST-$\delta^{18}O$ long-term trend components

Figure 12 shows the Plio-Pleistocene cross plots among the $\delta^{18}O$, SST, and pCO_2 trend and orbital components.

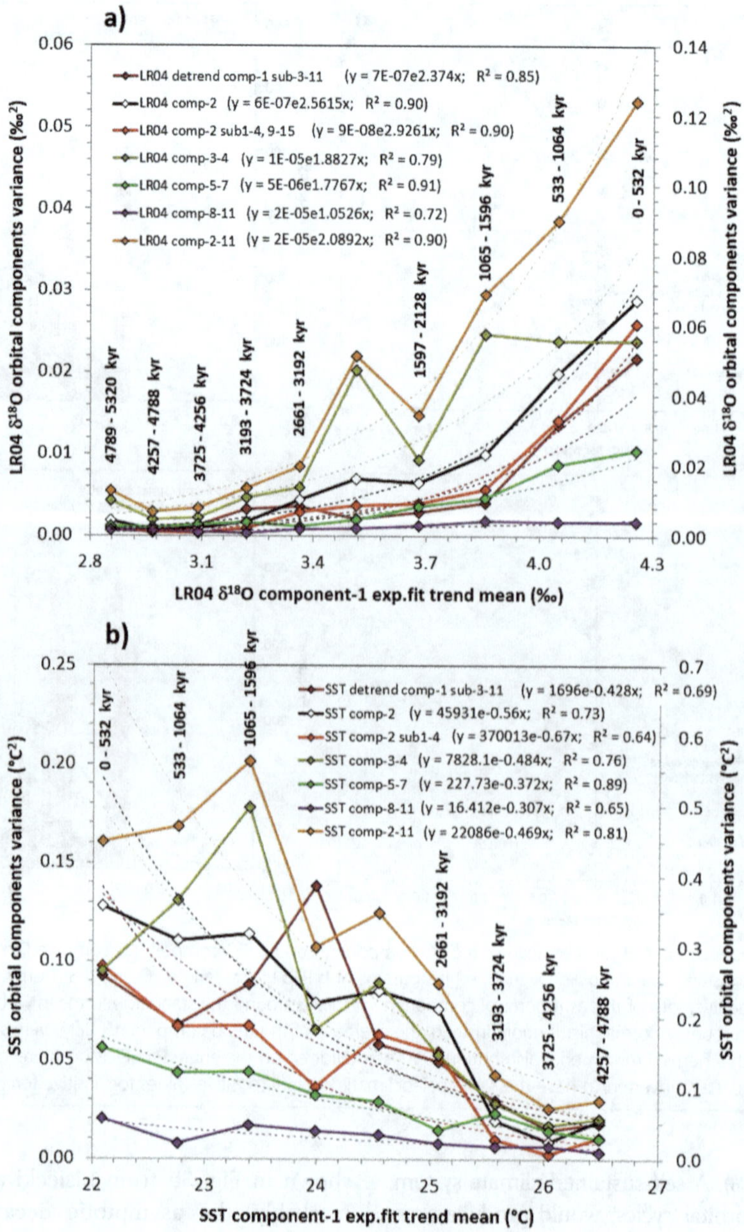

Fig. 10 Exponential relationships of the orbital variance of **a** $\delta^{18}O$ components and **b** SST components vs. the average of component-1 exponential fit trends by time segments binned at 532 kyr. The right *y*-axis shows the variance of comp-2-11 (orbitals) with a different scale to not compress the left *y*-axis values

These plots offer an overall view of the long-term trend links among the proxies. Although it is possible to express the trend relationship as exponential regressions, it was decided to use a simpler linear fitting since the correlation coefficients are similar, in order to quantify the trend rates of change. The $\delta^{18}O$ trend components exhibit a good linear interpolation vs. the boron-based pCO_2 trend of the Sites 668B-999A (Fig. 12a) and the alkenone-based pCO_2 trend of the Site 999A (Fig. 12b); differently, the $\delta^{18}O$ orbital components are unrelated to the same pCO_2 long-term trends. Similar relationships are shown by the $\delta^{18}O$ trend components and the alkenone-based SST trends of the Site 846 (Fig. 12c) and Site 999A (Fig. 12d). The tropical SST trend components of the Site 846 are linearly related to the boron-based atmospheric pCO_2 trend of the Sites 668B-999A (Fig. 12e) and the alkenone-based pCO_2 trend of the Site 999A (Fig. 12f); again, the SST orbital components are unrelated to the same pCO_2 long-term trends.

However, in the plots with more data points, the amplitude increase of the orbital components is clearly visible as pCO_2 and SST decrease, corroborating the analysis of Figs. 10 and 11. These results show that the

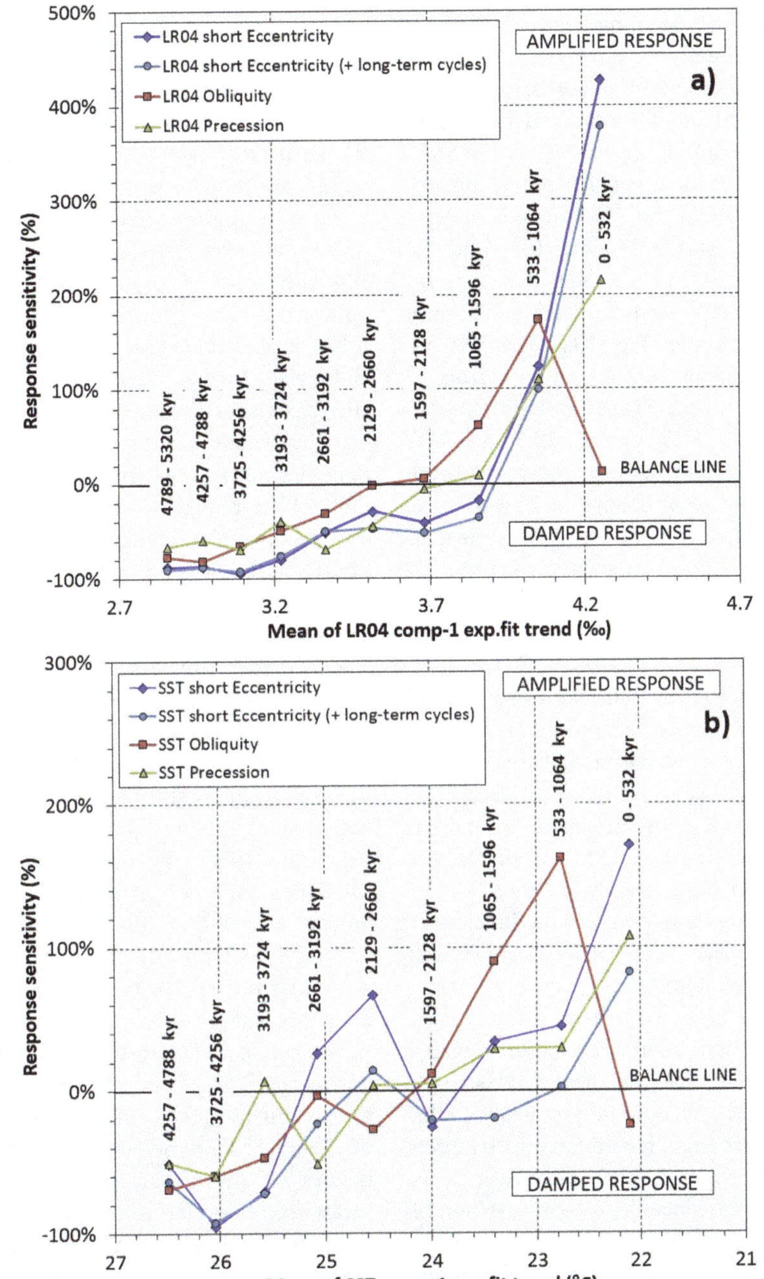

Fig. 11 Analysis of the order of magnitude in **a** $\delta^{18}O$ and **b** SST response modulation relative to orbital forcing (standardized response sensitivity R_S, see text) as a function of the LT-MCS. The balance line (0%) indicates that the variance of response matches the variance in forcing, suggesting a self-sustained climate system, which is only paced by orbital cycles. Values much larger than zero would indicate a nonlinear reinforcement of the signal response. Negative R_S shows a response variance lower than orbital forcing, suggesting a damping mechanism for the climate system

climate sensitivity to orbital forcing is lower when CO_2 and SST are higher (Early Pliocene) and that it increases non-linearly when CO_2 and SST are low (Mid-Late Pleistocene), in agreement to Martinez-Boti et al. (2015), a feature linked to the LT-MCS as shown in the present study. Despite that Site 846 is considered not "typical" tropical SST record, Fig. 13a shows a very close link between its trend component and that of the global ΔSST stack (Martinez-Boti et al. 2015) identified by SSA and driving 76.2% of the variance (ΔSST comp-1), noteworthy the same magnitude of the Site 846 SST and global $\delta^{18}O$ trend components (76.6% and 76.4%, respectively). The inclusion of the Site 846 among the ten of the ΔSST stack should not significantly affect the global character of the signal. Moreover, Lisiecki (2010b) calculates 0.28 °C/Myr the upwelling cooling effect at Site 846 which roughly matches a 30% fraction of the total 1.0 °C/Myr trend (Herbert et al. 2010a), resulting about 70% of the Site 846 cooling trend to be due to non-regional effects, according to the results of the SSA.

Figure 13 shows similar links of Fig. 12 among the global ΔSST long-term trend and the pCO_2 trends at Sites 668B and 999A (Fig. 13c) and 999A (Fig. 13d) and the global $\delta^{18}O$ (Fig. 13b). Figure 13e, f exhibits the same relationships by converting the standardized ΔSST into SST (°C) using the function of Fig. 13a. The SST vs. boron-based CO_2 trend slopes correspond to a tropical rate of change of 25.6 µatm/°C (Fig. 12e), very close to the global converted ΔSST stack of 25.5 µatm/°C (Fig. 13e). At 25 °C and atmospheric pressure, these values are equivalent to a rate of change of 26.4 ppm/°C and 26.3 ppm/°C, respectively. Coherently, as shown in Fig. 14b, the two SST trends are subparallel in time, although the Site 846 is cool-shifted. Considering the alkenone-based CO_2 trend slopes, these rates of changes are of 42.6 ppm/°C (Fig. 12f) and 39.2 ppm/°C (Fig. 13f), respectively. Such variability in rate of changes (~ 26 ppm/°C vs. ~ 41 ppm/°C) could be due to differences in the CO_2 proxy estimation (boron-based vs. alkenone-based). These slope data offer an overall view of the Plio-Pleistocene long-term climate sensitivity, and the oscillations around the linear trend should indicate that it can change for discrete time intervals. On the other hand, the climate sensitivity to orbital forcing change by time in response to the initial conditions set by LT-MCS (Tachikawa et al. 2014; Martinez-Boti et al. 2015; Lo et al. 2017; this study).

These results authorize to be considered of global in nature the non-Milankovitch trend links among pCO_2, SST, and $\delta^{18}O$. Considering that the main GHGs, such as carbon dioxide and methane, operate physically as the most effective positive (warming) climate forcings (Hansen et al. 2005; Hansen and Sato 2012; Van Der Meer et al. 2014), the hypothesis that the global long-term variation of atmospheric GHG concentration plays a relevant role in explaining the origin of the $\delta^{18}O$ and SST trend components, driving ~ 76% of the Plio-Pleistocene signal variance, appears plausible.

Figure 14 shows the correlation over time between the $\delta^{18}O$ enrichment trend components (LR04 comp-1 subcomp-1 and LR04 comp-1 exponential fit) and the CO_2 reduction trend of the ODP Sites 668B and 999A (Fig. 14a) and the SST cooling trends of the global ΔSST stack, ODP Sites 846 and 999A (Fig. 14b).

During the Zanclean, the pCO_2 alkenone ranged from 450 to 330 ppm, the highest values of the entire Plio-Pleistocene; the tropical SST was higher than the Plio-Pleistocene average; and the global $\delta^{18}O$ was depleted more than the average (Fig. 14). Conversely, the mean orbital forcing was slightly negative (Fig. 9). These observations indicate Zanclean as a period of a higher than average positive climate state, supporting the hypothesis that the Zanclean warmer climate may have been driven by an important greenhouse effect due to higher atmospheric CO_2 (Bartoli et al. 2011). The lack of significant Northern Hemisphere ice growth and 100-kyr climate response prior to 3.3 Myr suggest that pCO_2 alkenone above 320–330 ppm is associated with warming sufficient to prevent significant Northern Hemisphere ice sheet growth even during cold orbital phases (Seki et al. 2010). The pCO_2 and SST show a significant reduction during the Zanclean (subtrend-I), supporting the hypothesis that the ONHG was triggered by the decreasing trend in atmospheric CO_2 (Pagani et al. 2010; Seki et al. 2010). Subtrend-II (Piacenzian) is also consistent with a reduction of pCO_2 alkenone and SST, suggesting an INHG triggered by a further decrease in atmospheric CO_2 according to Lunt et al. (2008), who stated that Greenland glaciation was primarily controlled by a decrease in atmospheric CO_2 during the Late Pliocene. Nevertheless, the Piacenzian remains a period of a $\delta^{18}O$-SST higher than average positive climate state, with slightly negative or neutral mean orbital forcing (Fig. 9). For this reason, the Piacenzian, excluding the ONHG event, is still characterized by an average warming fingerprint due to a residual greenhouse effect. According to this hypothesis, Lunt et al. (2012), using a modeling approach, showed that on a global scale, the largest contributor to mid-Piacenzian warmth (3.26–3.03 Myr) is high CO_2. Subtrend-III (Gelasian-Calabrian p.p.) does not appear to comply with the pCO_2 alkenone data, probably due to the low-resolution proxy in this range, while it is more consistent with the decreasing boron-based pCO_2 and the SST subtrends. For the first time during the Plio-Pleistocene, the long-term $pCO_2$and SST are below or close the mean and the $\delta^{18}O$ is enriched more than the average (Figs. 14 and 9). The mean orbital forcing is neutral or slightly positive. The net result of these two unbalanced forcings may be a significant cooling effect and the start of an average "icehouse" world. Subtrend-IV (Calabrian-Middle Pleistocene p.p.) is consistent with a further reduction of pCO_2, reaching peak concentrations of 240–250 ppm (MIS-16), and with the continuation of the SST cooling trend. The long-term pCO_2, SST, and $\delta^{18}O$ are consolidated below the mean, while the mean orbital forcing appears slightly positive, resulting in a lower than average negative climate state, strengthening the icehouse effect. No CO_2 variation corresponds to the MPT, although it falls at the beginning of subtrend-IV after a long time interval of surface/deep-water cooling and ice volume growth. Finally, the MBO (Mid-Late Pleistocene) marks a mild recovery of the long-term pCO_2 clearly visible in the EPICA trend comp-1

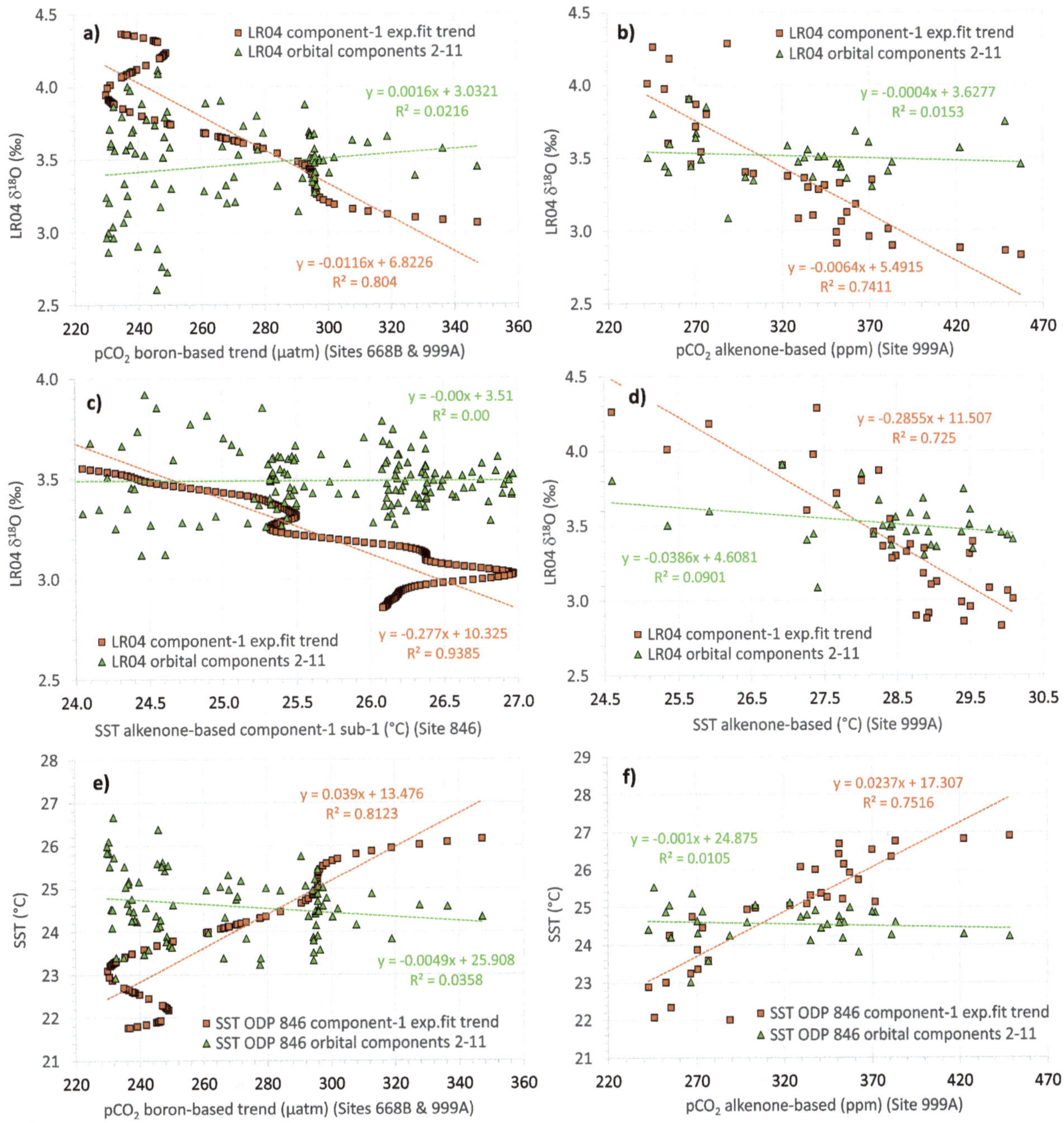

Fig. 12 Plio-Pleistocene cross plots between selected pCO_2 proxies and the $\delta^{18}O$ (**a**, **b**) and SST (**e**, **f**) SSA-components. Plio-Pleistocene cross plots between selected SST proxies and the $\delta^{18}O$ (**c**, **d**) SSA-components. A clear linear interpolation is shown between long-term trend components, whereas orbital components-2-11 are unrelated to the long-term trends. Proxy data: composite atmospheric boron-based pCO_2 proxy from ODP Sites 668B-999A (Hönisch et al. 2009; Bartoli et al. 2011), atmospheric alkenone-based pCO_2 and SST proxies from ODP Site 999A (Seki et al. 2010), alkenone-based SST proxy from ODP Site 846 (Herbert et al. 2010b), and LR04 $\delta^{18}O$ stack (Lisiecki and Raymo 2005b)

subcomp-3 and less distinctly in the CO_2 proxy records (see Additional file 1) (Lisiecki 2010b; Seki et al. 2010) and in the SST proxies (Fig. 14). The strong isotopic enrichment related to MIS-16 at the end of the subtrend-IV correlates well with the absolute CO_2 minimum in the EPICA comp-1 subcomp-3 at ~0.65 Myr. Considering that the $\delta^{18}O$ response lags CO_2 (Shackleton 2000; this study), the correlation between the CO_2 maximum in the EPICA comp-1 subcomp-3 at ~0.4 Myr and the related but slightly delayed isotopic depletion in the $\delta^{18}O$ comp-1 subcomp-1 is noteworthy (MBO). This observation supports the hypothesis that MIS-11, one of the most intense and prolonged interglacial oscillations of the Middle-Late Pleistocene, which is not proportional to eccentricity (Imbrie et al. 1993)

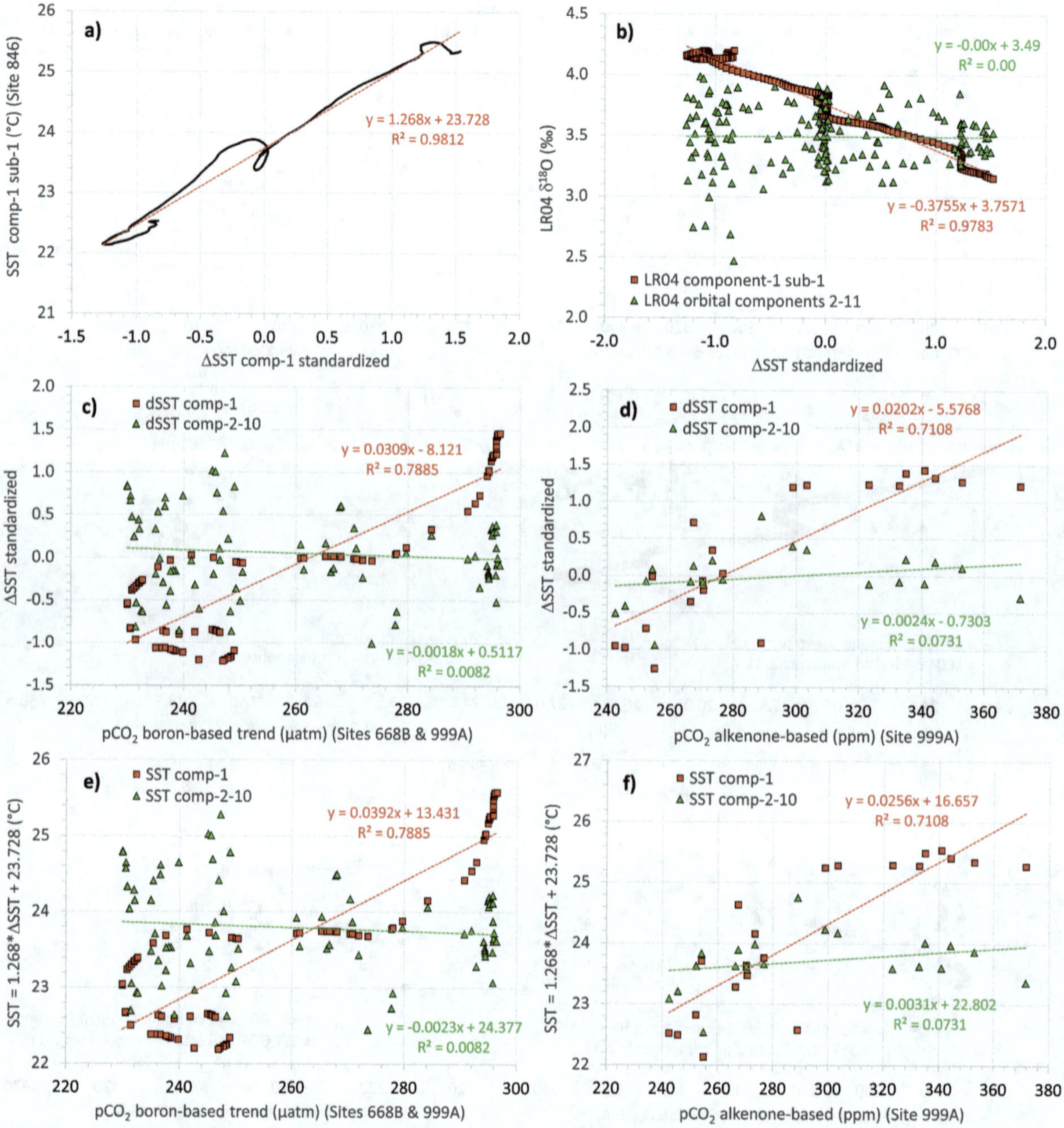

Fig. 13 Late Pliocene-Pleistocene cross plot between trend components of global ΔSST stack and tropical SST at Site 846 (**a**). Cross plots between the ΔSST stack and **b** $\delta^{18}O$, **c** boron-based, and **d** alkenone-based pCO_2 SSA-components. Cross plots between the converted ΔSST stack (°C by means of the function in **a**) and **e** boron-based pCO_2 and **f** alkenone-based pCO_2 trends. The long-term trend components are clearly linked, whereas orbital components are unrelated to the long-term trends. Proxy data: composite atmospheric boron-based pCO_2 proxy from ODP Sites 668B-999A (Hönisch et al. 2009; Bartoli et al. 2011), atmospheric alkenone-based pCO_2 and SST proxies from ODP Site 999A (Seki et al. 2010), alkenone-based SST proxy from ODP Site 846 (Herbert et al. 2010b), LR04 $\delta^{18}O$ stack (Lisiecki and Raymo 2005b), and global ΔSST stack (Martinez-Boti et al. 2015)

or cool insolation-induced interglacials (Berger et al. 2012), may be an astronomical event boosted by GHG mid-term recovery, which reached the highest value at ~400 kyr (Luthi et al. 2008; Berger and Yin 2012; Berger et al. 2016). The origin of this GHG mid-term recovery could be partly due to the long-term orbital cycles that reach a maximum relative (warming) at ~300–200 kyr (Figs. 2 and 6); the 400-kyr eccentricity cycle really reaches a maximum at about 200 kyr (Laskar et al. 2011). However, the timing of these astronomical events does not seem consistent with the CO_2 maximum of EPICA at ~400 kyr. In addition, we cannot exclude a contribution of the circum-Pacific explosive activity (Jicha et al. 2009). The Late Pleistocene was a period of quiescence of the circum-Pacific explosive activity,

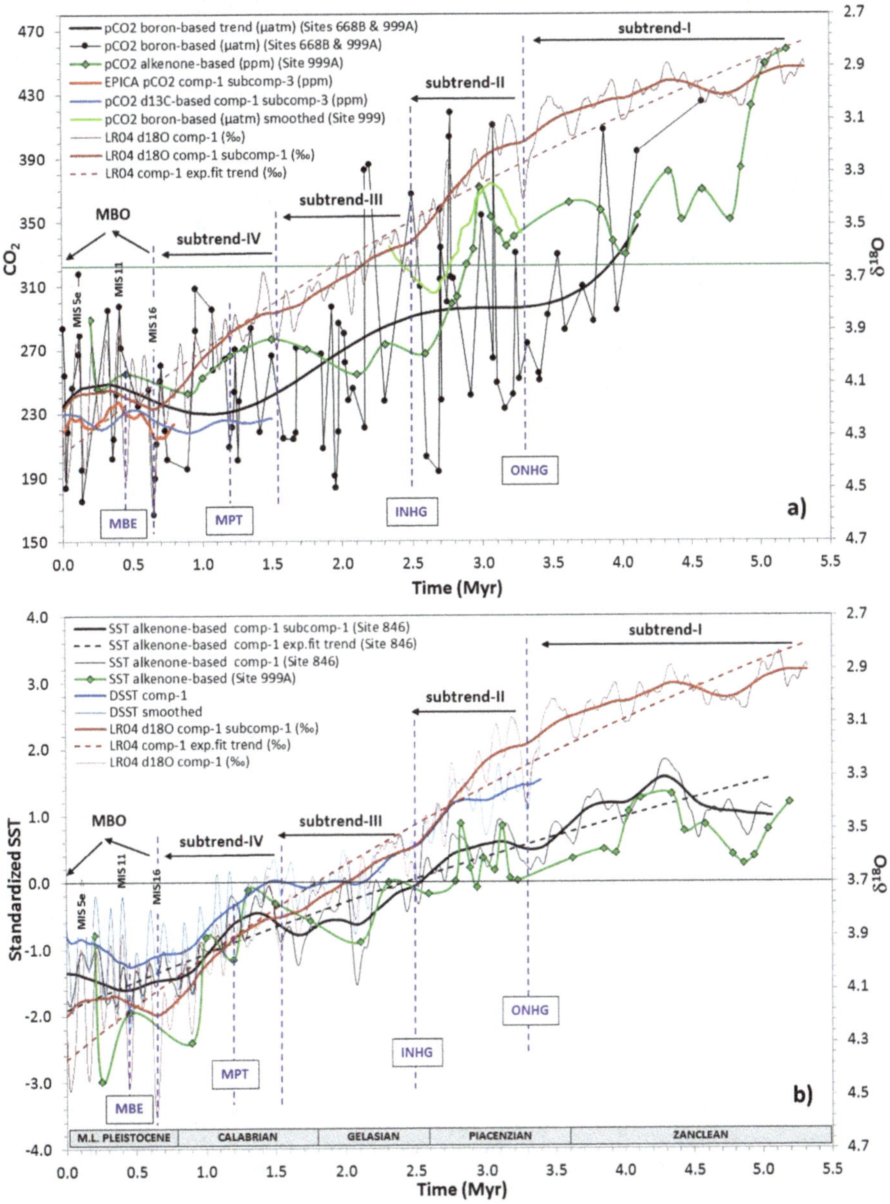

Fig. 14 Comparison over time of the long-term trend components of $\delta^{18}O$ and the **a** pCO_2 and **b** SST proxies. Composite atmospheric boron-based pCO_2 proxy from ODP Sites 668B-999A (Hönisch et al. 2009; Bartoli et al. 2011), atmospheric alkenone-based pCO_2 and SST proxies from ODP Site 999A (Seki et al. 2010), alkenone-based SST proxy from ODP Site 846 (Herbert et al. 2010b), EPICA pCO_2 record (Berger et al. 2016), $\delta^{13}C$-based pCO_2 proxy (Lisiecki 2010b), LR04 $\delta^{18}O$ stack (Lisiecki and Raymo 2005b), boron-based pCO_2 proxy from ODP Site 999 and global **ΔSST** stack (Martinez-Boti et al. 2015). The horizontal lines are the Plio-Pleistocene average of pCO_2 alkenone (ODP Site 999A) and the standardized mean of the SST. All records are plotted on their own age scale

which perhaps contributed to the mid-term CO_2 increment by a reduction in volcanic aerosols and the consequent minor algal ash-fertilization of the Pacific Ocean waters (Jicha et al. 2009).

Lagged patterns of $\delta^{18}O$ and SST orbital components

The phase relationships between the standardized La93 orbital forcings and the orbitally driven LR04-SST components are shown in Table 8 for three time intervals. The results indicate that the $\delta^{18}O$ and SST orbital components are all delayed on average compared to the related forcing of eccentricity, obliquity, and precession (with the small significant exception of the eccentricity in the pre-ONHG time due to very low SST signal coherency), although the patterns of coherency and phase lag are quite different across the time intervals.

These cross-spectral results are in general agreement with Imbrie et al. (1993), Shackleton (2000), Lisiecki and

Raymo (2005a), and Tziperman et al. (2006) and indicate that the climate system takes a certain amount of time to transfer the effect of orbital forcing to the global ice volume/deep water temperature and to the sea surface temperature. The coherency is generally lower in the tropical SST signal probably due to differences in the forcings controlled by high-latitude and low-latitude signals. The lagging of the $\delta^{18}O$ signal is usually larger than that of the SST (Table 8) because the ice volume component lags the deep water temperature signal in the $\delta^{18}O$ record (Lisiecki and Raymo 2005a; Herbert et al. 2010a). The coherency of the eccentricity signal increases from pre-ONHG time (incoherent) towards the Mid-Late Pleistocene, reflecting the progressive increase of eccentricity power both in global $\delta^{18}O$ and local tropical SST signals. The delta-lag in eccentricity between $\delta^{18}O$ and SST reaches its maximum during the Late Pleistocene (4.31–0.44 kyr), suggesting an increased contribution of the $\delta^{18}O$ ice volume component over the water temperature signal. The $\delta^{18}O$ and SST obliquity signals have the largest coherency and exhibit an average increasing trend in the lags towards the Mid-Late Pleistocene, starting from 4.10–3.46 kyr to 5.78–5.28 kyr and, finally, reaching values of 7.42–2.71 kyr (Table 8), in general agreement with Lisiecki and Raymo (2005a). The $\delta^{18}O$-SST delta-lag in obliquity is relatively small until

Table 8 Cross-spectral analysis results among the orbital forcings (Laskar et al. 1993) and the orbitally driven $\delta^{18}O$ and SST components for three time intervals. Data standardized (mean = 0, standard deviation = 1) and, where necessary, inverted to have the same paleoclimatic polarity (positive values for positive forcing and warming)

Time interval	Orbital forcing	Response component	Cross-spectrum freq. (kyr^{-1})	Cross-spectrum period (kyr)	Coherency	Phase shift		
Post-MPT start	Eccentricity	LR04 comp-2 subcomp-1-4, 9-15	0.010742	93	0.72	− 16.65°	− 4.31 kyr	$\delta^{18}O$ component lags eccentricity
		SST comp-2 subcomp-1-4	0.010742	93	0.70	− 01.70°	− 0.44 kyr	SST component lags eccentricity
	Obliquity	LR04 comp-3-4	0.024414	41	0.93	− 65.20°	− 7.42 kyr	$\delta^{18}O$ component lags obliquity
		SST comp-3-4	0.024414	41	0.68	− 23.78°	− 2.71 kyr	SST component lags obliquity
	Precession	LR04 comp-5-7	0.041992	24	0.78	− 74.50°	− 4.93 kyr	$\delta^{18}O$ component lags precession
		SST comp-5-7	0.041992	24	0.72	− 53.24°	− 3.52 kyr	SST component lags precession
ONHG-MPT start	Eccentricity	LR04 comp-2 subcomp-1-4, 9-15	0.010742	93	0.47	− 27.57°	− 7.13 kyr	$\delta^{18}O$ component lags eccentricity
		SST comp-2 subcomp-1-4	0.010742	93	0.45	− 21.37°	− 5.53 kyr	SST component lags eccentricity
	Obliquity	LR04 comp-3-4	0.024414	41	0.98	− 50.79°	− 5.78 kyr	$\delta^{18}O$ component lags obliquity
		SST comp-3-4	0.024658	41	0.95	− 46.85°	− 5.28 kyr	SST component lags obliquity
	Precession	LR04 comp-5-7	0.044678	22	0.89	− 57.37°	− 3.57 kyr	$\delta^{18}O$ component lags precession
		SST comp-5-7	0.042236	24	0.28	− 119.35	− 7.85 kyr	SST component lags precession
Pre-ONHG	Eccentricity	LR04 comp-2 subcomp-1-4, 9-15	0.009766	102	0.32	− 01.48°	− 0.42 kyr	$\delta^{18}O$ component lags eccentricity
		SST comp-2 subcomp-1-4	0.010742	93	0.13	+ 07.06°	1.83 kyr	SST component leads eccentricity
	Obliquity	LR04 comp-3-4	0.024414	41	0.95	− 36.02°	− 4.10 kyr	$\delta^{18}O$ component lags obliquity
		SST comp-3-4	0.024414	41	0.81	− 30.42°	− 3.46 kyr	SST component lags obliquity
	Precession	LR04 comp-5-7	0.042480	24	0.42	− 88.11°	− 5.76 kyr	$\delta^{18}O$ component lags precession
		SST comp-5-7	0.042480	24	0.59	− 87.60°	− 5.73 kyr	SST component lags precession

the beginning of the MPT and becomes large after this transition, probably following global ice volume growth. The cross-spectral results relative to precession are more variable. The precession lag is basically constant (5.76–5.73 kyr) during pre-ONHG time, then it becomes quite "unstable" (3.57–7.85 kyr) between ONHG and MPT and reaches lags of 4.93–3.52 kyr post-MPT. The minimum $\delta^{18}O$ phase lag relative to precession (3.57 kyr) during the ONHG-MPT time interval suggests the influence of a large global deep water temperature signal (Lisiecki and Raymo 2005a). During the same time interval, the precession's SST lag reached its maximum (7.85 kyr) which could be due to the effect of a deep water upwelling in the EEP (ODP Site 846). Table 9 exhibits, for the last 5 Myr, the cross-spectral analysis results between SST and LR04 orbital components, showing a slight $\delta^{18}O$ lagged response vs. SST, in agreement with Herbert et al. (2010a). The slightly earlier response of the $\delta^{18}O$ eccentricity component (0.05 kyr) is due to the pre-ONHG weak and unstable eccentricity signal. After the ONHG, the eccentricity response is strengthened and becomes more stable resulting in a slight $\delta^{18}O$ lag.

Discussion

Origin of the $\delta^{18}O$ and SST trend components

The LR04 trend component is related to the Plio-Pleistocene interval of the Cenozoic $\delta^{18}O$ enrichment trend starting at the end of the Early Eocene Climatic Optimum (EECO, 53–51 Myr), documenting the progressive development of the Antarctic and Boreal ice sheets (Zachos et al. 2001). The gradual climate cooling during the Cenozoic is generally correlated to a CO_2 concentration decrease in the atmosphere (Raymo and Ruddiman 1992; Zachos et al. 2001; Royer 2006; Zachos et al. 2008; Van de Wal et al. 2011; Zhang et al. 2013; Van Der Meer et al. 2014). Recent studies on Pliocene and Pleistocene alkenone $\delta^{13}C$ and boron-based pCO_2 proxies support this hypothesis (Seki et al. 2010; Pagani et al. 2010; Bartoli et al. 2011). Stap et al. (2014) find that the atmospheric temperature during the past 800 kyr has been controlled by a complex interaction of CO_2 and insolation, and both variables serve as thresholds for the NHG. Using CO_2 proxies and geochemical carbon cycle models, Royer (2006) shows that atmospheric CO_2 exhibits a pervasive and tight positive correlation with globally averaged surface temperatures, indicating that CO_2 played a strong role in driving the global climate for much of the Phanerozoic. The Cenozoic climatic record provides an important case study to understand the relationships between carbon cycling and climate (Zachos et al. 2001, 2008). In fact, much of the early Cenozoic was characterized by noticeably higher concentrations of GHGs and a much warmer mean global temperature. The most extreme event being between 53 and 51 Myr (EECO), when CO_2 concentration was on the order of 1000–1500 ppm, and the global temperatures reached a long-term Cenozoic maximum (Zachos et al. 2001, 2008; Beerling and Royer 2011). Zhang et al. (2013) estimate the new pCO_2 data for the Middle Miocene to be notably higher than published records with average pCO_2 concentrations in the range of 400–500 ppm, suggesting that CO_2 levels were highest during a period of global warmth associated with the Middle Miocene Climatic Optimum (MMCO, 17–14 Myr). Instead, during the mid-Pleistocene ice ages, CO_2 fell to 170–190 ppm, and the global temperature reached a long-term Cenozoic minimum (Beerling and Royer 2011; Berger et al. 2016). Considering a CO_2 variation from approximately 1000 ppm in the early Cenozoic to 170 ppm in the icy Pleistocene, the resulting negative climate forcing computed by Hansen and Sato (2012) for this CO_2 range exceeded 10 W/m^2, which is consistent with the observed strong Cenozoic cooling. These studies confirm the link among global temperature, the cryosphere, and atmospheric CO_2. The present study suggests a link between the

Table 9 Cross-spectral analysis results between orbitally driven $\delta^{18}O$ and SST components for the last 5 Myr. Data standardized (mean = 0, standard deviation = 1) and, where necessary, inverted to have the same paleoclimatic polarity (positive values for positive forcing and warming)

SST component	$\delta^{18}O$ component	Cross-spectrum freq. (kyr^{-1})	Cross-spectrum period (kyr)	Coherency	Phase shift		
SST comp-2 subcomp-1-4	LR04 comp-2 subcomp-1-4, 9-15	0.010498	95	0.89	0.20°	0.05 kyr	$\delta^{18}O$ leads SST eccentricity comp.
SST comp-2 subcomp-1-4 (post-ONHG)	LR04 comp-2 subcomp-1-4, 9-15 (post-ONHG)	0.010498	95	0.84	− 0.38°	− 0.10 kyr	$\delta^{18}O$ lags SST eccentricity comp.
SST comp-3-4	LR04 comp-3-4	0.024414	41	0.92	− 24.57°	− 2.80 kyr	$\delta^{18}O$ lags SST obliquity comp.
SST comp-5-7	LR04 comp-5-7	0.044800	22	0.46	− 16.40°	− 1.02 kyr	$\delta^{18}O$ lags SST precession comp.

atmospheric pCO_2 long-term variations and the SST-$\delta^{18}O$ trend components and highlights the primary role of the long-term change of the atmospheric GHG concentration in explaining the origin of the high-variance, highly significant $\delta^{18}O$ and SST trend components.

Factors controlling long-term atmospheric composition

Although it is generally accepted that the progressive Cenozoic cooling of the global climate resulted from a long-term decrease in atmospheric CO_2, details on how carbon cycle affects CO_2 and a clearer distinction of the involved time scales have not been fully resolved. Because of the relatively small size of the atmospheric reservoir, minimal imbalances between CO_2 sources and sinks cause rapid fluctuations in atmospheric CO_2 levels and global mean temperatures (Brady and Gislason 1997; Kent and Muttoni 2013). The results of the present study suggest ~76% of the Plio-Pleistocene climate variance could be driven by non-orbital carbon cycle. On time scales longer than one million years, atmospheric CO_2 levels are primarily controlled by the balance between the rate of volcanic and metamorphic sources from the Earth's interior and the rate of sink through the chemical weathering of silicate minerals and through the burial of organic carbon (Walker and Hays 1981; Berner et al. 1983; Raymo and Ruddiman 1992; Brady and Gislason 1997; Schrag et al. 2002; Hansen and Sato 2012). Plate tectonics may affect the primary inputs of atmospheric CO_2 through mantle outgassing at seafloor spreading ridges, volcanoes, and the metamorphism of carbonate rocks along subduction zones (Van Der Meer et al. 2014). Long-term sinks for CO_2 are controlled by the burial of organic matter and by the weathering of silicate rocks to form carbonates (Raymo and Ruddiman 1992; Berner and Kothavala 2001; Schrag et al. 2002; Van Der Meer et al. 2014). The weathering of silicate rocks is a chemical process whereby CO_2 consumption is buffered by the temperature dependence on the weathering rate, preventing the so-called "ice catastrophe" when temperatures fall (Walker and Hays 1981). As the atmospheric CO_2 concentration rises, the temperature and precipitation increase. They thereby enhance chemical weathering and atmospheric CO_2 depletion, preventing "greenhouse" disasters on geological timescales. Chemical weathering is also affected by changes in continental land area and paleoaltitude distribution (Berner et al. 1983; Schrag et al. 2002). The alteration of surficial marine basalts at low temperatures (< 40 °C) is another potentially important sink for atmospheric CO_2 over geologic time (Brady and Gislason 1997). The numerical modeling of the carbon cycle by Berner et al. (1983) suggests that changes in volcanic outgassing over the last 100 million years were the primary cause of the transition between the greenhouse climates of the Cretaceous and the ice ages of the Plio-Pleistocene. Their results indicate that the CO_2 content in the atmosphere is highly sensitive to changes in the seafloor spreading rate and continental land area. A tropical concentration of continental area may lead to a more efficient burial of organic carbon through increased tropical river discharge and the chemical weathering of silicate rocks (Schrag et al. 2002). By affecting CO_2 volcanic degassing, metamorphic-magmatic decarbonation and changes in the continental land area available for chemical weathering, plate tectonics is considered the main long-term controlling factor of the global climate (Berner et al. 1983; Van Der Meer et al. 2014). However, it is a matter of debate as to which main mechanism can adjust the balance between the CO_2 source through geotectonic volcanic degassing and the CO_2 sink by the chemical weathering of silicate rocks. Some studies show conflicting results regarding the role of the seafloor spreading rate, which is believed to be the major control of CO_2 degassing (Berner et al. 1983; Raymo and Ruddiman 1992). In fact, according to Rowley (2002), the seafloor spreading rate remained relatively constant over the past 180 Myr, while Cogné and Humler (2006) indicate a general increasing trend in the last 50 Myr. These results are discordant with the observed cooling trend of the Cenozoic climate. Recently, a study of carbon-cycle evolution indicated that increasing physical erosion in response to widespread mountain building is the most prominent driver of carbon cycling over the past 100 Myr (Li and Elderfield 2013). Raymo and Ruddiman (1992) proposed that the uplift of the Tibetan plateau was an important driving force behind the Cenozoic long-term cooling by atmospheric CO_2 consumption due to the increase of the chemical weathering of silicate rocks. However, the Cenozoic long-term shift in Earth's climatic state could be really related to regional transient differences in volcanic emissions by plate tectonic activity, which were particularly high during, e.g., the Late Paleocene and Early Eocene epochs (North Atlantic rifting and India-Asian northward migration) (Raymo and Ruddiman 1992; Zachos et al. 2001; Mosar et al. 2002; Zachos et al. 2008; Le Breton et al. 2012). From this viewpoint, a recent study on the Himalayan-Tibetan orogenic system shows a multiple collision model in which a continent-continent collision began around the Eocene/Oligocene boundary (34 Myr) (Aitchison et al. 2007), promoting the Early Oligocene glaciation (Zachos et al. 2001). Corroborating this hypothesis, the North Atlantic seafloor spreading rate (NASSR) exhibits a clear decreasing trend from high values in the Early Eocene (55–50 Myr) to a very low rate for most of the Oligocene (Mosar et al. 2002; Le Breton et al. 2012). On the other hand, recent studies seem to converge towards a significant 250–200-Myr period variation of the seafloor spreading rate

(Becker et al. 2009; Coltice et al. 2013). Becker et al. (2009) proposed a spreading rate long-term slowdown since 140 Myr with minor cycles superimposed over time, such as a relative increase at ~ 50 Myr and a minimum at ~ 32 Myr. Even the timing of the global average spreading rate (GASR) calculated by Conrad and Lithgow-Bertelloni (2007) appears to support the hypothesis of a direct geotectonic "forcing" to the long-term composition of atmospheric GHGs. On a more detailed timescale, a study in the North Atlantic shows the consistency of the NASSR relative maximum at ~ 15 Myr and the MMCO (Merkouriev and DeMets 2014). The correlation of the NASSR minimum at approximately 1.2–1.0 Myr (the lowest value in the last 20 Myr) and the CO_2 proxy minimum at 0.9–0.7 Myr (Fig. 14a), which correlate to the strong LT-LR04 cooling at the end of subtrend-IV, seems plausible. Indeed, the contribution of the important negative forcing created by aerosols (Hansen et al. 2005) emitted into the atmosphere during phases of explosive volcanic paroxysm in the circum-Pacific (Jicha et al. 2009) must also be taken in account. In fact, hundreds of explosive eruptions occurred during an extremely vigorous period of circum-Pacific volcanism beginning in the Late Eocene, which likely led to the production of sulfur aerosols in the stratosphere and net cooling. The feedback mechanism related to the enormous supply of ash that fertilized the Pacific Ocean waters might have caused algal growth and CO_2 uptake and thereby contributed to Oligocene glaciation at approximately 34 Myr (Jicha et al. 2009). It is interesting to observe how the Plio-Pleistocene in the circum-Pacific was a period of intense explosive volcanic activity characterized by three activity peaks at approximately 4.8, 2.8, and 1.5 Myr (Jicha et al. 2009). We cannot exclude that these events contributed in part to the three most pronounced segments of $\delta^{18}O$ enrichment highlighted in the LT-LR04 record (subtrend-I, subtrend-II, and subtrend-IV), culminating in the ONHG, INHG, and MIS-16 events, respectively. Moreover, the Late Pleistocene was a period of quiescence of the circum-Pacific explosive activity, which may have contributed to the mid-term CO_2 increment and $\delta^{18}O$ impoverishment that occurred at approximately 400–300 kyr, a possible cause of the MBO. A further contribution to the global emission of CH_4 and CO_2 may also come from the thermogenic cracking of organic carbon-rich sediments in large igneous provinces (LIPs), which are associated with geotectonic spreading areas or subduction zones (Svensen et al. 2004). Finally, another climate-controlling factor is oceanic paleocirculation. The "Panama Hypothesis" states that the gradual closure of the Panama seaway, between 13 and 2.6 Myr led to a decreased mixing of the Atlantic and Pacific water masses and a strengthening of the Atlantic thermohaline circulation, increased temperatures and precipitation in Northern Hemisphere high latitudes, and culminated with the INHG during the Pliocene, at approximately 3.2–2.7 Myr (Bartoli et al. 2005; Lunt et al. 2008). However, Lunt et al. (2008), using a more sophisticated general circulation model (GCM), demonstrated that the ice volume difference between the "closed" and "open" seaway model configurations is small and concluded that it was not a major forcing mechanism. This view accommodates $\delta^{18}O$ and SST components-8-11 of inter-tropical semi-precessional origin related to the oceanic heat transport towards high latitudes (Hagelberg et al. 1994; Berger et al. 2006), which contributes very little to the total variance of the paleoclimatic response (~ 0.6–0.5%). In summary, the geotectonic forcing appears a plausible hypothesis to explain the long-term changes in atmosphere composition (GHGs and volcanic aerosols) through direct (CO_2 outgassing and explosive volcanism) and indirect (orography and erosion, paleogeographic configuration, oceanic paleocirculation, and ocean fertilization) control by plate tectonics.

Nonlinear response and mean climate state

An interesting result of the present study shows that the nonlinear increase in signal amplitude from the Early Pliocene to the Late Pleistocene is observed in all $\delta^{18}O$ and SST astronomical components and is exponentially linked to the LT-MCS (Figs. 10 and 11). These results reinforce the notion that the climate sensitivity of orbital forcings may be dependent on the mean climate state in agreement with Lisiecki and Raymo (2007) and Hansen et al. (2013), with possible threshold effects typical of nonlinear systems (Rial et al. 2004). Because the Plio-Pleistocene orbital forcings do not exhibit a similar increase in amplitude, it is believed that the process controlling the nonlinear behavior of the signal response originates inside the climate system. The net balance of positive and negative feedback mechanisms is the most likely process behind most of the nonlinearities in the climate (Berger and Loutre 1997; Archer et al. 2004; Rial et al. 2004; Lisiecki and Raymo 2007; Brovkin et al. 2007; Hansen et al. 2007; Hansen et al. 2011; Martinez-Boti et al. 2015).

The most important positive feedback mechanisms that are able to amplify the climate response to primary forcing are the CO_2 exchange between the ocean and the atmosphere, the water vapor feedback, the methane release from permafrost peat bogs and from hydrates, all mechanisms related to greenhouse gases, and the surface albedo feedback. Hansen et al. (2007) revealed a remarkable correspondence between the Vostok Antarctic temperature and GHG climate forcing, confirming that the temperature causes gas changes over several hundred years and suggesting that increasing warming temperature causes a net release of GHGs from oceans, soils, and the biosphere. In the EPICA ice cores, atmospheric greenhouse gases have strongly covaried with

Antarctic temperature and $\delta^{18}O$ benthic throughout eight glacial cycles over the last 800 thousand years (Jouzel et al. 2007; Luthi et al. 2008; Loulergue et al. 2008; Berger et al. 2016). Even Schneider et al. (2013) at EPICA Dome C, during the glacial inception at the end of MIS-5e, confirm a 5-kyr lag in the CO_2 decline with respect to temperatures. They explain this lag by terrestrial carbon release and carbonate compensation. Brovkin et al. (2007), by simulating the effects of changes in atmospheric and oceanic circulations in response to glacial changes in radiative forcing, find that the net effect of five feedback mechanisms related to oceanic circulation and marine biogeochemistry is able to explain a 65 ppmv drop in glacial atmospheric CO_2, compared to 80–90 ppmv recorded in the ice cores (approximately 70–80% of the total drop). Cooler oceanic water can absorb more CO_2 than warmer water. As the ocean warms, it releases CO_2 to the atmosphere, providing an amplifying climate feedback that causes further warming. If the atmosphere is warmed, the saturation vapor pressure increases, and the amount of water vapor in the atmosphere will tend to increase, leading to further warming, because water vapor is an important greenhouse gas. Liptak and Strong (2016), in a modeling study of the Arctic Sea, confirm the existence of a positive thermodynamically driven sea ice-atmosphere feedback up to 35%. A recent study shows that permafrost is a very large carbon pool stored as peat and methane in circumpolar areas (Tarnocai et al. 2009). Lawrence et al. (2008) suggest that a rapid melting of Arctic Sea ice may start a feedback loop that rapidly melts Arctic permafrost, triggering further warming feedback by the release of methane and carbon dioxide. Methane hydrates are a form of water ice saturated by a large amount of methane within its crystal structure. Extremely large deposits of methane hydrates have been found within marine sediments along continental margins of the Earth (Dickens 2003). The release of this trapped methane has the potential to cause a significant increase in temperature as a feedback mechanism triggered by bottom water warming. This hypothesis has been proposed as a possible explanation of early Cenozoic hyperthermals, such as the Paleocene-Eocene Thermal Maximum (PETM) at approximately 55 Myr, a transient warming that shows characteristics of an extreme event of short-term positive feedback (Dickens 2003; Zachos et al. 2008). If the planet becomes warmer, the ice sheet area tends to decrease, exposing a darker surface of land and open water that absorbs more sunlight, causing additional heating. Instead, during times of global cooling, additional ice increases the reflectivity, which reduces the absorption of solar radiation and results in more cooling (albedo effect). According to Hansen et al. (2013) and Hansen (2013), the net effect of fast positive feedback is to amplify the global temperature response by a factor of two to three, and this reinforcement is state dependent. Even Menviel et al. (2008) state that the atmospheric CO_2 response is a delicately balanced sum of the terrestrial and marine inventory changes, emphasizing the importance of the initial climate state on the CO_2 response to large-scale ocean circulation changes. However, exponential growth cannot continue indefinitely because some negative feedback processes act as balancing mechanisms. The new findings of Early Pliocene asymptotic decays of R_s (Fig. 11) could represent the evidence of a damping effect of carbon cycle by negative feedback processes. The continental weathering is a chemical process of atmospheric CO_2 consumption that requires tens of thousands of years to act (Walker and Hays 1981; Schrag et al. 2002; Hansen 2013). Instead, the iceberg cooling effect is a fast negative feedback that comes into play as iceberg discharge reaches a rate that cools the regional ocean surface (Hansen and Sato 2012). The interaction between vegetation and climate is another interesting component in climate system dynamics, but the question of the overall role of the biogeophysical and biogeochemical feedbacks is not yet solved because both feedbacks can be positive or negative (Claussen 2009). Nevertheless, shifts in vegetation patterns have presumably provided a negative feedback in the climate system due to their influence on carbon cycle and great areal changes. During glacial inception, forests were reduced in their areal extent leading to an increase in atmospheric CO_2 concentration, while during deglaciation, atmospheric CO_2 concentration should have decreased due to the buildup of forests (Claussen 2009). Palynological studies attest to major changes in vegetation ecosystems both at glacial and interglacial time scales (Ning et al. 1993; Elenga et al. 2000; Bertini 2010). During a glacial climate, tropical rainforests reduce in their extent to tropical seasonal forests in tropical lowlands and to xerophytic taxa in tropical highlands (Elenga et al. 2000). Boreal forests regress equatorward with a compression and fragmentation of the forest zones, while cold steppe vegetation expands (Ning et al. 1993; Bertini 2010). However, after the Zanclean and towards the Late Pleistocene, a progressive impoverishment of the subtropical forest ecosystem occurs with an increase in herbaceous taxa following the long-term cooling and drying climate trend (Ning et al. 1993; Bertini 2010). Another negative feedback mechanism is the increase in precipitation rate in a greenhouse climate that may intensify weathering rates as fresh water is flushed more rapidly through soil and sedimentary rocks (Schrag et al. 2002). Removal of CO_2 is accomplished through burial of organic carbon, and increasing river discharge may lead to a more efficient burial of organic carbon

(Schrag et al. 2002). The oceanic carbon pump by primary productivity can act as possible damping mechanism, but the complex feedback between climate and ocean require a better understanding. In fact, on orbital time scale, export production by barite accumulation rates was generally higher during cold periods (Ma et al. 2015). This interpretation could imply a positive feedback on carbon cycle that reinforces the cooler orbital response, but it is difficult to predict the net changes in CO_2 efflux (Ma et al. 2015). On the other hand, based on the long-term trends, export production was on average higher during the warmer Late Pliocene than the colder Pleistocene (Ma et al. 2015). On the contrary, the depositional conditions of the Plio-Pleistocene Mediterranean sapropels exhibit warm and wet precession-related changes of phytoplankton productivity by river runoff increasing and bottom water anoxia (Emeis 1998; Gallego-Torres et al. 2007; Moller et al. 2012), suggesting a negative feedback on carbon cycle (Schrag et al. 2002) and a possible damping effect on warming climate response.

On a geological time scale, almost all these mechanisms are sufficiently "fast" (less than 100 kyr) and can act over an orbital short-term hemicycle like precession or "semi-precession." Being geologically fast processes, the feedback response should have in theory a similar frequency modulation as the primary astronomical forcing, although the processes should exhibit a slight delay compared to the latter and to the initial temperature change induced by insolation components (Berger and Loutre 1997; Ruggieri et al. 2009; Hansen et al. 2011). The δ^{18}O and SST lags shown in the present study support this hypothesis. Moreover, the amplitude of the feedback response should be linked to the orbital hemicycle duration and to the scale effect of the surfaces/volumes involved in the processes (Schrag et al. 2002; Menviel et al. 2008; Hansen et al. 2013), a possible explanation of the nonlinear features shown in Figs. 10 and 11. Therefore, the net cumulative effect of positive and negative fast feedback mechanisms, added to the initial temperature change, could lead to the exponential response of each astronomically driven response as a function of the LT-MCS at a given time (initial condition), the duration of the orbitally paced feedback processes, and the synergy of scale-dependent processes. The different growth rates of the nonlinear orbital responses (Fig. 10) and the asymmetry in R_s (Fig. 11) suggest a high-efficiency paced system depending on whether feedbacks amplify ($R_s > 0$) or suppress ($R_s < 0$) response changes in the orbital forcings with a state-dependent magnitude related to the duration time of the feedback processes. Interestingly, the results of the present study indicate an amplified orbital response during times of a global icehouse state (Mid-Late Pleistocene, with the remarkable exception of obliquity), which can reach a magnitude of three to five times global δ^{18}O or two to three times tropical SST forcing, and a maximum of two times the damped response for all signals during the time of a global greenhouse state (Early Pliocene). In this study, it is proposed that orbital response be considered basically as composite feedback responses paced by orbital and damped or amplified in a range of − 100 to + 400% the forcing, as function of the LT-MCS at a given time (initial conditions), the hemicycle duration, and the synergy of the feedback processes. Therefore, the sensitivity of the climate system to orbital forcings could depend on the long-term climate factors in terms of average atmospheric GHG composition, the average concentration of volcanic aerosols, the global mean temperature, the cryosphere average volume, and, presumably, the global forest cover and ocean primary productivity, factors affecting high-efficiency feedback mechanisms related to carbon cycle and ice-albedo. This interpretation may explain the energy excess "paradox" of the astronomically paced signals compared to the small energy of the orbital forcing, especially concerning the eccentricity bands (Imbrie et al. 1993; Wunsch 2004; Berger et al. 2005; Berger et al. 2012). The progressive increase in the strength of the 95-kyr δ^{18}O response from the Zanclean (very weak) to the Mid-Late Pleistocene may be interpreted as the effect of increased sensitivity to eccentricity due to the growing trend of the global ice volume following the cooling trend. In this time interval, the LT-LR04 component undergoes an average enrichment from 2.99‰ (Zanclean) to 4.18‰ (Middle Pleistocene-Holocene) due to long-term changes in atmosphere composition, and the climate system may become sensitive to the weak eccentricity insolation by overcoming the critical condition of global mean temperature and average ice volume, a hypothesis already formulated by Raymo (1998) and many others (Berger et al. 1999; Raymo et al. 2006; Berger and Yin 2012). The models of Imbrie et al. (2011) and Abe-Ouchi et al. (2013) also support this hypothesis. Their studies show that ice volume plays a very important role in pacing the orbital response of the climate system. Considering orbital signals basically as feedback responses paced by orbitals may explain why the ~ 100-kyr eccentricity-driven response is so strong post-MPT, when the obliquity "ice killing," which prevents a long-life ice sheet, was strongly mitigated by damping effect (Fig. 11). Also, the hysteresis loop of the North American ice sheet may have contributed to lead a long-life ice sheet. After the inception of the ice sheet, its mass balance remains mostly positive through several precession cycles, whose amplitudes decrease towards an eccentricity minimum (Abe-Ouchi et al. 2013). The larger the ice sheet grows and extends towards lower latitudes, the smaller is the insolation required to make the mass balance negative. A long-life ice sheet sensitive to eccentricity during a time of a global icy state may have activated a synergistic response of positive feedback processes modulated on ~ 100-kyr eccentricity response times. The negligible response of $\delta^{18}O$ in the eccentricity band during the Zanclean could be due to the higher mean temperature and the lower mean global ice volume, which would fail to move permanently to the climate signal due to obliquity "ice killing," a powerful forcing during this period. Considering the asymptotic decay of the $\delta^{18}O$ and SST R_s up to a maximum of − 100% towards the Early Pliocene (Fig. 11), a working hypothesis could be a damping effect due to the action of negative feedbacks during the time of higher global average temperature, such as, for instance, a higher global forest coverage, a greatest ocean primary productivity, and a more efficient burial of organic carbon. Zanclean was a period of global greenhouse state, characterized by high atmospheric GHGs and mean global temperature (Bertini 2010; Seki et al. 2010; Bartoli et al. 2011). During Zanclean, northern and central Italy were covered by vegetation dominated by hygrophilous and thermophilous forest taxa typical of a humid subtropical to warm-temperate climate. Precipitation was sufficiently high for the persistence of a "broad-leaved evergreen/warm temperate mixed forest" up to 3.5 Myr (Bertini 2010). Therefore, a wide forest cover, along with a possible high ocean primary productivity and organic carbon burial, could increase the quantitative role of the CO_2-related negative feedbacks (Emeis 1998; Schrag et al. 2002; Gallego-Torres et al. 2007; Claussen 2009; Moller et al. 2012; Ma et al. 2015) causing a damping effect of the orbitally paced responses. In the Mid-Late Pleistocene, an icy state leads to a significant reduction in global forest cover, making these processes quantitatively less important but emphasizing the vegetation-snow albedo feedback in synergy with the sea-ice albedo feedback which tends to amplify Northern Hemisphere and global mean temperature changes (Claussen 2009).

Mid-Late Pleistocene obliquity damped responses

To attempt the interpretation of these nonlinear features (Figs. 10 and 11), some possible mechanisms are discussed in this section as working hypotheses for future studies. The global $\delta^{18}O$ and tropical SST sharply decline to near-zero in obliquity R_s during the Mid-Late Pleistocene suggesting an attenuation mechanism of the obliquity driving force and, consequently, a reduction of the feedback amplification processes linked to the mean icy state. Understanding this phenomenon is important because the post-MPT obliquity forcing attenuation could have contributed to the strengthening of the eccentricity response, favoring a long-life ice growth. Lisiecki and Raymo (2007) state that the change in $\delta^{18}O$ glacial dynamics at ~ 1.4 Myr is associated with an abrupt decline in 41-kyr power and a decrease in modulation sensitivity to obliquity. Interestingly, the shape of the $\delta^{18}O$ and SST components-3-4 after the MPT (Figs. 4c and 8c), characterized by the envelope of two to three low-amplitude obliquity cycles in a weak ~ 90–93-kyr framework, suggests a connection between eccentricity and obliquity, similar to that proposed by Huybers (2007) of obliquity-cycle skipping. The obliquity-pacing results indicate that the 100-kyr variability could result from the skipping of one or two obliquity beats, corresponding to 80-kyr or 120-kyr glacial cycles that, on average, give 100-kyr periodicity, resulting from long-term changes in the climate system, such as ice sheet growth (Huybers 2007). Why does obliquity-cycle skipping occur? Climate friction is a dissipative feedback between obliquity variations and climate which may cause a secular drift of the spin axis (Laskar et al. 2004). In response to obliquity forcing, glacial and interglacial conditions drive the redistribution of the ice/water mass and the isostatic adjustment to the surface loading, affecting the dynamical ellipticity of the Earth (obliquity–oblateness feedback) (Rubincam 1993, 1995; Levrard and Laskar 2003; Laskar et al. 2004). Delayed responses in both climatic and viscous relaxation processes may introduce a secular term in the obliquity evolution (Rubincam 1995). Because both the Earth's ice load history and viscoelastic structure are not strongly constrained, it is difficult to produce accurate predictions of the coupled response of the entire system (Laskar et al. 2004). Therefore, simplifying assumptions have to be made to estimate the magnitude and the direction of the secular drift (increase or decrease in the obliquity). Specifically, the sign of the secular obliquity change is particularly sensitive to the obliquity-lag estimates and changes of oblateness (Williams et al. 1998; Levrard and Laskar 2003). Levrard and Laskar (2003) showed that climate friction impact is positive and likely negligible over the last 3 Myr, based on the model assumptions that the ice sheet response time-lag could not be longer than the obliquity period and explicitly assumed it to be 8 kyr. However, the secular change of obliquity becomes negative for low obliquity lag response (< 4.2 kyr) and high positive change of oblateness (Levrard and Laskar 2003). Response lags are difficult to estimate. According to Lisiecki and Raymo (2005a), the $\delta^{18}O$ lag in obliquity reaches values of ~ 6.8 kyr in the Pleistocene. The results of the present study indicate a $\delta^{18}O$ obliquity lag of 7.4 kyr post-MPT (Table 8), close to the value used by Levrard and Laskar (2003). In contrast, the obliquity lag estimation of Hilgen et al. (1993) provides a lower value of 5.6 kyr with respect to the Late Pleistocene if the phase relation with precession is kept constant. Considering the margin of error of these estimates, it is currently not possible to establish whether the obliquity lags may be lower post-MPT to justify the observed obliquity damping effect. Furthermore, other simplified assumptions in the dissipative equations model, together with age model bias, could affect the final results. A recent study by Adhikari and Ivins (2016) using space geodetic and satellite gravimetric present-day data has shown that global-scale continent-ocean mass transport changes (terrestrial water storage and global cryosphere) explain nearly the entire amplitude (83 ± 23%) and mean directional shift (within 5.9° ± 7.6°) of the observed polar motion on a very short present-day time scale. The study of Adhikari and Ivins (2016) suggests a high sensitivity of the poles to the mass transport changes and to the isostatic rebound. In conclusion, new studies are needed to clarify the effective role of climate friction or to identify other mechanisms that might be responsible for this unusual transition in obliquity response.

Tropical SST coupled with benthic $\delta^{18}O$ cycles

Despite the differences in signal-to-noise ratio and oceanographic context between the LR04 $\delta^{18}O$ and the ODP Site 846 SST records (global benthic stack and upwelling tropical single record), the SSA results show the tropical SST tightly coupled with benthic $\delta^{18}O$ over the past five million years, both at long and orbital time scales, in agreement with the original work of Herbert et al. (2010a). Some differences certainly exist (e.g., a strong cooling event at ~ 2.1 kyr or MBO warming less marked in the SST signal; the tropical trend cooling slightly negative shifted compared to global ΔSST, nonetheless subparallel in time) (Fig. 14b), but they may be explained in terms of temporal and regional upwelling effects at the ODP Site 846. These similarities suggest common causes for both the $\delta^{18}O$ long-term trend and cycles and their expression in the tropical SST. The most likely candidate for some orbitally paced cycles lies in high-latitude oceanographic processes tied to the global carbon cycle (Liu and Herbert 2004; Herbert et al. 2010a). These processes drive the oceanic circulation through changes in deep water properties, affecting vertical density stability and upwelling in the tropical oceans. Changes in the atmospheric GHG composition controlled by insolation processes in the high latitudes may also act to imprint obliquity on the tropical Pacific where the local direct effect of obliquity is weak (Liu and Herbert 2004; Medina-Elizalde and Lea 2005; Masson-Delmotte et al. 2006). The $\delta^{18}O$-SST delta-lag in obliquity became large after the beginning of the MPT, suggesting a signal of global ice volume growth (Table 8). Such observations, together with the high coherence between SST and $\delta^{18}O$ obliquity responses (Table 9) may suggest remote connections from high latitudes towards the tropics. These interpretations could explain why the tropical SST obliquity component contains a small variance in comparison to that of $\delta^{18}O$ (5.5% vs. 9.9%). The smaller coherency between tropical SST and $\delta^{18}O$ precession components (Table 9) might indicate that the response in the EEP was caused by low-latitude precession forcing and regional effects and was less coupled to the high-latitude ice volume (Liu and Herbert 2004). Tachikawa et al. (2014) found residual SST variability characterized by a 23-kyr periodicity in the western Pacific warm pool (WPWP) region. They concluded a wind-driven northward surface current advection towards the WPWP to be responsible for the regional SST precessional periodicity, attributed to meridional shifts of the Intertropical Convergence Zone. Even the "half-precession" cycle of equatorial origin (Berger et al. 2006), isolated in this study by SSA from both tropical SST and global $\delta^{18}O$ records, may be transmitted to high latitudes via advective transport (Hagelberg et al. 1994). The last interglacial half-precession cycle from the Northwestern Chinese Loess Plateau, interpreted as a direct response to low-latitude forcing through its modulation on the East Asian summer monsoon (Sun and Huang 2006), supports this interpretation. The coherency increase of the short eccentricity's $\delta^{18}O$ and SST signals and the maximum delta-lag in eccentricity between $\delta^{18}O$ and SST (4.31 vs. 0.44 kyr) towards the Late Pleistocene (Table 8) may reflect the progressive rise of eccentricity power and an increased contribution of the high-latitude $\delta^{18}O$ ice volume component over the water temperature signal. Berger et al. (2006) show the presence of significant 100-kyr and 11-kyr cycles in the amplitude of the seasonal cycle in the intertropical regions, but the amplitude decreases rapidly when moving away from the equator. The small differences in variance between the short eccentricity's SST and $\delta^{18}O$ components (5.1% vs. 6.5%) might depend on these considerations. Conversely, the big difference in variance of the closely enveloped long-term cycles (Earth's orbit secular frequencies, obliquity modulation cycles, and long eccentricity cycles) between SST (10.3%) and $\delta^{18}O$ (4.2%) components could suggest a greater sensitivity of the tropics to these long-term cycles. The tropical oceans should be shielded from feedback processes that produce large temperature sensitivity in the high latitudes, such as ice-albedo feedbacks, sea ice-atmosphere feedback, and permafrost feedback (Lawrence et al. 2008; Tarnocai et al. 2009; Herbert et al. 2010a; Liptak and Strong 2016). With the exception of obliquity, the marked difference in eccentricity and precession R_s in the Late Pleistocene between $\delta^{18}O$ (200–400%) and SST (100–170%) (Fig. 11) may be due to a lower synergy of the positive feedback mechanisms in tropical regions. In summary, both remote connections from high-latitude and direct low-latitude orbital forcings could couple Plio-Pleistocene global $\delta^{18}O$ and tropical SST orbital cycles through oceanic and atmospheric circulations. Finally, the remarkable constancy (~ 76%) of the Plio-Pleistocene trend variance among global $\delta^{18}O$, local tropical SST (Site 846), and global SST (ΔSST stack) might reflect the latitude-independent nature of the long-term factors controlling the atmosphere composition and global mean temperature.

Conclusions

The main conclusions of this study can be summarized as follows:

1) A detailed evaluation is provided of the variance paced by long-period orbital modulation, short eccentricity, obliquity, precession, and half-precession cycles, estimated for the $\delta^{18}O$ at approximately 4.2%, 6.5%, 9.9%, 2.0%, and 0.6%, respectively, and for the tropical SST at about 10.3%, 5.1%, 5.5%, 1.6%, and 0.5%, respectively, both resulting in total orbital contributions of ~ 23.0%. On long time scales, the Plio-Pleistocene astronomical forcing could be considered a random walk at zero-sum in which the orbital responses oscillate around a basically time-invariant quasi-normal distributed mean, supporting the hypothesis that orbital forcings did not have a major role in guiding the Plio-Pleistocene climate system response.
2) The exponential growth of the orbital variance towards the Late Pleistocene is function of the LT-MCS at a given time and regards all $\delta^{18}O$ and SST components with an increasing rate related to the orbital period, from half-precession to eccentricity. The asymmetry of the R_s suggests an amplified orbital response during times of a global icehouse state (Mid-Late Pleistocene, with the exception of obliquity), which can reach a magnitude of + 200 to + 400% global $\delta^{18}O$ forcing, or + 100 to + 170% the tropical SST forcing, and a minimum up to – 100% the damped response for all signals during time of a global greenhouse state (Early Pliocene).
3) It is proposed that orbital responses be considered as composite feedback lagged responses paced by orbitals and damped or amplified in a range of – 100 to + 400% the forcing, depending on the initial conditions set by LT-MCS, the hemicycle duration, and the synergy of the feedback processes related to carbon cycles and albedo effects. This interpretation may explain the energy excess "paradox" of the astronomically paced signals compared to the small energy of the orbital forcing, especially concerning the eccentricity bands.
4) The asymptotic decay of the $\delta^{18}O$ and SST R_s up to – 100% towards the Early Pliocene could be due to a damping effect of negative feedbacks during a time of higher mean global temperature. Zanclean was a period of global greenhouse state, and a wide forest cover along with a possible high ocean primary productivity and a more efficient burial of organic carbon could have increased the quantitative role of the CO_2-related negative feedbacks, causing a damping effect of the orbitally paced responses.
5) The global $\delta^{18}O$ and tropical SST anomalous sharply decline to near-zero in obliquity R_s during the Late Pleistocene suggesting an attenuation mechanism of the obliquity driving force and a reduction of the related feedback amplification processes. The post-MPT obliquity damping may have contributed to the strengthening of the eccentricity response by mitigating the obliquity "ice killing," favoring a 100-kyr long-life hysteresis ice growth volume. A long-life ice sheet sensitive to eccentricity during the time of a global icy state may have activated a synergistic response of positive feedback processes modulated on short eccentricity response times. In the Mid-Late Pleistocene, an icy state leads to a significant reduction in global forest cover making its negative feedbacks quantitatively less important but emphasizing the positive feedback processes, such as the vegetation-snow albedo in synergy with the sea-ice albedo and permafrost feedbacks. However, the origin of the post-MPT obliquity attenuation mechanism is an open question.
6) The tropical SST orbital cycles are tightly coupled with the global benthic $\delta^{18}O$ cycles over the past five million years, suggesting both Plio-Pleistocene oceanic and atmospheric remote connections from high-latitude and direct low-latitude orbital forcings.
7) The tropical SST, the global SST, and the global $\delta^{18}O$ trend components, each explaining ~ 76% of the Plio-Pleistocene climate variance, could be related to the long-term pCO_2 proxies, suggesting non-orbital latitude-independent control factors. These trend components significantly change the climate mean using arbitrary time intervals, suggesting the Plio-Pleistocene mean climate state was changed primarily by long-term carbon cycle controlling the Earth's atmospheric composition. The geotectonic forcing appears to be a plausible hypothesis to explain the long-term changes in atmospheric composition (GHGs and volcanic aerosols) through direct (CO_2 outgassing and explosive volcanism) and indirect (orography and erosion, paleogeography, oceanic paleocirculation, and ocean fertilization) control by plate tectonics.
8) The NHG gradually developed in four phases (subtrends) of progressive lowering of the average temperatures and ice volume growth, triggered by long-term changes in the atmospheric composition. Nevertheless, transient interaction effects with very long-term orbital cycles, expression of the chaotic behavior of the solar system, may have contributed

to setting these transitions. After MIS-16, the cooling trend was broken by a wide swing of $\delta^{18}O$ depletion and SST recovery, a relative stasis of ice growth and temperature cooling during the last 600 kyr (MBO), which could be related to mid-term GHG recovery.

The idea of the geotectonic forcing is the most attractive hypothesis by a long time, and the recent improvements in the seafloor spreading rate evaluation lead in the right direction. However, further studies are needed to better estimate high-resolution seafloor spreading rate and to assess the role of the volcanic CO_2 degassing and the explosive volcanic aerosols in the budget of the carbon cycle. Even much higher resolution CO_2 proxies, covering a long time period, are necessary to better address carbon cycle. Nevertheless, the relevance of the primary long-term climate system, controlling the Earth's atmospheric composition and modifying the average global climate, must be underlined. Finally, singular spectrum analysis provides a valuable tool in cyclostratigraphy with the remarkable advantage of separating full-resolution time series by variance strength.

Abbreviations

ANOVA: Analysis of variance; EECO: Early Eocene Climatic Optimum; EEP: Eastern Equatorial Pacific; EPICA: European Project for Ice Coring in Antarctica; FFS: Fourier frequency spectrum; FFT: Fast Fourier transform; GASR: Global average spreading rate; GCM: General circulation model; GHG: Greenhouse gas; GSSP: Global Stratotype Section and Point; HT846: Herbert et al. (2010b) sea surface temperature of ODP Site 846; INHG: Intensification of Northern Hemisphere Glaciation; LGM: Last Glacial Maximum; LIP: Large igneous province; LT-HT846: Long-term HT846 SST component; LT-LR04: Long-term LR04 $\delta^{18}O$ component; LT-MCS: Long-term mean climate state; MBE: Mid-Brunhes Event; MBO: Mid-Brunhes Oscillation; MIS: Marine Isotope Stage; MMCO: Middle Miocene Climatic Optimum; MPT: Mid-Pleistocene Transition; NASSR: North Atlantic seafloor spreading rate; NHG: Northern Hemisphere Glaciation; ODP: Ocean Drilling Program; ONHG: Onset of the Northern Hemisphere Glaciation; PETM: Paleocene-Eocene Thermal Maximum; R_S: Response sensitivity; SSA: Singular spectrum analysis; SST: Sea surface temperature; TISA: Time-integral squared amplitude; WPWP: Western Pacific warm pool

Acknowledgements

Special thanks to Gian Battista Vai for the useful suggestions and for the encouragement to publish the study. I am grateful to Francesco Mulargia for the methodological review. Thanks also to Nicola Scafetta for the useful discussion. I thank Lorraine Lisiecki for kindly providing me the $\delta^{13}C$-based pCO_2 proxy data. Special thanks to Alessandra Negri for improving the English language and other constructive comments.

Funding

This research did not receive any specific grant from funding agencies in the public, commercial, or not-for-profit sectors.

Authors' contributions

The author read and approved the final manuscript.

Competing interests

The author declares that he has no competing interests.

References

Abe-Ouchi A, Saito F, Kawamura K, Raymo ME, Okuno J, Takahashi K, Blatter H (2013) Insolation-driven 100,000-year glacial cycles and hysteresis of ice-sheet volume. Nature 500. https://doi.org/10.1038/nature12374

Adhikari S, Ivins ER (2016) Climate-driven polar motion: 2003–2015. Sci Adv 2. https://doi.org/10.1126/sciadv.1501693

Aitchison JC, Ali JR, Davis AM (2007) When and where did India and Asia collide? J Geophys Res 112:B05423. https://doi.org/10.1029/2006JB004706

Archer D, Martin P, Buffett B, Brovkin V, Rahmstorf S, Ganopolski A (2004) The importance of ocean temperature to global biogeochemistry. Earth Planet Sci Lett 222(2004):333–348. https://doi.org/10.1016/j.epsl.2004.03.011

Bartoli G, Hönisch B, Zeebe RE (2011) Atmospheric CO2 decline during the Pliocene intensification of northern hemisphere glaciations. Paleoceanography 26:PA4213. https://doi.org/10.1029/2010PA002055

Bartoli G, Sarnthein M, Weinelt M, Erlenkeuser H, Garbe-Schonberg D, Lea DW (2005) Final closure of Panama and the onset of northern hemisphere glaciation. Earth Planet Sci Lett 237(2005):33–44. https://doi.org/10.1016/j.epsl.2005.06.020

Becker TW, Conrad CP, Buffett B, Dietmar MR (2009) Past and present seafloor age distributions and the temporal evolution of plate tectonic heat transport. Earth Planet Sci Lett 278(2009):233–242. https://doi.org/10.1016/j.epsl.2008.12.007

Beerling DJ, Royer DL (2011) Convergent Cenozoic CO_2 history. Nat Geosci 4, pp. 418-420.

Berger A, Li XS, Loutre MF (1999) Modelling northern hemisphere ice volume over the last 3 Myr. Quat Sci Rev 18(1999):1–11

Berger A, Loutre MF (1997) Long-term variations in insolation and their effects on climate, the LLN experiments. Surv Geophys 18:147–161

Berger A, Loutre MF, Mélice JL (2006) Equatorial insolation: from precession harmonics to eccentricity frequencies. Clim Past Discuss 2:519–533 https://doi.org/10.5194/cp-2-131-2006

Berger A, Mélice JL, Loutre MF (2005) On the origin of the 100-kyr cycles in the astronomical forcing, Paleoceanography, 20:PA4019, https://doi.org/10.1029/2005PA001173.

Berger A, Yin Q (2012) Modeling the interglacials of the last 1 million years. In: Berger et al (eds) Climate change: inferences from paleoclimate and regional aspects. Springer-Verlag, Wien 2012.

Berger A, Yin QZ, Herold N (2012) MIS-11 and MIS-19, analogs of our Holocene interglacial. In: 3rd International Conference on Earth System Modelling, Hamburg, https://meetingorganizer.copernicus.org/3ICESM/3ICESM-11.pdf

Berger A et al (2016) Past interglacials working group of PAGES, interglacials of the last 800,000 years. Rev Geophys 54:162–219. https://doi.org/10.1002/2015RG000482

Berger AL (1978) Long-term variations of daily insolation and quaternary climatic change. J Atmos Sci 35:2362–2367

Berner RA, Kothavala Z (2001) GEOCARB III: a revised model of atmospheric CO_2 over phanerozoic time. Am J Sci 301(2):182–204

Berner RA, Lasaga AC, Garrels RM (1983) The carbonate-silicate geochemical cycle and its effect on atmospheric carbon dioxide over the past 100 million years. Am J Sci 283:641–683

Bertini A (2010) Pliocene to Pleistocene palynoflora and vegetation in Italy - state of the art. Quat Int 225(2010):5–24

Blanc PL, Fontugne MR, Duplessy JC (1983) The time-transgressive initiation of boreal ice-cap: continental and oceanic evidence reconciled. Palaeogeogr Palaeoclimatol Palaeoecol 42:211–224

Boulila S, Galbrun B, Miller KG, Pekar SF, Browning JV, Laskar J, Wright JD (2011) On the origin of Cenozoic and Mesozoic "third-order" eustatic sequences. Earth Sci Rev 109:94–112

Brady PV, Gislason SR (1997) Seafloor weathering controls on atmospheric CO_2 and global climate. Geochimica Et Cosmochimica Acta 61(5):965–973

Brovkin V, Ganopolski A, Archer D, Rahmstorf S (2007) Lowering of glacial atmospheric CO2 in response to changes in oceanic circulation and marine biogeochemistry. Paleoceanography 22:PA4202. https://doi.org/10.1029/2006PA001380

Clark PU, Archer D, Pollard D, Blum JD, Rial JA, Brovkin V, Mix AC, Pisias NG, Roy M (2006) The middle Pleistocene transition: characteristics, mechanisms, and implications for long-term changes in atmospheric pCO2. Quat Sci Rev. https://doi.org/10.1016/j.quascirev.2006.07.008

Claussen M (2009) Late Quaternary vegetation-climate feedbacks. Clim Past 5: 203–216, https://doi.org/10.5194/cp-5-203-2009

Cogné JP, Humler E (2006) Trends and rhythms in global seafloor generation rate. Geochem Geophys Geosyst 7:Q03011. https://doi.org/10.1029/2005GC001148

Coltice N, Seton M, Rolf T, Müller RD, Tackley PJ (2013) Convergence of tectonic reconstructions and mantle convection models for significant fluctuations in seafloor spreading. Earth Planet Sci Lett 383(2013):92–100. https://doi.org/10.1016/j.epsl.2013.09.032

Conrad CP, Lithgow-Bertelloni C (2007) Faster seafloor spreading and lithosphere production during the mid-Cenozoic. Geology 35(1):29–32. https://doi.org/10.1130/G22759A

Dickens GR (2003) Rethinking the global carbon cycle with a large, dynamic and microbially mediated gas hydrate capacitor. Earth Planet Sci Lett 213(2003):169–183. https://doi.org/10.1016/S0012-821X(03)00325-X

Elenga H et al (2000) Pollen-based biome reconstruction for southern Europe and Africa 18,000 years ago. J Biogeogr 27:621–634

Elsner JB, Tsonis AA (1996) Singular spectrum analysis: a new tool in time series analysis. Springer

Emeis K.-C., Sakamoto T. (1998) - The Sapropel Theme of Leg 160. In: Robertson A.H.F., Emeis K.-C., Richter C. & Camerlenghi A. (Eds.). Proceedings of the ocean drilling program, scientific results. 160

Gallego-Torres D, Martínez-Ruiz F, Paytan A, Jiménez-Espejo FJ, Ortega-Huertas M (2007) Pliocene–Holocene evolution of depositional conditions in the eastern Mediterranean: role of anoxia vs. productivity at time of sapropel deposition. Palaeogeogr Palaeoclimatol Palaeoecol 246(2007):424–439. https://doi.org/10.1016/j.palaeo.2006.10.008

Ghil M, Allen RM, Dettinger MD, Ide K, Kondrashov D, Mann ME, Robertson A, Saunders A, Tian Y, Varadi F, Yiou P (2002) Advanced spectral methods for climatic time series. Rev Geophys 40(1):3.1–3.41. https://doi.org/10.1029/2001RG000092

Gibbard PL, Head MJ (2009) IUGS ratification of the Quaternary system/period and the Pleistocene series/epoch with a base at 2.58 Myr. Quaternaire 20(4):411–412

Hagelberg TK, Bond G, deMenocal P (1994) Milankovitch band forcing of sub-Milankovitch climate variability during the Pleistocene. Paleoceanography 9(4):545–558. https://doi.org/10.1029/94PA00443

Hansen J. (2013) - An old story, but useful lessons. http://www.columbia.edu/~jeh1/mailings/2013/20130926_PTRSpaperDiscussion.pdf

Hansen J, Sato M (2012) Paleoclimate implications for human-made climate change. In: Berger A, Mesinger F, Sijacki D (eds) Climate change. Inferences from paleoclimate and regional aspects. Springer-Verlag, Wien 2012, p 350, https://doi.org/10.1007/978-3-7091-0973-1_2

Hansen J, Sato M, Kharecha P, Russell G, Lea DW, Siddall M (2007) Climate change and trace gases. Phil Trans R Soc A 365:1925–1954. https://doi.org/10.1098/rsta.2007.2052

Hansen J, Sato M, Kharecha P, von Schuckmann K (2011) Earth's energy imbalance and implications. Atmos Chem Phys 11:13421–13449, https://doi.org/10.5194/acp-11-13421-2011

Hansen J, Sato M, Russell G, Kharecha P (2013) Climate sensitivity, sea level and atmospheric carbon dioxide. Phil Trans R Soc A 371:20120294. https://doi.org/10.1098/rsta.2012.0294

Hansen et al (2005) Efficacy of climate forcings. J Geophys Res 110:D18104. https://doi.org/10.1029/2005JD005776

Hassani H (2007) Singular spectrum analysis: methodology and comparison. J Data Sci 5(2007):239–257

Head M.J. & Gibbard P.L. (2005) - Early–Middle Pleistocene transitions. An overview and recommendation for the defining boundary. From: Head MJ & Gibbard PL (eds). Early–middle Pleistocene transitions: the land–ocean evidence. Geological Society, London, Special Publications, 247, 1–18. 0305–8719/05/$15© The Geological Society of London 2005

Head MJ, Pillans B, Farquhar SA (2008) The Early–Middle Pleistocene transition: characterization and proposed guide for the defining boundary. Episodes 31(2):255–259

Herbert TD, Peterson LC, Lawrence KT, Liu Z (2010a) Tropical ocean temperatures over the past 3.5 million years. Science 328:1530–1534. https://doi.org/10.1126/science.1185435

Herbert T.D., Peterson L.C., Lawrence K.T. & Liu Z. (2010b) - Plio-Pleistocene tropical alkenone SST reconstructions LAST UPDATE: 6/2010, NOAA/NCDC Paleoclimatology Program. ftp://ftp.ncdc.noaa.gov/pub/data/paleo/contributions_by_author/herbert2010/herbert2010.txt

Hilgen FJ, Lourens LJ, Berger A, Loutre MF (1993) Evaluation of the astronomically calibrated time scale for the Late Piocene and the earliest Pleistocene. Paleoceanography 8:549–565

Hönisch B, Hemming NG, Archer D, Siddall M, McManus JF (2009) Atmospheric carbon dioxide concentration across the mid-Pleistocene transition. Science 324:1551. https://doi.org/10.1126/science.1171477

Huybers P (2007) Glacial variability over the last two million years: an extended depth-derived agemodel, continuous obliquity pacing, and the Pleistocene progression. Quat Sci Rev 26(2007):37–55

Imbrie J, Berger A, Boyle EA, Clemens SC, Duffy A, Howard WR, Kukla G, Kutzbach J, Martinson DG, Mcintyre A, Mix AC, Molfino B, Morley JJ, Peterson LC, Pisias NG, Prell WL, Raymo ME, Shackleton NJ, Toggweile JR (1993) On the structure and origin of major glaciation cycles 2. The 100,000-year cycle. Paleoceanography 8(6):699–735

Imbrie JZ, Imbrie-Moore A, Lisiecki LE (2011) A phase-space model for Pleistocene ice volume. Earth Planet Sci Lett 307:94–102

Jicha BR, Scholl DW, Rea DK (2009) Circum-Pacific arc flare-ups and global cooling near the Eocene Oligocene boundary. Geology 37(4):303–306. https://doi.org/10.1130/G25392A.1

Jouzel J et al (2007) Orbital and millennial Antarctic climate variability over the past 800,000 years. Science 317:793–796. https://doi.org/10.1126/science.1141038

Joyce JE, Tjalsma LRC, Prutzman JM (1990) High-resolution planktic stable isotope record and spectral analysis for the last 5.35 M.Y.: Ocean Drilling Program site 625 northeast Gulf of Mexico. Paleoceanography 5:507–529

Kent DV, Muttoni G (2013) Modulation of Late Cretaceous and Cenozoic climate by variable drawdown of atmospheric pCO_2 from weathering of basaltic provinces on continents drifting through the equatorial humid belt. Clim Past 9:525–546. https://doi.org/10.5194/cp-9-525-2013

Kominz M, Pisias N (1979) Pleistocene climate: deterministic or stochastic? Science 204(Issue 4389):171–173. https://doi.org/10.1126/science.204.4389.171

Laskar J, Fienga A, Gastineau M, Manche H (2011) La2010: a new orbital solution for the long-term motion of the Earth. Astron Astrophys 532:A89. https://doi.org/10.1051/0004-6361/201116836

Laskar J, Joutel F, Boudin F (1993) Orbital, precessional and insolation quantities for the Earth from −20 Myr to +10 Myr. Astron Astrophys 270:522–533

Laskar J, Robutel P, Joutel F, Gastineau M, Correia ACM, Levrard B (2004) A long term numerical solution for the insolation quantities of the Earth. Astron Astrophys 428:La2004

Lawrence DM, Slater AG, Tomas RA, Holland MM, Deser C (2008) Accelerated Arctic land warming and permafrost degradation during rapid sea ice loss. Geophys Res Lett 35:L11506. https://doi.org/10.1029/2008GL033985

Le Breton E, Cobbold PR, Dauteuil O, Lewis G (2012) Variations in amount and direction of seafloor spreading along the Northeast Atlantic Ocean and resulting deformation of the continental margin of Northwest Europe. Tectonics 31:TC5006. https://doi.org/10.1029/2011TC003087

Levrard B, Laskar J (2003) Climate friction and the Earth's obliquity. Geophys J Int 154:970–990

Li G, Elderfield H (2013) Evolution of carbon cycle over the past 100 million years. Geochim Cosmochim Acta 103(2013):11–25. https://doi.org/10.1016/j.gca.2012.10.014

Liptak J, Strong C (2016) A modeling investigation of the Arctic Sea ice-atmosphere feedback. Clim Dyn 47(Issue 7):2471–2480

Lisiecki LE (2010a) Links between eccentricity forcing and the 100,000-year glacial cycle. Nat Geosci 3. https://doi.org/10.1038/NGEO828

Lisiecki LE (2010b) A benthic $\delta^{13}C$-based proxy for atmospheric pCO_2 over the last 1.5 Myr. Geophys Res Lett 37:L21708. https://doi.org/10.1029/2010GL045109

Lisiecki LE, Raymo ME (2005a) A Pliocene-Pleistocene stack of 57 globally distributed benthic δ18O records. Paleoceanography 20:PA1003. https://doi.org/10.1029/2004PA001071

Lisiecki L.E. & Raymo M.E. (2005b) Appendix 1 - Global Plio-Pleistocene stack of benthic oxygen isotope records. DOI:https://doi.org/10.1594/PANGAEA.701576. DOI:https://doi.org/10.1029/2004PA001071. In supplement to: Lisiecki & Raymo (2005)

Lisiecki LE, Raymo ME (2007) Plio–Pleistocene climate evolution; trends and transitions in glacial cycle dynamics. Quat Sci Rev 26(2007):56–69. https://doi.org/10.1016/j.quascirev.2006.09.005

Liu Z, Herbert TD (2004) High-latitude influence on the eastern equatorial Pacific climate in the early Pleistocene epoch. Nature 427:720–723

Lo L, Chang S, Wei K, Lee S, Ou T, Chen Y, Chuang C, Mii H, Burr GS, Chen M, Tung Y, Tsai M, Hodell DA, Shen C (2017) Nonlinear climatic sensitivity to greenhouse gases over past 4 glacial/interglacial cycles. Nat Sci Rep 7:4626. https://doi.org/10.1038/s41598-017-04031-x

Loulergue L, Schilt A, Spahni R, Masson-Delmotte V, Blunier T, Lemieux B, Barnola JM, Raynaud D, Stocker TF, Chappellaz J (2008) Orbital and millennial-scale features of atmospheric CH4 over the past 800,000 years. Nature 453. https://doi.org/10.1038/nature06950

Lourens LJ, Hilgen FJ (1994) Long-period orbital variations and their relation to third-order eustatic cycles and the onset of major glaciations 3.0 million years ago. In: Lourens LJ (ed) Astronomical forcing of Mediterranean climate during the last 5.3 million years. Thesis Utrecht University, ISBN 90-393-0754-7

Lunt DJ, Foster GL, Haywood AM, Stone EJ (2008) Late Pliocene Greenland glaciation controlled by a decline in atmospheric CO_2 levels. Nature 454: 1102–1106. https://doi.org/10.1038/nature07223

Lunt DJ, Haywood AM, Schmidt GA, Salzmann U, Valdes PJ, Dowsett HJ, Loptson CA (2012) On the causes of mid-Pliocene warmth and polar amplification. Earth Planet Sci Lett 321-322(2012):128–138. https://doi.org/10.1016/j.epsl.2011.12.042

Luthi D, Le Floch M, Bereiter B, Blunier T, Barnola JM, Siegenthaler U, Raynaud D, Jouzel J, Fischer H, Kawamura K, Stocker TF (2008) High-resolution carbon dioxide concentration record 650,000–800,000 years before present. Nature 453. https://doi.org/10.1038/nature06949

Ma Z, Ravelo AC, Liu Z, Zhou L, Paytan A (2015) Export production fluctuations in the eastern equatorial Pacific during the Pliocene-Pleistocene: reconstruction using barite accumulation rates. Paleoceanography 30. https://doi.org/10.1002/2015PA002860

Martinez-Boti MA, Foster GL, Chalk TB, Rohling EJ, Sexton PF, Lunt DJ, Pancost RD, Badger MPS, Schmidt DN (2015) Plio-Pleistocene climate sensitivity evaluated using high-resolution CO2 records. Nature 518. https://doi.org/10.1038/nature14145

Masson-Delmotte et al (2006) Past temperature reconstructions from deep ice cores. Relevance for future climate change. Clim Past 2:145–165 www.clim-past.net/2/145/2006/

Medina-Elizalde M, Lea DW (2005) The mid-Pleistocene transition in the tropical Pacific. Science 310:1009–1012

Menviel L, Timmermann A, Mouchet A, Timm O (2008) Meridional reorganizations of marine and terrestrial productivity during Heinrich events. Paleoceanography 23:PA1203. https://doi.org/10.1029/2007PA001445

Merkouriev S, DeMets C (2014) High-resolution Quaternary and Neogene reconstructions of Eurasia-North America plate motion. Geophys J Int. https://doi.org/10.1093/gji/ggu142

Moller T, Schulz H, Hamann Y, Dellwig O, Kucera M (2012) Sedimentology and geochemistry of an exceptionally preserved last interglacial sapropel S5 in the Levantine Basin (Mediterranean Sea). Mar Geol 291-294(2012):34–48. https://doi.org/10.1016/j.margeo.2011.10.011

Mosar J, Lewis G, Torsvik TH (2002) North Atlantic Sea-floor spreading rates: implications for the Tertiary development of inversion structures of the Norwegian–Greenland Sea. J Geol Soc London 159(2002):503–515

Mudelsee M, Raymo ME (2005) Slow dynamics of the Northern Hemisphere glaciation. Paleoceanography 20. https://doi.org/10.1029/2005PA001153

Ning S, Jia-Xin C, Königsson LK (1993) Late Cenozoic vegetational history and the Pliocene-Pleistocene boundary in the Yushe Basin, SE Shanxi, China. Grana 32(4-5):260–271. https://doi.org/10.1080/00173139309429990

Pagani M, Liu Z, LaRiviere J, Ravelo AC (2010) High Earth-system climate sensitivity determined from Pliocene carbon dioxide concentrations. Nat Geosci 3. https://doi.org/10.1038/NGEO724

Raymo ME (1994) The initiation of northern hemisphere glaciation. Annu Rev Earth Planet Sci 22:353–383

Raymo ME (1998) Glacial puzzles. Science 281(5382):1467–1468

Raymo ME, Lisiecki LE, Nisancioglu KH (2006) Plio-Pleistocene ice volume, Antarctic climate, and the global $\delta^{18}O$ record. Science 313:492–495

Raymo ME, Ruddiman WF (1992) Tectonic forcing of late Cenozoic climate. Nature 359:117–122

Rial JA, Pielke RA Sr, Beniston M, Claussen M, Canadell J, Cox P, Held H, de Noblet-Ducoudré N, Prinn R, Reynolds JF, Salas JD (2004) Nonlinearities, feedbacks and critical thresholds within the Earth's climate system. Clim Change 65(1–2):11–38

Rowley DB (2002) Rate of plate creation and destruction 180 Myr to present. Geol Soc Am Bull 114:927–933

Royer DL (2006) CO_2-forced climate thresholds during the Phanerozoic. Geochim Cosmochim Acta 70(2006):5665–5675

Rubincam DP (1993) The obliquity of Mars and climate friction. J Geophys Res 98: 10827–10832

Rubincam DP (1995) Has climate changed the Earth's tilt ? Paleoceanography 10(3):365–372

Ruggieri E, Herbert T, Lawrence KT, Lawrence CE (2009) Change point method for detecting regime shifts in paleoclimatic time series. Application to $\delta^{18}O$ time series of the Plio-Pleistocene. Paleoceanography 24:PA1204. https://doi.org/10.1029/2007PA001568

Schneider R, Schmitt J, Köhler P, Joos F, Fischer H (2013) A reconstruction of atmospheric carbon dioxide and its stable carbon isotopic composition from the penultimate glacial maximum to the last glacial inception. Clim Past 9: 2507–2523. https://doi.org/10.5194/cp-9-2507-2013

Schrag DP, Berner RA, Hoffman PF, Halverson GP (2002) On the initiation of a snowball Earth. Geochem Geophys Geosyst 3(6). https://doi.org/10.1029/2001GC000219

Seki O, Foster GL, Schmidt DN, Mackensen A, Kawamura K, Pancost RD (2010) Alkenone and boron-based Pliocene pCO_2 records. Earth Planet Sci Lett 292(2010):201–211. https://doi.org/10.1016/j.epsl.2010.01.037

Shackleton NJ (2000) The 100,000-year ice-age cycle identified and found to lag temperature, carbon dioxide, and orbital eccentricity. Science 289:1897–1902

Shackleton NJ, Opdyke ND (1977) Oxygen isotope and paleomagnetic evidence for early northern hemisphere glaciation. Nature 270:216–219

Stap LB, de Boer B, Ziegler M, Bintanja R, Lourens LJ, van de Wal RSW (2016) CO_2 over the past 5 million years: continuous simulation and new $\delta^{11}B$-based proxy data. Earth Planet Sci Lett 439(2016):1–10. https://doi.org/10.1016/j.epsl.2016.01.022

Stap LB, van de Wal RSW, de Boer B, Bintanja R, Lourens LJ (2014) Interaction of ice sheets and climate during the past 800 000 years. Clim Past 10:2135–2152. https://doi.org/10.5194/cp-10-2135-2014

Sun J, Huang X (2006) Half-precessional cycles recorded in Chinese loess: response to low-latitude insolation forcing during the last interglaciation. Quat Sci Rev 25(9–10):1065–1072. https://doi.org/10.1016/j.quascirev.2005.08.004

Svensen H, Planke S, Malthe-Sørenssen A, Jamtveit B, Myklebust R, Eidem TR, Rey SS (2004) Release of methane from a volcanic basin as a mechanism for initial Eocene global warming. Nature 429:542–545

Tachikawa K, Timmermann A, Vidal L, Sonzogni C, Timm OE (2014) CO_2 radiative forcing and intertropical convergence zone influences on western Pacific warm pool climate over the past 400 ka. Quat Sci Rev 86(2014):24–34. https://doi.org/10.1016/j.quascirev.2013.12.018

Tarnocai C, Canadell JG, Schuur EAG, Kuhry P, Mazhitova G, Zimov S (2009) Soil organic carbon pools in the northern circumpolar permafrost region. Global Biogeochem Cycles 23:GB2023. https://doi.org/10.1029/2008GB003327

Tziperman E, Raymo ME, Huybers PJ, Wunsch C (2006) Consequences of pacing the Pleistocene 100 kyr ice ages by nonlinear phase locking to Milankovitch forcing. Paleoceanography 21(PA4206):1–11. https://doi.org/10.1029/2005PA001241

Van de Wal RSW, de Boer B, Lourens LJ, Kohler P, Bintanja R (2011) Reconstruction of a continuous high-resolution CO_2 record over the past 20 million years. Clim Past 7:1459–1469. www.clim-past.net/7/1459/2011/. https://doi.org/10.5194/cp-7-1459-2011

Van Der Meer D.G., Zeebe R.E., van Hinsbergen D.J.J., Sluijs A., Spakman W. & Torsvik T.H. (2014) - Plate tectonic controls on atmospheric CO_2 levels since the Triassic. PNAS Early Edition, www.pnas.org/cgi/doi/10.1073/pnas.1315657111

Vautard R, Ghil M (1989) Singular spectrum analysis in nonlinear dynamics, with applications to paleoclimatic time series. Physica D 35(1989):395–424 North-Holland, Amsterdam

Walker JCG, Hays PB (1981) A negative feedback mechanism for the long-term stabilization of Earth's surface temperature. J Geophys Res 86(C10):9776–9782

Westerhold T, Röhl U, Frederichs T, Agnini C, Raffi I, Zachos JC, Roy H, Wilkens RH (2017) Astronomical calibration of the Ypresian timescale: implications for seafloor spreading rates and the chaotic behavior of the solar system? Clim Past 13:1129–1152 https://doi.org/10.5194/cp-13-1129-2017

Williams DM, Kasting JF, Frakes LA (1998) Low-latitude glaciations and rapid changes in the Earth's obliquity explained by obliquity–oblateness feedback. Nature 396:453–455

Wunsch C (2004) Quantitative estimate of the Milankovitch-forced contribution to observed Quaternary climate change. Quat Sci Rev 23(2004):1001–1012. https://doi.org/10.1016/j.quascirev.2004.02.014

Zachos JC, Dickens GR, Zeebe RE (2008) An early Cenozoic perspective on greenhouse warming and carbon-cycle dynamics. NATURE, Vol. 451,17 January 2008. https://doi.org/10.1038/nature06588

Zachos JC, Pagani M, Sloan L, Thomas E, Billups K (2001) Trends, rhythms, and aberrations in global climate 65 Myr to present. Science 292, pp.686-693.

Zagwijn WH (1974) The Pliocene--Pleistocene boundary in western and southern Europe. BOREAS. Vol. 3:75–97

Zhang Y.G., Pagani M., Liu Z., Bohaty S.M. & DeConto R. (2013) - A 40-million-year history of atmospheric CO_2. PHIL. TRANS. R. SOC. A, 371: 20130096. https://doi.org/10.1098/rsta.2013.0096

13

Weakening of rainfall intensity on wet soils over the wet Asian monsoon region using a high-resolution regional climate model

Hiroshi G. Takahashi[1,2*] and Jan Polcher[3]

Abstract

This study estimated the sensitivity of rainfall characteristics (rainfall amount, rainfall frequency, rainfall intensity, and rainfall extremes based on 30-min intervals) to land-surface conditions over Southeast Asia, which has a wet land surface during the rainy season. To obtain the regional difference in sensitivity and simulate basic cloud-precipitation systems, we used a high-resolution regional climate model. To extract the systematic signals of sensitivity and exclude random errors, a series of six sensitivity experiments, which were driven by a reanalysis dataset and the observed sea surface temperature (SST), were conducted over the Indochina Peninsula. In our experiments, soil moisture was prescribed at 0.20, 0.25, 0.30, 0.35, 0.40, and 0.45 m^3 m^{-3} over the whole domain and during the whole calculation period. More experiments would allow us to divide the responses into systematic signals and random noise. The slope of a meteorological variable as a function of the six prescribed soil moisture values was defined as the sensitivity. It was found that the sensitivity of rainfall frequency to soil moisture was positive overall, whereas the sensitivity of rainfall intensity was negative overall, although evapotranspiration (sensible heat flux) increased (decreased) in a manner similar to the increase in soil moisture over the whole domain. The sensitivity of rainfall amount to an increase in soil moisture was dependent on the location. This implies that the response of rainfall characteristics to soil moisture is not simple, suggesting that changes in rainfall characteristics are not solely determined by evapotranspiration. In addition, the sensitivity of rainfall characteristics displayed remarkable regional characteristics. The characteristics described above were noticeable over the inland flat plains. We also discussed the mechanism in the response of rainfall characteristics to soil moisture. The coupling of an increase in water vapor in the planetary boundary layer and a decrease of sensible heat flux can explain the response. The increase in water vapor in the planetary boundary layer was associated with a reduction of the development of deep convections and an increase of boundary layer clouds.

Keywords: Rainfall characteristics, Asian monsoon, Land surface condition, Soil moisture, Regional climate model, Soil-rainfall feedback

Introduction

The land-atmosphere coupling is a significant component of the climate system. Changes in surface conditions can alter the regional climate. In terms of the effects of human activities on regional climate changes, changes in land-surface conditions by human activities, such as deforestation and urbanization, and conversion to cropland, are major drivers.

Changes in land-surface conditions, such as soil moisture, can change the balances of water and energy at the Earth's surface. Many previous studies using global and regional climate models have examined the realistic or potential impacts of deforestation on the regional and global climates (e.g., Henderson-Sellers and Gornitz (1984); Shukla et al. (1990); Polcher and Laval (1994); Pielke (2001); Takata et al. (2009)). There is a general agreement that deforestation leads to reduced rainfall

*Correspondence: hiroshi3@tmu.ac.jp
[1]Department of Geography, Tokyo Metropolitan University, 1-1 Minamioosawa, Hachioji, Tokyo 192-0397, Japan
[2]Japan Agency for Marine-Earth Science and Technology, Yokohama 236-0001, Japan
Full list of author information is available at the end of the article

and evaporation and increased surface temperature in regions where soil moisture changes have occurred. The land-atmosphere coupling strength has been investigated in terms of the improvement of seasonal forecasts; for example, Koster et al. (2004) found that the key soil moisture regions for seasonal forecasts are semi-arid regions (boundary regions between wet and dry regions). However, these previous studies used coarse-resolution climate models.

Recent studies have used high-resolution observations or high-resolution regional climate models to explore the feedback between soil moisture and precipitation. Taylor et al. (2012) showed that a dry surface resulted in a large amount of rainfall over the drier regions of West Africa. Taylor (2015) demonstrated the role of the horizontal non-uniformity of soil moisture on an initiation of convection over Europe. These results were supported by the direct observation of the mesoscale circulations driven by heterogeneous soil moisture conditions (Taylor et al. 2007). As a climate modeling study, Froidevaux et al. (2014) suggested that a dry soil patch activates an initiation of convection and a wet soil patch generates precipitation due to a convectively unstable condition. The importance of surface flux heterogeneity was also examined by a large-eddy simulation (Rochetin et al. 2017). In addition, Tuttle and Salvucci (2016) suggested the different response of precipitation to soil moisture due to the background regional aridity. Thus, understanding of feedback over drier regions has gradually improved on mesoscale, whereas over wetter regions it is still limited. Over wetter regions, a few numerical studies using a high-resolution climate model (Takahashi et al. 2010; Sugimoto and Takahashi 2017) have investigated changes in rainfall due to changes in soil moisture. This study focused on the wetter Asian monsoon region, because of a lack of previous studies with a focus on wetter climates.

It is well known that global warming can induce an increase in rainfall amount and a decrease in weaker rainfall events (e.g., IPCC 2013), which suggests a change in rainfall characteristics. The changes in rainfall characteristics and extremes are significant issues associated with climate change, and with changes in floods and droughts. Changes in soil moisture greatly modify the budgets of energy and water at the surface, which can result in changes in rainfall characteristics. Previous numerical studies over wet regions (Takahashi et al. 2010; Sugimoto and Takahashi 2017) considered changes in rainfall characteristics due to regional changes in soil moisture. An increase in soil moisture tends to result in a weakening of rainfall intensity over the wet Asian monsoon region. The effect of an increase in soil moisture on rainfall frequency has been found to differ markedly among regions. Although many numerical studies have shown the simulated evidence, it is generally difficult to obtain observational evidence. For example, Ono and Takahashi (2016) and Takahashi (2016) showed a similar tendency of the effect of land-surface conditions on rainfall intensity over the wet Asian monsoon region from observed datasets. However, more studies are required to fully understand the mechanisms of these responses.

For example, a wetter surface increases evaporation and water vapor in the lower troposphere and decreases sensible heating in turn the height of planetary boundary layer. This may increase the boundary layer clouds. However, more sensible heating can increase the height of planetary boundary layer and may increase rainfall intensity when the lower troposphere is sufficiently wet. Because both mechanisms are possible, more studies are necessary to understand the impact of soil moisture on precipitation characteristics.

Moreover, regional characteristics of the sensitivity of clouds and precipitation to soil moisture need to be evaluated to understand the impact of soil moisture changes by human activities, because human-induced land-surface changes are generally smaller than the resolution of global climate models (GCMs), i.e., typically 100 km. In addition, it is very difficult to understand the physical process of the impact of soil moisture on regional climate using GCMs because of their coarse horizontal resolution. Moreover, only a high-resolution regional climate simulation can include realistic orographic effects.

The main aim of this study is to understand the sensitivity of rainfall characteristics to soil moisture over wet Asian monsoon region. We focus on the wet Southeast Asian monsoon region where our understanding of the impact of soil moisture on rainfall characteristics remains inadequate. This study tries to understand the impact of soil moisture on rainfall characteristics according to a framework of one-way impacts from soil moisture to the atmosphere on a fine horizontal scale to better understand the physical processes involved. "Methods/experimental" section describes the model setting and data used in this study. The "Result" section presents our results regarding the response of rainfall characteristics to soil moisture. Finally, the mechanisms of the response of rainfall characteristics to soil moisture are discussed in the "Discussion" section, and our conclusions are presented in the "Conclusion" section.

Methods/experimental

Model setting

This study used a high-resolution regional climate model to determine the sensitivity of rainfall characteristics to land surface conditions. First, the spatial scale of the changes in land use due to human activities was generally less than 100 km. In addition, we examined regional differences of sensitivity associated with topography and land-sea distribution. Second, because this study focused on the sensitivity of rainfall characteristics, a basic rainfall system should be simulated. Particularly over the tropical Asian monsoon region, convective rainfall systems with a diurnal precipitation cycle are dominant (e.g., Takahashi et al. 2010).

For the purposes of our study, we used a non-hydrostatic regional climate model, the Advanced Research Weather Research and Forecasting (WRF) modeling system (Skamarock et al. 2008) Version 3.4.1. First, we conducted control (CTL) experiments. We used the atmospheric conditions from the ERA-Interim product (Dee et al. 2011; http://www.ecmwf.int/en/research/climate-reanalysis/era-interim) of the European Centre for Medium-Range Weather Forecasts (ECMWF) and sea surface temperature (SST) from the Optimum Interpolation Sea Surface Temperature, version 2 (OISST2; Reynolds et al. 2002; http://www.esrl.noaa.gov/psd/data/gridded/data.noaa.oisst.v2.html) as the initial and boundary conditions. To investigate the impact of soil moisture in a simple manner, we used monthly mean SST data. The monthly SST data were linearly interpolated with time. The interpolated SST data were prescribed every 6 h. In the CTL experiments, the soil moisture values were predicted. In the main experiments, soil moisture was fixed at a specific value (see "Sensitivity experiments under the prescribed soil moisture" section).

The Southeast Asian monsoon region, which has a relatively wet climate, was selected as the computational target region. The model domain is shown in Fig. 1. A portion of the Indochina Peninsula in Southeast Asia has some of the heaviest precipitation levels of any area in the world (e.g., Takahashi 2016). We conducted 2-month time-slice experiments for the first half of the summer monsoon season (Takahashi and Yasunari 2006) extending from June to July (JJ). We selected the 2 months of June

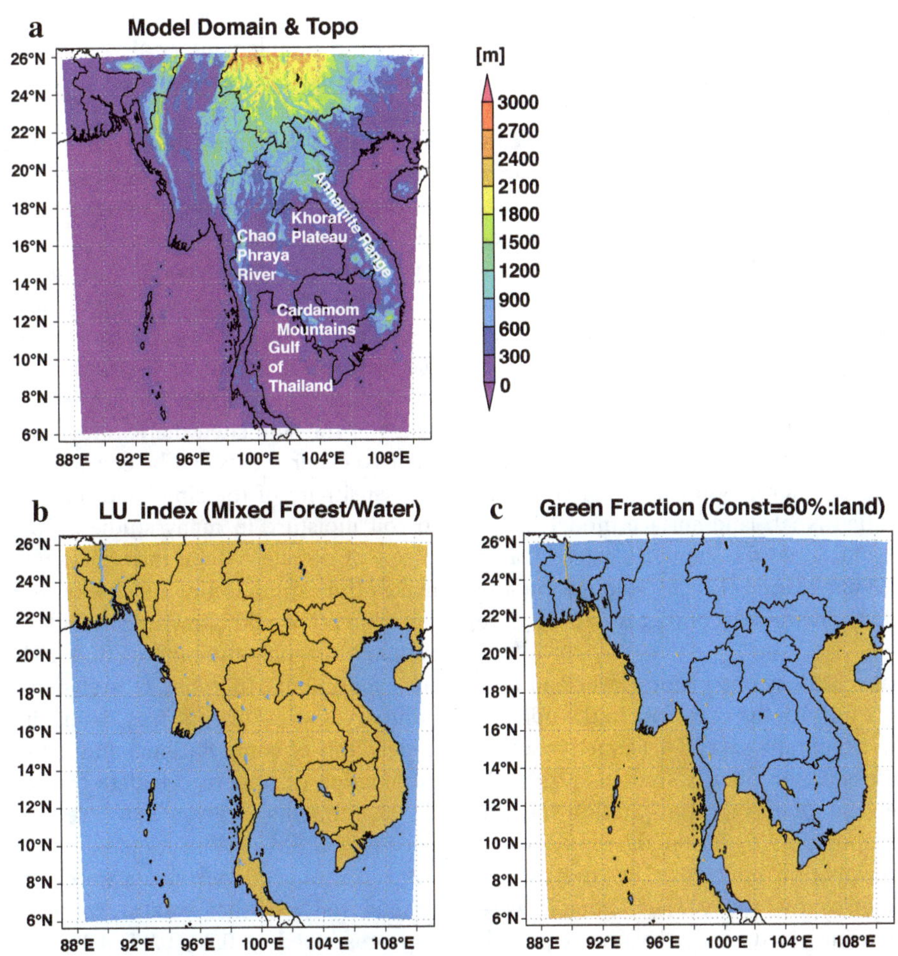

Fig. 1 **a** Study area, geographical names and topography in Domain 1. **b** Landuse index used in this simulation. For simplicity, all land uses were converted to a specific land use category except for water bodies. **c** Spatial distribution of the vegetation fraction, which was used in the land surface model (LSM). Differences in this parameter can affect the Bowen ratio under the same near-surface atmospheric, and soil moisture conditions. For simplicity, all values over land were unified to 0.6, which indicated that 60% of the grid area was covered by vegetation

2010 and July 2010 as the simulation periods to represent a typical summer monsoon year. We also conducted the same experiments during JJ of 2011 and 2012. Even under the atmospheric conditions in the flood year of 2011 (e.g., Takahashi et al. 2015), the basic results were the same as in 2010. In addition, because 2012 was a typical monsoon year, the results were basically the same as in 2010. Thus, we only show the results for 2010 in this paper. These simulation periods extended from 1 June 2010 to 1 August 2010, 1 June 2011 to 1 August 2011, and 1 June 2012 to 1 August 2012. The first 2 days of each simulation were a spin-up period and were not included in the analysis.

The model configuration details are provided here. As mentioned above, to resolve basic rainfall systems, 8-km horizontal spacing was used. Due to the huge computational cost, a relatively low resolution was selected. We confirmed that the 8-km spacing experiments could capture the basic rainfall systems over the tropics, such as the diurnal rainfall cycles over the Southeast Asian monsoon region (Sugimoto and Takahashi 2016). The domain has 27 terrain-following vertical levels. Moreover, because cumulus convective parameterization controls not only convective systems but also the related atmospheric circulation through thermodynamic effects, sensitivity experiments may depend almost exclusively on cumulus convective parameterization. Because test simulations with cumulus convective parameterization showed an unrealistic spatial distribution in rainfall amount over and around the Indochina Peninsula, we did not apply cumulus convective parameterization in the domain, referred to as the previous studies on the warm seasons when convective rainfall is predominant (e.g., Takahashi et al. 2009); Takahashi et al. (2015); Sugimoto and Takahashi (2016); Dado and Takahashi (2017)). The Mellor-Yamada-Janjić planetary boundary layer scheme (Janjić 1994), the single-moment six-class microphysics scheme (Hong and Lim 2006), the unified Noah land surface model (LSM; Chen and Dudhia 2001), and the Rapid Radiative Transfer Model for GCM application (RRTMG) longwave and short-wave radiation schemes (Iacono et al. 2008) were used.

The unified Noah LSM has one canopy layer and four soil layers down to 2 m, with the root zone in the upper 1 m. The surface skin temperature was determined by a surface energy balance equation for the combined ground-vegetation surface. The total evapotranspiration was the sum of the direct evaporation (E_{dir}) from the top soil layer, evaporation of precipitation intercepted by the canopy (E_c), and transpiration by the canopy (E_t). E_{dir} and E_c were calculated by a Penman-based energy balance approach that included a stability-dependent aerodynamic resistance. The canopy resistance was calculated with empirical equations (Jacquemin and Noilhan 1990). The dry wilting points over this study region were approximately from 0.01 to 0.14 $m^3\ m^{-3}$ as volumetric soil moisture content. The saturation volumetric soil moisture content was less than 0.45 $m^3\ m^{-3}$. These values were dependent on the dominant soil type of the grid point.

To analyze the sensitivity of rainfall characteristics to land-surface conditions, we recorded every 30 min only for the accumulated rainfall amount. The other simulated meteorological variables were recorded every 3 h.

Observed dataset

To evaluate the spatial pattern of simulated rainfall over the Southeast Asian monsoon region, this study used the Tropical Rainfall Measuring Mission (TRMM) 3B42 (Huffman et al. 2007) dataset from 1998 to present. The dataset was partly evaluated by rain gauge observations. We used the data in 2010, 2011, and 2012. The spatial resolution of TRMM3B42 is 0.25° in longitude and latitude. Although other rain-gauge-based datasets are available, their quality is dependent on the input dataset. Because few observations are used over mountain ranges, the performance of rain-gauge-based datasets is highly dependent on a complementary algorithm used in the dataset. Thus, we used the TRMM 3B42 dataset in this study.

Sensitivity experiments under the prescribed soil moisture

To isolate the one-way impact of soil moisture on regional climate including rainfall characteristics through changes in the Bowen ratio, we conducted a series of sensitivity experiments. In general, most previous studies have estimated the sensitivity of soil moisture from two experiments under different soil moisture conditions, for example, soil moisture values as fixed at 0.05 $m^3\ m^{-3}$ and 0.45 $m^3\ m^{-3}$ over the target area of interest. The differences in the simulated meteorological variables between the two experiments can be considered to correspond to the sensitivity of the simulated meteorological variables to soil moisture in many studies. However, the chaotic nature of climate systems introduces random noise, which makes the estimation of sensitivity more difficult. More experiments would allow us to divide the responses into systematic signals and random noise. Thus, we conducted six experiments to estimate systematic sensitivity. The number of experiments was determined based on the upper limit of computational and data-storage costs. An explanation of how to calculate the sensitivity from the six experiments is given in " Prescribed soil moisture sensitivity (PSMsens)" section.

A series of six experiments were conducted with prescribed soil moisture values. Volumetric soil moisture was prescribed at 0.20, 0.25, 0.30, 0.35, 0,40, and 0.45 $m^3\ m^{-3}$ as the volumetric soil moisture content over the whole domain and during the whole calculation period, where the experiments were referred to as SM020, SM025, SM030, SM035, SM040, and SM045, respectively.

The four soil layers in the LSM were prescribed at the same value. The aim was to determine the atmospheric response to the soil moisture through the changes in the Bowen ratio. Note that the feedback of rainfall to soil moisture was switched off to isolate the one-way impact of soil moisture on the atmosphere. Over wet climate regimes, the shortage of soil moisture for evapotranspiration is not important, unlike in dry climate regimes.

A soil moisture value of 0.45 $m^3\ m^{-3}$ is typical just after a major rainfall event, and 0.20 $m^3\ m^{-3}$ is typical for the dry season or a rainfall break over the target region of this study. Thus, a soil moisture range of 0.2 $m^3\ m^{-3}$ to 0.45 $m^3\ m^{-3}$ was used, which is typical of a relatively wet Asian monsoon region.

In addition, for simplicity, we unified the land use index as "Mixed forest" in the unified Noah LSM of the WRF except for water bodies (Fig. 1b). Moreover, the fraction of vegetation coverage was fixed at 60% over the whole domain except for water bodies (Fig. 1c), which meant that 60% of the grid area was covered by vegetation. This setting can yield the spatially uniform impact of soil moisture on the atmosphere except for the impact of topography and the land-sea contrast.

Prescribed soil moisture sensitivity (PSMsens)

To determine the mean changes in rainfall characteristics, we calculate the prescribed soil moisture sensitivity (PSMsens) using the following steps. First, monthly mean values of meteorological variables including rainfall characteristics were calculated. The definitions of rainfall characteristics are provided in "The definition of rainfall characteristics" section. For the meteorological variables, excluding the rainfall characteristics, a monthly value was simply averaged from 3-h outputs for a June-July (JJ) mean value in each grid for each experiment. For rainfall characteristics, we calculated a monthly value based on 30-min outputs. At the spatial scale, we transformed the original output (8-km horizontal resolution) into 0.1° in longitude and latitude. The effect of the transformation on analysis results was negligibly small.

Second, we calculated the slopes of meteorological variables, including the rainfall characteristics in each grid as a function of the prescribed soil moisture from the JJ averages of the six sensitivity experiments (SM020, SM025, SM030, SM035, SM040, and SM045). The slopes quantified the increase or decrease in meteorological variables due to changes in the soil moisture. A positive (negative) PSMsens indicates an increase (decrease) in a meteorological variable with an increase in soil moisture. In the figures, we unified the PSMsens values to yield into a change per 0.1-volumetric-soil-moisture increase. These values were based on the prescribed ranges of soil moisture. We defined this PSMsens because the importance of soil moisture shortage for evapotranspiration is small over wet climate regimes.

In the estimation of slopes, to reduce the adverse effects of outliers, we adapted the Theil-Sen method, which is based on direct estimation of the median of pairwise slopes. This is because the estimation of a slope is generally strongly affected by outliers when the least square method is applied. In this step, we also calculated the Pearson correlation coefficient in each grid between meteorological variables and the prescribed soil moisture values. The correlation coefficients were used to evaluate whether an estimated slope was systematic or not. We plotted only the systematic sensitivities, which were statistically significant at the 90% significance level (four degrees of freedom), determined by Student's t test. From here, PSMsens indicates the slope of a meteorological variable as a function of soil moisture.

The definition of rainfall characteristics

To understand the sensitivity of rainfall characteristics, we defined rainfall frequency, rainfall intensity, and rainfall extremes, based on the previous studies of numerical experiments and analysis of observed datasets (e.g., Sugimoto and Takahashi 2017; Takahashi 2016). The four parameters: rainfall amount, rainfall frequency, rainfall intensity, and rainfall extremes, were calculated from the simulated 30-min rainfall amount. The high temporal resolution of the output of numerical experiments in this study allowed a more accurate estimation of rainfall characteristics than 1 h or 3 h outputs. The total rainfall amount was defined as the accumulated rainfall amount over the analysis period of 2 months. The rainfall frequency was calculated from the rainfall occurrence, which was determined as a grid-scale 30 min rainfall amount was larger than 0.05 mm 30 min^{-1} (0.1 mm h^{-1}). The rainfall frequency was defined as the number of grid-scale rainfall occurrences divided by the total number of 30 min simulation slots, expressed as a percentage. We confirmed that our conclusion was unchanged when we used different threshold values. The rainfall intensity was calculated as the total rainfall amount divided by the rainfall frequency. This can also be also considered a conditional rainfall rate. In addition, we defined the 30-min rainfall extreme as the maximum 30-min rainfall amount within the simulation period at each grid, to better understand the changes in rainfall intensity due to the changes in land-surface conditions. To determine the rainfall extremes, we used 2832 30-min samples (59 days ×24 h ×2 (30 min)) for each grid.

Changes in rainfall amount can be derived from changes in rainfall frequency and rainfall intensity. To identify the contributions of the changes in rainfall frequency R_F and rainfall intensity R_I to rainfall amount R_A, we conducted an order estimation. Rainfall amount was defined as follows:

$$R_A = R_F R_I \quad (1)$$

The changes in rainfall amount, rainfall frequency, and rainfall intensity were expressed as δR_A, δR_F, and δR_I, respectively.

$$R_A + \delta R_A = (R_F + \delta R_F)(R_I + \delta R_I) \quad (2)$$

Typical values of δR_F and δR_I range from 0.01 to 0.1. Thus, $\delta R_F \delta R_I$ is negligibly small. Therefore, the changes in rainfall amount may be rewritten as follows:

$$\delta R_A \approx R_F \delta R_I + \delta R_F R_I \quad (3)$$

or

$$\frac{\delta R_A}{R_A} \approx \frac{\delta R_F}{R_F} + \frac{\delta R_I}{R_I} \quad (4)$$

Thus, the changes in rainfall amount can be approximated as a simple linear combination, which can be expressed as a percentage.

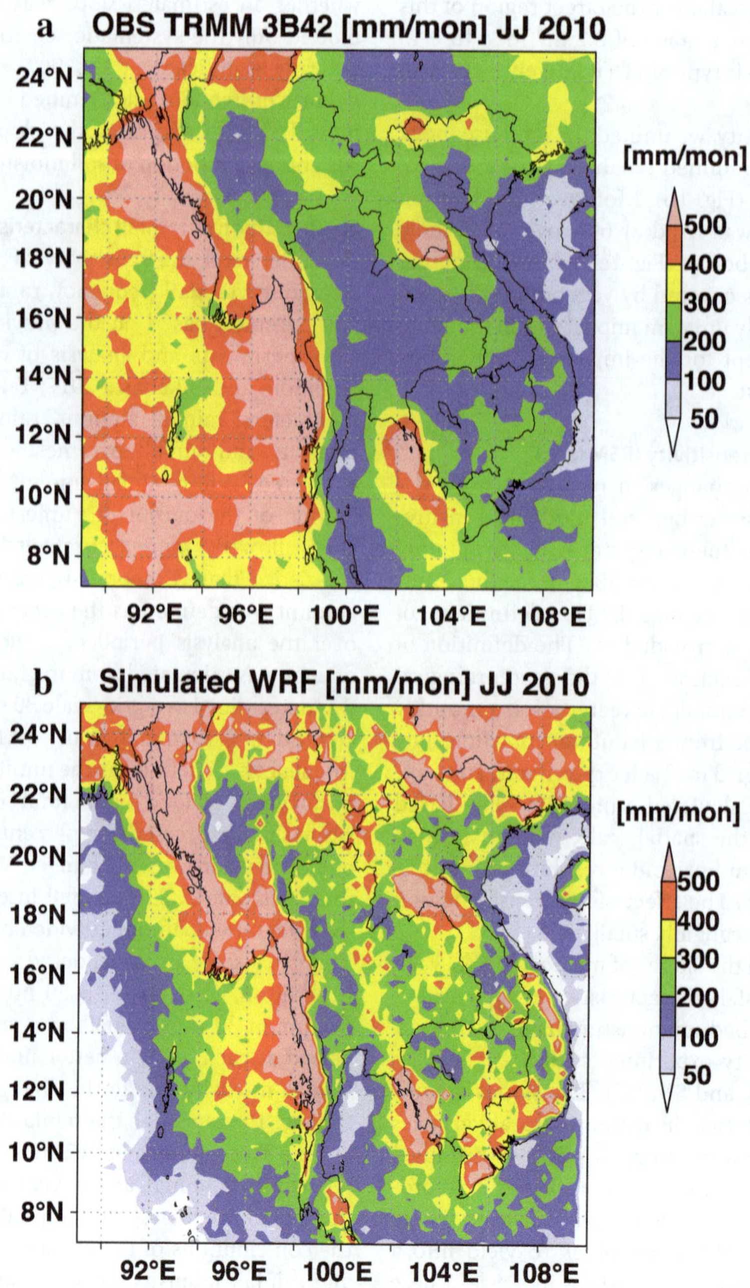

Fig. 2 a Observed (TRMM-3B42) monthly rainfall amount over the study region in June and July of 2010. **b** Same as the upper panel of Fig. 2 but for simulated monthly rainfall. The unit is mm month^{-1}

Result

CTL performance

Because this study focused on rainfall activity over the wet Asian monsoon region, we evaluated the rainfall amount for 2 months for comparison with observed data (Fig. 2).

The spatial pattern of the simulated rainfall amounts in June and July 2010 was very similar to the observed rainfall of June and July in 2010 from the TRMM3B42. Specifically, spatial peaks in rainfall amount were found along the western coast of the Indochina Peninsula and the western slopes of the Cardamom Mountains and Annamite Range in both the simulation and observations. Differences between the simulations and observations were most commonly found over higher altitude regions in mountain ranges and may have been induced by excessive condensation in the microphysics scheme over the higher altitude regions. This can be understood by the overestimation of orographic rainfall in these numerical experiments. A notable overestimation was also found over the mountainous region in the northern part of the domain, although we could not find any plausible cause for this. It should also be noted that the observations can also have biases and errors. There was a clear underestimation of the simulated rainfall over the center of the Bay of Bengal, which may be due to the lateral boundary effects of regional models. In addition, there was a clear overestimation of the estimated rainfall of TRMM-3B42, which was determined from a comparison with the more accurate TRMM-2A25 product climatology (e.g., Fig. 1 of Takahashi 2016). Nevertheless, the simulated rainfall over land regions, which was the focus of this study, was acceptable for the analysis.

The impact of soil moisture on rainfall amount

We investigated the PSMsens of rainfall amount (Fig. 3a). Overall, the PSMsens of rainfall amount shows positive and negative areas, which indicates that rainfall amount increased and decreased with increasing soil moisture. The PSMsens of rainfall amount depended on the location. A positive PSMsens of rainfall amount was observed

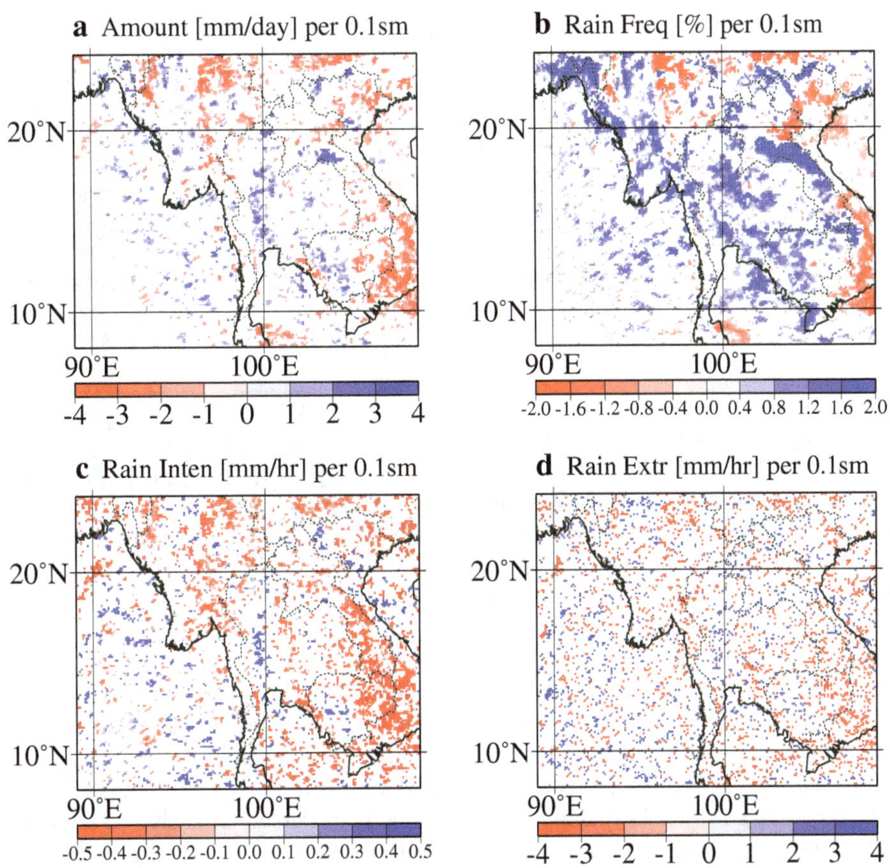

Fig. 3 Spatial map of prescribed soil moisture sensitivity (PSMsens (slope values)) of **a** rainfall amount, **b** rainfall frequency, **c** rainfall intensity, and **d** rainfall extremes as a function of the prescribed soil moisture value. All variables of rainfall characteristics were based on a 30-min time interval and 59-day data. We used the SM020, SM025, SM030, SM035, SM040, and SM045 experiments for the calculation of systematic sensitivity. The blue color indicates an increase in rainfall amount, rainfall frequency, rainfall intensity, and rainfall extremes with soil moistening. The units are mm day^{-1} per 0.1 increase in soil moisture, % per 0.1 increase in soil moisture, mm h^{-1} per 0.1 increase in soil moisture, and mm h^{-1} per 0.1 increase in soil moisture, respectively

over low, flat land, and basin-shaped valley regions, whereas a negative PSMsens was observed over mountainous regions. PSMsens signals were weak along coastal regions.

Concurrent with the changes in rainfall amount, the latent heat flux and sensible heat flux from the land surface increased and decreased over the whole calculation domain, respectively, because of the increase in soil moisture (Fig. 4). The PSMsens of latent heat flux was positive, indicating that evapotranspiration increased as a function of soil moisture, although the PSMsens values of latent heat flux from the surface differed slightly among regions. The changes in rainfall amount will be discussed in the "Discussion" section in terms of the changes of the radiation budget at the surface.

It was clear that the spatial distribution of the PSMsens of rainfall amount was not simple (Fig. 3), although the sensitivity of evapotranspiration to soil moisture was simple, being positive over the whole calculation domain (Fig. 4a). This implies that the relationship between rainfall amount and evapotranspiration is complicated and may interact with the lower atmospheric structure including the planetary boundary layer.

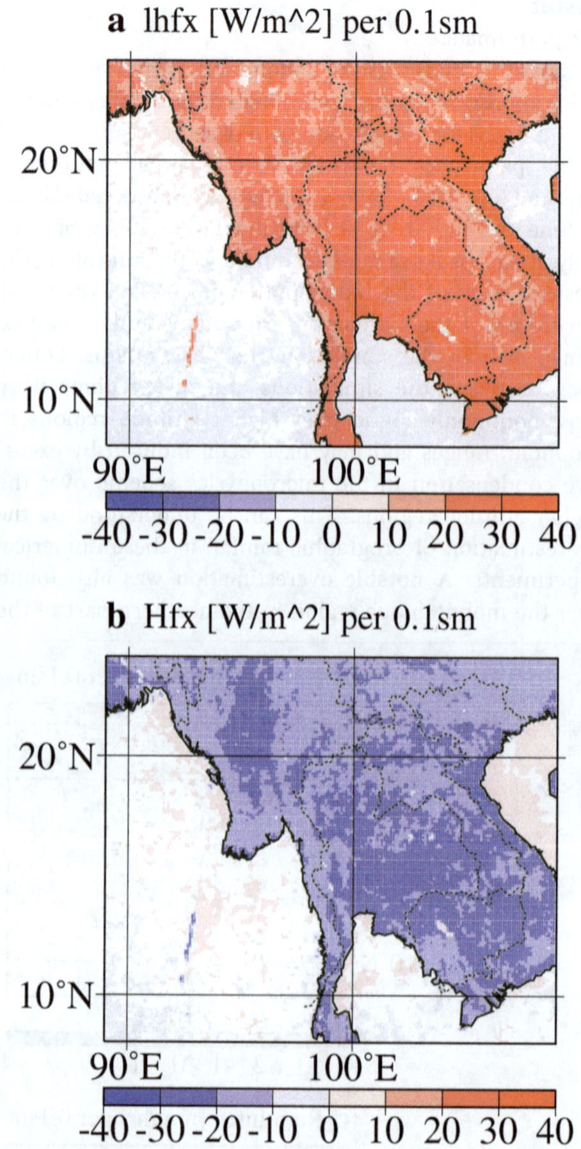

Fig. 4 Same as Fig. 3 but for **a** latent heat and **b** sensible heat fluxes. The units are W m^{-2} per 0.1 increase in soil moisture, respectively

The sensitivity of rainfall characteristics

Changes in soil moisture can lead to subsequent changes in rainfall characteristics, such as an increase in heavy rainfall. Even if the changes in rainfall amount are relatively small, there can be changes in rainfall characteristics. To understand the changes in rainfall characteristics, we calculated the PSMsens of rainfall frequency, rainfall intensity, and rainfall extremes.

Rainfall frequency

In a series of sensitivity experiments over the wet Asian monsoon region, positive PSMsens of rainfall frequency was observed over most parts of the calculation domain (Fig. 3b). This indicates that wetter soil conditions resulted in higher rainfall frequency. The spatial pattern of positive PSMsens of rainfall frequency shows peaks over the coastal regions facing the Bay of Bengal and Gulf of Thailand, and over the inland flat plain and basin-shaped valley regions close to mountain ranges, such as west of the Annamite Range. On the other hand, negative PSMsens of rainfall frequency were also observed over the limited regions, such as mountainous regions and the eastern coast of the Indochina Peninsula (having a dry season in June and July, e.g., Nguyen-Le et al. 2014).

We also produced a histogram of the PSMsens of rainfall frequency (Fig. 5a). The sensitivity values of rainfall frequency were mostly positive, which implies that the coupling of the increase in evapotranspiration and decrease in sensible heat flux resulted in an increase in rainfall frequency in this wet region.

Rainfall intensity

We also investigated the PSMsens of rainfall intensity (Fig. 3c). A negative PSMsens of rainfall intensity was observed over most of the calculation domain, which indicates that rainfall intensity weakened due to increased soil moisture in this wet region. Some positive PSMsens of rainfall intensity were observed over basin-shaped valleys, such as Chao Phraya River Basin and the northern Vietnam regions. No clear signals were observed along the coastal regions, which can explain the weak signals in rainfall amount along the coastal region, although the signals in rainfall frequency were clear.

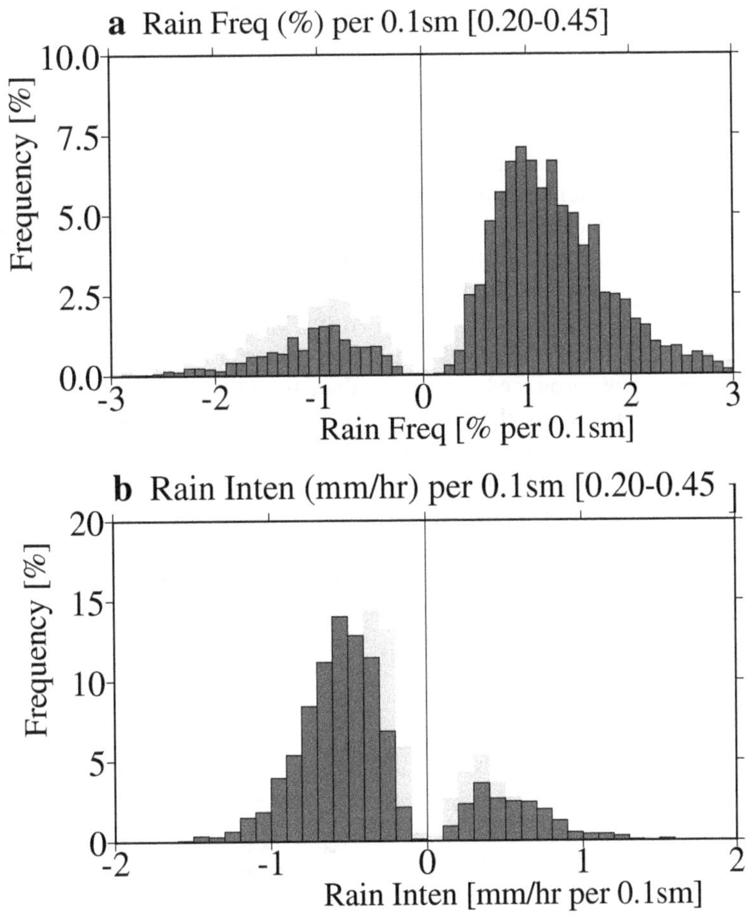

Fig. 5 Histogram of PSMsens (slope values) of **a** rainfall frequency and **b** rainfall intensity over the land regions. The dark shading represents the histogram for lowland regions (elevation is below 300 m). The light shading denotes the histogram for all land regions. The horizontal axes in **a** and **b** are mm h^{-1} per 0.1 increase in soil moisture and % per 0.1 increase in soil moisture, respectively. The vertical axes are the frequency of each bin, expressed as a percentage

The histogram of PSMsens of rainfall intensity clearly showed that rainfall intensity mostly weakened due to increased soil moisture, with the soil moisture ranging from 0.2 m^3 m^{-3} to 0.45 m^3 m^{-3} (Fig. 5b). This indicated that the coupling of increased evapotranspiration and decreased sensible heat flux resulted in a weakening in rainfall intensity in this region, although the sign of the sensitivity of rainfall intensity was dependent on the location. This response of rainfall intensity to the coupling of increased evapotranspiration and decreased sensible heat flux appears opposite over the mountain ranges compared to that over lowland regions. The mechanism underlying the changes in rainfall frequency and rainfall intensity is discussed in the "Discussion" section.

Rainfall extremes

We also investigated the PSMsens of rainfall extremes (Fig. 3d). A negative PSMsens of rainfall extremes was observed over most of the calculation domain, which indicated that rainfall extremes weakened due to the increased soil moisture in this wet region. Note that the spatial pattern of the PSMsens of rainfall extremes was very similar to that of rainfall intensity (Fig. 3c). When we compared the PSMsens signals of rainfall extremes with those of rainfall intensity, the signals of PSMsens of rainfall extremes were found to be relatively scattered. It was noteworthy that the PSMsens of rainfall extremes was not similar to that of rainfall amount, but was similar to that of rainfall intensity, which implies that the sensitivity of rainfall extremes to land-surface conditions cannot be understood only from rainfall amount. Thus, it is very important to examine not only rainfall amount but also rainfall characteristics over many regions.

Contribution of changes in rainfall characteristics to rainfall amount

To understand the contribution of rainfall frequency and rainfall intensity to rainfall amount, we first expressed the changes in rainfall amount as a percentage. Changes in rainfall amount were found over the flat plains and valleys close to the mountain ranges, and over the mountain ranges. Increases in rainfall amount were calculated over the flat plain and valley regions. Decreases in rainfall amount were observed over the mountain ranges. As an exception, a decrease in rainfall amount was observed along the eastern coast of the Indochina Peninsula, where a climatological dry season occurs in June and July. We examined the contribution of these changes in rainfall frequency and rainfall intensity to rainfall amount.

To quantify the contribution of rainfall frequency and rainfall intensity to rainfall amount, we produced maps of PSMsens, expressed as a percentage relative to the average of a set of six runs (Fig. 6). As shown in the "The definition of rainfall characteristics" section, the changes in rainfall amount can be approximated by a linear combination of changes in rainfall frequency and rainfall intensity, which were expressed as a percentage.

Rainfall frequency tended to contribute to the increase in rainfall amount as a whole (Fig. 6b). Clear signals were observed over the flat plain and valley regions. From a quantitative perspective, the increases in rainfall frequency were of the order of several percentage points, which indicates that the changes in rainfall frequency contributed quantitatively to modifications in rainfall amount over large areas.

The changes in rainfall intensity were mostly negative (Fig. 6c). Weakening signals were mainly observed over the mountain ranges. Strengthening signals were found over the flat plains and valleys, for example Chao Phraya River Basin. From a quantitative perspective, the absolute values of the signals exceeded 15%, which was larger than the changes in rainfall frequency. When the signs of rainfall frequency and rainfall intensity were different, the contribution of rainfall intensity to rainfall amount quantitatively exceeded that of rainfall frequency.

Discussion

Surface energy budget and rainfall amount

To understand the changes in rainfall amount, we investigated the PSMsens of the energy budget at the surface.

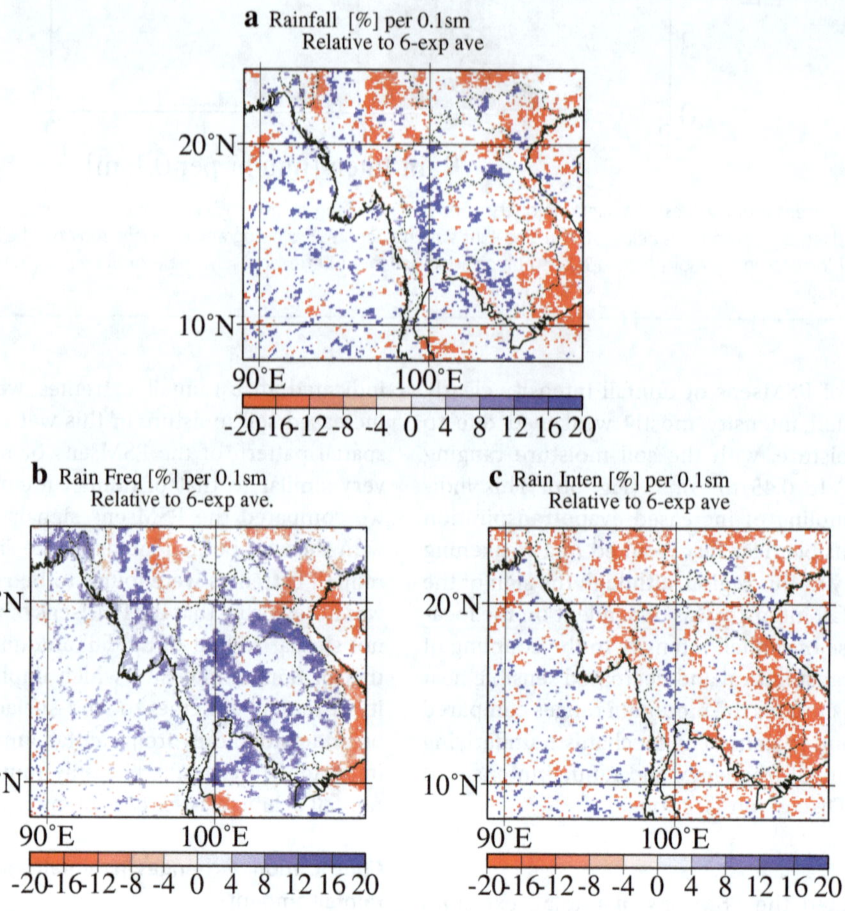

Fig. 6 Same as Fig. 3 but for **a** rainfall amount, **b** rainfall frequency, and **c** rainfall intensity, expressed as a percentage. A detailed explanation is provided in section "The definition of rainfall characteristics"

We investigated the changes in surface energy budget derived from the law of energy conservation. The sum of the absorbed shortwave and downward longwave radiation is defined as the radiative input to the surface. The input is redistributed to upward longwave radiation (surface longwave emission), sensible and latent heat fluxes, and ground heat flux. We calculated net radiation *Rn*, defined as the sum of the sensible and latent heat fluxes, and the ground heat flux (e.g., Takahashi et al. 2015).

The PSMsens of net radiation at the surface was positive over almost the entire land region (Fig. 7a). The increase in net radiation at the surface can be explained by the decrease in upward longwave radiation due to decreased land-surface skin temperature (Fig. 7b). Under wet conditions, the land-surface skin temperature is reduced due to evaporative cooling at the surface, which is also associated with changes of the Bowen ratio. At the same time, downward longwave radiation can decrease due to a decrease in surface air temperature, whereas it can increase due to an increase in water vapor in the planetary boundary layer (Fig. 7c). The decrease in upward longwave radiation at the surface due to the decrease in surface skin temperature was much larger than the increase in downward longwave radiation at the surface, which mostly increased net radiation at the surface.

When net radiation at the surface increases, the long-term and global mean rainfall amount should increase in terms of the global energy budget. However, at the regional scale, the energy balance at the surface cannot simply explain the changes in rainfall amount, which may be modified by the land-sea distribution and topography through the changes in atmospheric circulation, water vapor transport, development of the planetary boundary layer, and cloud-precipitation activity.

We focused on the regional characteristics of net radiation at the surface. Changes in the downward shortwave radiation associated with cloud changes had a large impact (Fig. 7d). The negative PSMsens of downward shortwave radiation at the surface can indicate an increase in low-level clouds, which have negative radiative forcing. In addition, the PSMsens of downward shortwave radiation at the surface also had clear regional characteristics.

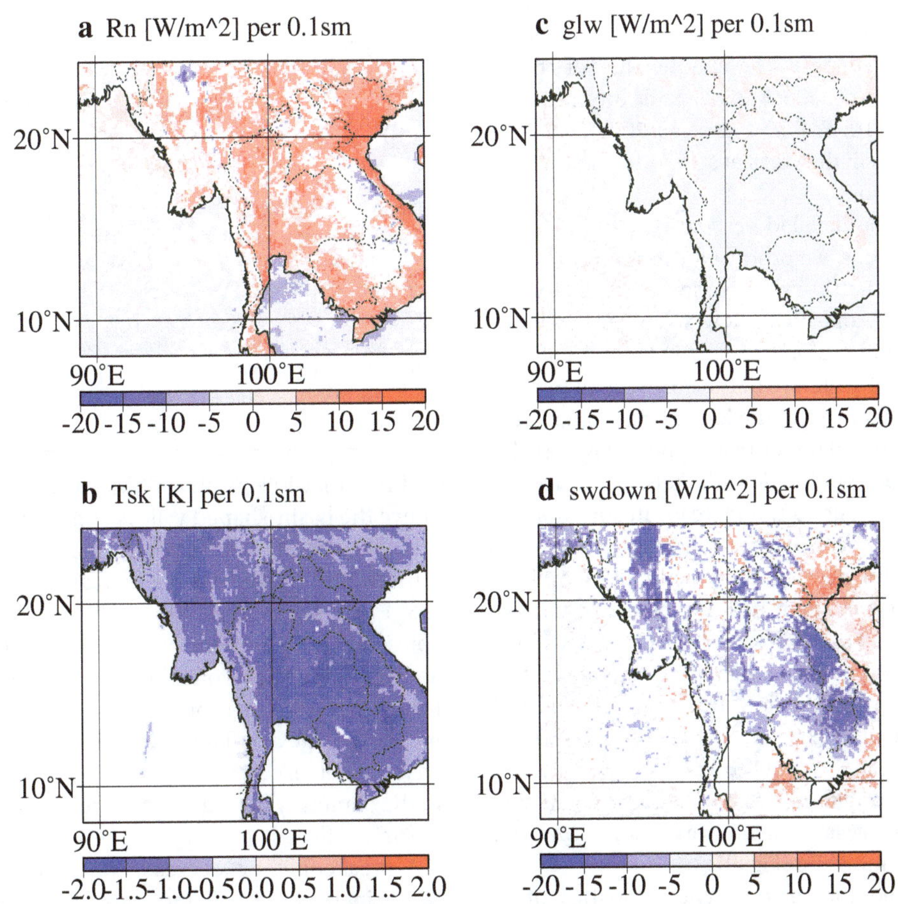

Fig. 7 Same as Fig. 3 but for **a** net radiation at the surface, **b** surface skin temperature, **c** downward longwave radiation at the surface, and **d** downward shortwave radiation at the surface. The units are W m^{-2} per 0.1 increase in soil moisture, K per 0.1 increase in soil moisture, and W m^{-2} per 0.1 increase in soil moisture, and W m^{-2} per 0.1 increase in soil moisture, respectively

Negative spatial peaks in the PSMsens of downward shortwave radiation at the surface were calculated over the inland plains near mountain ranges, such as the eastern Korate Plateau, northern Myanmer, and a plain in Bangladesh. The following subsection discusses the changes in low-level clouds.

Mechanism of changes in rainfall characteristics due to an increase in soil moisture

This subsection discusses the coupling of the positive PSMsens of rainfall frequency and the negative PSMsens of rainfall intensity. The rainfall characteristics could be associated with the atmospheric instability (e.g., vertical structure of water vapor) and trigger of convections. An increase in sensible heat flux can induce more intense development of the planetary boundary layer, which generates convective clouds and precipitation (Ek and Holtslag 2004; Taylor and Ellis 2006; Hohenegger et al. 2009; Sugimoto and Ueno 2010; Taylor et al. 2012; Froidevaux et al. 2014). The predominant triggers of the convective systems in this region were the cold air pools of the previous convective systems (Kataoka and Satomura 2005; Takahashi et al. 2010), which were difficult to analyze statistically. On the other hand, sufficient water vapor in the lower troposphere supports the maintenance of convective systems (Satomura 2000; Kataoka and Satomura 2005; Takahashi et al. 2010; Froidevaux et al. 2014). Thus, we focused on changes in the vertical structures of hydrometeors and water vapor.

To understand the detailed vertical structure of cloud and rainfall conditions, we produced a longitude–vertical (in pressure coordinates) cross section of total hydrometeors (both liquid and solid clouds and precipitation) and specific humidity along the 18°N latitudinal band (Fig. 8). The vertical structure of the PSMsens of hydrometeors showed an increase in the lower troposphere and a decrease in the mid and upper troposphere (Fig. 8a). This indicated an increase in low-level clouds and a decrease in deep convections due to the increase in soil moisture. In the lower troposphere, an increase in water vapor was also distinct (Fig. 8b). Because an increase in soil moisture induced an increase in latent heat flux and a decrease in sensible heat flux, changes in vertical structure of hydrometeors and water vapor are explained by the decrease in planetary boundary layer height due to the decrease in sensible heat flux. In other words, the increases in low-level clouds and water vapor are due to weakening of vertical mixing. In addition, we confirmed that sensitivity in planetary boundary layer height was clearly negative to an increase in soil moisture (Fig. 9).

We also assessed the regional characteristics of the PSMsens in hydrometeors and water vapor. Figure 8 shows that the negative PSMsens of hydrometeors was very distinct above mountain ranges (around 98° E, 101° E,

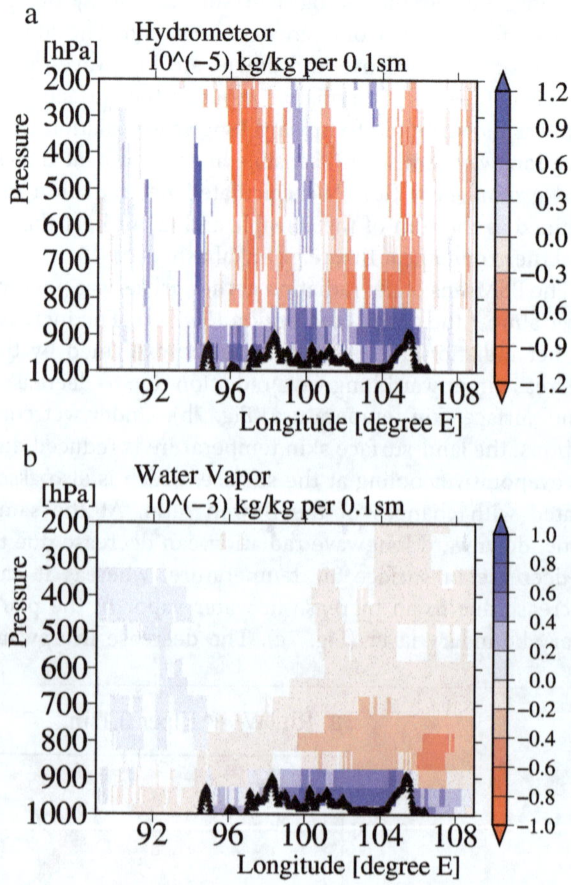

Fig. 8 Longitude-vertical (in pressure coordinates) cross section of PSMsens (slope values) in **a** hydrometeors and **b** water vapor along the 18° N latitudinal band. A hydrometeor was defined as the total amount of cloud and precipitations in both the liquid and solid phases. The black triangles indicate the topography along the 18° N latitudinal band. Both horizontal axes in **a** and **b** are longitude in °E. The unit of hydrometeors (water vapor) is 10^{-5} kg kg^{-1} per 0.1 increase in soil moisture (10^{-3} kg kg^{-1} per 0.1 increase in soil moisture)

and 105° E), while positive PSMsens signals were observed above the basin-shaped valleys and the flat plains (around 100° E, 104° E). The major increase in low-level clouds (105° E, 18° N) and minor increases in low-level clouds (96° E, 18° N) and (100° E, 18° N) shown in Fig. 8a were consistent with the decrease in downward shortwave radiation at the surface along the 18°N line (Fig. 7). This spatial feature clearly corresponded to the changes in rainfall intensity (Fig. 3c). Thus, the general changes in rainfall characteristics can primarily be explained by the changes in the planetary boundary layer height due to the changes in surface fluxes.

A weakening or decrease of deep convections due to an increase in soil moisture may also contribute to the vertical structure of hydrometeors and water vapor. Weakening of vertical and horizontal mixing by deep convections may interact with the vertical structure of

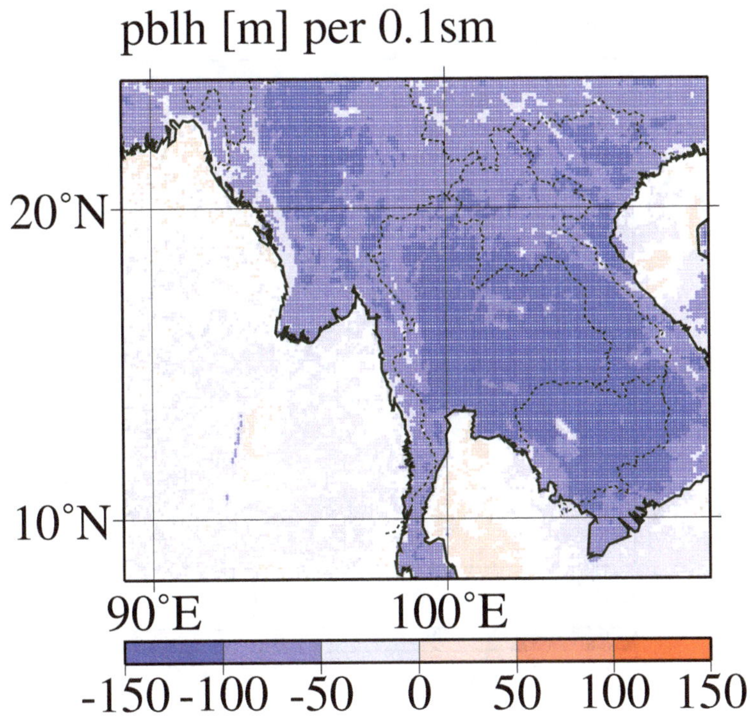

Fig. 9 Same as Fig. 3 but for planetary boundary layer height. The units are m per 0.1 increase in soil moisture

hydrometeors and water vapor. There may be a positive feedback mechanism between active deep convection and the increase in planetary boundary layer height.

We also examined changes in low-level circulations. Figure 10 shows the weakening of surface winds from ocean to land due to an increase in soil moisture (Fig. 10), which is also explained by the weakening of sensible heating over the land regions. We considered two major factors. One was the weakening of the land-ocean thermal contrast, which was obvious from Fig. 11. The other was

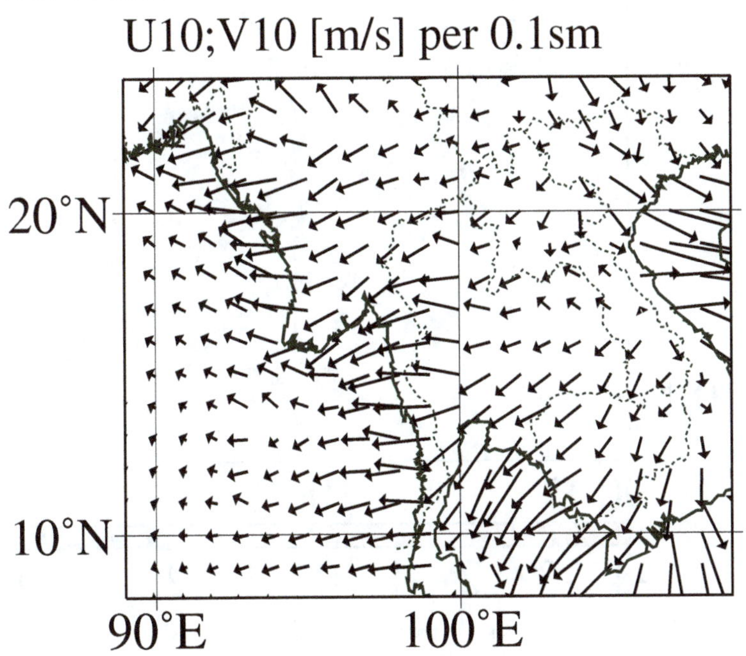

Fig. 10 Same as Fig. 3 but for winds at 10 m above the surface. The unit is m s^{-1} per 0.1 increase in soil moisture

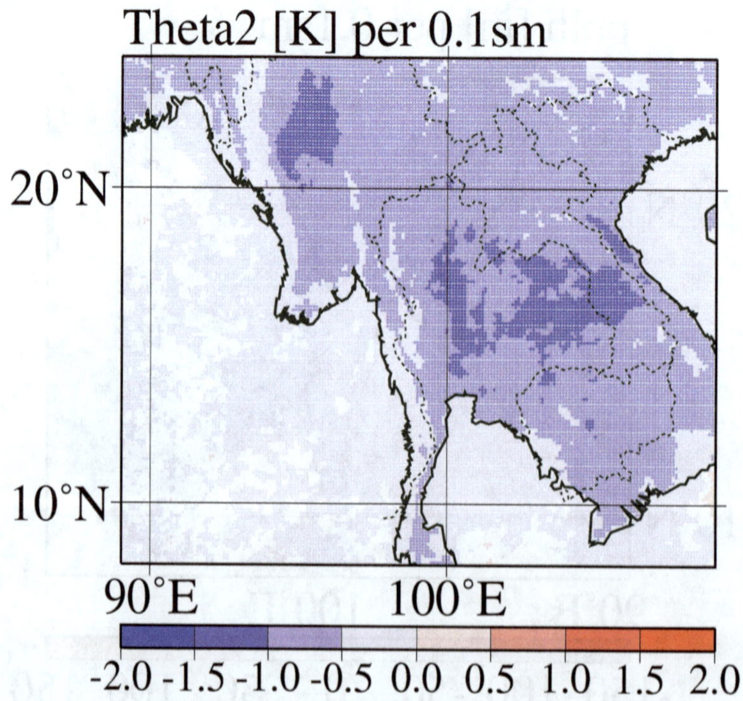

Fig. 11 Same as Fig. 3 but for potential temperature at 2 m above the surface. The unit is K per 0.1 increase in soil moisture

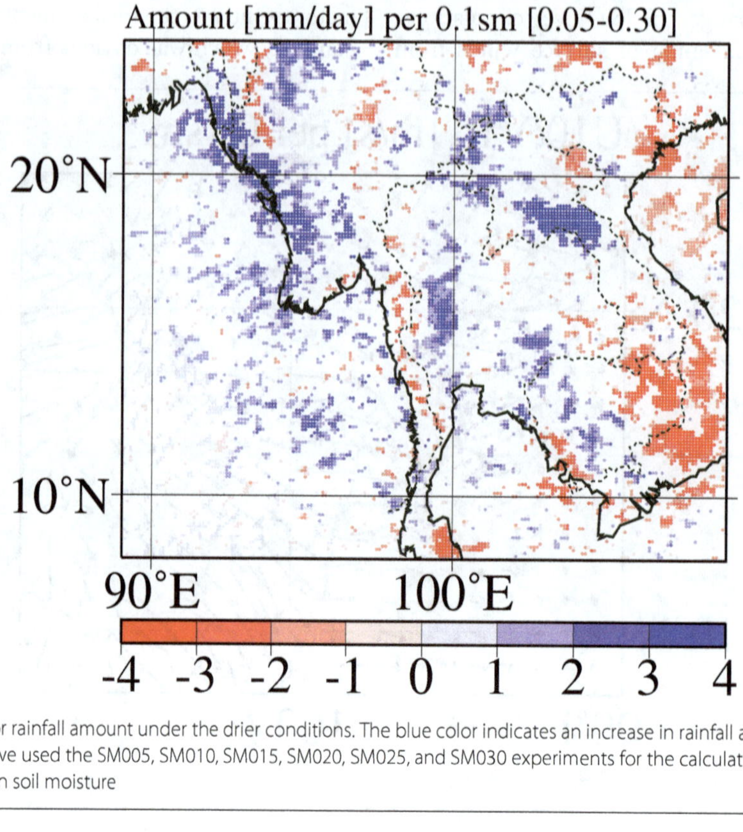

Fig. 12 Same as Fig. 3 but for rainfall amount under the drier conditions. The blue color indicates an increase in rainfall amount with soil moistening. Under the drier conditions, we used the SM005, SM010, SM015, SM020, SM025, and SM030 experiments for the calculation of PSMsens. The unit is mm day^{-1} per 0.1 increase in soil moisture

the decrease in the downward transport of momentum from the free troposphere into the boundary layer due to the decrease in the planetary boundary layer height (e.g., Stull 2006), which was also identified in previous studies (e.g., Takahashi et al. 2013; Dado and Takahashi 2017).

The relationship between the changes in circulation and rainfall characteristics was associated with the weakening of local circulations. However, local circulations could also be affected by rainfall and cloud activity. Thus, they strongly interact with each other, and the cause-effect relationship is not clear.

In addition, because of the weakening effect of vertical water vapor transport due to soil moistening, precipitable water over downstream regions of the Indochina Peninsula decreased. Thus, the downstream effect of soil moisture may not be negligible. However, because this may be associated with the size of the domain of the regional climate model experiments, an additional investigation of this effect is required.

The sensitivity under drier conditions

In this subsection, we discuss the PSMsens under the drier conditions in the range of 0.05 m^3 m^{-3} to 0.30 m^3 m^{-3}, because the sensitivity can differ between dry and wet climate conditions, and between dry and wet seasons (Tuttle and Salvucci 2016; Sugimoto and Takahashi 2017). Three additional experiments were conducted. The experimental settings were basically the same, but prescribed soil moisture values were fixed at 0.05, 0.10, and 0.15 m^3 m^{-3} over the whole land region during the calculation period. To calculate the PSMsens in the drier case, the six experiments with soil moisture values fixed from 0.05 to 0.30 m^3 m^{-3} by 0.05 m^3 m^{-3}, were used. The wetter three experiments (SM020, SM025, and SM030) were overlapped to calculate the PSMsens under both wet and dry conditions.

The spatial pattern of the PSMsens of rainfall amount was basically the same as in the wet case (Fig. 12). In particular, the positive PSMsens of rainfall amount was clearly enhanced under the drier conditions, whereas

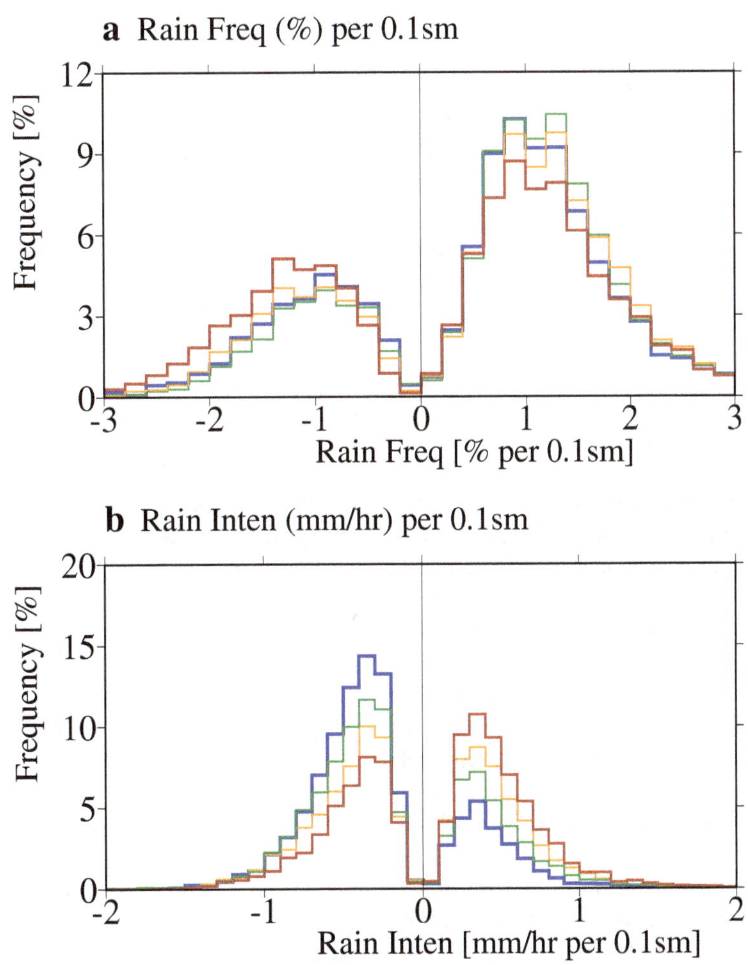

Fig. 13 Same as Fig. 5 but for PSMsens in **a** rainfall frequency and **b** rainfall intensity from different experiments over the land regions. The blue color indicates the same as in Fig. 5 (SM020 to SM045), green is SM015 to SM040, orange is SM010 to SM035, and red is SM005 to SM030. The horizontal axes in **a** and **b** are % per 0.1 increase in soil moisture and mm h^{-1} per 0.1 increase in soil moisture. The vertical axes are the frequency of each bin, expressed as a percentage

the negative PSMsens was similar to that under the wet conditions (Fig. 12). Changes in the PSMsens of rainfall frequency and rainfall intensity are summarized in Fig. 13. All slopes were calculated based on the six experiments, although the experiments overlapped. For example, the red histogram in the figure was calculated from the experiments SM005, SM010, SM015, SM020, SM025, and SM030 (see the caption of Fig. 13).

The PSMsens of rainfall frequency had more positive values when calculated under the wetter conditions, whereas somewhat more negative values were observed under drier conditions (Fig. 13a). However, the positive PSMsens of rainfall frequency was dominant under all conditions. This result suggests that the PSMsens of rainfall frequency is mostly positive, which indicates that an increase in soil moisture contributes to an increase in rainfall frequency over this region.

For rainfall intensity, the PSMsens had more negative values under wetter conditions (Fig. 13a), while more positive values were calculated under the drier conditions. In addition, the sign of PSMsens of rainfall intensity changed according to differences in land surface wetness conditions. This suggests that rainfall intensity is suppressed under the wetter conditions due to the increase in soil moisture, whereas it may be enhanced under the drier conditions even with an increase in soil moisture. This difference may be explained by the richness of water vapor in the lower troposphere. To understand the threshold, regional characteristics, and related mechanisms, further experiments are required, including experiments over the different regions.

Conclusion

To investigate the impact of land-surface conditions (soil moisture) on rainfall characteristics over the wet Asian monsoon region, we conducted numerical experiments using a high-resolution regional climate model. Specifically, sensitivity experiments were conducted that prescribed soil moisture values of 0.2, 0.25, 0.3, 0.35, 0.4, and 0.45 $m^3\ m^{-3}$ over the whole calculation domain. We focused on not only rainfall amount, but also rainfall frequency, rainfall intensity, and rainfall extremes based on 30-min intervals. We discussed how the increase in soil moisture changes rainfall characteristics. A slope of a meteorological variable as a function of the prescribed soil moisture values was defined as the sensitivity, which is referred to as PSMsens.

The results showed that an increase in soil moisture resulted in an increase or decrease in rainfall amount depending on the location. It was noteworthy that rainfall amount did not simply increase, although evapotranspiration simply increased over the whole domain. This implies that the relationship between rainfall amount and evapotranspiration is not simple.

Rainfall frequency mostly increased due to an increase in soil moisture, which was consistent with the increased boundary layer clouds. The increase in low-level clouds was associated with increased water vapor in the lower troposphere, which was explained by an increase in latent heat flux and decrease in sensible heat flux. The decrease in sensible heat flux resulted in the weakening of vertical mixing in the lower troposphere, which also contributed to the increase in water vapor in the lower troposphere. In addition, the increase in low-level clouds was consistent with a decrease in shortwave radiation at the surface.

On the other hand, rainfall intensity and rainfall extremes decreased overall with an increase in soil moisture. The weakening of rainfall intensity and rainfall extremes was consistent with a reduction of the development of deep convections. The decrease in rainfall intensity and rainfall extremes can be due to the decrease in the planetary boundary layer height, which resulted from the decrease in sensible heat flux. The regional characteristics of the sensitivity in rainfall intensity and rainfall extremes indicated a decrease over all land regions, including mountain ranges except for 100-km-scale basin-shaped valleys, for example Chao Phraya River Basin and Khorat Plateau. The regional characteristics of rainfall intensity and rainfall extremes can be associated with a weakening of development of the planetary boundary layer.

In addition, the weakening of large-scale low-level circulations at the calculation domain scale contributed to the decrease of horizontal transport in water vapor from ocean to land. This may contribute by compensating for the increase in evapotranspiration. The weakening of the large-scale low-level circulations was probably associated with the weakening of the thermal contrast between land and ocean. However, the weakening of low-level circulations could also be associated with the vertical transport of momentum due to changes in the planetary boundary layer.

Abbreviations

CTL: Control experiment; GCM: Global climate model; IPCC: Intergovernmental panel on climate change; JJ: June-July; PSMsens: Prescribed soil moisture sensitivity; RCM: Regional climate model; SM0XX: Soil moisture fixed sensitivity experiment at 0.XX level; SST: Sea surface temperature; TRMM: Tropical rainfall measuring mission

Acknowledgements

We thank Dr. Shiori Sugimoto for discussing the interpretation of results. This study is partly supported by the 8th Japan Aerospace Exploration Agency (JAXA) Precipitation Measuring Mission (PMM) Project No. 309. We are also grateful for the constructive comments from the editor and reviewers in further improving this manuscript.

Funding

This work was partly supported by the "Green Network of Excellence (GRENE)" program by the Ministry of Education, Culture, Sports, Science, and Technology (MEXT), Japan. This study is partly supported by the 8th Japan Aerospace Exploration Agency (JAXA) Precipitation Measuring Mission (PMM) Project No. 309.

Authors' contributions
HGT proposed the topic and conceived and designed the study. HGT carried out the experimental study. HGT and JP analyzed the data and helped in the interpretation. HGT and JP wrote the manuscript. Both authors read and approved the final manuscript.

Competing interests
The authors declare that they have no competing interests.

Author details
[1]Department of Geography, Tokyo Metropolitan University, 1-1 Minamioosawa, Hachioji, Tokyo 192-0397, Japan. [2]Japan Agency for Marine-Earth Science and Technology, Yokohama 236-0001, Japan. [3]Laboratoire de Météorologie Dynamique (LMD), Institut Pierre Simon Laplace (IPSL), Ecole Polytechnique, Centre National de la Recherche Scientifique (CNRS), Paris, France.

References

Chen F, Dudhia J (2001) Coupling an advanced land surface-hydrology model with the Penn State-NCAR MM5 modeling system. Part I: Model implementation and sensitivity. Mon Wea Rev 129(4):569–585

Dado JMB, Takahashi HG (2017) Potential impact of sea surface temperature on rainfall over the western Philippines. Progress Earth Planetary Sci 4(1):23

Dee D, Uppala S, Simmons A, Berrisford P, Poli P, Kobayashi S, Andrae U, Balmaseda M, Balsamo G, Bauer P, et al (2011) The ERA-Interim reanalysis: Configuration and performance of the data assimilation system. Q J R Meteorol Soc 137(656):553–597

Ek M, Holtslag A (2004) Influence of soil moisture on boundary layer cloud development. J Hydrometeorol 5(1):86–99

Froidevaux P, Schlemmer L, Schmidli J, Langhans W, Schär C (2014) Influence of the background wind on the local soil moisture–precipitation feedback. J Atmos Sci 71(2):782–799

Henderson-Sellers A, Gornitz V (1984) Possible climatic impacts of land cover transformations, with particular emphasis on tropical deforestation. Clim Chang 6(3):231–257

Hohenegger C, Brockhaus P, Bretherton CS, Schär C (2009) The soil moisture–precipitation feedback in simulations with explicit and parameterized convection. J Climate 22(19):5003–5020

Hong SY, Lim JOJ (2006) The WRF single-moment 6-class microphysics scheme (WSM6). J Korean Meteor Soc 42(2):129–151

Huffman GJ, Bolvin DT, Nelkin EJ, Wolff DB, Adler RF, Gu G, Hong Y, Bowman KP, Stocker EF (2007) The TRMM multisatellite precipitation analysis (TMPA): Quasi-global, multiyear, combined-sensor precipitation estimates at fine scales. J Hydrometeorol 8(1):38–55

Iacono MJ, Delamere JS, Mlawer EJ, Shephard MW, Clough SA, Collins WD (2008) Radiative forcing by long-lived greenhouse gases: Calculations with the AER radiative transfer models. J Geophys Res 113(D13). https://doi.org/10.1029/2008JD009944

IPCC (2013) Climate Change 2013: The Physical Science Basis: Working Group I Contribution to the Fifth Assessment Report of the Intergovernmental Panel on Climate Change. Cambridge University Press, Cambridge, United Kingdom and New York

Jacquemin B, Noilhan J (1990) Sensitivity study and validation of a land surface parameterization using the HAPEX-MOBILHY data set. Boundary-Layer Meteorol 52(1):93–134

Janjic ZI (1994) The step-mountain eta coordinate model: Further developments of the convection, viscous sublayer, and turbulence closure schemes. Mon Wea Rev 122(5):927–945

Kataoka A, Satomura T (2005) Numerical simulation on the diurnal variation of precipitation over northeastern Bangladesh: A case study of an active period 14-21 June 1995. SOLA 1:205–208

Koster RD, Dirmeyer PA, Guo Z, Bonan G, Chan E, Cox P, Gordon C, Kanae S, Kowalczyk E, Lawrence D, et al (2004) Regions of strong coupling between soil moisture and precipitation. Science 305(5687):1138–1140

Nguyen-Le D, Matsumoto J, Ngo-Duc T (2014) Climatological onset date of summer monsoon in Vietnam. Int J Climatol 34(11):3237–3250

Ono M, Takahashi HG (2016) Seasonal transition of precipitation characteristics associated with land surface conditions in and around Bangladesh. J Geophys Res Atmos 121(19):11190–11200

Pielke RA (2001) Influence of the spatial distribution of vegetation and soils on the prediction of cumulus convective rainfall. Rev Geophys 39(2):151–177

Polcher J, Laval K (1994) The impact of African and Amazonian deforestation on tropical climate. J Hydrol 155(3-4):389–405

Reynolds RW, Rayner NA, Smith TM, Stokes DC, Wang W (2002) An improved in situ and satellite SST analysis for climate. J Climate 15(13):1609–1625

Rochetin N, Couvreux F, Guichard F (2017) Morphology of breeze circulations induced by surface flux heterogeneities and their impact on convection initiation. Q J R Meteorol Soc 143(702):463–478

Satomura T (2000) Diurnal variation of percipitation over the Indo-China Peninsula: Two-dimensional numerical simulation. J Meteor Soc Japan 78(4):461–475

Shukla J, Nobre C, Sellers P, et al (1990) Amazon deforestation and climate change. Science 247(4948):1322–1325

Skamarock WC, Coauthors (2008) A Description of the Advanced Research WRF Version 3. NCAR Technical Note. NCAR/TN-475+STR, https://doi.org/10.5065/D68S4MVH. https://opensky.ucar.edu/islandora/object/technotes:500

Stull R (2006) The atmospheric boundary layer. In: Wallace JM, Hobbs PV (eds). Atmospheric Science: an Introductory Survey, 2nd edn. Elsevier Inc., USA

Sugimoto, S, Takahashi HG (2016) Effect of spatial resolution and cumulus parameterization on simulated precipitation over South Asia. SOLA 12A:7–12

Sugimoto S, Takahashi HG (2017) Seasonal differences in precipitation sensitivity to soil moisture in Bangladesh and surrounding regions. J Climate 30(3):921–938. https://doi.org/10.1175/JCLI-D-15-0800.1

Sugimoto S, Ueno K (2010) Formation of mesoscale convective systems over the eastern Tibetan Plateau affected by plateau-scale heating contrasts. J Geophys Res Atmos 115:16105

Takahashi HG (2016) Seasonal and diurnal variations in rainfall characteristics over the tropical Asian monsoon region using TRMM-PR data. SOLA 12A:22–27

Takahashi HG, Adachi SA, Sato T, Hara M, Ma X, Kimura F (2015) An oceanic impact of the Kuroshio on surface air temperature on the pacific coast of Japan in summer: Regional H_2O greenhouse gas effect. J Climate 28(18):7128–7144. https://doi.org/10.1175/JCLI-D-14-00763.1

Takahashi HG, Fujinami H, Yasunari T, Matsumoto J (2010) Diurnal rainfall pattern observed by Tropical Rainfall Measuring Mission Precipitati on Radar (TRMM-PR) around the Indochina peninsula. J Geophys Res Atmos 115:07109

Takahashi HG, Fujinami H, Yasunari T, Matsumoto J, Baimoung S (2015) Role of tropical cyclones along the monsoon trough in the 2011 Thai flood and interannual variability. J Clim 28(4):1465–1476

Takahashi HG, Ishizaki NN, Kawase H, Hara M, Yoshikane T, Ma X, Kimura F (2013) Potential impact of sea surface temperature on winter precipitation over the japan sea side of japan: a regional climate modeling study. J Meteor Soc Japan 91(4):471–488. https://doi.org/10.2151/jmsj.2013-404

Takahashi HG, Yasunari T (2006) A climatological monsoon break in rainfall over Indochina—A singularity in the seasonal march of the Asian summer monsoon. J Clim 19(8):1545–1556

Takahashi HG, Yoshikane T, Hara M, Takata K, Yasunari T (2010) High-resolution modelling of the potential impact of land surface conditions on regional climate over Indochina associated with the diurnal precipitation cycle. Int J Climatol 30(13):2004–2020

Takahashi HG, Yoshikane T, Hara M, Yasunari T (2009) High-resolution regional climate simulations of the long-term decrease in September rainfall over Indochina. Atmos Sci Lett 10(1):14–18

Takata K, Saito K, Yasunari T (2009) Changes in the Asian monsoon climate during 1700–1850 induced by preindustrial cultivation. Proc Natl Acad Sci 106(24):9586–9589

Taylor CM (2015) Detecting soil moisture impacts on convective initiation in Europe. Geophys Res Lett 42(11):4631–4638

Taylor CM, de Jeu RA, Guichard F, Harris PP, Dorigo WA (2012) Afternoon rain more likely over drier soils. Nature 489(7416):423–426

Taylor CM, Ellis RJ (2006) Satellite detection of soil moisture impacts on convection at the mesoscale. Geophys Res Lett 33(3):03404

Taylor CM, Parker DJ, Harris PP (2007) An observational case study of mesoscale atmospheric circulations induced by soil moisture. Geophys Res Lett 34(15):15801

Tuttle S, Salvucci G (2016) Empirical evidence of contrasting soil moisture–precipitation feedbacks across the United States. Science 352(6287):825–828

Permissions

All chapters in this book were first published in PEPS, by Springer; hereby published with permission under the Creative Commons Attribution License or equivalent. Every chapter published in this book has been scrutinized by our experts. Their significance has been extensively debated. The topics covered herein carry significant findings which will fuel the growth of the discipline. They may even be implemented as practical applications or may be referred to as a beginning point for another development.

The contributors of this book come from diverse backgrounds, making this book a truly international effort. This book will bring forth new frontiers with its revolutionizing research information and detailed analysis of the nascent developments around the world.

We would like to thank all the contributing authors for lending their expertise to make the book truly unique. They have played a crucial role in the development of this book. Without their invaluable contributions this book wouldn't have been possible. They have made vital efforts to compile up to date information on the varied aspects of this subject to make this book a valuable addition to the collection of many professionals and students.

This book was conceptualized with the vision of imparting up-to-date information and advanced data in this field. To ensure the same, a matchless editorial board was set up. Every individual on the board went through rigorous rounds of assessment to prove their worth. After which they invested a large part of their time researching and compiling the most relevant data for our readers.

The editorial board has been involved in producing this book since its inception. They have spent rigorous hours researching and exploring the diverse topics which have resulted in the successful publishing of this book. They have passed on their knowledge of decades through this book. To expedite this challenging task, the publisher supported the team at every step. A small team of assistant editors was also appointed to further simplify the editing procedure and attain best results for the readers.

Apart from the editorial board, the designing team has also invested a significant amount of their time in understanding the subject and creating the most relevant covers. They scrutinized every image to scout for the most suitable representation of the subject and create an appropriate cover for the book.

The publishing team has been an ardent support to the editorial, designing and production team. Their endless efforts to recruit the best for this project, has resulted in the accomplishment of this book. They are a veteran in the field of academics and their pool of knowledge is as vast as their experience in printing. Their expertise and guidance has proved useful at every step. Their uncompromising quality standards have made this book an exceptional effort. Their encouragement from time to time has been an inspiration for everyone.

The publisher and the editorial board hope that this book will prove to be a valuable piece of knowledge for researchers, students, practitioners and scholars across the globe.

List of Contributors

Keiji Horikawa, Jing Zhang and Tomohiro Kodaira
Graduate School of Science and Engineering for Research, University of Toyama, Gofuku 3190, Toyama 930-8555, Japan

Masafumi Murayama
Center for Advanced Marine Core Research, Kochi University, B200 Monobe, Nankoku 783-8502, Japan

Kumiko Tsujimoto
Graduate School of Environmental and Life Science, Okayama University, 3-1-1 Tsushima-naka, Kita-ku, Okayama 700-8530, Japan

Tetsu Ohta
Independent Consultant, Okayama, Japan

Kentaro Aida
Center for Research in Isotopes and Environmental Dynamics (CRiED), University of Tsukuba, 1-1-1 Tennodai, Tsukuba, Ibaraki 305-8577, Japan

Katsunori Tamakawa
International Centre for Water Hazard and Risk Management (ICHARM) of the Public Works Research Institute, 1-6 Minamihara, Tsukuba, Ibaraki 305-8516, Japan

Monichoth So Im
Department of Meteorology, Ministry of Water Resources and Meteorology, 364 Monivong Boulevard, Chamkarmorn, Phnom Penh, Cambodia

Hiroaki Ueda
Faculty of Life and Environmental Sciences, University of Tsukuba, 1-1-1 Tennoudai, Tsukuba, Ibaraki 305-8506, Japan

Youichi Kamae
Faculty of Life and Environmental Sciences, University of Tsukuba, 1-1-1 Tennoudai, Tsukuba, Ibaraki 305-8506, Japan
Scripps Institution of Oceanography, University of California San Diego, 9500 Gilman Drive, MC 206, La Jolla, CA 92093-0206, USA

Toshi Kawana
College of Geoscience, School of Life and Environmental Sciences, University of Tsukuba, 1-1-1 Tennoudai, Tsukuba, Ibaraki 305-8506, Japan

Megumi Oshiro
Graduate School of Life and Environmental Sciences, University of Tsukuba, 1-1-1 Tennoudai, Tsukuba, Ibaraki 305-8506, Japan

Hisayuki Kubota
Faculty of Science, Hokkaido University, Kita 10 Nishi 8, Kita-ku, Sapporo 060-0810, Japan

Ryuichi Shirooka
Department of Coupled Ocean-Atmosphere-Land Processes Research, Japanese Agency for Marine-Earth Science and Technology (DCOP, JAMSTEC), Yokosuka, Japan

Jun Matsumoto
Department of Coupled Ocean-Atmosphere-Land Processes Research, Japanese Agency for Marine-Earth Science and Technology (DCOP, JAMSTEC), Yokosuka, Japan
Department of Geography, Graduate School of Urban Environmental Sciences, Tokyo Metropolitan University, Tokyo, Japan

Esperanza O. Cayanan and Flaviana D. Hilario
Philippine Atmospheric, Geophysical and Astronomical Services Administration (PAGASA), Quezon City, Philippines

Ikumi Akasaka
Department of Geography, Senshu University, 2-1-1 Higashi-Mita, Tama-ku, Kawasaki, Kanagawa 214-8580, Japan

Hisayuki Kubota
Hokkaido University, Kita 10, Nishi 8, Kita-ku, Sapporo, Hokkaido 060-0808, Japan

Jun Matsumoto
Department of Geography, Tokyo Metropolitan University, 1-1 Minami-Osawa, Hachioji, Tokyo 192-0397, Japan
Department of Coupled Ocean-Atmosphere-Land Processes Research, Japanese Agency for Marine-Earth Science and Technology, Yokosuka, Kanagawa 237-0061, Japan

Esperanza O. Cayanan, Rosalina G. de Guzman and Flaviana D. Hilario
Philippine Atmospheric, Geophysical and Astronomical Services Administration, Science Garden Complex, Diliman, 1101 Quezon City, Philippines

Luc Beaufort
Aix Marseille University, CNRS, IRD, Collège de France, CEREGE, Avenue Louis Philibert, BP8013545 Aix-en-Provence, Cedex 04, France

Michaël Grelaud
Institute of Environmental Science and Technology, Autonomous University of Barcelona (UAB), Bellaterra 08193, Spain

Masaki Satoh Naomi Kuba1 and Ying-Wen Chen
Atmosphere and Ocean Research Institute, The University of Tokyo, 5-1-5 Kashiwanoha, Kashiwa, Chiba 277-8568, Japan

Akira T. Noda, Tatsuya Seiki, Chihiro Kodama and Yohei Yamada
Japan Agency for Marine-Earth Science and Technology, 3173-15, Showa-machi, Kanazawa-ku, Yokohama, Kanagawa 236-0001, Japan

Yousuke Sato
Department of Applied Energy, Graduate School of Engineering, Nagoya University, Furo cho, Chikusa-ku, Nagoya, Aichi 464-8603, Japan
RIKEN Center for Computational Science, 7-1-26, Minatojima-minami-machi, Chuo-ku, Kobe, Hyogo 650-0047, Japan

Shin-Ya Ogino, Pei-Ming Wu and Shuichi Mori
Japan Agency for Marine–Earth Science and Technology (JAMSTEC), Natsushima–cho, Yokosuka 237-0061, Japan

Manabu D. Yamanaka
Japan Agency for Marine–Earth Science and Technology (JAMSTEC), Natsushima–cho, Yokosuka 237-0061, Japan
Professor Emeritus, Kobe University, Rokkodai–cho, Nada, Kobe 657-8501, Japan

Hamada Jun-Ichi and Jun Matsumoto
Japan Agency for Marine–Earth Science and Technology (JAMSTEC), Natsushima–cho, Yokosuka 237-0061, Japan
Research Center for Climatology and Department of Geography, Tokyo Metropolitan University, Minami–Osawa, Hachioji 192-0397, Japan

Fadli Syamsudin
Center for Regional Resources Development Technology (PTPSW), Agency for the Assessment and Application of Technology (BPPT), Jl. Raya Puspiptek, South Tangerang 15314, Indonesia

Masahiro Tanoue
Department of Civil Engineering, Shibaura Institute of Technology, 3-7-5 Toyosu, Koto-ku, Tokyo 135-8548, Japan

Kimpei Ichiyanagi
Faculty of Advanced Science and Technology, Kumamoto University, 2-39-1, Kurokami, Chuo-ku, Kumamoto 860-8555, Japan

Kei Yoshimura
Institute of Industrial Science Chiba Experimental Station, The University of Tokyo, 5-1-5 Kashiwanoha, Kashiwa-city, Chiba 277-8574, Japan

Masashi Kiguchi
Institute of Industrial Science, The University of Tokyo, 4-6-1 Komaba, Meguro-ku, Tokyo 153-8505, Japan

Toru Terao
Faculty of Education, Kagawa University, 1-1, Saiwaicho, Takamatsu, Kagawa 760-8522, Japan

Taiichi Hayashi
Center for Southeast Asian Studies, Kyoto University, 46, Shimoadachi-cho, Yoshida, Sakyo-ku, Kyoto 606-8501, Japan

Julie Mae B. Dado
Department of Geography, Tokyo Metropolitan University, Minami-Osawa 1-1, Hachioji, Tokyo 192-0397, Japan

Hiroshi G. Takahashi
Department of Geography, Tokyo Metropolitan University, Minami-Osawa 1-1, Hachioji, Tokyo 192-0397, Japan
Japan Agency for Marine-Earth Science and Technology, 3173-25, Showa-machi, Kanazawa-ku, Yokohama, Kanagawa 236-0001, Japan

Masakazu Yoshimori
Faculty of Environmental Earth Science, Hokkaido University, Kita 10, Nishi 5, Kita-ku, Sapporo 060-0810, Japan
Arctic Research Center, Hokkaido University, Sapporo, Japan

Masahiro Watanabe and Akira Oka
Atmosphere and Ocean Research Institute, The University of Tokyo, Kashiwa, Japan

Ayako Abe-Ouchi
Atmosphere and Ocean Research Institute, The University of Tokyo, Kashiwa, Japan
Japan Agency for Marine-Earth Science and Technology, Yokohama, Japan

Hideo Shiogama
Center for Global Environmental Research, National Institute for Environmental Studies, Tsukuba, Japan

Rumi Ohgaito
Japan Agency for Marine-Earth Science and Technology, Yokohama, Japan

Youichi Kamae
Faculty of Life and Environmental Sciences, University of Tsukuba, Tsukuba, Japan

Paolo Viaggi
Museo Geologico Giovanni Capellini, SMA, Bologna University, Via Zamboni 63, I-40126 Bologna, Italy

Jan Polcher
Laboratoire de Météorologie Dynamique (LMD), Institut Pierre Simon Laplace (IPSL), Ecole Polytechnique, Centre National de la Recherche Scientifique (CNRS), Paris, France

Index

P

R

S

T

U

W